River Science

Research and Management for the 21st Century

River Science

Research and Management for the 21st Century

EDITED BY

David J. Gilvear
Plymouth University, UK

Malcolm T. Greenwood
Loughborough University, UK

Martin C. Thoms
The University of New England, Australia

Paul J. Wood
Loughborough University, UK

WILEY Blackwell

Registered office: John Wiley & Sons, Ltd, The Atrium, Southern Gate, Chichester, West Sussex, PO19 8SQ, UK

Editorial offices: 9600 Garsington Road, Oxford, OX4 2DQ, UK
The Atrium, Southern Gate, Chichester, West Sussex, PO19 8SQ, UK
111 River Street, Hoboken, NJ 07030-5774, USA

For details of our global editorial offices, for customer services and for information about how to apply for permission to reuse the copyright material in this book please see our website at www.wiley.com/wiley-blackwell.

Library of Congress Cataloging-in-Publication Data

Names: Gilvear, D. (David), editor.
Title: River science : research and management for the 21st century / edited by David J. Gilvear, Malcolm T. Greenwood, Martin C. Thoms, Paul J. Wood.
Description: Hoboken, NJ : John Wiley & Sons, 2016. | Includes bibliographical references and index.
Identifiers: LCCN 2015050981 | ISBN 9781119994343 (cloth)
Subjects: LCSH: Stream ecology. | Rivers.
Classification: LCC QH541.5.S7 R595 2016 | DDC 577.6/4–dc23 LC record available at http://lccn.loc.gov/2015050981

A catalogue record for this book is available from the British Library.

Wiley also publishes its books in a variety of electronic formats. Some content that appears in print may not be available in electronic books.

Cover design by: ©Angela Gurnell

Set in 9.5/13pt MeridienLTStd by SPi Global, Chennai, India

Printed in Singapore by C.O.S. Printers Pte Ltd

1 2016

Dedication

This volume is dedicated to Geoff Petts – vice chancellor, professor, river scientist, teacher, colleague and friend, whose inspiration and fortitude in bringing together the many elements fundamental to our understanding of river science have been a platform for many; without his visionary ideas river science would not be as advanced as it is today.

Contents

Part 2
Contemporary river science

List of contributors

Michelle L. Anderson
Flathead Lake Biological Station, University of Montana, Polson, Montana, USA

Patrick D. Armitage
Freshwater Biological Association, Wareham, Dorset, UK

Claudio Baigún
Instituto de Investigaciones e Ingeniería Ambiental, Universidad de San Martín, Argentina

James Banning
College of Natural and Health Sciences, Department of Biology, University of Tampa, Tampa, USA

Thomas S. Bansak
Flathead Lake Biological Station, University of Montana, Polson, Montana, USA

Darrell Brown
School of Business, Portland State University, Portland, USA

Andrew F. Casper
Illinois River Biological Station, University of Illinois, Havana, USA

Heejun Chang
Department of Geography, Portland State University, Portland, USA

Samantha D. Chilcote
Flathead Lake Biological Station, University of Montana, Polson, Montana, USA

Adrian L. Collins
Sustainable Soils and Grassland Systems Department, Rothamsted Research, North Wyke, UK

Nathan R. De Jager
US Geological Survey, Upper Midwest Environmental Sciences Center, Wisconsin, USA

Michael D. Delong
Large River Studies Center, Biology Department, Winona State University, Minnesota, USA

Veronica Dujon
Department of Sociology, Portland State University, Portland, USA

David Ervin
Department of Economics, Department of Environmental Science and Management, and Institute for Sustainable Solutions, Portland State University, Portland, USA

Ian Foster
Department of Environmental & Geographical Sciences, University of Northampton, Northampton, UK and Department of Geography, Rhodes University, Grahamstown South Africa

Joseph Flotemersch
National Exposure Research Laboratory, U.S. Environmental Protection Agency, Ecological Exposure Research Division, Cincinnati, Ohio, USA

Laura Gangi
School of Geography, Earth and Environmental Sciences, University of Birmingham, Birmingham, UK; Chair of Hydrology, Faculty of Environment and Natural Resources, University of Freiburg, Freiburg, Germany; Forschungszentrum Juelich GmbH, Juelich, Germany

David J. Gilvear
School of Geography, Earth and Environmental Sciences, University of Plymouth, Plymouth, UK

Elise F. Granek
Department of Environmental Science and Management, School of the Environment, Portland State University, Portland, USA

Ken J. Gregory CBE,
Visiting Professor,
University of Southampton and
Emeritus Professor
University of London, UK

Angela M. Gurnell
School of Geography, Queen Mary University of London, London, UK

David M. Hannah
School of Geography, Earth and Environmental Sciences, University of Birmingham, Birmingham, UK

Matthew J. Hill
Centre for Hydrological and Ecosystem Sciences, Department of Geography, Loughborough University, Loughborough, UK

Peter Hunter
School of Natural Sciences, University of Stirling, Stirling, Scotland

James A. Gore
College of Natural and Health Sciences, Department of Biology, University of Tampa, Tampa, USA

Malcolm T. Greenwood
Centre for Hydrological and Ecosystem Science, Department of Geography, Loughborough University, Loughborough, UK

Andy Large
School of Geography, Politics and Sociology, Newcastle University, Newcastle upon Tyne, UK

Ian Maddock
Institute of Science and the Environment, University of Worcester, Worcester, UK

Jenny Mant
The River Restoration Centre, Cranfield University, Cranfield, UK

Kate L. Mathers
Centre for Hydrological and Ecosystem Sciences, Department of Geography, Loughborough University, Loughborough, UK

Jonathan Millett
Centre for Hydrological and Ecosystem Sciences, Department of Geography, Loughborough University, Loughborough, UK

Alexander M. Milner
School of Geography, Earth and Environmental Sciences, University of Birmingham, Birmingham, UK

Victoria S. Milner
Institute of Science and the Environment, University of Worcester, Worcester, UK

John M. Nestler
IIHR Hydroscience and Engineering, University of Iowa, Iowa City, USA

Malcolm Newson
Tyne Rivers Trust, Corbridge, Northumberland, UK

Melissa Parsons
Riverine Landscapes Research Laboratory, Geography and Planning, University of New England, Armidale, Australia

Brian L. Reid
Flathead Lake Biological Station, University of Montana, Polson, Montana, USA

Michael Reid
Riverine Landscapes Research Laboratory, Geography and Planning, University of New England, Armidale, Australia

Murray Scown
Riverine Landscapes Research Laboratory, Geography and Planning, University of New England, Australia

Vivek Shandas
Toulan School of Urban Studies and Planning, Portland State University, Portland, USA

Jack A. Stanford
Flathead Lake Biological Station, University of Montana, Polson, Montana, USA

Michael Stewardson
Department of Infrastructure Engineering, Melbourne School of Engineering, The University of Melbourne, Victoria, Australia

Martin C. Thoms
Riverine Landscapes Research Laboratory, Geography and Planning, University of New England, Australia

Desmond E. Walling
Department of Geography, College of Life and Environmental Sciences, University of Exeter, Exeter, UK

Markus Weiler
Chair of Hydrology, Faculty of Environment and Natural Resources, University of Freiburg, Freiburg, Germany

Robert L. Wilby
Centre for Hydrological and Ecosystem Science, Department of Geography, Loughborough University, Loughborough, UK

Paul J. Wood
Centre for Hydrological and Ecosystem Science, Department of Geography, Loughborough University, Loughborough, UK

J. Alan Yeakley
Department of Environmental Science and Management, School of the Environment, Portland State University, Portland, USA

Preface

Ken J. Gregory
Visiting Professor University of Southampton and, Emeritus Professor University of London, London, UK

When I was appointed to the Chair of Physical Geography in the University of Southampton in 1976 I asked my Exeter research students if they wished to move with me or preferred to stay at the University of Exeter. The one research student who decided to move was Geoff Petts – surprising in some ways because he had already completed two years research so the move would be for his final writing-up year. Although I thought that it was a good idea to get experience of two universities, I had not influenced Geoff's decision, but later realised that this was typical of his subsequent career – the ability to see the potential as opportunities became available.

A foundation

Geoff had graduated from the University of Liverpool in 1974 with a joint honours degree in Physical Geography and Geology. The NERC studentship at the University of Exeter that we had obtained for research on river channel adjustments downstream from reservoirs was the second of a series awarded for investigations of river channel adjustments arising from a range of different causes. The empirical approach employed used field measurements of channel capacities downstream from dams in 13 areas throughout England and Wales to compare with the dimensions of unregulated channels. At that time there had been comparatively few such investigations, and indeed the effects of human activity on river channels had not been explicitly explored until classic papers by Wolman (1967a,b; see Gregory, 2011), although scour below dams had been surveyed by engineers as a necessary input to dam construction. The Tone had been investigated (Gregory and Park, 1974) but the results obtained by Geoff from a range of UK areas greatly extended understanding of changes that could occur. Areas studied included the Derbyshire Derwent where, in addition to comparing the size of channels downstream from reservoirs with channel size along unregulated rivers showing that capacities were reduced to c. 40% of the expected size, Geoff also demonstrated how a bench formed within the channel had produced the reduction in capacity and that dendrochronology could be used to date trees that had grown on the benches. This allowed confirmation that the reductions in capacity had occurred at dates corresponding to reservoir construction.

This research (Petts, 1978) was one of a series of NERC studentship investigations which deliberately focused on the national picture so that instead of concentrating on a single field area, then very popular with the growth of process-based investigations, the intention was to address large-scale problems by employing empirical measurements from several different areas of Britain. Such an approach was demanding for a research student, but Geoff demonstrated his ability to apply himself to the opportunity, assembling the literature context from the international publications, undertaking field surveys upstream and downstream from reservoirs in different areas of the country, then proceeding to identify the significance of event effectiveness, of sediment availability, of vegetation indicators, culminating in establishing the appropriate elements of a general model including relaxation paths of complex response. This resulted in an impressive array of papers dealing with the channel change effects downstream of the Derwent dams (Petts, 1977), with the application of complex response to channel morphology adjustments (Petts, 1979), with the range of channel changes in regulated rivers (Petts, 1982) and with implications for stream habitats (Petts, 1980a) introducing a link with aquatic ecology that was subsequently to feature throughout Geoff's later research. At a time when specific applications of research results were not often considered, he appreciated the potential significance of the research results for management (Petts, 1980b) which were considered in relation to long-term consequences (Petts, 1980c).

Having established his publication record so effectively, Geoff then had the vision to produce a book *Impounded Rivers* (Petts, 1984a) – which he described as the 'outcome of seven years of research and discussion with friends and professional colleagues'. This book was notable in that it contained hydrology, water quality, morphological effects, ecological aspects including vegetation and macroinvertebrates as well as fisheries, thus providing a truly multi-disciplinary approach to management problems and prospects that were the subject for the final chapter. This book demonstrated the value of providing a context and approach, which we would now refer to as holistic, to succeed the preceding engineering emphasis. In the final part of the preface to his book, Geoff made a plea for a long-term perspective in river management (Petts, 1984a, xv), a theme which he has pursued in much of his later work.

Explanation of the detail of his early research is necessary because it shows how these foundations were fundamental for the way in which he has been able to develop his career. After gaining his PhD he was first appointed in 1977 to the Dorset Institute of Higher Education (later to become part of the University of Bournemouth), but then in 1979 was appointed as lecturer in geography University of Loughborough where he remained until 1994, being senior lecturer (1986–89), Professor of Physical Geography (1989–94) and head of Geography (1991–94). In 1994 he was appointed Professor of Physical Geography University of Birmingham becoming Director of Environmental Science and Management (1994–97), he founded the University's Centre for Environmental Research and Training (CERT) in 1996, became Director of Environmental Science and Training in 1997, Head of the School of Geography and Environmental Science from 1998–2001, and then Pro-Vice Chancellor from 2001–07. With this background and progression it was perhaps inevitable that a move to lead an institution would follow, and in 2007 Geoff became Vice Chancellor and Rector of the University of Westminster. On taking up his post he said 'I am particularly looking forward to working in partnership with staff, students and other stakeholders to grow the University's contributions to the emerging economic, social and environmental demands of urban life in London and other cities across the globe'.

Research development and impacts

A career involving progressively greater amounts of administration, at Loughborough,

Birmingham and Westminster, could have led to a decline of further research, publication and scientific impact, but Geoff has proved to be one of those individuals who maintains his academic contacts. His contributions can be encapsulated in terms of his developing research on flow regulation, the books and contributions in edited volumes that he has produced, and the establishment of the journal *Regulated Rivers*. Furthermore, by pursuing these three themes he has produced enlightening general perspectives, has established collaboration with many other scientists, including many international colleagues, especially European.

Research on flow regulation continued with investigations of a number of other areas leading to the context of flow regulation impacts, progressing research towards other themes. Further investigations of morphological change included the lowland English river Ter, Essex (Petts and Pratts, 1983) and the Rheidol in Wales (Petts and Greenwood, 1985). Whereas ecological changes had previously often been analysed independently from morphological changes, Geoff was involved in research combining the two (e.g., Petts and Greenwood, 1985) and also provided important dimensions such as timescales for ecological change (Petts, 1987). Although changes in water quality and reduced sediment transport downstream of dams had previously been investigated, Geoff was involved with analysis of monitored results from a controlled release from Kielder reservoir on the North Tyne (Petts *et al.*, 1985), analysed sedimentation along the Rheidol (Petts, 1984) and bar development along the North Tyne (Petts and Thoms,1987). Although ecology and its relation to morphological changes had been major sections of his book (Petts, 1984) other aspects were subsequently explored including invertebrate faunas (Petts and Greenwood, 1985), the macroinvertebrate response and physical habitat change to river regulation on the River Rede (Petts, Armitage and Castella, 1993), and the effects of water abstractions on invertebrate communities in UK streams (Castella *et al.*, 1995). Such specific investigations allowed elaboration of more general ecological concerns such as a perspective on the abiotic processes sustaining the ecological integrity of running waters (Petts, 2000), dams and geomorphology (Petts and Gurnell, 2005), a scientific basis for setting minimum ecological flows (Petts *et al.*, 1995), and reservoir operating rules to sustain environmental flows in regulated rivers (Yin, Xin'an *et al.*, 2011).

Flow regulation research led to evaluations of water resources such as the case of Lake Biwa, Japan (Petts, 1988), in turn leading naturally to concern for management problems such as the management of fish populations in Canada (Petts *et al.*, 1989), advancing science for water resources management (Petts *et al.*, 2006), linking hydrology and biology for assessing water needs for riverine ecosystems (Petts *et al.*, 2006), the role of ecotones in aquatic landscape management (Petts, 1990), and sustaining the ecological integrity of large floodplain rivers (Petts, 1996).

Such general themes inevitably meant that Geoff was able to make a very significant contribution in text books and edited volumes – both influential in shaping the development of a subject at a particular stage of its research development. Since *Regulated Rivers* (Petts, 1984a) Geoff has been involved in writing and editing more than 20 books. Texts contributing to the advancement of understanding of rivers include *Rivers and Landscape* (Petts and Foster, 1985), *The Rivers Handbook* Volume I (Calow and Petts, 1992), Volume II (Petts and Calow, 1994) and *Fluvial Hydrosystems* (Petts and Amoros, 1996). Such volumes demonstrated the benefits of multi-disciplinary approaches and Geoff Petts has galvanised the production of edited volumes that have been significant in bringing research results together at a time when branches of disciplines are evolving and hybrid approaches are being articulated. Thus *Regulated Rivers in the UK* (Petts and Wood, 1988) demonstrated the state of the art in relation to river regulation effects, *Historical Analysis of Large Alluvial Rivers in Western Europe* (Petts *et al.*, 1989) achieved a similar result for channel changes in Europe, and *Global Perspectives on River Conservation* (Boon *et al.*, 2000) provided a timely world approach to an inter-disciplinary field.

A recurrent theme emerging from these publications has been the commitment that Geoff has shown to multi-disciplinary approaches, and an outstanding contribution was the way in which his idea for a journal led to *Regulated Rivers* – first published in 1987. This interdisciplinary journal, for which Geoff still continues as Managing Editor, evolved from *Regulated Rivers: Research and Management* (1987–2001) to *River Research and Applications*, having achieved its stated aim to become an international journal dedicated to the promotion of basic and applied scientific research on rivers. In 2010 it appeared as 10 issues with 1314 pages developing from the four issues per year with 375 pages in 1987. It is now an established international journal ranked second in the Science Watch list for Water Resources 1981–2009.

Such progress in publication and research has positioned Geoff to make significant general contributions including changing river channels: the geographical tradition in which he compared the geographical approach with the geological and engineering traditions and advocated a return to large rivers and linking geomorphology and ecology (Petts, 1995). Other position statements have included advancing science for water resources management (Petts *et al.*, 2006), research progress and future directions for dams and geomorphology (Petts and Gurnell, 2005), instream-flow science for sustainable river management (Petts, 2009), and our collaborative proposal for restructuring physical geography (Gregory *et al.*, 2002). The direction of Geoff's scientific contributions has necessitated collaborative work and multi-authored publications – a necessary characteristic of research publication since the days of Geoff's first research. Collaboration with research students, with research grant investigators and with associates from international organisations has been very beneficial and, for example, collaboration with Angela Gurnell has been reflected in publications in the fields of glacial geomorphology, fluvial geomorphology in Italy, including the Tagliamento – encompassing the intriguing paper on trees as riparian engineers (Gurnell and Petts, 2006). The substantial range of associates in the past is testified to by the authors of the chapters of this volume, combining intersection with the phases of his career and the range of disciplines transected in that career.

Recognition

With research output of more than 20 books and 100 scientific papers and as founder and Editor-in-Chief of the international journal *River Research and Applications* it is appropriate that there has already been significant acknowledgement and recognition of the contributions that Geoff has made. He has been a member of several scientific advisory committees including the

International Council for Science (ICSU) Scientific Committee on Water Research; UNESCO IHP Eco-Hydrology Programme; and US Department of the Interior, Fish and Wildlife Service, Long-term Monitoring Programme for the upper Mississippi River. He has been invited to give numerous keynote addresses, he was Director of the International Water Resources Association (1992–94), a Council Member of the Freshwater Biological Association (2000–03), and was appointed Vice President of the new International Society for River Science which was launched in 2006. In 2007, he was awarded the Busk Medal of the Royal Geographical Society for his contributions to inter-disciplinary research on river conservation; in conferring the award the President commended the way in which he had forged inter-disciplinary links between geographers, civil engineers, biologists, ecologists and conservationists. His track record makes it very appropriate that he received a Lifetime Achievement Award from the International Society for river science in 2009.

Following such recognition it is equally appropriate that this collection of essays is published to honour the contribution that Geoff has made, particularly when he has done much to influence the progress of river science with responsibility for founding the journal *Regulated Rivers*. His essential characteristics that have pervaded his academic career include dynamism, opportunism, vision and a multi-disciplinary focus. It is his particular combination of attributes and skills that have enabled him to make a lasting contribution. In his speech receiving the Busk Medal at the RGS in September 2007, Geoff acknowledged his parents giving him a subscription to the *Geographical Magazine* – which he said meant that his 'future was set by the excitement of the topics being reported'. He has managed to continue and convey that excitement throughout his work and it has spilled over in his other interests, particularly hockey and cricket later supplemented, (or succeeded?), by golf and fishing.

Geoff's research began with river regulation: careers such as his can include changes which are analogous to construction of a dam which retains most of the discharge so that relatively little research is published after administration and leadership begin to dominate. However, as Geoff's career has been regulated, he has continued to research and publish and to influence the development of river science in a variety of ways. It is therefore excellent that Geoff's contribution has provided the raison d'etre for this book and that the editors have been so effective in organising such an illustrious list of authors and managing the production of such a timely volume.

References and bibliography

(includes Geoff Petts' publications referred to in the text, which are a selection from his total ouptut)

Boon, P.J., Davies B., and Petts, G.E. (eds) 2000. *Global Perspectives on River Conservation*. Wiley, Chichester.

Calow, P. and Petts, G.E. (eds.) 1992. *Rivers Handbook*. Volume 1, Blackwell Scientific, Oxford.

Castella, E., Armitage, P.D., Bickerton, M. and Petts, G.E. 1995. The ordination of differences: assessing the effects of water abstractions on invertebrate communities in UK streams. *Hydrobiologia*, 308, 167–182.

Gregory, K.J. 2011. Wolman MG (1967) A cycle of sedimentation and erosion in urban river channels. *Geografiska Annaler* 49A: 385–395. *Progress in Physical Geography* 35, 833–844.

Gregory, K.J. and Park, C.C. 1974. The adjustment of stream channel capacity downstream of a reservoir. *Water Resources Research* 10, 870–873.

Gregory K.J., Gurnell A.M. and Petts, G.E. 2002. Restructuring physical geography. *Transactions of the Institute of British Geographers* 27, 136–154.

Gurnell, A.M. and Petts, G.E. 2006. Trees as riparian engineers: the Tagliamento River, Italy. *Earth Surface Processes and Landforms* 31, 1558–1574.

Petts, G.E. 1977. Channel response to flow regulation: the case of the River Derwent, Derbyshire. In: Gregory, K.J. (ed.) *River Channel Changes*, Wiley, Chichester, 368–385.

Petts, G.E. 1978. The adjustment of river channel capacity downstream from reservoirs in Great Britain. Unpublished PhD thesis, University of Southampton.

Petts, G.E. 1979. Complex response of river channel morphology subsequent to reservoir construction. *Progress in Physical Geography* 3, 329–362.

Petts, G.E. 1980a. Implications of the fluvial process – channel morphology interaction below British reservoirs for stream habitat. *The Science of the Total Environment* 16, 149–163.

Petts, G.E. 1980b. Morphological changes of river channels consequent upon headwater impoundment. *Journal of the Institution of Water Engineers and Scientists* 34, 374–382.

Petts, G.E. 1980c. Long-term consequence of upstream impoundment. *Environmental Conservation* 7, 325–332.

Petts, G.E. 1982. Channel changes within regulated rivers. In: Adlam, B.H., Fenn, C.R. and Morris, L. (eds) *Papers in Earth Studies*. Geobooks, Norwich, 117–142.

Petts, G.E. 1984a. *Impounded Rivers: Perspectives for Ecological Management*. Wiley, Chichester.

Petts, G.E. 1984b. Sedimentation within a regulated river: Afon Rheidol, Wales. *Earth Surface Processes and Landforms* 9, 125–143.

Petts, G.E. 1987. Timescales for ecological change in regulated rivers. In: Craig, J. and J.B. Kemper (eds) *Regulated Streams: Advances in Ecology*. Plenum, New York, 257–266.

Petts, G.E. 1988. Water management: the case of Lake Biwa, Japan. *Geographical Journal* 154, 367–376.

Petts, G.E. 1995. Changing river channels: the geographical tradition. In: Gurnell, A.M. and Petts, G.E. (eds) *Changing River Channels*, Wiley, Chichester, 1–23.

Petts, G.E. 1996. Sustaining the ecological integrity of large floodplain rivers. In: Anderson, M.D., Walling, D.E. and Bates P. (eds) *Floodplain Processes*. Wiley, Chichester, 535–551.

Petts, G.E. 2000. A perspective on the abiotic processes sustaining the ecological integrity of running waters. *Hydrobiologia* 422/423, 15–27.

Petts, G.E. 2009. Instream-flow science for sustainable river management. *Journal of the American Water Resources Association* 45, 1071–1086.

Petts, G.E. and Amoros, C. 1996. (eds) *Fluvial Hydrosystems*. Chapman and Hall, London.

Petts, G.E. and Calow, P. (eds.) 1994. *Rivers Handbook*, Volume 2. Blackwell Scientific, Oxford.

Petts, G.E. and Foster, I.D.L. 1985 *Rivers and Landscape*. Arnold, London.

Petts, G.E. and Greenwood, M.T. 1985. Channel changes and invertebrate faunas below Nant-y-Moch Dam, River Rheidol, Wales, UK. *Hydrobiologia* 122, 65–80.

Petts, G.E. and Gurnell, A.M. 2005. Dams and geomorphology: research progress and future directions. *Geomorphology* 71, 27–47.

Petts, G.E. and Pratts, J.D. 1983. Channel changes following reservoir construction on a lowland English river, *Catena*, 10, 77–85.

Petts, G.E. and Thoms, M.C. 1986. Channel sedimentation below Chew Valley Lake, Somerset, UK. *Catena* 13, 305–320.

Petts, G.E. and Thoms, M.C. 1987. Bar development in a regulated gravel-bed river: River North Tyne, UK. *Earth Surface Processes and Landforms* 12, 433–440.

Petts, G.E. and R.J. Wood (eds) 1988. *Regulated Rivers in the UK*. Special Issue of Regulated Rivers, Wiley, Chichester.

Petts, G.E., Armitage, P.D. and Castella, E. 1993. Physical habitat change and macroinvertebrate response to river regulation: the River Rede, UK. *Regulated Rivers* 8, 167—178.

Petts, G.E., Foulger, T.R., Gilvear, D.J., *et al.* 1985, Wave-movement and water-quality changes during a release from Kielder Reservoir, North Tyne River, UK. *Journal of Hydrology* 80, 371–389

Petts, G.E., Imhof, J., Manny, B.A., *et al.* 1989. Management of fish populations in large rivers: a review of tools and approaches", *Canadian Journal of Fisheries and Aquatic Science*, 578–588.

Petts, G.E., Maddock, I., Bickerton, M.A. and Ferguson, A. 1995. The scientific basis for setting minimum ecological flows. In: Harper, D. and Ferguson, A. (eds) *The Ecological Basis for River Management*. Wiley, Chichester, 1–18.

Petts, G.E., Morales, Y. and Sadler, J.P. 2006. *Linking hydrology and biology in assessing water needs for riverine ecosystems*. Invited Commentary *Hydrological Processes*.

Petts, G.E., Nestler, J. and Kennedy, R. 2006. Advancing science for water resources management. *Hydrobiologia* 565, 277–288.

Petts, G.E., Roux, A.L. and Moller, H. (eds) 1989. *Historical Analysis of Large Alluvial Rivers in Western Europe*. Wiley, Chichester.

Wolman, M.G. 1967a: A cycle of sedimentation and erosion in urban river channels. *Geografiska Annaler* 49A, 385–395.

Wolman, M.G. 1967b: Two problems involving river channel changes and background observations. In: *Quantitative Geography Part II Physical and Cartographic Topics, Northwestern Studies in Geography* 14, 67–107.

Yin, Xin'an, Zhang, Yang Z-F and Petts, G.E. 2011. Reservoir operating rules to sustain environmental flows in regulated rivers. *Water Resources Research* 47, W08509, doi 10.1029/2010WR009991.

CHAPTER 1

An introduction to river science: research and applications

Martin C. Thoms[1], David J. Gilvear[2], Malcolm T. Greenwood[3] and Paul J. Wood[3]

[1] *Riverine Landscapes Research Laboratory, Geography and Planning, University of New England, Australia*

[2] *School of Geography, Earth and Environmental Sciences, Plymouth, UK*

[3] *Centre for Hydrological and Ecosystem Science, Department of Geography, Loughborough University, Loughborough, UK*

Introduction

River science is a rapidly developing interdisciplinary field of study focusing on interactions between the physical, chemical and biological components within riverine landscapes (Thoms, 2006; Dollar *et al.*, 2007) and how they influence and are influenced by human activities. These interactions are studied at multiple scales within both the riverscape (river channels, partially isolated backwaters and riparian zones) and adjacent floodscape (isolated oxbows, floodplain lakes, wetlands and periodically inundated flat lands). It is an exciting and robust field of study because of the integrative nature of its approach towards understanding complex natural phenomena and its application to the management of riverine landscapes.

The modern era of river science is a challenging one because climate, landscapes and societies are changing at an ever-increasing rate. Thus, our use, perceptions and values related to riverine landscapes are also changing. The twenty-first century will be different to the twentieth century both in terms of the way in which we undertake research and manage rivers. Increasing globalisation and data availability will allow unique opportunities for sharing of information and experiences, at unparalleled rates. Therefore, we can expect an exponential upward trajectory in societies' understanding of rivers and their appreciation of them as one of the globe's key ecosystems. This will be especially true as the goods and services that rivers provide, in particular the demand for water as the resource, becomes scarcer in many regions. Water security is predicted to become a key global issue in the twenty-first century (Gleick, 2003). Thus river ecosystems and their associated landscapes are likely to be viewed and valued by society in the same way that the importance of tropical rainforests, as a regulator of climate change, became evident in the twentieth century.

Rivers and their associated landscapes are ubiquitous global features, even in the driest and coldest regions of the world (Hattingh and Rust, 1999; Bull and Kirby, 2002; Doran *et al.*, 2010). The physical, geochemical and ecological characteristics of the world's riverine landscapes are as

River Science: Research and Management for the 21st Century, First Edition.
Edited by David J. Gilvear, Malcolm T. Greenwood, Martin C. Thoms and Paul J. Wood.

diverse as the peoples of the world and their cultural origins (Miller and Gupta, 1999; Cushing *et al.*, 2006). Many rivers meander slowly through lowland regions, with some never making their way to the sea, while those that do so often rush down steep rocky gorges or flow hidden beneath the ground within alluvial aquifers or limestone caves. Some rivers flow in multiple channels and others exist as a series of waterholes connected by intermittent channels for most of the time. Some rivers only flow after prolonged rainfall and some flow all year round with little variation in water levels.

Human societies and populations have been drawn to these landscapes for millennia because of the provision of important resources, like water for human survival, irrigation, power, navigation, food and timber. The flat fertile lands of river floodplains have drawn people to them for agriculture and have been used by them as important transport routes, even in contemporary societies where road, rail and air freight may be more rapid. However, rivers and their floodplains also present challenges to those that choose to inhabit these landscapes because of their propensity to flood, erode their banks as well as to contract and even become dry during extended periods of drought (Lake, 2009; Pennington and Cech, 2010). The prosperity of human societies is closely linked to natural variations in the character and behaviour of riverine landscapes both regionally and over time, in many parts of the world (cf. Petts *et al.*, 1989; Wohl, 2011). Past civilisations have waxed and waned, and even disappeared, as result of the unpredictable and highly variable nature of riverine landscapes (e.g., Schumm, 2005).

Riverine landscapes and their associated ecosystems are the foundation of our social, cultural and economic wellbeing. The degraded condition of many of the world's rivers and floodplains is a testament to our failure to understand these complex systems and manage them wisely. The exponential increase in the number of riverine studies, from various regions, highlights the growing stresses placed on river systems in response to demands made directly upon them and their surrounding catchments. A recent assessment of the worlds 100 most-populated river basins, by The World Resources Institute, found 34 of these basins displayed high to extreme levels of stress, while only 24 had minimal levels of stress. This was primarily a result of water related pressures in these basins. These rivers flow through countries with a collective GDP of $US 27 trillion (World Resources Institute, 2014). Similarly, other studies with a more regional focus, demonstrate the impact of inappropriate activities on the health and/or condition of river systems. The Sustainable Rivers Audit undertaken in the Murray Darling Basin, Australia, for example, found rivers in 21 of the 23 sub-basins were in poor to very poor condition in terms of their hydrology, physical form, vegetation, fish and macroinvertebrate communities, because of changes in hydrological regimes, land use and inappropriate channel management (Murray-Darling Basin Authority, 2013). River science is the interdisciplinary study of these complex biophysical systems and seeks to understand the drivers that influence pattern and process within these critically important systems. In order to minimise future river catastrophes and degradation, river science should underpin our approach to their management and the setting of policy regarding these landscape scale systems.

Many animal and plant communities depend upon riverine landscapes and their associated ecosystems for some or all of their lifecycle. Most rely on riverine landscapes as a source of water and nutrients. The strong linkage between rivers, humans and biological communities is strongest where human societies are also heavily dependent upon riverine landscapes for food and where fish is a major component of their diet. In many of these locations the concept of a 'healthy river' was, or remains, culturally important and an intuitive component of human survival (Kelman, 2006). Given the dependency on rivers and their health or productivity by humans and organisms, it is surprising that the subject of river science as a discipline in its own right has only emerged in recent years. The journal *River Research and Applications* and its predecessor *Regulated Rivers: Research and Management*, the pre-eminent scientific publication devoted to river ecosystems, only commenced publishing in 1987. In part, this is a reflection and response to the distancing of many human societies from riverine landscapes and the ecosystem goods and services, and environmental hazards that are an inherent component of these natural landscapes. Historically a gulf between river scientists and river managers has existed resulting in a lag between the advancement of the science and improved river management (Cullen, 1996; Parsons *et al.*, Chapter 10 in this volume): this lag, in part, still exists today.

The development of the discipline of river science

River science is a relatively recent discipline compared to the traditional academic disciplines of biology, chemistry, geology, mathematics and physics. However, river science does have a recognisable lineage within some disciplines, most notably biology, geology, geomorphology, hydrology and limnology. One of the first to document interactions between humans and their environment was George Marsh in 1864 (Lowenthal, 2000). Marsh highlighted the links between the collapse of civilisations through environmental degradation, most notably catchment land-use changes and the resource condition of catchment ecosystems, including its soil and water resources. It is no exaggeration to say that *Man and Nature* (Marsh, 1864) helped launch the modern conservation movement and helped many to recognise the damage that societies across the globe were doing to the natural environment. It also challenged society to behave in more responsible ways toward the earth and its natural systems. *Man and Nature* (Marsh, 1864) stands next to *Silent Spring* (Carson, 1962) and *A Sand County Almanac* (Leopold, 1949) by any measure of historic significance within the modern conservation movement (Lowenthal, 2000).

Three merging paths of activity have advanced our understanding of rivers as ecosystems and their role within the broader landscape since the publication of Marsh (1864). The first path was the articulation of conceptual constructs of the study of rivers and their landscapes. This began with the seminal paper by Hynes (1975) 'The stream and its valley', which acknowledged that hill slopes and fluvial processes are primary drivers of lotic ecosystems. It also provided a frame of reference for adopting a catchment-scale approach to the study of lotic systems and the coupling of hydrology, geomorphology and ecology to advance our understanding of rivers as natural complex systems. Another catalyst for scientific coupling was publication of the *River Continuum Concept* – (RCC) (Vannote *et al.*, 1980) that elegantly if not explicitly, linked hydrological, geomorphological and ecological components of a river system within the context of the longitudinal profile of a river. This was notable in that it took a source to mouth perspective, and indirectly – via reference to the concept of stream ordering (Horton, 1945) – a stream network perspective. The RCC provided the impetus for a relatively rapid progression in the conceptual understanding of river ecosystems; with the publication of the *Serial Discontinuity Concept* (SDC) by Ward and Stanford (1983), the *Flood Pulse Concept* (FPC) by Junk *et al.* (1989) and the *Patch Dynamics Concept* (PDC) by Townsend (1989). The research of Stanford and Ward (1993) on *hyporehos-stream linkages* also reinvigorated research in the field of surface and sub-surface linkages pioneered in the 1970s (e.g., Williams and Hynes, 1974) and provided a clear vertical dimension to our conceptual understanding of lotic systems. Later, the *Fluvial Hydrosystem Concept* of Petts and Amoros (1996) provided one of the first larger scale frameworks with which to view riverine landscapes; an approach carried forward by Dollar *et al.* (2007) and others. Both Petts and Amoros (1996) and Dollar *et al.* (2007) sought to describe patterns in riverine landscape in four dimensions (sensu Ward 1989) and at different scales to establish relationships between the physical character of riverine landscapes and their ecological functioning. The spatial arrangement of both physical and ecological elements within riverine landscapes is largely determined by the flow and sediment (both organic and inorganic) regimes. Functional and genetic links between adjoining components of the riverine landscape often result in clinal patterns conceptualised as continua. However, the integrity of river systems depends on the dynamic interactions of hydrological, geomorphological and biological processes acting in longitudinal, lateral and vertical dimensions over a range of temporal scales. Thus, resultant interactions may also produce riverine landscape mosaics rather than a system solely characterised by gradients. This was one of the central themes explored in the *River Ecosystem Synthesis* (RES) of Thorp *et al.* (2008). As a collective, all of these concepts and theories highlight the need for cross-disciplinary thinking and the importance of multiple scales of investigation for the research and management of riverine landscapes.

The second path was the establishment of the series of symposia under the banner 'International Symposium on Regulated Rivers', formerly established in 1985 (cf. Craig and Kemper, 1987), although the original meeting was held in 1979 as a special symposium at the North American Benthological Society meeting in Erie, Pennsylvania, USA, and was called *The* [First] *International Symposium on Regulated Streams* (later referred to as FISORS). Subsequent successful meetings have been held in Australia, Europe and North America. The International Symposium on Regulated Rivers series ended in Stirling, Scotland in 2006 (Gilvear *et al.*, 2008). After which it became the biennial conference of the *International Society for River Science* (ISRS). The inaugural meeting of the ISRS was held in Florida in 2009 with subsequent meetings in Berlin, Beijing and La Crosse, Wisconsin, USA in 2015. It was at the meeting in Florida that ISRS became a formal society, with its members focused on the interdisciplinary study of riverine landscapes and its applications to management and policy.

Closely associated with the symposium series was the launch of the journal *Regulated Rivers: Research and Management* in 1987; and this can be considered the third path of convergence in River Science. The journal changed its name in 2002 to *River Research and Applications* (RRA) and became the official journal of ISRS. This name change reflected the need for scientific coupling of traditional disciplines and marked the increased acceptance that River Science required contributions from hydrology, stream ecology, fluvial geomorphology and river engineering to be directed at the subject of understanding river ecosystems and their landscapes. Both ISRS and the journal have explicitly welcomed and encouraged interdisciplinary research and have resulted in an increase to the growing body of knowledge on river ecosystems.

The discipline of river science has in a relatively short period of time grown from its pioneering stage to become established within the community and has reached relative maturity. This is reflected in a meta-analysis of 1506 research publications within the journal *River Research and Applications* and its former iteration, *Regulated Rivers: Research and Management*, from herein termed *River Research and Applications* (RRA). Since the first publication in 1987, each manuscript was assessed in terms of its disciplinary focus. The nine disciplinary areas were: (i) catchment geomorphology; (ii) biology; (iii) chemistry; (iv) ecology; (v) engineering; (vi) fluvial geomorphology; (vii) hydrology; (viii) management; and (ix) policy. The spatial scale of each study was assigned to either the entire fluvial network, river zone, reach or site scale. In addition, the focus and approach of each study was determined as being in-channel, riparian, floodplain, drainage network or the entire system and if it was empirical, modelling or conceptual in nature.

A summary of the meta-analysis RRA research publications assessed is presented in Figure 1.1. There are three salient points emerging from this analysis. First, the number of papers appearing in RRA increased significantly between 1987 and 2013 (Figure 1.1a); (22 in 1987 to a maximum of 137 in 2012). This was also accompanied by increase in the number of RRA journal issues in 1987–2014 from four to ten. However, the number of manuscripts per volume also changed significantly in 2000; in that period the journal changed focus from largely managed and regulated rivers to a river science/river ecosystems focus. An average of 37 research manuscripts per volume were published in the 1987–99 period compared to 73 in 2000–13 (Figure 1.1a). Moreover, there was a steady increase of six additional published manuscripts per volume from 2000–13 contrasting with a relatively stable number of manuscripts per volume 1987–99. Second, a wide ranging set of disciplines has contributed to RRA but the relative contribution of the different disciplines has changed over time (Figure 1.1b). The disciplines of biology (31.8%), ecology (15.5%), geomorphology (15.6%) and hydrology (14.3%) were the major contributors to the journal, in terms of published articles, in 1987–99 compared to 2000–2013, where the disciplines of ecology (34.3%), geomorphology (22.7%) hydrology (14.5%) and management (15.9%) were the dominant contributors. Furthermore, multi-disciplinary studies became more prevalent, rising from 41.1% (1987–99) to 65.1% (2000–13). Third, the spatial scale, locational focus and research approach of the published studies also changed over the same period (Figure 1.1c). In terms of scale, the majority of published studies in 1987–13 were undertaken at the reach (63.8%) or site scales (21.8%). However, following 2000 there was an increase in the spatial scale at which researchers undertook stream and river studies. The number of studies conducted at larger river zone and network scales increased from 4.2% in 1987–99, to 17.7% in 2000–13 and from 1.7% in 1987–99 to 5.7% in 2000–13). Accompanying this was a decrease in site-based studies from 36.3% in 1987–99 to 7.3% in 2000–13. In addition, the number of studies undertaken over multiple spatial scales in 1987–13 increased steadily from a relative contribution of 2% in 1987 to 18% in 2013. Over the same period the locational

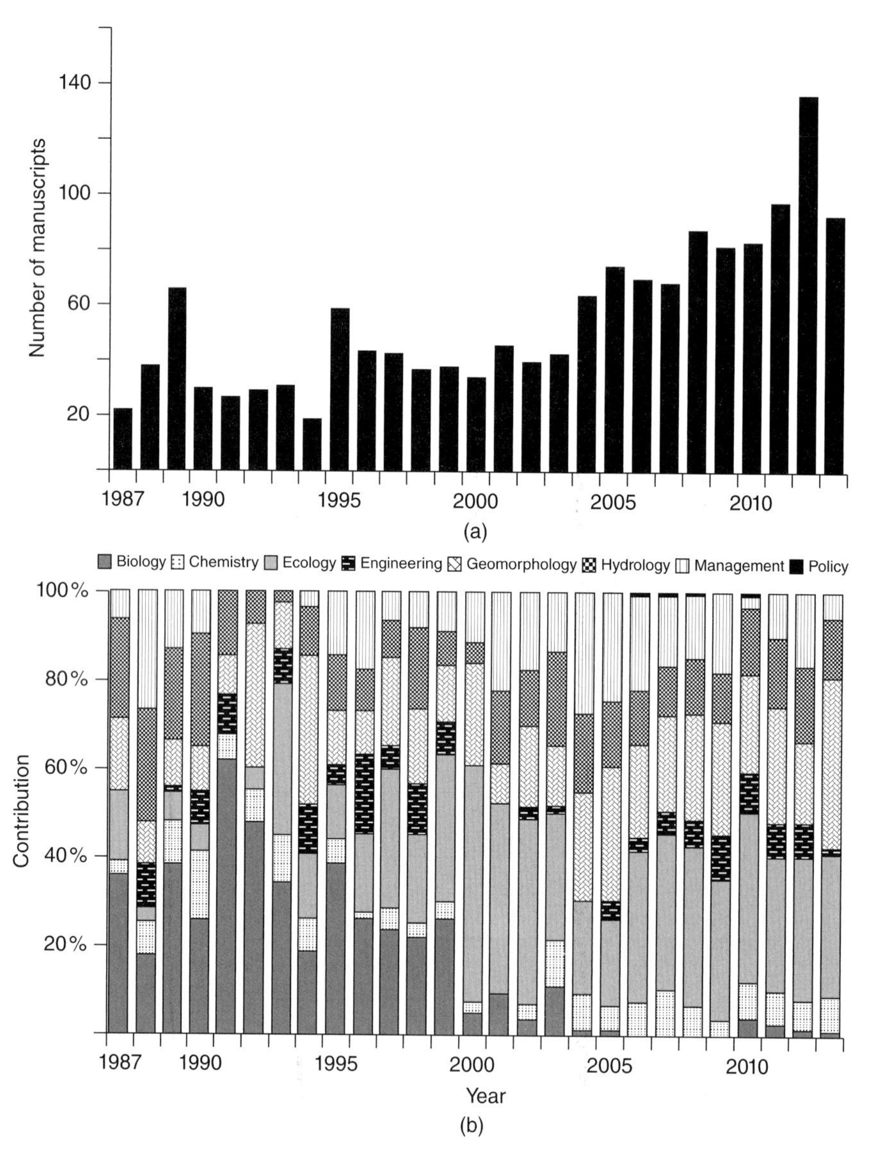

Figure 1.1 Meta-analysis of published research manuscripts in the journals *Regulated Rivers: Research and Management* and *River Research and Applications* for the period 1987–2013. (a) The annual number of publications; (b) the relative composition the various disciplinary foci; and (c) the scale of focus of the various published studies.

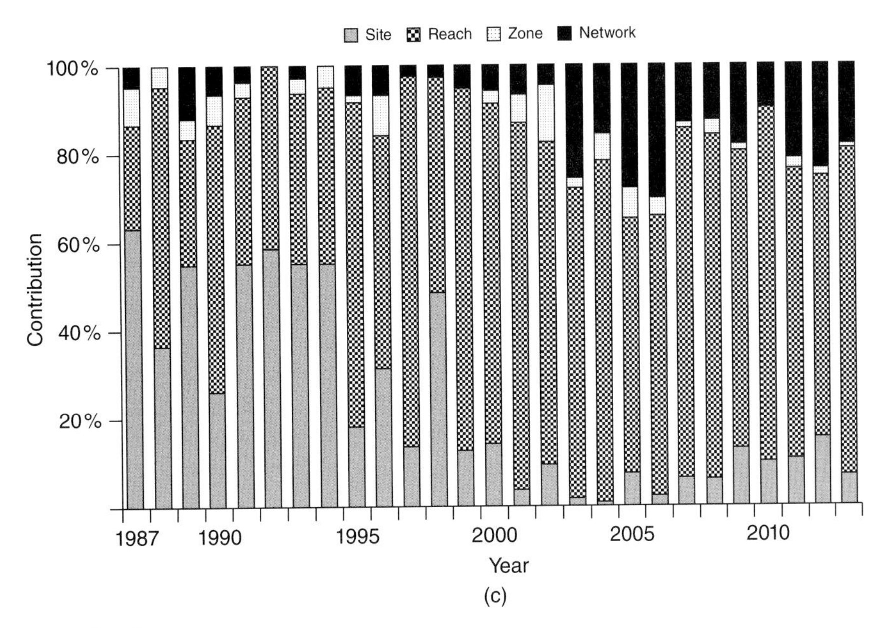

Figure 1.1 (*Continued*)

focus of the studies also changed from being dominated by in-channel focused (76% of studies in 1987–99 to 60% in 2000–13) to having a greater emphasis on entire systems, that is a combined in-channel, riparian and floodplain focus (6.9% of studies in 1987–99 compared to 20.5% in 2000–13). Finally, research publications in RRA are essentially empirical in nature, representing on average 91% of the published studies. This has only changed slightly with conceptual and modelling studies increasing in 2000–13 to contribute 13% of the total published papers.

River science continues to expand from descriptive studies of the physical or biological structure of river channels to a field which includes, among other things, biophysical processes involving conceptual and mathematical modelling, empirical investigations, remote sensing and experimental analysis of these complex process–response systems. These studies are being conducted at both greater (e.g., catchment – continental) and smaller (e.g., fine sediment biochemical processes) scales and more importantly span multiple scales. Through the emergence of a systems approach within science during the 1970s more broadly, an inevitable convergence of individual disciplines towards river science occurred; although the term *river science* would not come into contemporary use until the early twenty-first century.

The domain of river science

To quote Burroughs (1886) and direct it to riverscapes: 'one goes to rivers only for hints and half-truths ... their facts are often crude until you have observed them in many different ways and then absorbed and translated these'. Ultimately it is not so much what we see in rivers, rather what we see suggests. The discipline of river science allows those engaged with it to observe rivers, their associated landscapes and ecosystems through a multitude of lenses. Thus, it embraces a continuum of ideas, concepts and approaches, from those having a biotic focus (e.g., aquatic ecology, genetics, physiology) at one end of the spectrum to those with an abiotic focus, most notably hydrology, geomorphology and engineering at the other. Spanning these are those areas of landscape and community ecology and biogeography to mention but a few. Figure 1.2 schematically represents the development of River Science over time. Over the last 45 years, from its foundations in hydrology, geomorphology, ecology and engineering, new disciplines have emerged and coalesced to form the modern day science of rivers. During this time the focus of attention has also shifted to areas outside of the channel bed to the floodplain and hyporheos and from the reach scale to the river network. Closer to the corners of this conceptual diagram of river science are the more singular disciplinary foci, whilst those towards the central regions represent the greater inter-disciplinary elements. The content critical to the subject of river

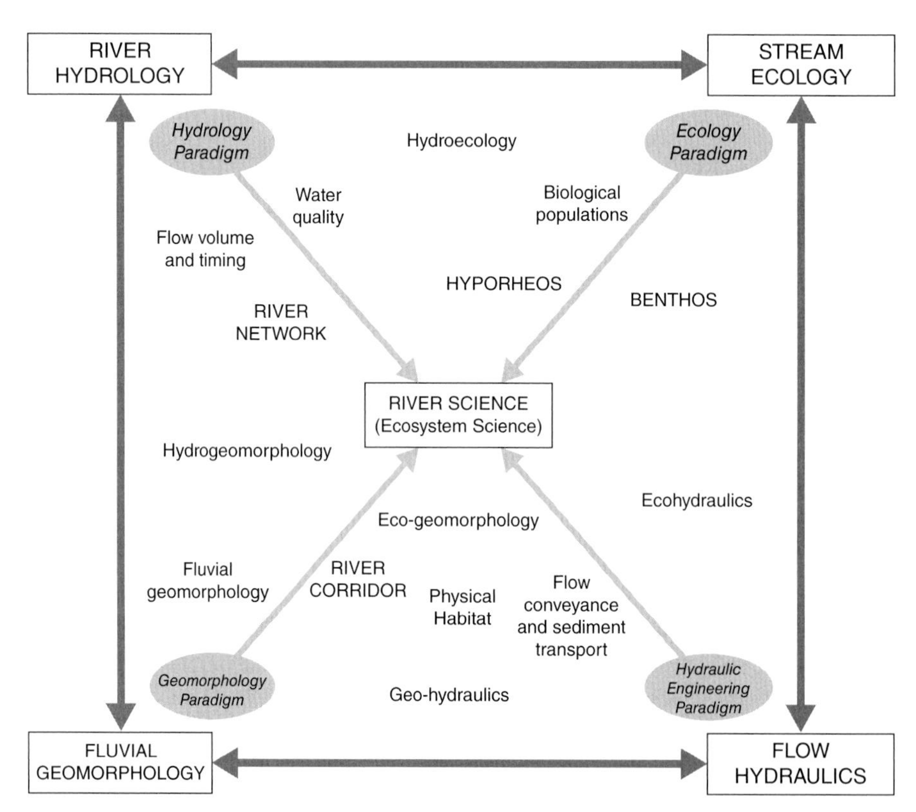

Figure 1.2 The evolution of river science over time from its foundations within river hydrology, fluvial geomorphology, flow hydraulics and stream ecology. The arrows that flow towards the centre of the page, from their subject specific paradigm, are conceptual timelines converging on the subject of river science and its focus on ecosystem science. In two-dimensional space a selection of disciplines and fields of enquiry (shown in lower case font) that emerged over time are shown to illustrate the conceptual development of river science as a subject. The widening of the focus of river science beyond the channel margins is illustrated in the diagram by differing components of river ecosystems (shown in upper case font) with their location reflecting the larger disciplinary area from which they emerged.

science, in terms of understanding river ecosystems, is clearly represented within the chapters in this volume.

Chapters in this volume and book structure

This volume is a reflection of, and a tribute to, the emergence of the discipline of river science and the recognition that it helps to provide an holistic approach through which to study, manage and conserve lotic ecosystems in the contemporary social, political and environmental landscape. Our aim for this edited book was to produce a volume which brings together the multiple strands of research that represent this rapidly developing arena of research (natural science, social sciences, engineering and environmental policy), that would provide a benchmark text for those familiar and new to the concept of river science. In addition, the volume represents a resource that will be valuable to researchers, practitioners, environmental regulators and those engaged in the development or implementation of policy. The volume was also specifically prepared as an acknowledgement of the ongoing commitment to river science provided by Professor Geoffrey Petts, editor in chief of *River Research and Application* over 30 years. To achieve this goal, recognised international research leaders within the field of river science were asked to position their contributions within the context of the historical development of the field, identify key research challenges for the future and highlight the wider societal implications of the research. The volume encompasses a range of chapters illustrating the dynamic nature of riverine processes (Gangi *et al.*, Chapter 14; Gurnell, Chapter 7; Milner *et al.*, Chapter 8; Nestler *et al.*, Chapter 5; Scown *et al.*, Chapter 6; Walling and Collins, Chapter 3) how riverine landscapes support natural ecosystem functioning (Delong and Thoms, Chapter 2; Milner, Chapter 12; Stanford *et al.*, Chapter 13) and how this knowledge can be used to inform policy and management practices (Foster and Greenwood, Chapter 4; Gilvear *et al.*, Chapter 9; Gore *et al.*, Chapter 15; Mant *et al.*, Chapter 16; Wilby, Chapter 18). The chapters clearly illustrate the relevance of river science to all parts of contemporary society, from the scientific community through to those living alongside rivers, of the physical, economic, cultural and spiritual benefits and risks associated with our ongoing relationship with rivers (Parsons *et al.*, Chapter 10; Wood *et al.*, Chapter 11; Yeakley *et al.*, Chapter 17). Collectively, the chapters demonstrate the growing maturity of river science and its central place in the management and conservation of rivers across the globe.

The book is comprised of two sections: Part 1 provides an overview of some fundamental principles of river science (Chapters 2–10), from its early development within the confines of traditional academic disciplines through to contemporary interdisciplinary research, which transcends traditional disciplinary boundaries and addresses research questions at multiple spatial (site through to catchment) and temporal scales (days to millennia) and also within the context of an ecosystems framework. Part 2 (Chapters 11–18) comprises a range of case studies, which illustrate how contemporary river science continues to address fundamental research questions regarding the organisation and functioning of river systems, how anthropogenic activities modify these systems and how we may ultimately manage, conserve and restore riverine ecosystems to sustain natural functioning and ecosystem health, and also to support the needs of an ever thirsty society for water, energy and the services that rivers provide.

We realise that a book of this nature could never realistically hope to cover all aspects of contemporary river science. Indeed, we are conscious that this volume only touches on the burgeoning body of research centred on the biogeochemistry of riverine ecosystems, such as nutrient spiralling (von Schiller *et al.*, 2015) and the processing, storage and transport of dissolved organic matter (DOM) and dissolved organic carbon (Singh *et al.*, 2014). We also recognise that the current volume only touches on issues associated with the impacts of, and future threats posed by, invasive/non-native species on lotic ecosystems across the globe (Scott *et al.*, 2012). In addition, the chapters exclusively address the upper and middle reaches of riverine catchments and they do not consider the interface between what many consider the end of the river, the brackish/estuarine system (Jarvie *et al.*, 2012). It is hoped that by following both the themes and topics illustrated in this volume, together with new initiative ideas, an in-depth and broadening knowledge of river science will be established.

References

Bull LL, Kirby MJ, (eds) (2002). *Dryland Rivers: Hydrology and Geomorphology of Semi Arid Channels*. Wiley and Sons, Chichester.

Burroughs J, (1886). *Signs and Seasons*. Riverside Press, Cambridge.

Carson R, (1962). *Silent Spring*. Mariner Books, Houghton Mifflin Company, Boston.

Craig JF, Kemper JB (eds) (1987). *Regulated Streams: Advances in Ecology: Proceedings of the Third International Symposium on Regulated Streams*, Held August 4–8, 1985, in Edmonton, Alberta, Canada. Plenum Publishing Company Limited, University of California.

Cullen PW, (1996). *Science brokering and managing uncertainty*. Proceedings of The Great Barrier Reef: Science, Use and Management, Vol. 1, pp. 309–318. CRC Reef Research Centre, Townsville.

Cushing CE, Cummins KW, Minshall GW, (2006). *River and Stream Ecosystems of the World*. University of California Press, London.

Dollar, ESJ, James CS, Rogers KH, Thoms MC, (2007). A framework for interdisciplinary understanding of rivers as ecosystems. *Geomorphology* 89: 147–162.

Doran PT, Lyons B, McKnight DM, (eds) (2010). *Life in Antarctic Deserts and Other Cold Dry Environments*. Astrobiological Analogs, Cambridge University Press, Cambridge.

Gilvear D, Wilby N, Kemp P, Large A, (2008). Special Issue: riverine ecohydrology: advances in research and applications. Selected papers from the Tenth International Symposium on Regulated Streams. Stirling, August 2006. *River Research and Applications* 24: 473–475.

Gleick PH, (2003). Global freshwater resources: soft-path solutions for the 21st century. *Science* 302: 1524–1527.

Hattingh J, Rust IC, (1999). Drainage evolution and morphological development of the late Cenozoic Sundays River, South Africa. In: Miller AJ, Gupta A, (eds) *Varieties of Fluvial Form*. Wiley and Sons, Chichester. pp. 145–166.

Horton RE, (1945). Erosional development of streams and their drainage basins: hydrophysical approach to quantitative morphology. *Geological Society of America Bulletin* 56: 275–370.

Hynes HBN, (1975). The stream and its valley. *Verhandlungen Internationale Vereinigung für Theoretiche und Angewandte Limnologie* 19: 1–15.

Jarvie HP, Jickells TD, Skeffington RA, Withers PJA, (2012). Climate change and coupling of macronutrient cycles along the atmospheric, terrestrial, freshwater and estuarine continuum. *Science of the Total Environment* 434: 252–258.

Junk WJ, Bayley PB, Sparks RE, (1989). The flood pulse concept in river- floodplain systems. *Special Publication Canadian Journal of Fisheries and Aquatic Sciences* 106: 110–127.

Kelman M, (2006). *A River and its City: The Nature of Landscape in New Orleans*. University of Californian Press, Berkeley.

Lake PS, (2009). *Drought and Aquatic Ecosystems: Effects and Responses*. John Wiley and Sons, Chichester.

Leopold A (1949). *A Sand County Almanac*. Ballantine Books, New York.

Lowenthal D, (2000). *George Perkins Marsh, Prophet of Conservation*. Columbia University Press, Washington.

Marsh GP, (1864). *Man and Nature; or Physical Geography as Modified by Human Action*. Sampson Low, Son and Marston, London.

Miller AJ, Gupta A, (eds) (1999). *Varieties of Fluvial Form*. John Wiley and Sons, Chichester.

Murray–Darling Basin Authority, (2013). Sustainable Rivers Audit report 2 (2008–2010). Murray–Darling Basin Authority.

Pennington KL, Cech TV, (2010). *Introduction to Water Resources and Environmental Issues*. Cambridge University Press, Cambridge.

Petts GE, Amoros C, (eds) (1996). *Fluvial Hydrosystems*. Chapman and Hall, London.

Petts GE, Möller H, Roux AL, (eds) (1989). *Historical Change of Large Alluvial Rivers: Western Europe*. John Wiley and Sons, Chichester.

Schumm SA, (2005). *River Variability and Complexity*. Cambridge University Press, Cambridge.

Scott SE, Pray CL, Nowlin WH, Zhang Y, (2012). Effects of native and invasive species on stream ecosystem functioning. *Aquatic Sciences* 74: 793–808.

Singh S, Inamdar S, Mitchell M, (2014). Changes in dissolved organic matter (DOM) amount and composition along nested headwater stream locations during baseflow and stormflow. *Hydrological Processes* 29: 1505–1520.

Stanford JA, Ward JV, (1993). An ecosystem perspective of alluvial rivers: connectivity and the hyporheic corridor. *Journal of the North American Benthological Society* 12: 48–60.

Thoms MC, (2006). An interdisciplinary and hierarchical approach to the study and management of river ecosystems. *International Association of Hydrological Sciences* 301: 170–179.

Thorp, JH, Thoms MC, Delong MD, (2008). *The Riverine Ecosystem Synthesis*. San Diego, California, Elsevier.

Townsend CR, (1989). The patch dynamics concept of stream community ecology. *Journal of the North American Benthological Society* 8: 36–50.

Vannote RL, Minshall GW, Cummins KW, *et al.* (1980). The river continuum concept. *Canadian Journal of Fisheries and Aquatic Sciences*, 37: 130–137.

von Schiller D, Bernal S, Sabater F, Martí E, (2015). A round-trip ticket: the importance of release processes for in-stream nutrient spiralling. *Freshwater Science* 34: 20–30.

Ward JV, (1989). The four-dimensional nature of lotic ecosystems. *Journal of the North American Benthological Society* 8: 2–8.

Ward JV, Stanford JA, (1983). The serial discontinuity concept of lotic ecosystems. In: Fontaine TD, Bartell SM, (eds) *Dynamic of Lotic Ecosystems*. Ann Arbor Science, Ann Arbor, MI; 347–356.

Williams DD, Hynes HBN, (1974). Occurrence of benthos deep in substratum of a stream. *Freshwater Biology* 4: 233–255.

Wohl, E, (2011). What should these rivers look like? Historical range of variability and human impacts in the Colorado Front Range, USA. *Earth Surface Processes and Landforms* 36: 1378–1390.

World Resources Institute (2014). World's 18 Most Water-Stressed Rivers. http://www.wri.org/blog/2014/03/world%E2%80%99s‐18‐most‐water‐stressed‐rivers. (Accessed 26 May 2015).

PART 1
Fundamental principles of river science

CHAPTER 2

An ecosystem framework for river science and management

Michael D. Delong[1] and Martin C. Thoms[2]

[1] *Large River Studies Center, Biology Department, Winona State University, Minnesota, USA*

[2] *Riverine Landscapes Research Laboratory, Geography and Planning, University of New England, Australia*

Introduction

River science, the interdisciplinary study of fluvial ecosystems, focuses on interactions between the physical, chemical and biological structure and function of lotic and lentic components within riverine landscapes (Thoms, 2006; Dollar *et al.*, 2007). These interactions are studied at multiple spatiotemporal scales within both the riverscape (river channels, partially isolated backwaters, and riparia of small streams to large rivers) and the surrounding floodscape (isolated oxbows, floodplain lakes, wetlands, and periodically inundated drylands). River science continues to expand from descriptive studies of the physical or biological structure of river channels to a field which includes, among other things, biophysical processes involving conceptual and mathematical modelling, empirical investigations and experimental analysis of these complex process–response systems. This emergence has also seen river scientists contributing effectively at the turbulent boundary of science, management and policy (Cullen, 1990, Parsons *et al.*, Chapter 10).

Successful interdisciplinary science requires the merger of two or more areas of understanding into a single conceptual–empirical structure (Pickett *et al.*, 1994; Thoms, 2006). Implicit to this process is the development, testing and application of new ideas, as well as the continued integration of concepts, paradigms and information from emerging sub-disciplines and other scientific fields that operate across a range of domains, scales and locations. The progression of science is a dynamic process influenced by current and historical developments, with the accumulation of knowledge within formal logical frameworks. Such frameworks are often built from direct observations which are synthesised within hypotheses and then empirically tested (Graham and Dayton, 2002). Frameworks are useful tools for achieving integration of different disciplines and have been used in many areas of endeavour. A framework is neither a model nor a theory; models describe how things work, theories explain phenomena, whereas frameworks show how facts, hypotheses and models may be linked (Pickett *et al.*, 1999) Frameworks, therefore, provide a way

River Science: Research and Management for the 21st Century, First Edition.
Edited by David J. Gilvear, Malcolm T. Greenwood, Martin C. Thoms and Paul J. Wood.

of ordering phenomena, thereby revealing patterns of structure and function (Rapport, 1985). The continued development of river science and the exchange of its endeavours with management require a diversity of frameworks, study designs and research questions and a commitment to the dual process of developing and testing theories and their application into the domain of management. While these steps are important to any field of enquiry, it is also crucial to challenge concepts and the prevailing 'wisdom' so as to: avoid stasis; accurately integrate fundamental knowledge within applied policies; and to ensure transfer of reliable information to future generations of scientists.

There is an intimidating array of 'models', 'concepts' and 'theories' on the nature of fluvial ecosystems to consider when developing a framework for research, rehabilitation or management. As the chapter title implies, a critical element of developing a viable framework is the incorporation of an ecosystem approach, the value of which has been described as having three dimensions of influence (Pickett and Cadenasso, 2002). First is that its basic definition is inclusive and free of limiting assumptions; second, its ability to be expressed in a range of models that articulate the components, interactions, extent and boundaries of the ecosystem under investigation; and, finally the powerful influence it can have in social discourse through its metaphorical strengths.

An ecosystem is a spatially explicit unit of the Earth that includes all of the organisms, along with all of the abiotic components within its boundaries (Tansley, 1935). This definition establishes that there is a clear spatial (and temporal) dimension to an ecosystem. Moreover, 'spatially explicit' and 'within its boundaries' infer that an ecosystem approach is not just limited to the ecosystem level of organisation; it can be used consider biotic–abiotic interactions across many levels of organisation. From this perspective, a river basin, a lateral channel or a single rock can be viewed as an ecosystem if appropriate boundaries and scale are applied (sensu Likens 1992). An ecosystem approach allows for examination of form and processes across different disciplines through consideration of both biotic and abiotic interactions, thereby providing the holistic approach needed for an applicable framework.

To understand the behaviour and begin to manage rivers as ecosystems requires a holistic, interdisciplinary approach that simultaneously considers their physical, chemical and biological components (Dollar *et al.*, 2007; Thoms, 2002; Thorp *et al.*, 2008). Interdisciplinary research is fraught with many problems including different approaches and conceptual tools, hence disciplinary paradigms lose their usefulness in the interdisciplinary arena. Development and use of common frameworks can alleviate this. The objectives of this chapter are to:

- provide a historical overview of different models of river ecosystems, including their genesis, strengths, limitations and potential to aid in interdisciplinary science and management of river ecosystems;
- outline a conceptual framework for use in the research and management of river ecosystems; and,
- highlight the use of such a framework in the research and management of riverine landscapes.

To accomplish these objectives, we propose a shift in how river networks are viewed for research and management. To truly continue forward, it is essential that we look where we have been by examining past models and, from there, ascertain the best approach for achieving a framework that fits the criteria described previously in this chapter.

A brief history of models that have contributed to our understanding river ecosystems

Fish and biocoenotic zones

One of the earliest efforts toward a general model of the structure and function of river ecosystems emanated from Europe during the latter part of the nineteenth century. Its focus was on the classification of river networks with the division of the network into five 'fish zones' (Hynes, 1970; Hawkes, 1975). These zones, which were named for dominant species of fish within a given river, were fixed in their longitudinal location and had abrupt transitions from one zone to the next. In addition, locations of these non-repeatable sections were considered highly predictable from upstream to downstream. This was later modified to include physical and chemical characteristics of each zone (e.g., Huet, 1959; Aarts and Nienhuis, 2003). Testing this model outside the region of its development highlighted several limitations to the zonation of river networks by fish zones (Aarts and Nienhuis, 2003). The primary limitation was that discontinuities in river basin geomorphology interfered with the expected pattern of fish zones, often resulting in the repeated occurrence of zones throughout a river (e.g., Tittizer and Krebs, 1996). Fish zones were later represented as biocoenotic zones, where the intent was to consider all aquatic organisms (Illies and Botosaneaunu, 1963), and later hydrological characteristics (Arts and Nienhuis, 2003). Despite these changes, other problems with the model became evident, specifically: (i) the predicted sequence of zones differed from one river to the next; (ii) some zones were absent from rivers; and (iii) some zones repeated along the downstream gradient of rivers. With few exceptions (e.g., Aarts and Neinhuis, 2003), fixed/biocoenotic zones are now rarely seen in the literature.

River continuum concept

Biocoenotic zonation was replaced by the river continuum concept – RCC (Vannote *et al.*, 1980). The RCC addressed limitations of biocoenotic zones by attempting to explain longitudinal changes in ecosystem form and function. It was designed to reflect gradual downstream transitions that had been observed in studies that found conflicts with the abrupt changes prescribed by fixed/biocoenotic zonation. The central premise of the RCC was that hydrological and geomorphological conditions change predictably from headwaters to terminus within a river network and with these come concomitant shifts in ecological processes and community structure. The RCC was simplified to describe ecological changes relative to stream order as the basis for defining the location of physical and ecological components longitudinally. One component that remained consistent between the RCC and biocoenotic zones was that both emphasised the longitudinal dimension and asserted there were predictable, fixed (in terms of location along the longitudinal gradient) zones with specific physical and ecological attributes, hence the RCC provided a model that was more broadly applicable than the taxon-specific methodology of biocoenotic zones. Moreover, relating expected ecological and physical conditions to stream order made it readily applicable to both researchers and managers.

Testing of the RCC began immediately after its publication and it still serves as a useful null hypothesis for river ecosystem studies. While some studies found support for the RCC (i.e., Culp and Davies, 1982; Cushing *et al.*, 1983; Minshall *et al.*, 1983), many questioned its general applicability. Townsend (1989), in a review of aquatic ecological concepts, stated in regard to the applicability of the RCC that it '... is remarkable primarily because it is not usually realized and cannot provide a world-wide generalization'. This shortcoming has been observed in studies of community structure (e.g., Winterbourn *et al.*, 1981; Perry and Schaeffer, 1987) as well as predictions on trophic dynamics in streams and rivers (e.g., Lewis *et al.*, 2001, Delong and Thorp, 2006; Lau *et al.*, 2009). The value of using production/respiration ratios as a measure of trophic status has also been called into question given that river networks are largely heterotrophic (P/R < 1) because of microbial production that is typically independent of metazoan production (Thorp and Delong, 2002; Marcarelli *et al.*, 2011). Thus its usefulness in underpinning a framework on how to approach the interdisciplinary study of riverine landscapes and their management is limited.

Studies contradicting the RCC typically tied their findings to differences in local lithology, geomorphology and hydrology. While based on the hypothesis of gradual changes in stream characteristics, the RCC suffered one of the same limitations as fixed/biocoenotic zones; specifically, it did not account for differences in geomorphology and the repeatability of 'zones' within and among river networks. A subsequent revision of the RCC by Minshall *et al.* (1985) acknowledged the need to account for climate, local lithology, and geomorphology in its predictions: 'Further reflection (on the classic view of rivers) indicates that the ideal rarely is so clearly achieved' (Minshall *et al.*, 1985). It was suggested in this revision that expected differences in ecological and physical conditions should be viewed on a sliding scale where, as an example, a braided fifth-order channel might be better explained by viewing it as five third-order channels (Minshall *et al.*, 1985). In essence, the modifications of the ecological predictions of the RCC were to consider deviations created by geomorphology, lithology, tributaries and climate on a sliding scale that was still based on stream order.

The frequent inability to get a fit between ecological structure and processes and the conceptual basis of the RCC can be linked to the hydrogeomorphic concepts on which it is based. The hydrogeomorphic basis of the RCC is drawn from a suite of studies described by Leopold *et al.* (1964) on the longitudinal morphology of river channels. While these studies did describe conditions where a continuum of fluvial processes and morphology could occur, they emphasised that these circumstances were not applicable to all rivers and in fact were rare (Leopold *et al.*, 1964). In addition, much of the underlying physical basis of the RCC relies on stream order, which does not provide a meaningful template for describing hydrogeomorphic processes within river systems (Gregory and Walling 1973). The lack of a hydrogeomorphic continuum was further emphasised by Statzner and Higler (1985), who examined hydrological data of the rivers used by Minshall *et al.* (1983) and demonstrated no uniform longitudinal pattern to measures of hydraulic stress. Large-scale hydraulic discontinuities do occur in rivers (Statzner and Higler, 1985) and the simplicity of the relationship between physical and biological gradients within river networks is overstated in the RCC. Also lacking from the physical component of the RCC is consideration of the stochastic nature of rainfall and runoff patterns that have a tendency to create hydrological discontinuities.

The apparent lack of congruence in the physical template of the RCC and associated ecological discrepancies *does not* provide a viable working model for scientists and managers except in its usefulness as a starting null hypothesis. The intent of the RCC was to provide a cohesive basis for the study of river networks through the integration of physical and biological gradients (Minshall *et al.*, 1985). This is reflected by Cushing *et al.* (1983), who stated that 'streams are best viewed as gradients, or continua, and that classification systems which separate discrete reaches are of little ecological value'. Furthermore, the emphasis on longitudinal change in physical structure and associated ecological processes was done in the absence of scale. Minshall *et al.* (1985) does address spatial and temporal heterogeneity in the context of its influence on the habitat templet (sensu Southwood 1977) but the RCC does not account for how physical and biological structure at smaller spatial scales may shape structural organisation at larger spatial scales (e.g., Boys and Thoms, 2006) or the influence of physical and ecological change occurring at multiple spatial and temporal scales.

Riverine ecosystem synthesis

Neither biocoenotic zonation nor the RCC has the potential to provide a basis upon which river research or management can be placed. While many reasons have been provided (Poole, 2002; Thorp *et al.*, 2008), chief among these is the failure to recognise that physical conditions are not always highly predictable on a longitudinal gradient and that a given set of hydrogeomorphic conditions can be repeated at multiple locations within a river network. Recognition of these attributes emerged in the late twentieth century as scientists came to appreciate river networks as a mosaic of patches existing at multiple spatial and temporal scales, with the type and arrangement of these physical patches influencing ecological form and function (e.g., Thoms, 2006; Townsend, 1989).

Physical patch structure of rivers was increasingly emphasised around the turn of the twenty-first century, and a series of concepts such as the process domain concept (Montgomery, 1999), river discontinuum (Poole, 2002) and hydrogeomorphic zones (Thoms, 2006) emerged. These various concepts put forward the view that rivers resemble a mosaic of physical patches operating at multiple spatiotemporal scales, where patches can be defined by their hydrological, sedimentological and morphological attributes independent of location within the stream network. The concepts of Poole (2002) and Thoms (2006) went further to note that patches can be found at multiple locations within a stream network where similar hydrological and geomorphological conditions exist. Development of the hydrogeomorphic mosaic is based on well-established principles of fluvial geomorphology and landscape ecology and complements the independent work of Townsend (1989) who suggested that a unifying stream framework based on the patchy nature of rivers would provide a more realistic and generalised means of examining ecological processes than continuum/clinal based concepts.

The riverine ecosystem synthesis (RES), integrates the hydrological and geomorphological constructs of the hydrogeomorphic mosaic perspective with expected ecological responses to the physical mosaic of river networks (Thorp *et al.*, 2008). This is where the RES departs from other concepts and models. While biocoenotic zonation and the RCC emphasised longitudinal patterns and were limited to what could best be considered fixed large-scale patches, the conceptual approach of the RES recognises that hydrogeomorphic–ecological linkages operate at multiple scales. Additionally, the RES does not have a preconceived bias of 'I am in "X" stream order, therefore I should expect "Y" physical conditions and "Z" ecological processes'. Instead, the RES calls for an analytical approach to allow for self-emergence of where you are and what should be expected. More importantly, this concept departs from the location-specific approaches that have constrained the advancement of river science and broader applications of what we learn (sensu Fisher 1997).

Patches are hierarchically organised in time and space within the RES, with each patch type possessing intrinsic hydrological, sedimentological and geomorphological attributes. Ecological traits, in turn, are also hierarchically organised, thus allowing for integration of hydrological, geomorphological and ecological character appropriate for the scale of interest in research and management. Patches vary in hydrological variability and physical complexity, including potential differences in number and permanency of lateral channels, spatial diversity of current velocities, temporal variability in flow/flood pulse rate and extent, substrate size and variability, chemical characteristics, riparian–channel interactions and riverscape–floodscape exchanges. Included at larger spatial scales are functional process zones (FPZs), which are repeatable along the longitudinal dimension and only partially predictable longitudinally, especially when comparing among ecoregions.

The hydrogeomorphic patch approach contrasts sharply from the longitudinal perspective by recognising that rivers are more than a single thread passing through a terrestrial landscape (c.f., Ward and Tockner 2001). This view has been emphasised by others through observation of the heterogeneous and discontinuous nature of river systems (Fausch *et al.*, 2002; Ward *et al.*, 2002; Thorp *et al.*, 2008; Carbonneau *et al.*, 2012). It is for this reason that a foundational property of the RES is recognition that river networks must be viewed as mosaics consisting of patches of differing size, quality and character as a function of their hydrological and geomorphological condition (Wiens, 2002; Thorp *et al.*, 2008). Additionally, the hydrological and geomorphological attributes of these patches will shape ecological structure and processes within these patches. The character of patches, therefore, establishes the basis on which the structure and function of river systems should be considered in research and management.

The scalar nature of patches leads to an additional key point on which the RES is based; specifically, the acknowledgement that river networks are comprised of hierarchically arranged patches that are formed by hydrological and geomorphological processes. Patches are not isolated entities functioning wholly independently of their surroundings. Ecological and hydrogeomorphic characteristics of patches are also shaped by their association with adjacent patches. The type and arrangement of smaller patches within any portion of the river network gives rise to distinctive, larger-scale patches with their own inherent qualities. The location of a patch, regardless of its scale, will be based on its hydrological and geomorphological character, giving their location low predictability along the longitudinal gradient of the network (Poole, 2002). A further advantage of the hierarchical nature of patches is that it provides a mechanism for clearly defining boundaries. Clearly defined spatial and temporal boundaries allow for clearer definition of the processes, both physical and ecological, operating within a patch and to delineate flow pathways across patch boundaries (Cadenasso *et al.*, 2003; Strayer *et al.*, 2003).

Underlying concepts for the use of frameworks in River Science

The complexity of riverine landscapes challenges many traditional scientific approaches and methods (Dollar *et al.*, 2007). A river's multi-causal, multiple-scale character constrains the usefulness of conventional reductionist-falsification approaches, except when applied at very small scales and within limited domains (Thoms and Parsons, 2002). The complex character of rivers instead requires a more iterative process that is scale aware, akin to what Pickett *et al.* (2007) labelled the new philosophy of science. Frameworks for the successful interdisciplinary study have been proposed; most notably that by Thoms (2002) and Dollar *et al.* (2007). Here we review the underlying concepts of these, the majority of which are based upon hierarchy.

A hierarchy is a graded organisational structure. A hierarchical level (or holon) is a discrete unit within a system and the features of a level reflect both the level above it and those of the level below it within the hierarchy (Figure 2.1). Higher levels within a hierarchy exert some constraint on lower levels, especially the level immediately below (i.e., L-3 influences L-4 more than it does L-5; Figure 2.1). Conversely, lower levels influence the structure and functioning of those at higher levels, particularly the level immediately above. Therefore, downward constraints and upward influences explain the character most strongly at the adjacent levels, and this gives rise to emergent properties of the level of interest. It is also important to note that a level within a hierarchy is not a scale but may be characterised by a scale (O'Neill *et al.*, 1989).

Scale defines the physical dimension of an entity and Quinn and Keogh (2002) characterise scale in terms of grain and extent. Grain refers to the smallest spatial or temporal interval in an observation set and it has also been referred as the smallest scale of pattern to which an organism may respond (O'Neill *et al.*, 1989) or the smallest scale of influence of an ecosystem disturbance or process driver (Rogers, 2003). Extent is the total area or duration over which observations are made, the largest pattern to which an organism responds (i.e., the habitats used by a fish or the time over which a given habitat is used), or the largest scale at which a disturbance or process driver exerts influence on the system. Therefore, grain and extent define the upper and lower limits of resolution in the description of a level of organisation or an ecosystem. Assigning a scale to a hierarchical level of organisation provides contextual meaning and more importantly it determines the variables and units of measure that can be associated with each level of a particular hierarchy.

Hierarchical concepts are common in the sub-disciplines of river science – ecology, geomorphology and hydrology – with each sub-discipline having distinct levels of organisation. The familiar hierarchical levels of ecological organisation (organism, species, community, ecosystem) are also fundamental to ecological understanding

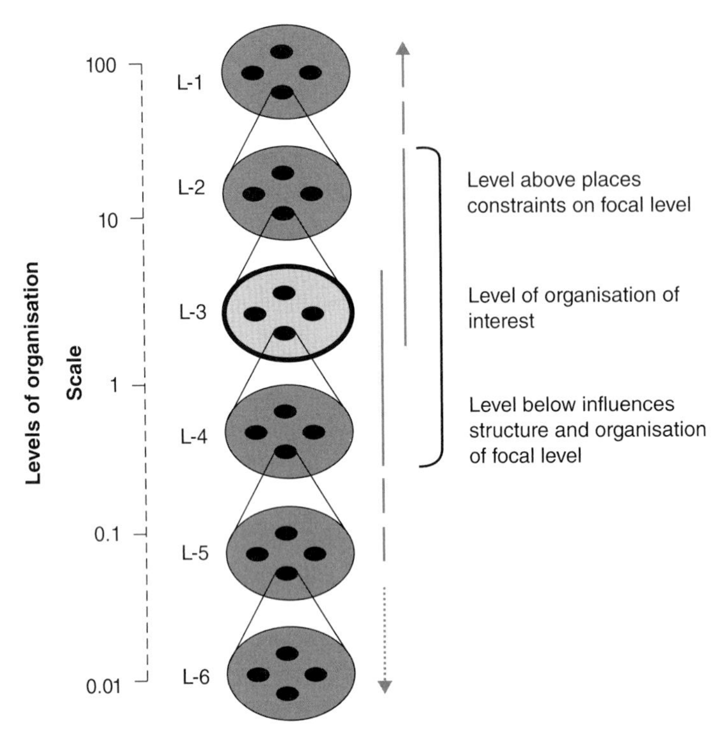

Figure 2.1 Conceptual diagram of hierarchical organisation demonstrating where one level of organisation is nested within the level immediately above. Increased dashing of arrows demonstrates that the decrease in the degree to which the level of organisation of interest is influenced by other levels within the hierarchy. In this example, the focus level, L-3, is most directly influenced by L-2 and L-4.

(Barrett *et al.*, 1997). While these levels of organisation or units are not scales (Petersen and Parker, 1998), they operate in characteristic spatial and temporal domains and are used to stratify components within the biological system. For example, physiology and behaviour are generally studied at the level of the individual, whereas species richness and diversity are studied at the community level and energy and nutrient fluxes are studied at the ecosystem level.

Linking between sub-disciplines of river science

Hierarchical interpretations within individual sub-disciplines are not generally applied in interdisciplinary applications because they are not often compatible in terms of variable representation and scale connotation (Thoms, 2002). The frameworks of Thoms (2002) and Dollar *et al.* (2007) provide individual disciplinary (or sub-system) hierarchical structures that are independent of scale and then use scale as the currency for linking them. Recognition of spatial and temporal scales inherent to the levels of organisation of a disciplinary hierarchy makes integration of multiple sub-systems possible (Dollar *et al.* 2007). This is an essential step in the development of a framework that incorporates an interdisciplinary approach (Pickett *et al.*, 1994). Integration of the scales allows researchers and managers to ask correct questions through recognition that there are causal linkages across geomorphological, hydrological and ecological frameworks. This perspective also acknowledges that structure and function of any kind are not limited to a single level or scale.

There are four steps in the application of the frameworks put forward by Thoms (2002) and Dollar *et al.* (2007). The first step requires identification of various sub-systems and the second focuses on describing the relevant levels of organisation that characterise the different sub-systems in the context of the issue/problem being addressed. The third step involves the identification of appropriate scales and variables within the different organisational levels and step four describes the process interactions between appropriately identified sub-system components.

Step one: sub-system identification

Many different disciplines study rivers, among these are hydrology, geomorphology, ecology, chemistry, engineering, social science and economics. For the purposes of this chapter we focus on the three primary sub-systems of ecology, hydrology and geomorphology. Geomorphology (in this context fluvial geomorphology) considers the landforms associated with river systems and the processes that form them. Hydrology focuses on the occurrence and movement of water through landscapes and river systems. Ecology considers the response of flora and fauna to changes in water supply, sediment movement and channel morphology along with changes in landscape character and changes in other biotic phenomena.

Step two: organisational hierarchies

Each of these three sub-systems can be represented as a hierarchical structure that makes explicit the different organisational levels. The identification and organisation of parallel organisational sub-system hierarchies constitute a substantive part of the framework. It is important to recognise that the three organizational hierarchies are dissimilar in terms of the nature of their levels within each hierarchy). Ecological levels are biological abstractions, geomorphological levels are physical entities while hydrological levels are variable descriptors. Despite the differences they are useful conceptual constructs for correlating the sub-systems through assigning scale.

Geomorphological factors sit within a hierarchy of influence, where larger-scale factors establish the conditions within which smaller-scale factors form. As a result, river systems can be divided into nested levels that encompass the relationships between a stream and its catchment at a range of spatial and temporal scales. Examples of this include: Frissell *et al.* (1986), van Niekerk *et al.* (1995), Montgomery and Buffington (1998), Petts and Amoros (1996) and Thoms *et al.* (2004). The uppermost level of organisation is the drainage basin and these typically have a wide spatial extent and are shaped by long-term geological and climatological processes, making them relatively resistant to change (Figure 2.2). Nested within a drainage basin are functional process zones (FPZs), which are lengths of river that have similar geological histories plus discharge and sediment regimes. They can be further defined by major breaks in slope and style of channel or floodplain (Thoms and Parsons, 2002). FPZs can be thought of as a longitudinal component of the river system except, in a marked departure from applying stream order, FPZs have limited predictability in location and can be found repeatedly along the length of a river (Thoms *et al.*, 2004, 2007). These geomorphological features allow for the application of an ecosystem framework because they,

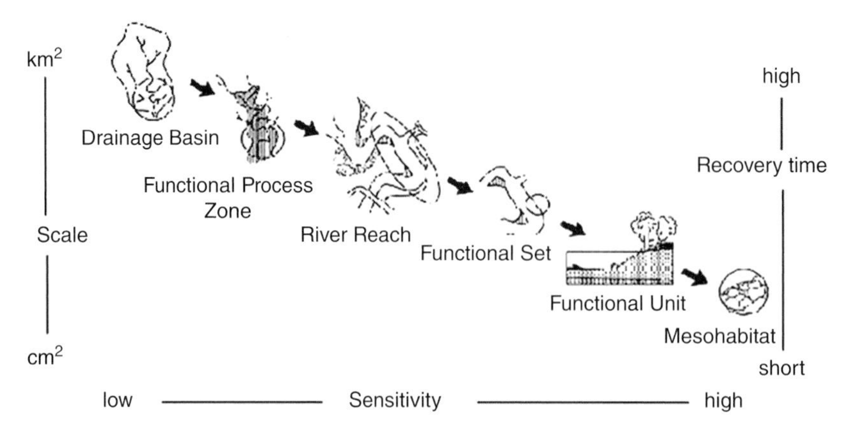

Figure 2.2 The geomorphological levels of organisation of riverine landscapes. Scale provides a relative measure of the spatial extent of each level. Modified from Petts and Amoros (1996).

like all other geomorphological levels of organisation, have a quantifiable spatial and temporal dimension and provide a clear demarcation of where you are in a river that cannot be accomplished when relying on location along a continuum. Reaches are repeatable sections within a given FPZ and are defined by their channel planform or bedform character. The critical point here is that a reach possesses a quantifiable spatial and temporal scale. All too frequently, a reach is an arbitrarily defined construct assigned to a single riffle–pool sequence, a section of river between bends, or a several kilometre long stream section. This approach leads to studies where the degree of success in linking appropriate measures among the hierarchies, or even to the results of other studies, may be limited because of a disregard for the true scalar nature of a reach. The geomorphological hierarchy continues down to the level of microhabitat (Thoms and Parsons, 2002; Dollar *et al.*, 2007).

Five levels of hydrological organisation have been identified as important in river ecosystem function (Thoms and Sheldon, 2000). Flow regime is a long-term statistical generalisation of flow behaviour or climate that can extend over hundreds of years (Figure 2.3). Flow history reflects the sequences of floods, droughts, connectivity and so on, over periods of 1–100 yr. The flood pulse represents the cycle of flooding typically over a period of < 1 yr. Next on the hierarchy is channel hydraulics, which is represented by measures of velocity and turbulence. The temporal scale of hydraulics is minutes to seconds that may influence bedform and boundary-sediment composition (Dollar *et al.*, 2007). The lowest hierarchical level is fluid mechanics, with a duration of seconds, which influences chemical processes and has the potential to impact which spaces are occupied by individual organisms. In addition to the temporal scales just described, spatial scales can also be applied to hydrological levels of organisation. The spatial dimensions of hydrology are reflective of their scope of influence, which can be viewed as complementary to the spatial scale of geomorphological hierarchy (Figure 2.3).

Hydrological models are commonly formulated and developed within a spatiotemporal hierarchical context, with different orders of complexity and different process descriptions at different scales.

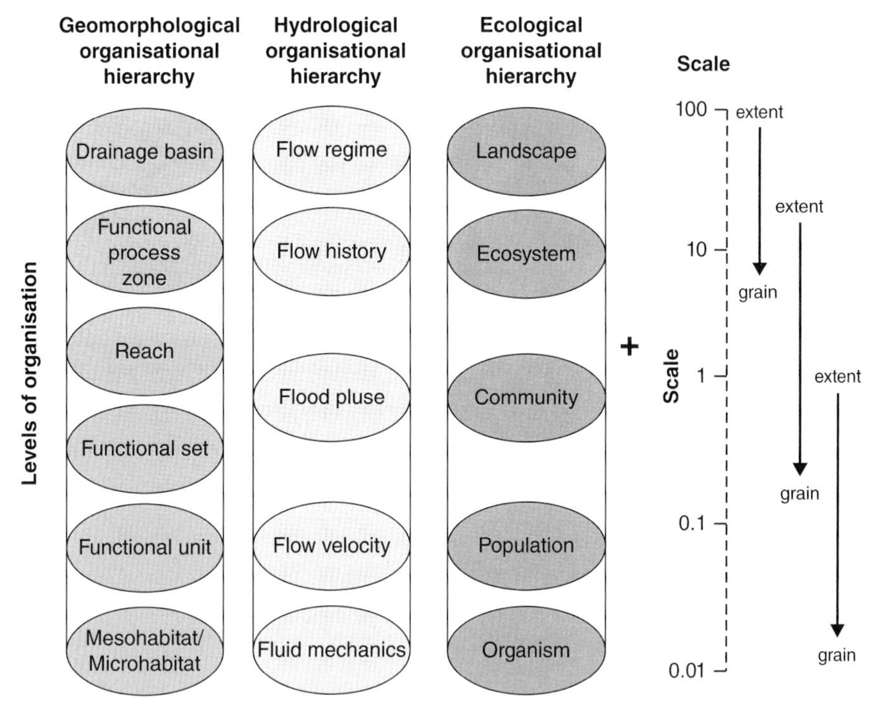

Figure 2.3 Organisational hierarchies in river science. To use this framework one must first define the relevant spatiotemporal dimension for the study or question. Scales for each hierarchy are then determined and allow the appropriate levels of organisation to be linked. The scale at the right demonstrates that linking levels across the three hierarchies may be vertical, depending on the nature of the question (modified from Dollar *et al.*, 2007; Thorp *et al.*, 2008).

Large-scale basin yield models commonly use statistically integrated representations of small-scale surface characteristics and processes that would be described explicitly in a small-scale urban flood-prediction model (Schubert *et al.*, 2008).

Step three: assignment of scales

The levels of organisation of the different sub-system hierarchies cannot be matched directly because they are scale independent and non-commensurate. For example, a baetid mayfly and a salmon both belong to the individual or organism level of the ecological hierarchy but each has a different perception in terms of grain and extent of patchiness within the landscape they inhabit. Each of these organisms responds to and influences a different range of geomorphological and flow characteristics. Likewise, an individual sand grain, a geomorphic unit, and a river reach will respond differently to a particular flow event, and the dynamics of each can also be described in terms of flow characteristics at different levels in the hydrological hierarchy. The three sub-systems can only be integrated to resolve a particular problem after appropriate scaling of the respective organisational structures thereby producing scaled hierarchies (Dollar *et al.*, 2007).

An important step here is identification of the spatiotemporal scales of the hierarchical sub-systems. In the case of an ecosystem framework for river science and management, scale is applied to the generic organisational hierarchies for geomorphology, hydrology and ecology (Figure 2.3). Integration of the three hierarchies requires them to be scaled according to the principles presented previously. In particular, different levels within each sub-systems' organisational hierarchy are related to the grain and extent of the proposed research or management activity. Assignment of scale to one level of organisation implies scales for all other levels, leading to a hierarchy of scales for a given sub-system. The same process of establishing scale is applied to the other two hierarchical sub-systems. Once spatial scales have been assigned, the temporal scales of sub-system components can also be distinguished, hence signifying their different frequencies of occurrence and/or rates of change. Processes at higher levels have lower rates and frequencies, and therefore operate more slowly and over larger spatial arenas than those at lower levels. Once scales have been set, decisions can be made that best address research and management goals at the appropriate scales (Figure 2.4).

Step four: component interactions

Interactions between the various sub-system hierarchies can be achieved in many ways and the use of flow chain models is a common approach (Shachak and Jones, 1995). Flow chain models have four basic components; the abiotic or biotic agent of change, or driver; the template or substrate upon which the driver acts; controllers of the driver; and an entity or process that responds to the driver (Shachak and Jones, 1995). Responders are sets of processes, organisms or parts of the physical environment and controllers act directly or indirectly on the agent. Organisms and processes respond differently to the sub-system in its different states. Floodwater, for example, acts as an agent of change by redistributing sediment. The pattern (product) of this redistribution may be controlled, in part, by large woody debris or outcropping bedrock in the channel (Dollar *et al.*, 2007). Flow chain models have been used in ecology to demonstrate changes in heterogeneity (Pickett *et al.*, 2003) and diversity (Shachak and Jones, 1995) and multi-level flow chain models allow for integration of time and space at multiple scales. Integration of scaled hierarchies with process models allows scientists and managers to view a landscape as a nested patch hierarchy rather than as a geographic arrangement of ecosystem components. This perspective allows the explicit recognition that agents of change act across a range of scales and helps in identifying those physical drivers of change, at appropriate scales. Biological response links in the hierarchy of flow chain models can also facilitate similar recognition and description of biological agents of change across the spectrum of scales.

The use and abuse of an interdisciplinary approach in the research and management of riverine landscapes

River science is an exciting and robust interdisciplinary scientific endeavour. Because of its fundamentally integrative nature, it has great potential for not only generating broad understanding about complex natural phenomena but also in providing relevant and critical guidance to decision makers seeking solutions to environmental problems and to river managers. Such understandings must be generated from sustained research at multiple scales and applied in multiple settings (Likens, 1992). Models, theories

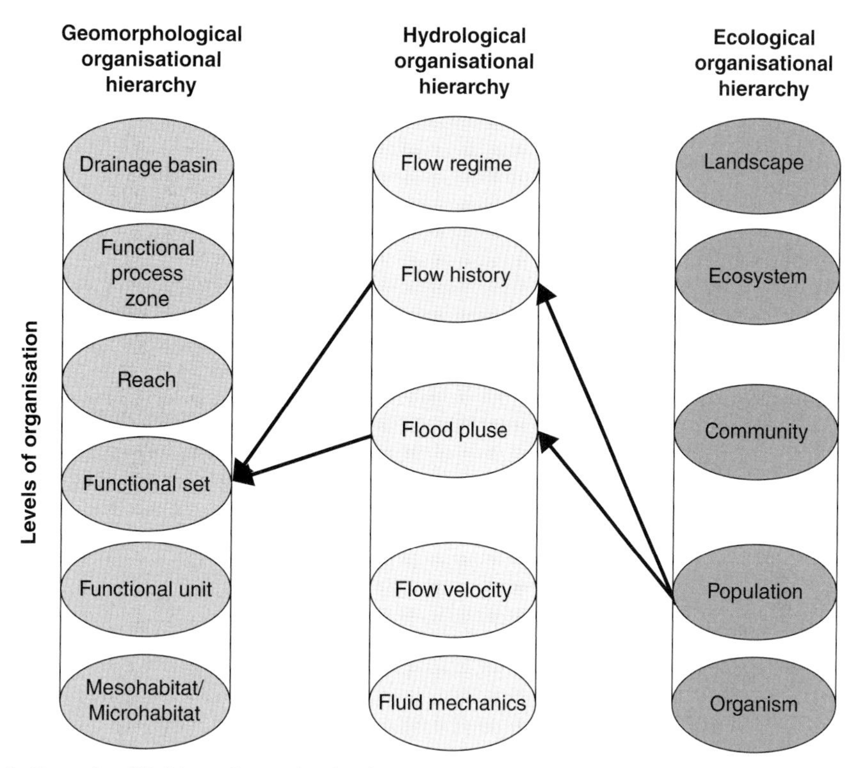

Figure 2.4 Example of linking relevant levels of organisation and scales for research or management. If the question relates to, in this example, the behavioural pattern of fish movements, the ecological level of organisation is the population level. Behavioural patterns would be influenced greatly by short-term (flood pulse) and longer-term (flow history) hydrological conditions. The most appropriate geomorphological scale at which to observe patch use and movement behaviour would be the functional set. The starting point for a study can be in any of the three levels of organisation.

and frameworks are only of academic value if not applied. In this section, the conceptual framework outlined in the previous sections is applied in a range of settings.

Monitoring and evaluation

The Sustainable Rivers Audit (SRA) is a long-term programme developed to assess river ecosystem health of the Murray–Darling Basin, Australia. Development of the SRA was strongly influenced by the Environmental Monitoring and Assessment Program (EMAP) of the U.S. Environmental Protection Agency, particularly for the development of study design and criteria for indicators (Davies *et al.*, 2010). The focus of the programme is to use a multidisciplinary approach to assess ecosystem health of river valleys within the Murray–Darling Basin. Surrogate measures of ecosystem processes (e.g., nutrient cycling, energy flow, sediment transport) derived from variables obtained under five themes – hydrology, fish, macroinvertebrates, vegetation and physical form – are used to generate indicators of ecosystem health. The expressed scale of measurement within each theme is the valley scale. Multiple scales were also considered in the design in recognition that some processes, or organisms, will respond across short- to long-term spatial and temporal scales

(Davies *et al.*, 2010). A pilot of the SRA was initiated in 2001 before the first formal audit over 2004–07.

Data for the hydrological, physical, fish and macroinvertebrate themes were all collected at the site scale. The exception was the vegetation theme where data were collected in a manner that was most appropriate for the FPZ to valley scale. Observations for the themes were aggregated in an attempt to 'scale up' to the valley scale. Further data aggregation was performed to distill the range of observations to create a single indicator score of ecosystem health. Regardless of whether the 'site scale' is considered to be a mesohabitat, functional unit or functional set, all of these are levels of organisation that would not exert an influence on ecological structure and function at the valley scale. There is a fundamental mismatch of scale in terms of the scale at which data were collected and the scale at which these data were applied for the river assessment; sensu hierarchy theory. In addition, there is further aggregation of data to single metrics of ecosystem health, further limiting the potential to elucidate the nature of responses of the theme-based variables, which are likely to be more sensitive to agents of change across multiple spatiotemporal scales. This latter element can only be divined through multivariate comparisons of predictor and response variables, even when comparisons are made for observed vs. expected (reference) expected conditions for response variables.

Issues of scale mismatch and inappropriate data aggregation are common to many monitoring and assessment schemes, including EMAP (Angradi *et al.*, 2006). The framework prescribed for the SRA is sound from the perspective of ecosystem structure; however, its utility is diminished by how scale is applied and data organisation. As outlined above, the levels of organisation immediately above and below place constraints and limits on the operational mechanisms of the level of interest. Levels of organisation that are farther removed exert less influence on structure and processes operating at the organisational level of interest. In this case, evaluation at the valley scale should also focus on the functional process zone and not the site level because the organisation and characteristics of FPZs will define process and structure of the river valley. Moving the focus from the site to FPZ level also changes the temporal scale because sensitivity to change in physical, chemical and ecological character are lower at larger spatial scales (Figure 2.2). The scope of observation of driver variables, therefore, must encompass longer temporal scales to fully assess their impact of ecosystem health.

Data for indicator variables must also be collected to reflect the appropriate scales at which the assessment is taking place. Aggregating data from multiple sites does not equate to measurements relatable to higher levels of organisation. Unlike the site scale, where small sample areas would be the determinant for a sample scheme, broader collection of data would, in the case of the SRA's intent, need to more closely reflect conditions at the valley and FPZ scale. Broad representation would not necessarily reflect every type of habitat present since this level(s) of organisation would not place limits on structure and organisation observed at the FPZ and valley scale (Thorp *et al.*, 2008). Sample themes should also reflect relevant levels of organisation. For example, direct measures for the generation of food web models and mass balance of nutrient processes are possible across multiple scales, including river valley and FPZs (sensu Woodward and Hildrew,

2002; Richardson *et al.*, 2004). Moreover, functional measures would provide a more representative comparison against short-term and long-term hydrological and geomorphological variables.

Ecosystem goods and services

Goods and services provided by river ecosystems have become an increasingly important part of the rationale in the maintenance of natural integrity and rehabilitation of riverine landscapes (Thorp *et al.*, 2010; Yeakley *et al.*, Chapter 17). Appreciation of scale is a critical component of determining the types of ecosystem services currently available as well as ascertaining the value of reestablishing a given type(s) of services. The latter situation, therefore, requires *a priori* development of a conceptual model built within an ecosystem framework as a basis for decision making. Reconnection of floodplains for water storage or nutrient sequestration provides an example of this. Functional process zones differ in the role of river–floodplain connectivity relative to floodplain geomorphology and hydrological patterns. When reconnection of a river channel to its floodplain is considered, the question might be 'how much spatial and temporal reconnection is needed to realise a cost-effective return on water storage?' Thorp *et al.* (2010) propose that benefits may reach an asymptote whereby there would be a point at which further setback of levees may lead to increased costs and loss of other services (e.g., agricultural land) while providing little additional gains in water storage. The decision-making process must, therefore, consider where services are currently provided, where in the river network these services can be reintroduced, and to what extent they can be reestablished while not adversely impacting other services. Also included in this would be recognition of the relevant scale(s) to address for optimising benefits.

Nutrient sequestration and removal, while an ecosystem-level process, can be evaluated at different spatial and temporal scales. Mass balance determination of inputs and outputs at the river valley scale are useful in determining basins critical to contributions of nitrogen. Such was the case in identifying the Upper Mississippi River Basin as the major source of nitrogen loading contributing to the hypoxia zone in the Gulf of Mexico (Rabalais and Turner, 2001). Smaller scale evaluation becomes important only when more specific information on location of nutrient input and removal are needed. Richardson *et al.* (2004) quantified nitrogen dynamics across a suite of functional sets in the Upper Mississippi River with backwaters identified as key locations for denitrification. They also determined that the capacity for denitrification in backwaters was limited because natural hydrological dynamics delivered nitrogen to backwaters in the spring when biological activity would be low. Richardson *et al.* (2004) further concluded that putting water into backwaters during biologically active periods would likely be of limited benefit because nitrogen content in river water would greatly exceed potential for denitrification. This was only possible through examination of nitrogen processes at the appropriate scale and integrating biological, geomorphological and hydrological conditions.

Research

The key attribute of bringing this approach into research is recognising that an ecosystem framework is not limited to addressing questions at the ecosystem level of organisation. Indeed, this framework can be implemented for any geomorphological, hydrological, or ecological level of organisation through the application of relevant spatial and temporal scales. Realisation of scale, in turn, allows for establishment of research boundaries that are identified through self-emergence of the inherent properties through multivariate analysis rather than arbitrarily designating a sample area. Reliance on the self-emergent properties of a study area or focal level of organisation is critical in that it minimises bias that commonly occurs when decisions are based solely on investigator-based criteria. The examples below help to demonstrate the applicability of an ecosystem approach. Moreover, these investigations highlight the benefits of interdisciplinary studies and the value added in considering the influence of levels of organisation above and below the focus level of a study for achieving greater insights on pattern and process. They also demonstrate the need to not rely solely on a reductionist approach with a single predictive variable or a suite of variables at a single scale. Study designs based on a holistic ecosystem framework increase the likelihood of generating meaningful causal relationships across and within ecological, geomorphological and hydrological realms.

Parsons and Thoms (2007) examined the association between large woody debris and channel morphology in the River Murray across multiple spatial scales. Reach was the highest spatial scale, which consisted of a 95-km section of the river. Eight functional sets were identified using variables describing channel planform. Functional units were identified within each functional set based on location longitudinally and laterally within the channel (Figure 2.5). Location of wood was identified through aerial photographs and field observations. The location of large wood in the channel was recorded as well as length, angle to flow, distance from bank, structural complexity and length. Initial analysis revealed no distinct organisation of wood at the reach scale, indicating that distribution was uniform. The degree of curvature of the channel was a key determinant of distribution of large wood in functional sets and eight of the twelve types of functional units possessed unique distributions of wood. They concluded that examination at only one scale would not have fully described the influence of channel morphology on the arrangement of large woody debris and, in fact, would have identified no pattern had the study been done only at the reach scale (Parsons and Thoms, 2007). The study also demonstrated the application of multivariate statistics in establishing boundaries for the hierarchical scales.

Southwell and Thoms (2011) took a multi-scale approach to examine the pattern of sediment character and nutrient concentrations of inset-floodplain surfaces (benches) within the paleochannel trough of the Barwon–Darling River, Australia. Two reaches were identified and, within each reach, benches were selected within confined and unconfined sections of the paleochannel trough and channel planform. Sediment samples were identified based on grain size and entropy analysis revealed five sediment texture groups. Total carbon, total nitrogen and total phosphorus were measured from subsamples of sediment samples. Southwell and Thoms (2011) observed no strong longitudinal or lateral gradient in either nutrient concentration or sediment texture. They did, however, note that nutrient concentrations were closely correlated with sediment texture and that elevation of inter-channel bench surfaces also influenced sediment character. They concluded that the absence of a

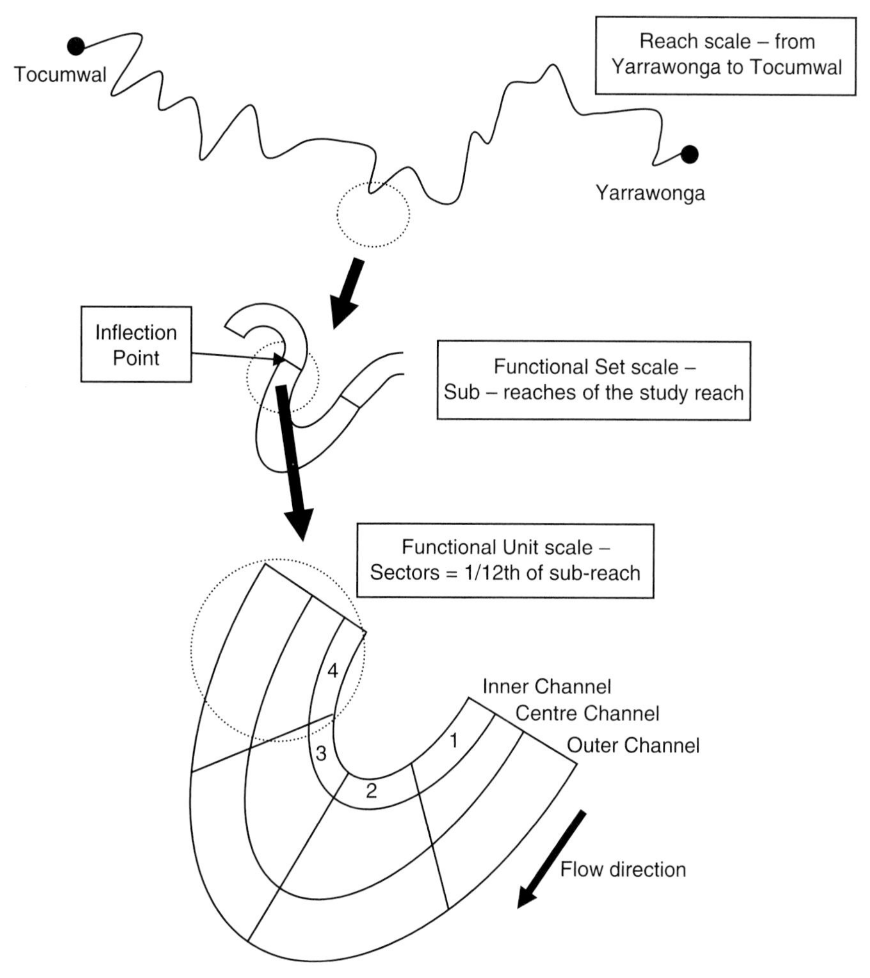

Figure 2.5 Example of the division of a study reach into different scales of measurement to examine the relationship between channel morphology and the location and abundance of large wood. As a perspective, the distance between Tocumwal and Yarrawongo is 46 km on a straight line. After Parsons and Thoms (2007).

longitudinal and lateral pattern in nutrient concentrations indicated that larger-scale geomorphic characteristics were not placing constraints on the spatial distribution of nutrients (Southwell and Thoms, 2011). Instead, arrangement of nutrients in the Barwon–Darling River was dictated by conditions at smaller spatial scales, specifically sediment texture and elevation. They also emphasise that the spatial mosaic evident in nutrient concentration across larger spatial scales highlights the patchy nature of river floodplains, thereby necessitating a scale-based analysis of form and processes.

Webb *et al.* (2011) considered the influence of hydrological and physical characteristics on fish communities within in-channel waterholes across multiple rivers in the Lower Balonne River system. Historical records from gauging stations closest to each site were used to generate a suite of hydrological variables reflective of flow regime, flow history and flow pulse. Fish were identified, measured for placement into size-classes, and counted to obtain community- and population-level measures. They determined that hydrological conditions at different temporal scales shaped different components in the organisation of fish assemblages (Webb *et al.*, 2011). Long-term hydrological measures acted as a determinant of community-level organisation, specifically, the number and types of species present as well as their persistence. Hydrological variables reflective of flow-pulse, on the other hand, acted on population-level traits – the size structure of individual populations. Moreover, they noted that long-term hydrological character operated at the spatial scale of the Lower Balonne system, whereas flow-pulse factors placed constraints at the waterhole scale. As was the case in the other examples, this study highlights the importance of incorporating multiple spatial and temporal spatial scales. It also accentuates the need to consider hydrology as something more than simply a measure of current velocity or annual discharge even for a study where these measures are scale appropriate.

Summary

River science is an exciting and robust field. Because of its fundamentally integrative nature there is great potential for generating broad understanding about complex natural phenomena. Such understanding, generated from sustained research at multiple scales, can also provide relevant and critical guidance to decision-makers seeking solutions to environmental problems and to managers of natural resources. The direction we have laid out for the future of river research and management is a marked departure from past approaches. Viewing river systems from an interdisciplinary perspective demands a philosophy of being scale aware and moves away from a reliance on the conventional reductionist falsification approach which limits our understanding of rivers as complex systems. Moreover, it is essential that we recognise the limitations of attempting to use conceptual models, rather than frameworks, as a basis for establishing the basis for research and management (Dollar *et al.*, 2007). Problems arise when conceptual models are too general (e.g., Vannote *et al.*, 1980), particularly when their lack of a scalar perspective leads to extrapolation of data among disparate regions which contribute to a lack of complete understanding of ecological and hydrogeomorphological character (Fausch *et al.*, 2002; Carboneau *et al.*, 2012).

Hierarchy and scale are fundamental tenets of hydrology, fluvial geomorphology and freshwater ecology. However, identification of appropriate scales or levels of organisation that link similar attributes across sub-disciplines is rarely attempted because of entrenched views within individual disciplines. Our framework matches a problem with a river system process in a manner whereby the appropriate causal explanations can be identified at the correct spatial and temporal scales. This, in turn, allows consideration of paradigms from different disciplines that may be descriptive, explanatory or experimental but which ultimately lead to multiscale predictions of pattern–process and process–pattern relationships. The primary components of the framework are:

(a) There should be an emphasis on the need to define the study domain in terms of its spatial and temporal dimensions.

(b) Ecological, geomorphological and hydrological complexity can only be deconstructed by research at multiple scales. Multi-scale studies provide a mechanism for embedding small scale understanding within the context of larger scale understanding.

(c) Studies at different scales are amenable to different approaches. At large scales there is seldom the luxury of replication and controls so that generalisation (pattern seeking) and causal explanation are more appropriate than experimental testing.

(d) The classic emphasis on falsifiability is too restrictive for river science because the prerequisites for its use – universality and simple causality – seldom apply in natural systems where organisms and their abiotic environment are characterised by multiple causality.

To achieve the full potential of an ecosystem framework, it is critical to train ourselves and our students to carry out interdisciplinary research. The first step is the recognition that multidisciplinary and interdisciplinary are not the same things. Integration of the sub-disciplines of river science and the emergence of new paradigms is required for it to be truly interdisciplinary and this includes the unravelling of their own key concepts and operational scales. We may attempt to do interdisciplinary research individually or through collaboration of ideas and data, but all too often the response is to retreat to the friendly confines of our own disciplines in the face of apparent complexity of interdisciplinary research. All too often integrating the different sub-disciplinary hierarchies has been a stumbling block for true interdisciplinary science. Integration of the sub-disciplines can be accomplished through the merging of the hierarchical levels of organisation and application of scale. Hierarchy theory has been around for some time, but its application has been restricted largely because of its apparent complexity. Many view hierarchy theory as highly complex and as more of an esoteric academic exercise. When, however, it is put into a framework such as that provided here, hierarchy becomes accessible and ripe for application. Complexity is something to embrace rather than from which to retreat. In embracing complexity, we can begin to make hierarchy, and River Science, something that is used and not just useful (Rogers, 2008).

References

Aarts, B. G. W. and P. H. Nienhuis. 2003. Fish zonations and guilds as the basis for assessment of ecological integrity of large rivers. *Hydrobiologia* 500:157–178.

Angradi, T. R., E. W. Schweiger, B. H. Hill, *et al.* 2006. *Environmental Monitoring and Assessment Program: Great River Ecosystems Field Operations Manual*. U. S. E. P. Agency, editor., Washington, DC.

Barrett, G. W., J. D. Peles and E. P. Odum. 1997. Transcending processes and the levels of organisation concept. *Bioscience* 47:531–535.

Boys, C. A. and M. C. Thoms. 2006. A hierarchical scale approach to the assessment of fish assemblages and their habitat associations in large dyland rivers. *Hydrobiologia* 572:11–31.

Cadenasso, M. L., S. T. A. Pickett, K. C. Weathers and C. G. Jones. 2003. A framework for a theory of ecological boundaries. *Bioscience* 53:750–758.

Carbonneau, P., M. A. Fonstad, W. A. Marcus and S. J. Dugdale. 2012. Making riverscapes real. *Geomorphology* 137:74–86.

Cullen, P. 1990. The turbulent boundary between science and water management. *Freshwater Biology* 24:201–209.

Culp, J. M. and R. W. Davies. 1982. Analysis of longitudinal zonation and the river continuum concept in the Oldham-South Saskatchewan River system. *Canadian Journal of Fisheries and Aquatic Sciences* 39:1258–1266.

Cushing, C. E., C. D. McIntire, J. R. Sedell *et al.* 1983. Relationships among chemical, physical, and biological indices along the river continua based on multivariate analyses. *Archiv für Hydrobiologie* 98:317–326.

Davies, P. E., J. H. Harris, T. J. Hillman and K. F. Walker. 2010. The Sustainable Rivers Audit: assessing river ecosystem health in the Murray-Darling Basin, Australia. *Marine and Freshwater Research* 61:764–777.

Delong, M. D. and J. H. Thorp. 2006. Significance of instream autotrophs in the trophic dynamics of a large river. *Oecologia* 147:76–85.

Dollar, E. S. J., C. S. James, K. H. Rogers and M. C. Thoms. 2007. A framework for interdisciplinary understanding of rivers as ecosystems. *Geomorphology* 89:147–162.

Fausch, K. D., C. E. Torgersen, C. V. Baxter and H. W. Li. 2002. Landscapes to riverscapes: bridging the gap between research and conservation of stream fishes. *Bioscience* 52:483–498.

Fisher, S. G. 1997. Creativity, idea generation, and the functional morphology of streams. *Journal of the North American Benthological Society* 16:305–318.

Frissell, C. A., W. J. Liss, C. E. Warren and M. D. Hurley. 1986. A hierarchical framework for stream habitat classification: viewing streams in a watershed context. *Environmental Management* 10:199–214.

Graham, M. H. and P. K. Dayton. 2002. On the evolution of ecological ideas: paradigms and scientific progress. *Ecology* 83:1481–1489.

Gregory, K. J. and D. E. Walling. 1973. *Drainage Basin Form and Process: A Geomorphological Approach*. Edward Arnold, London.

Hawkes, H. A. 1975. River zonation and classification. In: B. A. Whitton (ed.), *River Ecology*. Blackwell Science, Oxford, UK: pp. 312–374.

Huet, M. 1959. Profiles and biology of Western European streams as related to fish management. *Transactions of the American Fisheries Society* 88:155–163.

Hynes, H. B. N. 1970. *The Ecology of Running Waters*. Toronto Press, Toronto, Ontario.

Illies, J. and L. Botosaneaunu. 1963. Problémes et méthodes de la classification et de la zonation écologique des eaux courantes, considerées surtout du point de vue faunistique. *Verhandlungen der Internationalen Vereinigung für Theoretische und Angewandte Limnologie* 12:1–57.

Lau, S. C. P., K. M. Y. Leung and D. Dudgeon. 2009. What does stable isotope analysis reveal about trophic relationships and the relative importance of allochthonous and autochthonous resources in tropical streams? A synthetic study from Hong Kong. *Freshwater Biology* 54:127–141.

Leopold, L. B., M. G. Wolman and J. P. Miller. 1964. *Fluvial Processes in Geomorphology*. Freeman, San Francisco.

Lewis, W. M., Jr.,, S. K. Hamilton, M. A. Rodriquez, *et al.* 2001. Foodweb analysis of the Orinoco floodplain based on production estimates and stable isotope data. *Journal of the North American Benthological Society* 20:241–254.

Likens, G. E. 1992. *The Ecosystem Approach: Its Use and Abuse*. Ecology Institute. Oldendorf/Lube, Germany.

Marcarelli, A. M., C. V. Baxter, M. M. Mineu, and R. O. Hall. 2011. Quantity and quality: unifying food web and ecosytem perspectives on the role of resource subsidies in freshwaters. *Ecology* 92: 1215–1225.

Minshall, G. W., R. C. Peterson, K. W. Cummins, *et al.* 1983. Interbiome comparison of stream ecosystem dynamics. *Ecological Monographs* 53:1–25.

Minshall, G. W., K. W. Cummins, R. C. Petersen, *et al.* 1985. Developments in stream ecosystem theory. *Canadian Journal of Fisheries and Aquatic Sciences* 42:1045–1055.

Montgomery, D. R. 1999. Process domains and the river continuum concept. *Journal of the American Water Resources Association* 35:397–410.

Montgomery, D. R. and J. M. Buffington. 1998. Channel processes, classifcation, and response. In: R. J. Naiman and R. E. Bilby (eds), *River Ecology and Management: Lessons from the Pacific Coastal Ecoregion*. Springer-Verlag, Berlin: pp. 13–42.

O'Neill, R. V., A. R. Johnson and A. W. King. 1989. A hierarchical framework for the analysis of scale. *Landscape Ecology* 3:193–205.

Parsons, M. and M. C. Thoms. 2007. Hierarchical patterns of physical-biological association in river ecosystems. *Geomorphology* 89:127–146.

Perry, J. A. and D. J. Schaeffer. 1987. The longitudinal distribution of riverine benthos: a river dis-continuum? *Hydrobiologia* 148:257–268.

Peterson, D. L. and V. T. Parker. 1998. *Ecological Scale - Theory and Application: Complexity in Ecological Systems*. Columbia University Press, New York.

Petts, G. E. and C. Amoros. 1996. *Fluvial Hydrosystems*. Chapman & Hall, London.

Pickett, S. T. A. and M. L. Cadenasso. 2002. The ecosystem as a multidimensional concept: meaning, model, and metaphor. *Ecosystems* 5:1–10.

Pickett, S. T. A., J. Kolasa and C. G. Jones. 1994. *Ecological Understanding: The Nature of Theory and the Theory of Nature*, 1st edition. Academic Press, London.

Pickett, S. T. A., W. R. Burch and J. M. Grove. 1999. Interdisciplinary research: maintaining the constructive impulse in a culture of criticism. *Ecosystems* 2:302–307.

Pickett, S. T. A., M. L. Cadenasso and T. L. Benning. 2003. Biotic and abiotic variability as key determinants of savanna heterogeneity at multiple spatiotemporal scales. In: J. T. Du Toit, K. H. Rogers and H. C. Biggs (eds), *The Rivers of Kruger National Park*. South African Water Research Commission, Pretoria: pp. 22–40.

Pickett, S. T. A., J. Kolasa and C. G. Jones. 2007. *Ecological Understanding: The Nature of Theory and the Theory of Nature*, 2nd edition. Elsevier, London.

Poole, G. C. 2002. Fluvial landscape ecology: addressing uniqueness within the river discontiunuum. *Freshwater Biology* 47:641–660.

Quinn, G. P. and M. Keough. 2002. *Experimental Design and Data Analysis for Biologists*. Cambridge University Press, New York.

Rabalais, N. N. and R. E. Turner. 2001. Hypoxia in the northern Gulf of Mexico: Description, causes and change. *Coastal and Estuarine Studies* 58:1–36.

Rapport, A. 1985. Thinking about home environments: a conceptual framework. In: I. Altman and C. M. Werner (eds), *Home Environments*. Plenum Press, New York: pp. 255–286.

Richardson, W. B., E. A. Strauss, L. A. Bartsch, *et al.*. 2004. Denitrification in the Upper Mississippi Rive: rates, controls, and contribution to nitrate flux. *Canadian Journal of Fisheries and Aquatic Science* 61:1102–1112.

Rogers, K. H. 2003. Adopting a heterogeneity paradigm: implications for management of protected savannas. In: J. T. Du Toit, K. H. Rogers and H. C. Biggs (eds), *The Kruger Experience: Ecology and Management of Savanna Heterogeneity*. Island Press, Washington: pp. 41–58.

Rogers, K. H. 2008. Kilham Memorial Lecture: Limonlogy and the post-normal imperative: an African perspective. *Verhandlungen der Internationalen Vereinigung für Theoretische und Angewandte Limnologie* 30:171–185.

Schubert, J. E., B. F. Saunders, M. J. Smith and N. G. Wright. 2008. Unstructured mesh generation and landcover-based resistance for hydrodynamic modeling of urban flooding. *Advances in Water Resources* 31:1603–1621.

Shachak, M. and C. G. Jones. 1995. Ecological flow chains and ecological systems: concepts for linking species and ecosystem perspectives. In: C. G. Jones and J. H. Lawton (eds), *Linking Species and Ecosystems*. Chapman & Hall, London: pp. 280–294.

Southwell, M. and M. Thoms. 2011. Patterns of nutrient concentrations across multiple floodplain surfaces in a large dryland river system. *Geographical Research* 49:431–443.

Southwood, T. R. E. 1977. Habitat, the templet for ecological strategies? *Journal of Animal Ecology* 46:336–365.

Statzner, B. and B. Higler. 1985. Questions and comments on the river continuum concept. *Canadian Journal of Fisheries and Aquatic Sciences* 42:1038–1044.

Strayer, D. L., M. E. Power, W. F. Fagan, *et al.* 2003. A classification of ecological boundaries. *Bioscience* 53:723–729.

Tansley, A. G. 1935. The use and abuse of vegetational concepts and terms. *Ecology* 16:284–307.

Thoms, M. C. 2006. Variability in riverine ecosystems. *River Research and Applications* 22:115–121.

Thoms, M. C. and F. Sheldon. 2000. Lowland rivers: an Australian introduction. *Regulated Rivers: Research and Management* 16:375–383.

Thoms, M. C. and M. Parsons. 2002. Ecogeomorphology: an interdisciplinary approach to river science. *International Association of Hydrological Sciences* 27:113–119.

Thoms, M. C., S. M. Hill, M. J. Spry, *et al.* 2004. The geomorphology of the Darling River. In: *R. Breckwodt*, R. Boden and J. Andrews (eds), *The Darling*. The Murray Darling Basin Commission: pp. 68–105.

Thoms, M. C., S. Rayburg and M. Neave. 2007. The physical diversity and assessment of a large river system: The Murray-Darling Basin, Australia. In: A. Gupta (ed.), *Large Rivers*. Wiley, Chichester: pp. 587–608.

Thorp, J. H. and M. D. Delong. 2002. Dominance of autotrophic autochthonous carbon in food webs of heterotrophic rivers. *Oikos* 96:543–550.

Thorp, J. H., M. C. Thoms and M. D. Delong. 2008. *The Riverine Ecosystem Synthesis: Towards Conceptual Cohesiveness in River Science*. Academic Press, Amsterdam.

Thorp, J. H., J. E. Flotemersch, M. D. Delong, *et al.* 2010. Linking ecosystem services, rehabilitation, and river hydrogeomorphology. *Bioscience* 60:67–74.

Tittizer, T. and F. E. Krebs. 1996. *Ökosystemforschung: Der Rhein und seine Auen - Eine Bilanz*. Springer, Berlin.

Townsend, C. R. 1989. The patch dynamics concept of stream community ecology. *Journal of the North American Benthological Society* 8:36–50.

Van Niekerk, A. W., G. L. Heritage and B. P. Moon. 1995. River classification for management: the geomorphology of the Sabie River in the Eastern Transvaal. *South African Geographical Journal* 77:68–76.

Vannote, R. L., G. W. Minshall, K. W. Cummins, *et al.* 1980. The river continuum concept. *Canadian Journal of Fisheries and Aquatic Sciences* 37:130–137.

Ward, J. V. and K. Tockner. 2001. Biodiversity: towards a unifying theme for river ecology. *Freshwater Biology* 46:807–819.

Ward, J. V., K. Tockner and D. B. Arscott. 2002. Riverine landscape diversity. *Freshwater Biology* 47:517–539.

Webb, M., M. Reid and M. Thoms. 2011. The influence of hydrology and physical habitat character on fish assemblages at different temporal scales. *River Systems* 19:283–299.

Wiens, J. A. 2002. Riverine landscapes: taking landscape ecology into the water. *Freshwater Biology* 47:501–516.

Winterbourn, M. H., J. S. Rounick and B. Cowie. 1981. Are New Zealand stream ecosystems really different? *New Zealand Journal of Marine and Freshwater Research* 15:321–328.

Woodward, G. and A. G. Hildrew. 2002. Food web structure in riverine landscapes. *Freshwater Biology* 47:777–798.

CHAPTER 3

Fine sediment transport and management

Desmond E. Walling[1] and Adrian L. Collins[2]
[1] Department of Geography, College of Life and Environmental Sciences, University of Exeter, Exeter, UK
[2] Sustainable Soils and Grassland Systems Department, Rothamsted Research, North Wyke, UK

Background and context

Traditionally, studies of sediment transport by rivers have distinguished the coarse bedload component from the finer suspended load. The latter component is often further subdivided into a coarser fraction, designated the suspended bed material load, and a finer fraction termed the wash load (Shen, 1981).The wash load is commonly assumed to be derived from the catchment surface, to be rapidly transported through the channel system and to have limited interaction with the channel bed. As such it was generally seen by hydraulic engineers as having limited importance for river morphology and river management. By virtue of its source outside the river channel and the fact that most rivers can transport a much greater wash load than is actually transported, the wash load differs from the suspended bed material and bedload in that it is a non-capacity load that is supply controlled, rather than being controlled by the transport capacity of the river. This means that it is difficult to predict using hydraulic variables, and it was commonly excluded from theoretical treatments of sediment transport as being something that needed to be measured, should it prove important. Against this background, fine sediment transport by rivers traditionally received relatively little attention, compared with the coarser load, except where reservoir sedimentation was a potential problem or such information was used to assess rates of soil loss or land degradation (e.g., Graf, 1971; Shen, 1981).

Two developments changed this situation and directed increased attention to fine sediment transport by rivers. The first, which can be traced to the 1970s and 1980s, was the increasing recognition of the importance of fine sediment as a vector for the transfer of nutrients and contaminants through river systems (see Förstner and Muller, 1974; Golterman, 1977; Golterman *et al.*, 1983; Allan, 1986). Fine sediment particles are highly active chemically and act as a substrate for the adsorption of nutrients, particularly phosphorus (P), and many contaminants such as heavy metals, pesticides and other persistent organic pollutants (POPs). Sediment-associated transport can exert a key control on the transfer and fate of such substances within fluvial systems and an understanding of fine sediment transport and loads is an essential pre-requisite for understanding and controlling nutrient and contaminant fluxes and diffuse source

River Science: Research and Management for the 21st Century, First Edition.
Edited by David J. Gilvear, Malcolm T. Greenwood, Martin C. Thoms and Paul J. Wood.

pollution. This was well demonstrated by the pioneering work of the joint US–Canada International Commission on the Great Lakes (IJC) and its Pollution from Land Use Activities Reference Group (PLUARG) in the 1970s. This aimed to reduce eutrophication and pollution in Lake Erie and Lake Ontario and identified the need to reduce the mobilisation of sediment from agricultural land and its transport to the lakes (PLUARG, 1978).

The second development is linked to the above and reflects the growing recognition of the wider ecological importance of fine sediment in degrading aquatic and riparian ecology and habitats. This degradation is partly a response to the pollutants that are frequently associated with fine sediment, but can also reflect the physical impact of excessive amounts of fine sediment. The latter can, for example, involve reduced light transmission and smothering of the stream bed and aquatic vegetation. The silting of fish spawning gravels, which reduces the flow of water through the gravels and the supply of oxygen to the eggs (Heywood and Walling, 2007; Sear *et al.*, 2014), is another example. There are, however, many other ways in which fine sediment can impact adversely on aquatic ecology (see Chapman *et al.*, 2014; Collins *et al.*, 2011; Jones *et al.*, 2012a,2012b, 2014; Kemp *et al.*, 2011; Thompson *et al.*, 2014; Von Bertrab *et al.*, 2013; Wagenhoff *et al.*, 2013; Wood and Armitage, 1997).

The environmental problems outlined above highlight the potential role of fine sediment as a pollutant and this has been recognised in the EU within the Water Framework (European Parliament, 2000), Freshwater Fish (European Parliament, 2000) and Habitats Directives (European Council, 1992) and by the US Environmental Protection Agency (EPA) through the introduction of Total Maximum Daily Load (TMDL) standards (Hawkins, 2003). These problems have in turn directed increased attention to managing fine sediment mobilisation and transport and this has been coupled with a changing view of the significance of load magnitude. In the traditional hydraulic engineering context, linked to reservoir and channel sedimentation and land degradation, problems generally increased as sediment yields increased. In the wider ecological context, however, rivers with low sediment loads are often the most sensitive to small changes in fine sediment concentrations or load and such rivers can experience greater problems and necessitate more intensive management than those draining areas with higher sediment yields (Collins and Anthony, 2008a).

Key concepts

In seeking to develop an improved understanding of the fine sediment loads of rivers and to ultimately manage such loads, four key concepts can usefully be emphasised. These are, firstly, the non-capacity and supply-controlled nature of fine sediment transport, secondly, the significance of grain size, sediment composition and composite particles, thirdly, the importance of sediment source and finally the need to view the fine sediment load of a river as a component of the overall catchment sediment budget. These concepts will be briefly considered in turn.

Non-capacity supply controlled transport.

As indicated above, fine sediment or wash load transport differs from the transport of coarser sediment in that it cannot be treated as a capacity load. The supply is generally far more important than the transport capacity in determining the magnitude of the load. Such behaviour is clearly demonstrated by Figure 3.1. Figure 3.1a illustrates the variation of suspended sediment concentration during a sequence of storm hydrographs monitored at the outlet of the 46 km^2 catchment of the River Dart in Devon, UK. The data demonstrate that the sediment concentration and discharge peaks are out of phase and that the supply can be depleted and subsequently replenished during a sequence of events. Figure 3.1b presents the suspended sediment rating curve or plot of suspended sediment concentration versus discharge for the 262 km^2 catchment of the River Creedy at Cowley in Devon, UK. Suspended sediment concentrations can be seen to range over more than two orders of magnitude for a given water discharge or transport capacity and the sediment concentrations associated with a given discharge are significantly higher in summer than in winter and are generally higher on the rising stage than on the falling stage.

Sediment grain size, composition and composite particles

Recognition of the important role of fine sediment in the transport of nutrients and contaminants and its potential impact in degrading aquatic ecosystems has significantly expanded information requirements. In addition to information on the magnitude of fine sediment concentrations and loads, there is also a need for information on the properties, composition and structure of the sediment particles. Grain size composition exerts a key influence on sediment-associated transport, since clay- and fine silt-sized particles are generally more chemically active than larger particles (Horowitz, 1991). Likewise, the presence of organic matter, either as discrete particles, surface coatings or more complex associations with inorganic particles, can exert a key influence on the role of fine sediment as a substrate for contaminant transport. The complex nature of fine sediment transport is further emphasised by the fact that few particles will exist in isolation. Most will be transported as composite particles or flocs, comprising large numbers of smaller particles of mineral or organic matter and with highly complex structures (see Droppo, 2001; Droppo *et al.*, 2005). The individual components of flocs may be held together by several mechanisms, including electrochemical forces and sticky material and filaments associated with bacteria and extracellular polymeric substances (EPS). Figure 3.2 presents highly magnified images of several suspended sediment particles, which emphasise their complex structure. Traditional grain size analyses undertaken in the laboratory generally involve removal of organic matter and chemical and physical dispersion of the particles. The results may therefore bear little relation to the actual *in situ* or effective particle size of the particles transported by a river and any attempt to understand the hydrodynamic behaviour of suspended sediment particles must take this into account (Williams *et al.*, 2008).

The importance of sediment source

The need to understand sediment properties and the role of fine sediment in nutrient and contaminant transport necessarily directs attention to the importance of sediment source in influencing these key aspects. Source can be defined in terms of both spatial location within the upstream catchment (e.g., areas of contrasting geology or different sub-catchments) or source type, which reflect the processes responsible for sediment mobilisation and the related source areas. The latter could, for example, include

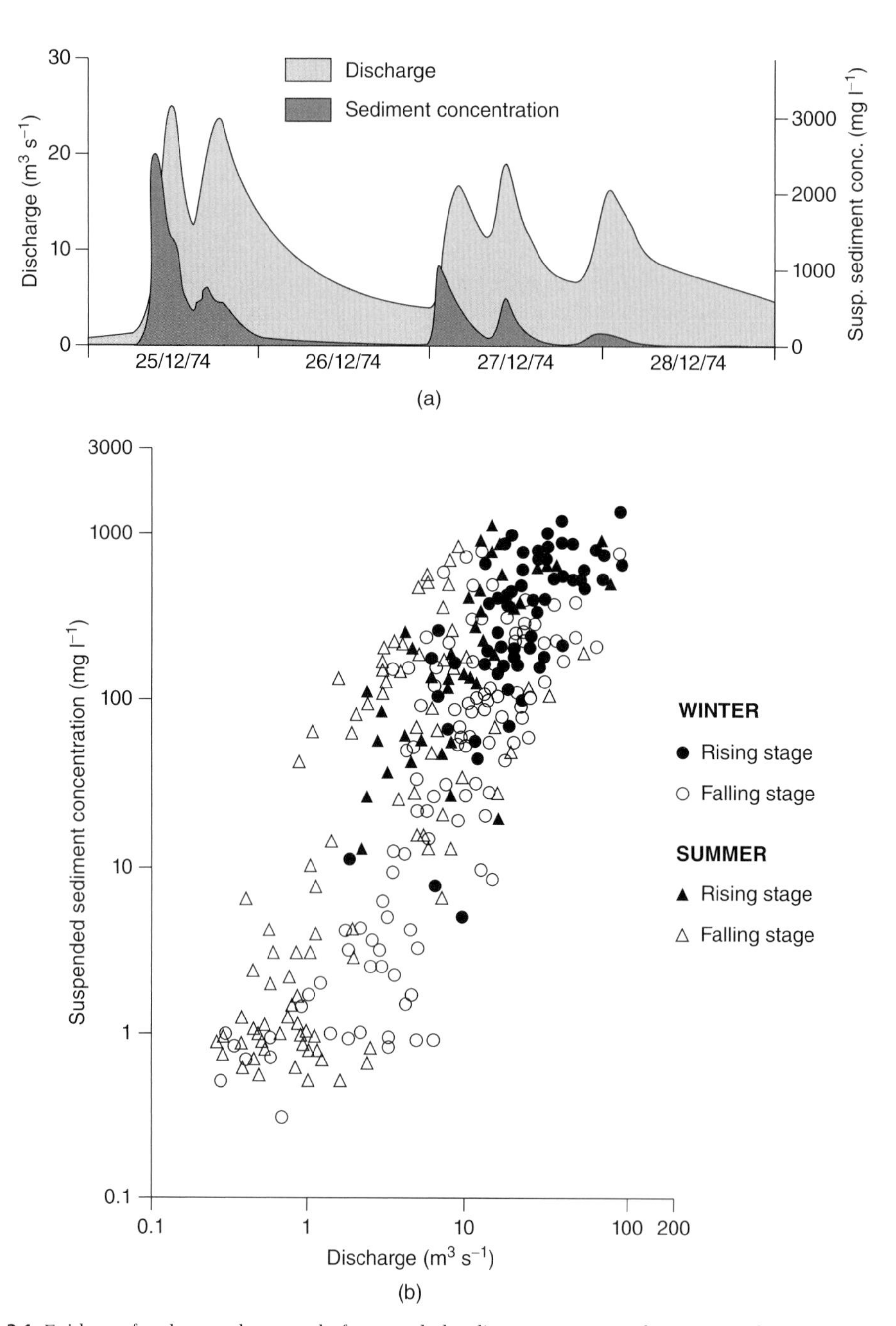

Figure 3.1 Evidence for the supply control of suspended sediment transport, showing (a) the variation of suspended sediment concentration through a sequence of storm hydrographs on the River Dart at Bickleigh, Devon, UK and (b) the relationship between suspended sediment concentraton and discharge for the River Creedy at Cowley, Devon, UK for the period 1972–74.

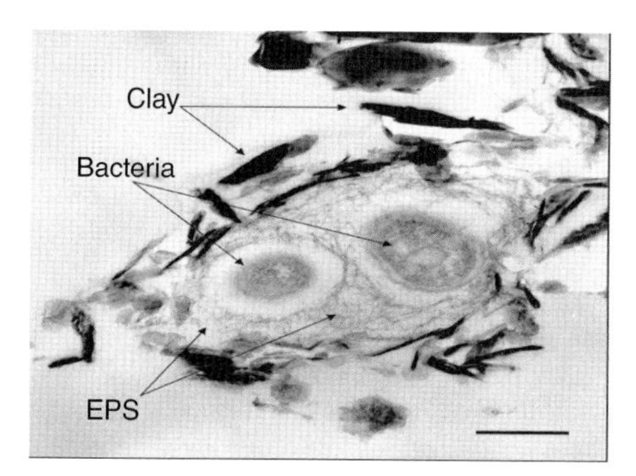

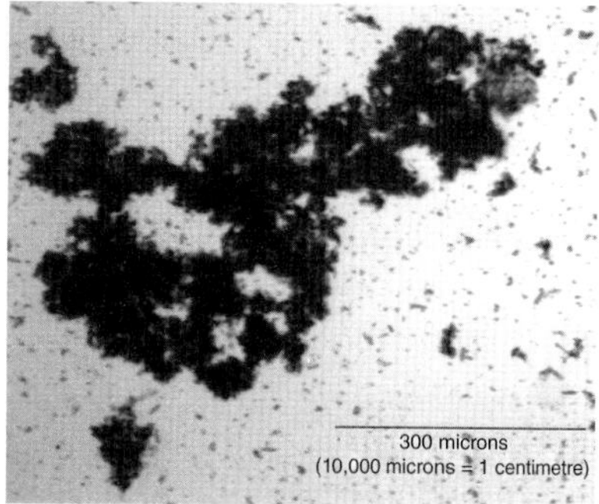

Figure 3.2 Micrographs of suspended sediment particles depicting (left) a small floc (scale bar = 0.5 μm) and (right) a group of larger flocs. (Source, Ian Droppo, Environment Canada.)

channel erosion, gully erosion and erosion of surface soils from areas under cultivation or pasture by sheet and rill erosion. In some catchments, roads and urban areas, point sources and effluent from sewage treatment can also represent important sources of fine particulates. Sediment sources can change both seasonally and between and during events and such changes can result in significant changes in sediment properties, including grain size (e.g., Ongley *et al.*, 1982). Information on the source of the sediment transported by a river is also likely to be of critical importance when developing sediment control or management strategies. To be effective and to maximise the return on expenditure, such strategies must target the most important sources (Gellis and Walling, 2011). Information on sediment source is difficult to obtain using traditional monitoring techniques, but recent advances in sediment source fingerprinting (Walling, 2013) have provided the means to obtain such information and this technique will be discussed further below.

Catchment sediment budgets

It is important to recognise that the fine sediment output from a catchment represents the result of a complex interaction of sediment mobilisation from a variety of sources within the catchment, the transfer of that sediment to and through the channel system, and the temporary and longer-term storage of the sediment as it moves through the sediment delivery continuum. This system must be understood if fine sediment transport and yields, and changes resulting from climate change or changing land use and land management, are to be successfully predicted. Much of the sediment mobilised from the upstream catchment area may not reach the catchment outlet. Equally, sediment yields could change as a result of remobilisation of stored sediment. The catchment sediment budget, as proposed by Trimble (1983) and illustrated in Figure 3.3 for the now classic example of Coon Creek, Wisconsin, USA, affords a valuable conceptual tool for representing this complex interaction of sources and sinks. In the 360 km^2 Coon Creek catchment, only ~5–7% of the sediment mobilised within the basin reached the basin outlet, with the remainder being stored within the catchment. Reduction in rates of soil loss from the agricultural areas in the catchment by about 25% after 1938, as a result of the implementation of soil conservation measures, was not reflected by reduced sediment output from the catchment. This was largely because of remobilisation of

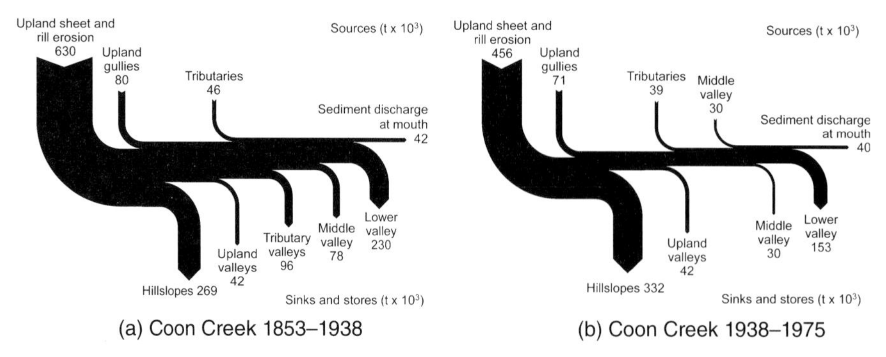

Figure 3.3 Sediment budgets for Coon Creek Wisconsin, for the periods 1853–1938 (a) and 1938–75 (b). (Based on data presented by Trimble, 1983.)

sediment from sinks in the middle valley. The sediment budget must be seen as key tool both for understanding sediment export from a catchment and, perhaps even more importantly, for supporting the design and implementation of effective sediment management programmes (Walling and Collins, 2008; Gellis and Walling, 2011).

Tools for meeting new information needs

Increased interest in the fine sediment loads of rivers has been paralleled by the development of a range of tools for meeting requirements for new information. These developments partly reflect the need to address new questions, but they are also a reflection of timely technological advances. They span improved monitoring techniques and equipment, sediment source fingerprinting, sediment tracing and modelling fine sediment yields across a range of temporal and spatial scales and for a range of purposes. These areas are reviewed below.

Monitoring techniques and equipment

The non-capacity and supply-controlled nature of fine sediment transport (e.g., Figure 3.1) means that carefully designed monitoring programmes are necessary to obtain reliable information on suspended sediment transport (Walling *et al.*, 1992). In large rivers, where discharge and sediment concentration change relatively slowly, a programme of daily sampling might be sufficient to define the record of variations in suspended sediment concentration or load. However, as the size of the catchment reduces and the response to rainfall becomes more flashy, the frequency of sampling needs to increase. Most of the suspended sediment load of a stream or river is transported during storm events and primarily by the large events. Typically, about 75% of the load is transported during about 5% of the time and it is critical that the sediment concentration record should be documented in detail during the key events. Suspended sediment samplers and sampling techniques are now well developed (see Gray *et al.*, 2008), but the need to visit the site can make it difficult to assemble a detailed record of suspended sediment concentration. The development of automatic samplers has provided a means of overcoming this problem, although problems can arise in ensuring that the sample collected is representative of the channel cross section. Such samplers can be programmed to collect suspended sediment samples when flow or concentration (turbidity) exceeds a pre-set threshold and to vary the sampling frequency according to the rate of change of flow or turbidity (e.g., Lewis, 1996). Recording turbidity meters, which now commonly employ optical backscatter (OBS) sensors, also offer a means of collecting a continuous surrogate record of suspended sediment concentration (e.g., Gray and Gartner, 2010; Schoellhamer and Wright, 2003) and are widely employed for monitoring suspended sediment transport. This approach is, however, heavily dependent on the existence of a well-defined calibration relationship between sediment concentration and turbidity and this relationship can be affected by changes in the grain size composition and colour of the sediment load (Sutherland *et al.*, 2000). The time integrating trap sampler developed at the University of Exeter (Phillips *et al.*, 2000; Russell *et al.*, 2000) is a very simple device which has met an important need for the automated collection of sizeable representative samples of suspended sediment for use in sediment fingerprinting investigations. Where large instantaneous samples of suspended sediment are required for subsequent analysis, continuous flow centrifuges have proved an effective means of dewatering and recovering the sediment (see Ongley and Blatchford, 1982).

The need for easily derived information on the grain size composition of suspended sediment samples has been addressed by the development of laboratory laser diffraction analysers. However, as indicated above, the grain size distribution measured in the laboratory may differ significantly from the *in situ* or effective distribution that exists in the river, due to the presence of flocs or composite particles, which are likely to be broken up during the laboratory measurements (Phillips and Walling, 1995). As a result, attention has been successfully directed to the *in situ* deployment of laser diffraction or scattering probes (e.g., Phillips and Walling, 1997; Gray *et al.*, 2004; Williams *et al.*, 2007). The current generation of LISST laser-based equipment developed by Sequoia Scientific specifically for river studies includes an *in situ* laser probe contained in a streamlined body (LISST-SL) and a portable battery powered streamside monitoring unit that pumps water directly from the river and which can be programmed to make measurements at intervals of between 5 minutes and 60 minutes (LISST-Streamside).

Sediment source fingerprinting

There is an increasing need for information on the source of the fine sediment transported by a river. Such information is essentially impossible to obtain using traditional monitoring techniques, but the development of sediment source fingerprinting techniques has provided a timely and effective means of meeting this need. Sediment source fingerprinting is founded on two key principles. Firstly, one or more diagnostic physical or chemical properties are used as fingerprints to discriminate the source materials associated with the potential fine sediment sources in a catchment. Secondly, comparison of the equivalent properties of the suspended sediment transported by a river with the fingerprints of the potential sources provides a means of establishing the relative contribution of the individual sources. Use of this approach can be traced back to the 1970s and the work of researchers such as Klages and Hsieh (1975), Wall and Wilding (1976) and Walling *et al.* (1979). However, the assessment of the relative importance of different sources provided by these early studies was essentially qualitative. Since then, the approach has been successfully developed and refined, with most emphasis being placed on determining the relative importance of different source types. Following Walling (2013), seven key developments which have been incorporated into current approaches, can be identified as follows:

(1) Use of multiple properties or composite fingerprints, involving a wide range of different physical and chemical properties, to strengthen the discrimination between different sources and to permit a greater number of potential sources to be identified. Sediment properties that have now been successfully used as source fingerprints include a wide range of geochemical parameters, isotopic signatures, radionuclides, sediment colour and spectral reflectance and compound specific stable isotopes (e.g., Collins *et al.* 2010a; Douglas *et al.*, 2003; Gibbs, 2008; Martínez-Carreras *et al.*, 2010; Tiecher *et al.*, 2015; Wallbrink *et al.*, 1998).

(2) Incorporation of statistical tests to confirm the ability of particular fingerprint properties to discriminate between potential sediment sources and to assist in the selection of the 'best' composite fingerprint (e.g., Collins *et al.*, 2012; Juracek and Ziegler, 2009; Laceby *et al.*, 2015; Motha *et al.*, 2003).

(3) Use of numerical mixing (or unmixing) models to provide quantitative assessments of the relative contribution of different potential sources (e.g., Collins *et al.*, 2010a; Fox and Papanicolaou 2008; Haddachi *et al.*, 2014; Lamba *et al.*, 2015; Lin *et al.*, 2015; Nosrati *et al.*, 2014; Palmer and Douglas, 2008).

(4) Use of specific size fractions to take account of contrasts in grain size composition between suspended sediment and catchment source materials, testing fingerprint properties for conservative behaviour and incorporation of grain size and organic matter enrichment/depletion effects into the mixing models used for source apportionment (e.g., Collins *et al.*, 1998, 2013a,b; Motha *et al.*, 2003, Russell *et al.*, 2001).

(5) Extension of the approach to consider a wider range of 'targets', in addition to samples of suspended sediment. These include surrogates for suspended sediment, such as floodplain surface scrapes and fine sediment deposits from river channels (e.g., Collins *et al.*, 2010a), particular 'problem sediments', such as interstitial fine sediment recovered from fish spawning gravels (e.g., Walling *et al.*, 2003) and recent fine sediment deposits from lakes and estuaries (e.g., Gibbs, 2008; Haiyan, 2015). In some studies attention has focused on the source of the organic material associated with the sediment (Collins *et al.*, 2013c, 2014).

(6) Extension of the approach to incorporate a temporal dimension and to document changes in sediment source through time. Such work has included both 'before and after' studies in experimental catchments where sediment control measures and changes in land management have been implemented (e.g. Merten *et al.*, 2010) and use of sediment cores collected from lakes and river floodplains to reconstruct longer-term changes in sediment source (e.g., Foster and Walling, 1994; Pittam *et al.*, 2009; Collins *et al.*, 2010b).

(7) Taking account of the uncertainty associated with source apportionment procedures. Incorporation of Monte Carlo procedures and Bayesian statistics into the mixing models used to determine the relative contributions of potential sources has permitted the uncertainty associated with source characterisation and other components of the source fingerprinting approaches to be propagated through the calculations (e.g., Franks and Rowan, 2000; Collins *et al.*, 2012, 2014; Laceby and Olley, 2015; Nosrati *et al.*, 2014; Palmer and Douglas, 2008; Pulley *et al.*, 2015). Sediment source fingerprinting techniques have now been widely applied in Europe, North America and Australia, to support investigations of fine sediment transport by rivers and the development and implementation of sediment management and control programmes. In Australia, a number of studies have been undertaken to establish the primary sources of the fine sediment transported to the coast adjacent to the Great Barrier Reef (GBR) (e.g., Douglas *et al.*, 2008; Hughes *et al.*, 2009; Wilkinson *et al.*, 2011). The GBR is currently under stress from terrestrially derived sediment and information on sediment source is a critical requirement for the design of catchment management programmes aimed at reducing land–sea sediment fluxes.

Tracing soil and sediment redistribution

Production of a contemporary sediment budget for a catchment, similar to that depicted in Figure 3.3, requires information on rates of gross and net soil loss from slopes and the deposition and storage of sediment as it is transported towards the stream and through the channel network. As with sediment source, such information is difficult to obtain using traditional monitoring and sediment tracing techniques have proved to be particularly useful for this purpose (Walling, 2006). Source fingerprinting techniques could be viewed as a tracing technique, but here attention will focus on the more direct use of fallout radionuclides to trace sediment movement and redistribution in catchments. This approach is founded on the existence of a number of natural and manmade radionuclides that reach the land surface as fallout, primarily as wet fallout in association with rainfall, and are rapidly and strongly fixed by the surface soil or sediment. By studying the post-fallout redistribution and fate of the selected fallout radionuclide, it is possible to obtain information on soil and sediment redistribution and, therefore, erosion and deposition rates.

The fallout radionuclide most widely used for this purpose is caesium-137 (^{137}Cs) (see IAEA, 2014; Walling, 2012; Zapata, 2002). Caesium-137 is a manmade radionuclide that was produced by the testing of thermonuclear weapons in the 1950s and early 1960s. Significant bomb-derived fallout occurred in most areas of the world during the period extending from the mid 1950s through to the 1970s, although the depositional fluxes were much greater in the northern than the southern hemisphere. In the absence of further bomb tests after the Nuclear Test Ban Treaty in 1963, fallout effectively ceased in the mid 1970s. However, in some areas of the world a further short-lived fallout input occurred in 1986 as a result of the Chernobyl accident.

Caesium-137 has a half-life of 30.2 years and much of the original fallout is likely to still remain within the upper horizons of the soils and sediments of a catchment. By investigating the current distribution of the radionuclide in the landscape, it is possible to obtain information on the net effect of soil and sediment redistribution processes operating over the past ca. ~50 years (i.e., since the main period of fallout) and thus quantify medium-term erosion and deposition rates. Mean soil redistribution rates over the past ~50 years are established by comparing the inventories measured at individual sampling points with the reference inventory for the study site, which represents the inventory found at a site which has experienced neither erosion nor deposition. Points with inventories less than the reference inventory are indicative of eroding areas, whereas those with inventories in excess of the reference value indicate deposition. The timescale will need to be modified where significant Chenobyl fallout has occurred. A range of conversion models have been developed for use in estimating erosion and deposition rates, based on the degree of departure of the measured inventory from the reference inventory (e.g., Walling and He, 1999a; Walling *et al.*, 2011; Li *et al.*, 2009). Using a similar approach, ^{137}Cs measurements have also been successfully used to document rates and patterns of overbank deposition on river floodplains over the past ~50 years (Golosov and Walling, 2014; Walling and He, 1997; Terry *et al.*, 2002)

Although most studies employing fallout radionuclides have been based on ^{137}Cs, both excess lead-210 ($^{210}Pb_{ex}$) and beryllium-7 (^{7}Be) have also been used in a similar manner (see IAEA, 2014; Mabit *et al.*, 2008, 2014; Walling, 2012). These two fallout radionuclides differ from ^{137}Cs in being of natural, geogenic and cosmogenic origin, respectively. Pb-210 has a similar half-life to ^{137}Cs (22.3 years) but that of ^{7}Be is very much shorter (53 days). By virtue of its ongoing fallout, $^{210}Pb_{ex}$ provides a means of assessing soil and sediment redistribution over periods of ~100 years, whereas ^{7}Be can be used at the timescale of individual events or a few weeks. Walling and He (1999b) report the successful use of $^{210}Pb_{ex}$ in soil erosion studies and He and Walling (1996) provide examples of its application for estimating rates of overbank sedimentation on floodplains. The use of ^{7}Be to document short-term soil redistribution rates is reported by Porto *et al.* (2014), Schuller *et al.* (2006) and Walling *et al.* (1999, 2009).

Most studies that have employed fallout radionuclides to document soil and sediment redistribution in catchments have focused on small areas such as individual fields or representative transects and have involved the collection of a substantial number of samples. Extrapolation of the results to larger areas can introduce problems due to restrictions on the number of samples that can be collected and analysed. Increased attention is therefore being directed to the problem of upscaling the approach (see Mabit *et al.*, 2007; Walling *et al.*, 2014). The approach recently documented by Porto *et al.* (2011) involves sampling an essentially random network of points distributed across a larger area and using the resulting information to provide a representative sample of erosion and deposition rates within the landscape of the study area. This will provide information on both the magnitude of erosion and deposition rates and the relative importance of zones experiencing erosion and deposition (see Figure 3.4).

Modelling sediment yields

There is a long tradition, particularly from engineering disciplines, of modelling the in-channel processes of scour, sediment transport and deposition in alluvial river systems, with the sediment transfer functions reflecting differing levels of complexity and corresponding data requirements. The US Bureau of Reclamation Generalized Stream Tube model for Alluvial River Simulation (GSTARS) is a well-known example of a sediment routing model used for practical engineering purposes (Yang *et al.*, 1998). However, such models focus on the coarser channel-derived sediment.

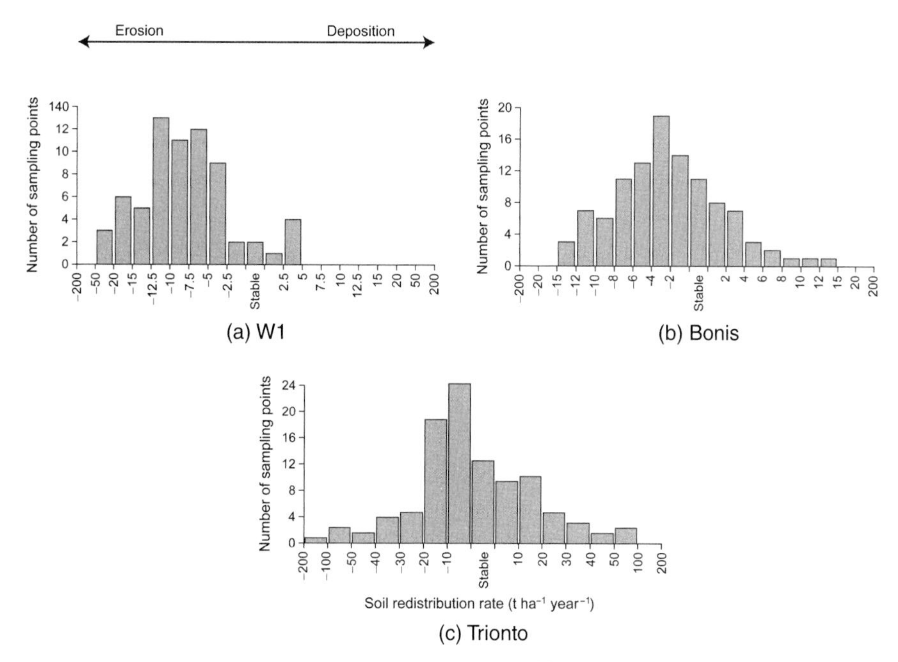

Figure 3.4 Distributions of soil redistribution rates derived from ^{137}Cs measurements in two small catchments (W1 = 0.015 km^2 and Bonis = 1.39 km^2) and an intermediate sized catchment (Trionto = 31.61 km^2) in southern Italy. Porto *et al.* 2011. Reproduced with permission from John Wiley & Sons.

The understanding and management of fine sediment problems requires models that characterise the linkages between the catchment surface and the channel network and can represent the influence of topography, soil type, land use and other factors on sediment mobilisation and delivery. The increased use of Geographical Information Systems (GIS) and digital elevation models (DEMs) has promoted the development and application of spatially distributed, process-based models of soil erosion and sediment delivery that capture many of the key controls involved. Well-known examples of such models include, amongst others, SHESED (Wicks and Bathurst, 1996), EUROSEM (Morgan *et al.*, 1998), WEPP (Nearing *et al.*, 1989) and SEDEM (Van Rompaey *et al.*, 2001). Another example from the UK is the PSYCHIC (Phosphorus and Sediment Yield Characterisation in Catchments) model (Davison *et al.*, 2008; Stromqvist *et al.*, 2008), which was designed specifically to assist catchment screening and the identification of pollution hotspots for informing mitigation planning. Its development reflects the increasing use of computer models to inform and support decisions on diffuse pollution issues and to target the implementation of abatement measures. The conceptual framework for PSYCHIC is based on the source–mobilisation–delivery transfer continuum. Mobilisation is conceptualised as initiating sediment redistribution locally at plot scale, whereas delivery represents a difference variable linking mobilisation and inputs to the river channel system. Sediment mobilisation is estimated using a modified form of the Morgan–Morgan–Finney soil erosion model (Morgan, 2001). Sediment delivery to river channels is determined by using connectivity factors based on the presence of drains predicted from the Hydrology

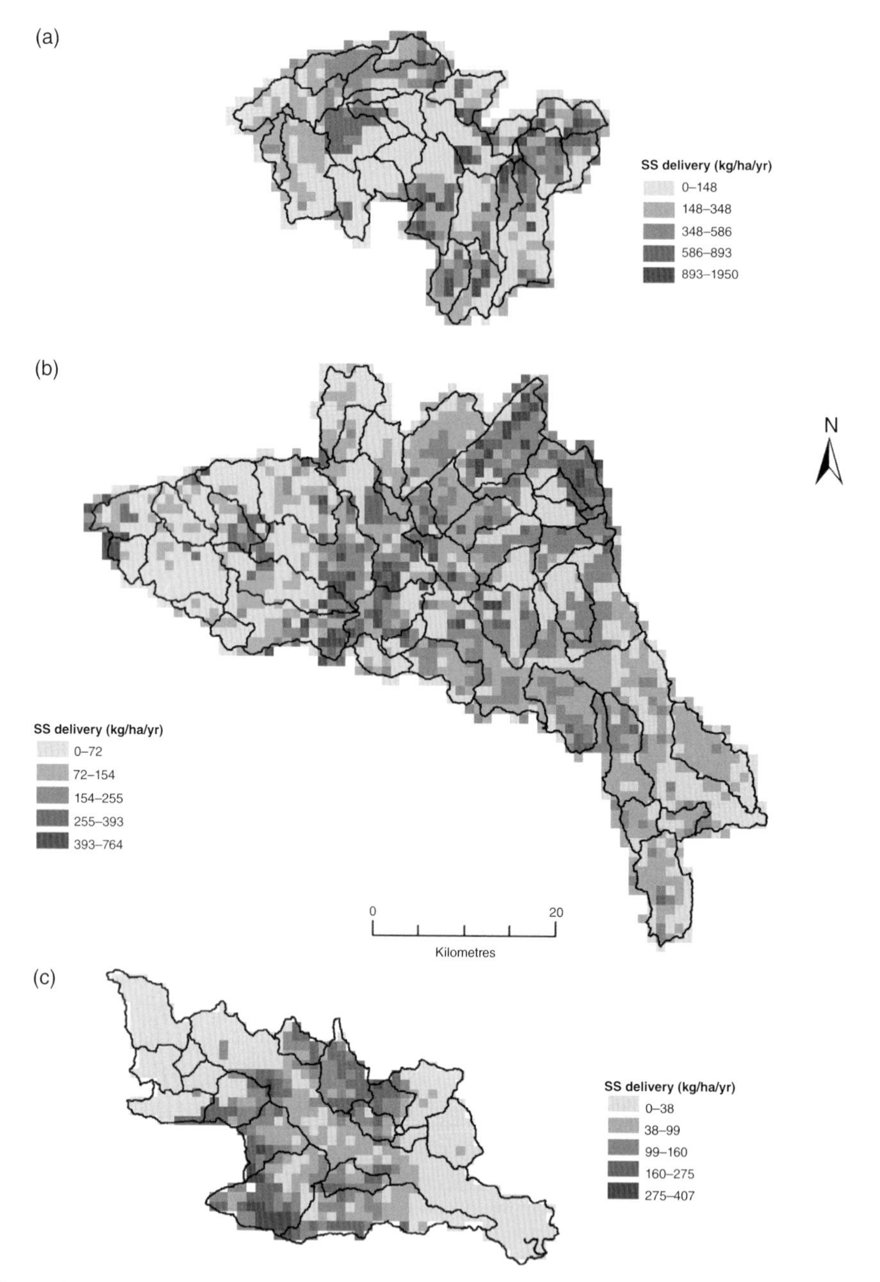

Figure 3.5 Sediment delivery to rivers predicted by PSYCHIC for the Derwent-Cocker (a), Teme (b) and Wensum (c) river catchments in England. (Based on Collins et al., 2007.)

of Soil Types (HOST) classification scheme and distance to watercourse. Figure 3.5, as an example, shows sediment delivery to streams within three contrasting river catchments in England, predicted using the PSYCHIC model. This version of the PSYCHIC model only represents sediment loss from agricultural land and does not include a channel erosion and routing function. For policy support purposes, the outputs of this model have been combined using GIS with estimates of sediment loss from additional sectors and sources, to simulate total sediment inputs to all rivers across England and Wales under current or future projected environmental conditions (see case study section and Collins *et al.*, 2009a).

Where larger river basins are involved, input data requirements and computational constraints are likely to limit the potential for applying a fully distributed and physically based approach to modelling and predicting sediment yields. In this situation, the functioning of the river basin must be simplified to incorporate the key processes and drivers of sediment yield and its area subdivided into small sub-units, which can be modelled using a lumped approach. The SedNet model, developed in Australia as a semi-lumped model for use in larger river basins (Wilkinson *et al.*, 2004, 2009), provides a good example of the potential of such models. Key features of the SedNet model are the sediment budget approach, the use of the river network to provide the basic structure and the estimation of mean annual sediment yields, rather than shorter-term yields. The network is subdivided into a series of individual links and the sediment budget is evaluated for each link, to estimate the output from the link into the next link downstream. Inputs to the link include hillslope and gully erosion from the catchment area draining to the link, bank erosion along the link and upstream inputs. Sinks within the link include overbank floodplain deposition and reservoir deposition. Within-channel storage is ignored as this is assumed to be negligible at the decadal timescale. Hillslope erosion from the catchment contributing to the link is, for example, estimated using the RUSLE model (Renard *et al.*, 1997) coupled with a sediment delivery ratio and bank erosion is modelled based on stream power and bank material properties. The model is particularly useful for management purposes, because it can provide information on the sediment yield from individual links, the contribution of each link to the sediment flux at the basin outlet and the relative importance of slope and channel (gully and bank) erosion. Such information is valuable for targeting remediation measures to reduce downstream sediment loads.

Management and policy

Since fine sediment plays a pivotal role in influencing the physical, chemical and biological integrity of aquatic ecosystems, the need to manage excess sediment stress on watercourses is integral to river catchment management and associated policy. With this recognition comes the need to assess environmental status for sediment and this, in turn, underscores the requirement for meaningful and practical sediment targets for informing compliance and gap analysis. Both water column and river substrate metrics have been proposed as river sediment targets (Collins *et al.*, 2011). Water column metrics include light penetration, turbidity, sediment concentration summary statistics and sediment regimes. Substrate metrics include embeddedness and riffle stability. However, establishing such metrics involves many problems including the uncertainty associated with toxicological dose–response experimental data. Furthermore, many of the thresholds reported in existing scientific and grey literature are based on correlative relationships that fail to capture the specific mechanisms controlling fine sediment impacts on aquatic habitats and are stationary in nature. A good example of the latter is the existing European Union Freshwater Fish Directive indicative target for annual mean suspended sediment concentration (25 mg l^{-1}) which up until 2013 was applied as a static global threshold in many Member States (Collins and Anthony, 2008b).

Against this background, the definition of meaningful fine sediment targets for informing river catchment management continues to attract debate from scientists, practitioners and policy-makers alike. The temporal windows representing the key life stages of sentinel species, such as the spawning and incubation season for salmonids, must be given greater emphasis in the identification of practical thresholds. Similarly, some consideration must be given to 'background' sediment inputs to watercourses for different physiographic settings, since no cost-effective mitigation programme should seek to address these natural levels of stress (cf. Foster *et al.*, 2011). Given the need to provide more meaningful fine sediment targets for individual contrasting catchments and to use those targets in analysing the gap between current sediment stress and good ecological condition for a range of biota, it can be argued that generic modelling toolkits capable of coupling sediment stress and its mitigation, with biotic endpoints, represent one pragmatic way forward for policy-makers working at strategic scales (Collins *et al.*, 2011). In this context, ongoing work in the UK funded by the Department for Environment, Food and Rural Affairs (Defra) is seeking to develop an integrated modelling toolkit for helping to revise fine sediment targets for individual river catchments across England and Wales. The Demonstration Test Catchment (DTC) platform (McGonigle *et al.*, 2014) established in 2009 and now in its second phase running till 2017, supported by the same body, is working to compile a robust evidence base on the impact of sediment mitigation measures from farm to catchment to national scale. Progress on these fronts is dependent on interdisciplinary working, whilst the capacity for managing excess fine sediment stress must be placed in the context of the need to maximise food production from agricultural land (Foresight, 2011; Pretty and Bharucha, 2014), which is frequently the dominant sediment source (Zhang *et al.*, 2014), for the purpose of securing food security.

Case studies

Establishing a catchment sediment budget

The use of both sediment source fingerprinting and sediment tracing techniques in tandem and in combination with information on the sediment flux at a catchment outlet provided by standard monitoring techniques can provide an effective and valuable basis for establishing a catchment sediment budget (e.g., Minella *et al.*, 2014; Walling *et al.*, 2001, 2002, 2006). Thus, for example, estimates of floodplain and channel storage can be added to the measured output flux to estimate the total sediment input to the channel system and information on the source of the sediment load can be used to estimate the primary source of this sediment input. If fallout radionuclides are used to document gross and net rates of soil loss from the slopes, comparison of these estimates with estimates of sediment input to the channel from slope sources, provides a means of obtaining a first order estimate of conveyance losses and storage associated with slope–channel transfer. This approach, coupled with additional measurements of channel storage using the approach reported by Lambert and Walling

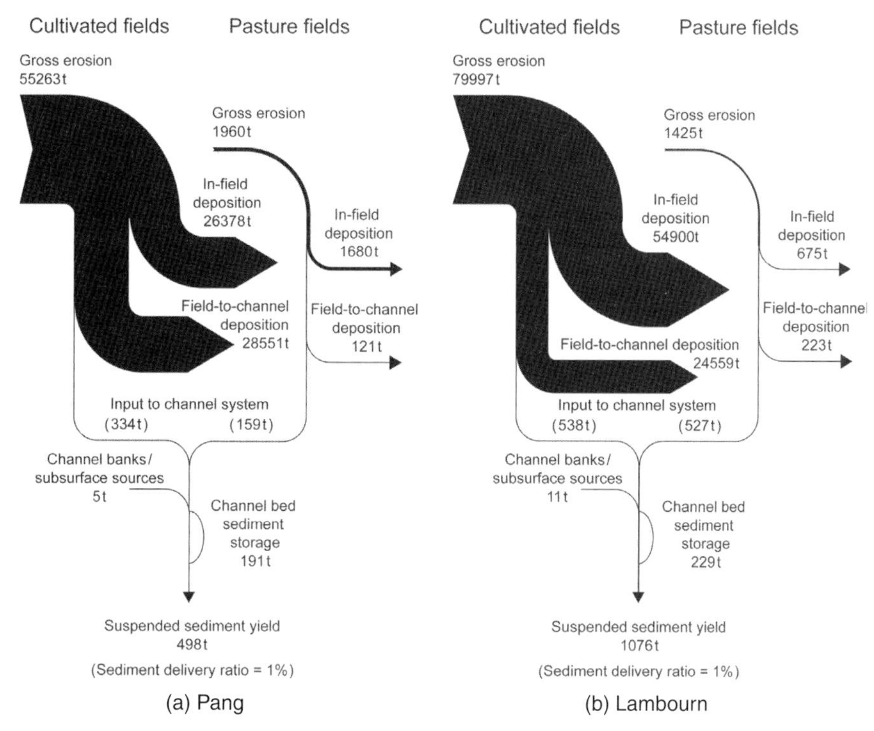

Figure 3.6 Catchment sediment budgets for (a) the Pang and (b) the Lambourn catchments in Berkshire, UK. The values indicated represent values of annual sediment flux and storage. Walling *et al.* 2006. Reproduced with permission from Elsevier.

(1988) was used by Walling *et al.* (2006) to establish tentative sediment budgets for the Pang (166 km^2) and Lambourn (234 km^2) catchments (see Figure 3.6).These two catchments, located on the chalk of southern England, formed part of the Lowland Catchment Research Programme (LOCAR) funded by the UK Natural Environment Research Council (see http://www.nerc.ac.uk/research/programmes/locar/). The location of the catchments on highly permeable strata and the resulting dominance of groundwater flow mean that storm runoff is limited and that little sediment reaches the catchment outlets. However, there is evidence of relatively high rates of sediment mobilisation and redistribution within the catchments, and their sediment budgets are dominated by slope and slope to channel sediment sinks.

Reduction of the sediment output from these catchments would clearly need to target the slopes of the cultivated areas, since these are the primary sediment source. A substantial reduction in sediment mobilisation from the cultivated slopes would, however, be required to reduce sediment output from the catchments, since only a small proportion of the soil eroded from the cultivated area reaches the channel system. However, a small increase in the conveyance loss or deposition associated with field–channel transfer could result in an appreciable reduction in the sediment input to the channel system and should thus be seen as a priority target for remedial measures. Equally, the importance of in-field and field–channel storage in reducing the sediment input to the channels means that any change in the functioning of these sinks or stores, resulting in reduced deposition or perhaps remobilisation of stored sediment, could potentially result in a major increase in the sediment outputs from the catchments in relative terms.

A national scale modelling assessment of fine sediment compliance with the EU Freshwater Fish Directive across England and Wales

A recent modelling study undertaken in the UK (Collins and Anthony, 2008b) provides a useful example of a national scale assessment of the gap between current and compliant sediment losses from the agricultural sector, based on the EU Freshwater Fish Directive (FFD) (78/659/EC) guideline standard (an annual mean concentration of 25 mg l^{-1}). The modelling methodology was founded on a statistical relationship between measured suspended sediment concentration and modelled total sediment inputs to watercourses from diffuse and point sources. Mean annual total suspended sediment loads for each Water Framework Directive (WFD) waterbody across England and Wales were estimated as the sum of the modelled individual loads for the diffuse agricultural and urban sectors, eroding channel banks and point source discharges. Diffuse agricultural sediment inputs for all rivers were calculated using the PSYCHIC process-based model (see above), which deploys 1 km^2 resolution statistical input information on a number of key environmental drivers, including climate, slope, soil types and characteristics, drainage density, land use and cropping and livestock density. National scale sediment contributions from diffuse urban sources were estimated using an Event Mean Concentration (EMC) approach based on the inter-quartile ranges of empirical data for sediment runoff from industrial areas, main roads and residential zones. The EMCs were combined with estimated mean annual runoff from urban areas derived using the Wallingford procedure (National Water Council, 1983). Corresponding total sediment inputs from eroding channel banks were estimated using a prototype national scale index based on the river regime (Gustard *et al.*, 1992), the duration of excess shear stress and channel density. Point source sediment loadings to all rivers across England and Wales were computed using a database of consented effluent discharges from sewage treatment works, but with a correction based on the relationship between measured and consented average suspended sediment concentrations.

The predicted mean annual total suspended sediment loads delivered to all rivers were coupled with corresponding flow regime distributions to estimate time-averaged suspended sediment concentrations. Structured regression modelling was used to optimise the relationship between modelled and measured time-averaged suspended sediment concentrations, for the purpose of estimating the annual mean suspended sediment concentration and the likelihood of 'good ecological status' (GES) due to sediment contributions from the agricultural sector alone (Figure 3.7). The findings suggested that on the basis of using the FFD to define GES for sediment, approximately 83% of the total catchment area across England and Wales appeared to require no further reduction in sediment loss to rivers from diffuse agricultural sources. Maps of compliance, however, will inevitably depend on the sediment thresholds used to define GES, and in recognition of the issues associated with the 'global' FFD guideline standard, alternative means of setting thresholds on a catchment-specific basis are currently

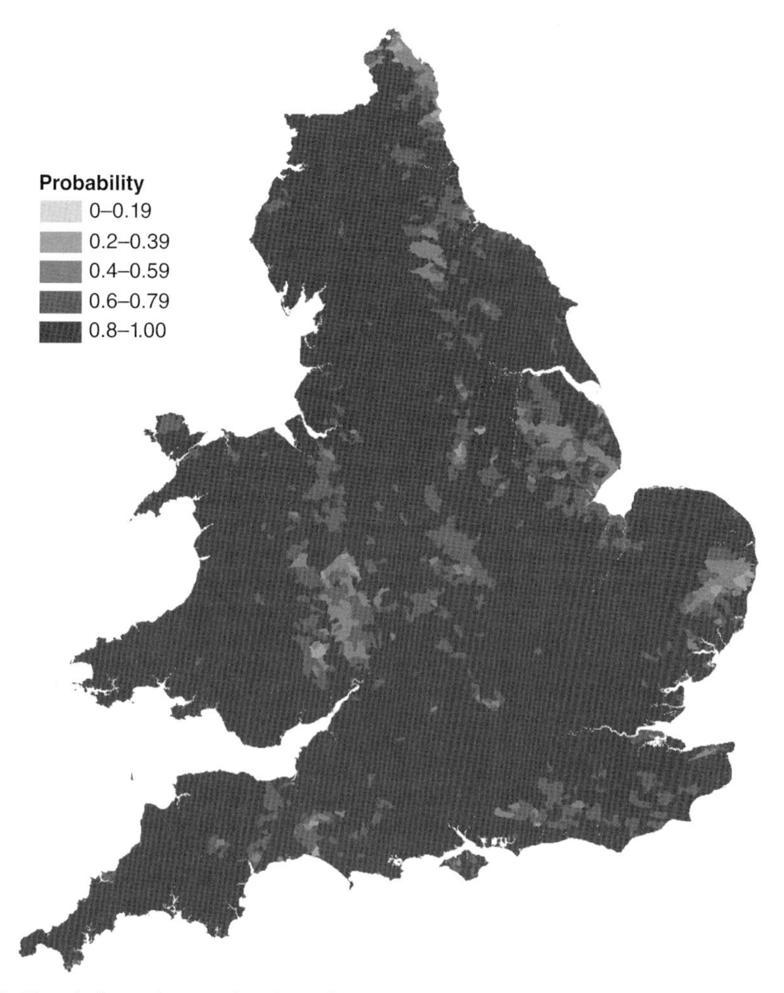

Figure 3.7 Likelihood of meeting 'good ecological status' (GES) for fine sediment across England and Wales, as defined by the EU Freshwater Fish Directive (FFD) guideline standard (Based on Collins and Anthony, 2008b.)

being investigated to inform catchment management for sediment across the UK.

Summary and the way forward

About 25 years ago the fine sediment loads of rivers were frequently seen as being of limited importance. They are now recognised as representing a key element of river behaviour with wide-ranging ecological and environmental significance and an important focus for catchment management programmes. The availability of new instrumentation to provide improved data on suspended sediment loads, the development of a range of techniques to document sediment sources and soil and sediment redistribution within catchments, as well as the development of improved catchment-based distributed models have resulted in important advances in our understanding of the fine sediment dynamics of catchments and our ability to predict their behaviour. The growing awareness of the environmental significance of fine sediment, and particularly its ecological importance, is directing increasing attention to sediment management in river catchments. The development and implementation of successful fine sediment management strategies will depend on the availability of a sound understanding of both sediment budgets and sediment-related stress and biotic impacts, as well as a reliable evidence base to support policy (cf. Collins *et al.* 2009b).

Looking to the future, there is a need to continue to improve our understanding of catchment sediment dynamics and their response to land use and climate change and our ability to model catchment behaviour. As management attracts greater attention, it is important that the available models should be capable of predicting catchment response under different management scenarios, in order to assess their likely impact and success. Sediment source tracing must be seen as providing key information for targeted management and there is a need to exploit the potential for further improvements in source discrimination, to identify source-specific inputs, and to progress its transfer from being a research tool to one that can be more easily and widely applied on a routine basis. To support policy-making it is important that further attention should be directed to establishing more meaningful sediment targets or metrics for assessing catchment compliance and this will require further research on the ecological impacts of fine sediment. In this context, attention should be directed to the relative roles of the organic and inorganic components of fine sediment loads in contributing to sediment-related stress. Developing effective strategies for controlling fine sediment loss to watercourses will require an improved empirical data base on the cost-effectiveness of mitigation options, set in the context of a competitive agricultural sector and the need to engage catchment stakeholders. In addition there is a need to develop and refine both farm-scale toolkits for guiding the selection and targeting of on-farm mitigation strategies and catchment-scale modelling frameworks for scaling up the likely benefits. The latter should incorporate the link between sediment stress and biotic impacts and thereby permit decision making to focus more directly on protecting aquatic ecosystems.

References

Allan, R.J. 1986. *The Role of Particulate Matter in the Fate of Contaminants in Aquatic Ecosystems.* Inland Waters Directorate, National Water Research Institute, Canada Centre for Inland Waters.

Chapman, J.M., Proulx, C.L., Veilleux, M.A.N., et al. 2014 Clear as mud – a meta-analysis of the effects of sedimentation on freshwater fish and the effectiveness of sediment-control measures. *Water Research* **56**, 190–202.

Collins, A.L. and Anthony, S.G. 2008a. Predicting sediment inputs to aquatic ecosystems across England and Wales due to current environmental conditions. *Applied Geography* **28**, 281–294.

Collins, A.L. and Anthony, S.G. 2008b. Assessing the likelihood of catchments across England and Wales meeting 'good ecological status' due to sediment contributions from agricultural sources. *Environmental Science and Policy* **11**, 163–170.

Collins, A.L., Walling, D.E. and Leeks, G.J.L. 1998. Use of composite fingerprints to determine the provenance of the contemporary suspended sediment load transported by rivers. *Earth Surface Processes and Landforms* **23**, 31–52.

Collins, A.L., Stromqvist, J., Davison, P.S. and Lord, E.I. 2007. Appraisal of phosphorus and sediment transfer in three pilot areas identified for the Catchment Sensitive Farming initiative in England: application of the prototype PSYCHIC model. *Soil Use and Management* **23**, 117–132.

Collins, A.L., Anthony, S.G., Hawley, J. and Turner, T. 2009a. The potential impact of projected change in farming by 2015 on the importance of the agricultural sector as a sediment source in England and Wales. *Catena* **79**, 243–250.

Collins, A.L., McGonigle, D.F., Evans, R., et al. 2009b. Emerging priorities in the management of diffuse pollution at catchment scale. *International Journal of River Basin Management* **7**, 179–185.

Collins, A.L., Walling, D.E., Webb, L. and King P. 2010a. Apportioning catchment scale sediment sources using a modified composite fingerprinting technique incorporating property weightings and prior information. *Geoderma* **155**, 249–261.

Collins, A.L., Walling, D.E. Stroud, R.W., et al. 2010b Assessing damaged road verges as a suspended sediment source in the Hampshire Avon catchment, southern United Kingdom. *Hydrological Processes* **24**, 1106–1122.

Collins, A.L. Naden, P.S., Sear, D.A., et al. 2011. Sediment targets for informing river catchment management: international experience and prospects. *Hydrological Processes* **25**, 2112–2129

Collins, A.L., Zhang, Y., McChesney, D., et al. 2012. Sediment source tracing in a lowland agricultural catchment in southern England using a modified procedure combining statistical analysis and numerical modelling. *Science of the Total Environment* **414**, 301–317.

Collins, A.L., Zhang, Y.S., Duethmann, D., et al. 2013a. Using a novel tracing-tracking framework to source fine-grained sediment loss to watercourses at sub-catchment scale. *Hydrological Processes* **27**, 959–974.

Collins, A.L., Zhang, Y., Hickinbotham, R., et al. 2013b. Contemporary fine-grained bed sediment sources across the River Wensum Demonstration Test Catchment, UK. *Hydrological Processes* **27**, 857–884.

Collins, A.L., Williams, L.J., Zhang, Y.S., et al. 2013c. Catchment source contributions to the sediment-bound organic matter degrading salmonid spawning gravels in a lowland river, southern England. *Science of the Total Environment* **456–457**, 181–195.

Collins, A.L., Williams, L.J., Zhang, Y.S., et al. 2014. Sources of sediment-bound organic matter infiltrating spawning gravels during the incubation and emergence life stages of salmonids. *Agriculture, Ecosystems and Environment* **196**, 76–93.

Davison, P.S., Withers, P.J.A., Lord, E.I., et al. 2008. PSYCHIC – a process-based model of phosphorus and sediment mobilisation and delivery within agricultural catchments. Part 1. Model description and parameterisation. *Journal of Hydrology* **350**, 290–302.

Douglas, G.B. Palmer, M. and Caitcheon, G., 2003. The provenance of sediments in Moreton Bay, Australia: a synthesis of major, trace element and Sr–Nd–Pb isotopic geochemistry, modelling and landscape analysis. *Hydrobiologia* **494**, 145–152

Douglas, G.B., Ford, P.W., Palmer, M.R., et al. 2008. Fitzroy River basin, Queensland, Australia. IV. Identification of flood sediment sources in the Fitzroy River. *Environmental Chemistry* **5**, 243–257.

Droppo I.G. 2001. Rethinking what constitutes suspended sediment. *Hydrological Processes* **15**, 1551–1564.

Droppo, I.G., Leppard, G.G., Liss, S.N. and Milligan, T.G. (eds) 2005. *Flocculation in Natural and Engineered Environmental Systems*. CRC Press, Boca Raton, FL.

European Council 1992. Council Directive 92/43/EEC of 21 May 1992 on the conservation of natural habitats and of wild fauna and flora. *Official Journal L 206*, 22/07/1992, pp, 7–50.

European Parliament 2000. Establishing a framework for community action in the field of water policy. *Directive EC/2000/60*, EU, Brussels.

Foresight 2011. *The Future of Food and Farming: Challenges and Choices for Global Sustainability*. The Government Office for Science, London.

Förstner, U. and Muller, G. 1974. *Schwermetalle in Fliissen und Seen als Ausdruck der Umweltverschmutzung*. Springer-Verlag, Berlin.

Foster, I.D.L. and Walling, D.E. 1994. Using reservoir deposits to reconstruct changing sediment yields and sources in the catchment of the Old Mill Reservoir, South Devon, UK, over the past 50 years. *Hydrological Sciences Journal* **39**, 347–368.

Foster, I.D.L., Collins, A.L., Naden, P.S., et al. 2011. The potential for palaeolimnology to determine historic sediment delivery to rivers. *Journal of Palaeolimnology* **45**, 287–306.

Fox, J.F. and Papanicolaou, A.N. 2008. An un-mixing model to study watershed erosion processes. *Advances in Water Resources* **31**, 96–108.

Franks, S.W. and Rowan, J.S. 2000. Multi-parameter fingerprinting of sediment sources: Uncertainty estimation and tracer selection. In: Bentley, L.R. et al. (eds) *Computational Methods in Water Resources XIII*. Balkema, Rotterdam: pp. 1067–1074.

Gellis, A.C. and Walling, D.E. 2011. Sediment source fingerprinting (tracing) and sediment budgets as tools in targeting river and watershed restoration programs. In: *Stream Restoration in Dynamic Fluvial Systems: Scientific Approaches, Analyses, and Tools*. American Geophysical Union, Geophysical Monograph Series: pp. 194, 263–291.

Gibbs, M. 2008. Identifying source soils in contemporary estuarine sediments: A new compound-specific isotope method. *Estuaries and Coasts* **31**, 344–359.

Golterman H.L. (ed.) 1977. *Interactions between Sediment and Fresh Water*. The Hague, Dr. W Junk Publishers.

Golterman, H.L., Sly, P.G, and Thomas, R.L. 1983, *Study of the Relationship between Water Quality and Sediment Transport*. UNESCO Technical Papers in Hydrology No. 26. UNESCO, Paris.

Golosov, V. and Walling, D.E. 2014. Using fallout radionuclides to investigate recent overbank sedimentation rates on river floodplains: an overview. In: *Sediment Dynamics from the Summit to the Sea*, IAHS Publication 387, IAHS Press, Wallingford, UK: pp. 228–234.

Graf, W.H. 1971, *Hydraulics of Sediment Transport*. McGraw Hill, New York.

Gray, J.R., Agrawal, Y.C. and Pottsmith, H.C. 2004. The LISST-SL streamlined isokinetic suspended-sediment profiler. *Proceedings of the Ninth International Symposium on River Sedimentation*, vols. **1–4**, 2549–2555.

Gray, J.R. and Gartner, J.W. (2010). Surrogate technologies for monitoring suspended-sediment transport in rivers. In: Poleto, C. and Charlesworth, S. (eds) *Sedimentology of Aqueous Systems*, London, Wiley-Blackwell: Chapter **1**, pp 3–45

Gray, J.R., Glysson, G.D. and Edwards, T.E. 2008. Suspended-sediment samplers and sampling methods. In: Garcia, M. (ed.) *Sedimentation Engineering – Processes, Measurement, Modeling and Practice*, American Society of Civil Engineers Manual **110**: pp. 320–339.

Gustard, A., Bullock, A. and Dixon, J.M., 1992. Low flow estimation in the United Kingdom. *Institute of Hydrology Report 108*, Wallingford, UK.

Haddachi, A., Olley, J. and Laceby, J.P. 2014. Accuracy of mixing models in predicting sediment source contributions. *Science of the Total Environment* **497–498**, 139–152.

Haiyan, F. 2015. Temporal variations in sediment source from a reservoir catchment in the black soil region, Northeast China. *Soil Tillage and Research* **153**, 59–65.

Hawkins, R.H. 2003. *Survey of Methods for Sediment TMDLs in Western Rivers and Streams of the the United States. Report to US EPA Office of Water*, Assessment and Watershed Protection Division, Washington DC.

He, Q. and Walling, D.E. 1996. Use of fallout Pb-210 measurements to investigate longer-term rates and patterns of overbank sediment deposition on the floodplains of lowland rivers. *Earth Surface Processes and Landforms* **21**, 141–154.

Heywood, M.J.T and Walling, D.E. 2007. The sedimentation of salmonid spawning gravels in the Hampshire Avon catchment, UK: implications for the dissolved oxygen content of intragravel water and embryo survival. *Hydrological Processes* **21**, 770–788.

Horowitz, A.J. 1991. *A Primer on Sediment-Trace Element Chemistry*. (Second edn) Lewis Publishers, Chelsea, Michigan.

Hughes, A.O., Olley, J.M., Croke, J.C. and McKergow, L.A. 2009. Sediment source changes over the last 250 years in a dry-tropical catchment, central Queensland, Australia. *Geomorphology* **104**, 262–275.

IAEA, 2014. *Guidelines for Using Fallout Radionuclides to Assess Erosion and Effectiveness of Soil Conservation Strategies*. IAEA-TECDOC 1741. IAEA, Vienna.

Jones, J.I., Murphy, J.F., Collins, A.L., et al. 2012a. The impact of fine sediment on macro-invertebrates. *River Research and Applications* **28**, 1055–1071.

Jones, J.I., Collins, A.L., Naden, P.S. and Sear, D.A. 2012b. The relationship between fine sediment and macrophytes in rivers. *River Research and Applications* **28**, 1006–1018.

Jones, J.I., Duerdoth, C.P., Collins, A.L., et al. 2014. Interactions between diatoms and fine sediment. *Hydrological Processes* **28**, 1226–1237.

Juracek, K.E. and Ziegler, A.C. 2009. Estimation of sediment sources using selected chemical tracers in the Perry lake basin, Kansas, USA. *International Journal of Sediment Research* **24**, 108–125.

Kemp, P., Sear, D., Collins, A., et al. 2011. The impacts of fine sediment on riverine fish. *Hydrological Processes* **25**, 1800–1821.

Klages, M.G. and Hsieh, Y.P. 1975. Suspended solids carried by the Galatin River of southwestern Montana: II. Using mineralogy for inferring sources. *Journal of Environmental Quality* **4**, 68–73.

Laceby, J.P. and Olley, J. 2015. An examination of geochemical modelling approaches to tracing sediment sources incorporating distribution mixing and elemental correlations. *Hydrological Processes* **29**, 1669–1685.

Laceby, J.P., McMahon, J, Evrard, O. and Olley, J. 2015. A comparison of geological and statistical approaches to element selection for sediment fingerprinting. *Journal of Soils and Sediments*. DOI 10.1007/s11368-015-1111-1119

Lamba, J., Karthikeyan, K.G. and Thompson, A.M. 2015. Apportionment of suspended sediment sources in an agricultural watershed using sediment fingerprinting. *Geoderma* **239–240**, 25–33.

Lambert, C.P. and Walling, D.E. 1988 Measurement of channel storage of suspended sediment in a gravel-bed river. *Catena* **15**, 65–80.

Lewis, J. 1996. Turbidity-controlled suspended sediment sampling for runoff-event load estimation. *Water Resources Research* **32**, 2299–2310.

Li. S., Lobb, D.A., Tiessen, K.H.D. and McConkey, B.G. 2009. Selecting and applying cesium-137 conversion models to estimate soil erosion rates in cultivated fields. *Journal of Environmental Quality* **39**, 204–219.

Lin, J., Huang, Y., Wang, M-K., et al. 2015. Assessing the sources of sediment transported in gully systems using a fingerprinting approach: an example from southeast China. *Catena* **129**, 9–17.

Mabit, L., Bernard, C. and Laverdiere, M.R. 2007. Assessment of erosion in the Boyer River watershed (Canada) using a GIS oriented sampling strategy and ^{137}Cs measurements. *Catena* **71**, 242–249.

Mabit, L., Benmansour, M. and Walling, D.E. 2008. Comparative advantages and limitations of the fallout radionuclides ^{137}Cs, ^{210}Pb$_{ex}$ and ^{7}Be for assessing soil erosion and sedimentation. *Journal of Environmental Radioactivity* **99**, 1799–1807.

Mabit, L., Benmansour, M., Abril, J.M. et al. 2014. Fallout lead-210 as a soil and sediment tracer in catchment sediment budget investigations: a review. *Earth-Science Reviews* **138**, 335–351.

Martínez-Carreras, N., Udelhoven, T., Krein, A., et al. 2010. The use of sediment colour measured by diffuse reflectance spectrometry to determine sediment sources: Application to the Attert River catchment (Luxembourg). *Journal of Hydrology* **382**, 49–63.

McGonigle, D.F., Burke, S.P., Collins, A.L. et al. 2014. Developing Demonstration Test Catchments as a platform for transdisciplinary land management research in England and Wales. *Environmental Science: Processes and Impacts* **16**, 1618–1628.

Merten, G.H., Minella, J.P.G., Moro, M., et al. 2010. The effects of soil conservation on sediment yield and sediment source dynamics in a catchment in southern Brazil. In: *Sediment Dynamics for a Changing Future*, Proceedings of the Warsaw Symposium, IAHS Publication no. 337, IAHS Press Wallingford, UK: pp. 59–67.

Minella, J.P.G., Walling, D.E. and Merten, G.H. 2014. Establishing a sediment budget for a small agricultural catchment in southern Brazil, to support the development of effective sediment management strategies. *Journal of Hydrology* **519**, 2189–2201.

Morgan, R.P.C. 2001. A simple approach to soil loss prediction: a revised Morgan-Morgan-Finney model. *Catena* **44**, 305–322.

Morgan, R.P.C., Quinton, J.N., Smith, R.E. et al. 1998. The European soil erosion model (EUROSEM): a dynamic approach for predicting sediment transport from fields and small catchments. *Earth Surface Processes and Landforms* **23**, 527–544.

Motha, J.A., Wallbrink, P.J., Hairsine, P.B. and Grayson, R.B. 2003. Determining the sources of suspended sediment in a forested catchment in southeastern Australia. *Water Resources Research* **39**, 1056–1069.

National Water Council 1983. *Design analysis of urban storm drainage; the Wallingford procedure volume 1; principles, methods and practice*. National Water Council, London, UK.

Nearing, M.A., Foster, G.R., Lane, L.J. and Finkner, S.C. 1989. A process-based soil erosion model for USDA-Water Erosion Prediction Project Technology. *Transactions American Society of Agricultural Engineers* **32**, 1587–1593.

Nosrati, K., Govers, G., Semmens, B.X. and Ward, E.J. 2014. A mixing model to incorporate uncertainty in sediment fingerprinting. *Geoderma* **217-218**, 173–180.

Ongley, E.D. and Blatchford, D.P. 1982. Application of continuous flow centrifugation to contaminant analysis of suspended sediment in fluvial systems. *Environmental Technology Letters* **3**, 219–228.

Ongley, E.D., Bynoe, M.C. and Percival, J.B. 1982. Physical and geochemical characteristics of suspended solids, Wilton Creek, Ontario. *Hydrobiologia* **91**, 41–57,

Palmer, M.J. and Douglas, G.B. 2008. A Bayesian statistical model for end member analysis of sediment geochemistry, incorporating spatial dependences. *Applied Statistics* **57**, 313–327.

Phillips J.M. and Walling D.E. 1995. Assessment of the effects of sample collection, storage and resuspension on the representativeness of measurements of the effective particle size distribution of fluvial suspended sediment. *Water Research* **29**, 2498–2508.

Phillips J.M. and Walling D.E. 1997. Calibration of a PAR-TEC 200 Laser back-scatter probe for *in-situ* sizing of fluvial suspended sediment. *Hydrological Processes* **12**, 221–231.

Phillips, J.M., Russell, M.A. and Walling, D.E. 2000. Time-integrated sampling of fluvial suspended sediment: a simple methodology for small catchments. *Hydrological Processes* **14**, 2589–2602.

Pittam, N.J., Foster, I.D.L. and Mighall, T.M. 2009. An integrated lake-catchment approach for determining sediment source changes at Aqualate Mere, Central England. *Journal of Paleolimnology* **42**, 215–232

PLUARG, 1978. *Environmental Management Strategy for the Great Lakes System*. International Joint Commission, Windsor, Ontario, Canada. http://agrienvarchive.ca/download/PLUARG_env_man_strat.pdf.

Porto, P. and Walling, D.E. 2014. Use of ^{7}Be measurements to estimate rates of soil loss from cultivated land: Testing a new approach applicable to individual storm events occurring during an extended period. *Water Resources Research* **50**, 8300–8313.

Porto, P., Walling, D.E. and Callegari, G 2011. Using ^{137}Cs measurements to establish catchment sediment budgets and explore scale effects. *Hydrological Processes* **25**, 886–900.

Pretty, J. and Bharucha, Z. 2014. Sustainable intensification in agricultural systems. *Annals of Botany*, 1–26.

Pulley, S., Foster, I. and Antunes, P. 2015. The uncertainties associated with sediment fingerprinting suspended and recently deposited fluvial sediment in the Nene river basin. *Geomorphology* **228**, 303–319.

Renard, K.G., Foster, G.R., Weesies, G.A., et al. 1997. *Predicting soil erosion by water: A guide to conservation planning with the Revised Universal Soil Loss Equation (RUSLE)*. US Department of Agriculture *Agricultural Handbook*. No. 703. US Department of Agriculture, Washington, DC.

Russell, M.A, Walling, D.E. and Hodgkinson, R.A. 2000. Appraisal of a simple sampling device for collecting time-integrated fluvial suspended sediment samples. In *The Role of Erosion and Sediment Transport in Nutrient and Contaminant Transfer*. IAHS Publication No.263, IAHS Press, Wallingford, UK, 119–127.

Russell, M.A., Walling, D.E. and Hodgkinson, R.A. 2001. Suspended sediment sources in two small lowland agricultural catchments in the UK. *Journal of Hydrology* **252**, 1–24.

Schuller, P., Iroumé, A., Walling, D.E., et al. 2006. Use of beryllium-7 to document soil redistribution following forest harvest operations. *Journal of Environmental Quality* **35**, 1756–1763.

Schoellhamer, D.H. and Wright, S.A. 2003. Continuous monitoring of suspended sediment discharge in rivers by use of optical back scatterance sensors. In: *Erosion and Sediment Transport Measurement: Technological and Methodological Advances*. IAHS Publication No. 283, IAHS Press, Wallingford, UK: pp. 28–36.

Sear, D.A., Pattison, I., Collins, A.L., et al. 2014. Factors controlling the temporal variability in dissolved oxygen regime of salmon spawning gravels. *Hydrological Processes* **28**, 86–103.

Shen, H.W. (ed.) 1981. *River Mechanics*. Water Resources Publications, Littleton, CO, USA.

Strőmqvist, J., Collins, A.L., Davison, P.S. and Lord, E.I. 2008. PSYCHIC – a process-based model of phosphorus and sediment transfers within agricultural catchments. Part 2. A preliminary evaluation. *Journal of Hydrology* **350**, 303–316.

Sutherland, T.F., Lane, P.M., Amos, C.L. and Downing, J. 2000. The calibration of optical backscatter sensors for suspended sediment of varying darkness levels. *Marine Geology* **162**, 587–597.

Terry, J.P., Garimella, S. and Kostaschuk, R.A. 2002. Rates of floodplain accretion in a tropical island river system impacted by cyclones and large floods. *Geomorphology* **42**, 171–182.

Thompson, J., Cassidy, R., Doody, D.G. and Flynn, R., 2014. Assessing suspended sediment dynamics in relation to ecological thresholds and sampling strategies in two Irish headwater catchments. *Science of the Total Environment* **468**, 345–357.

Tiecher, T., Caner, L., Minella, J.P.G., et al. 2015. Tracing sediment sources in a subtropical rural catchment of southern Brazil by using geochemical tracers and near-infrared spectroscopy, *Soil and Tillage Research 2015*, 10.1016/j.still.2015.03.001

Trimble, S.W. 1983. A sediment budget for Coon Creek basin in the Driftless Area, Wisconsin, 1853–1977. *American Journal of Science* **283**, 454–474.

Van Rompaey, A.J.J., Verstraeten, G., Van Oost, K., et al. 2001. Modelling mean annual sediment yield using a distributed approach. *Earth Surface Processes and Landforms* **26**, 1221–1236.

Von Bertrab, M.G., Krein, A., Stendera, S., et al. 2013. Is fine sediment deposition a main driver for the composition of benthic macroinvertebrate assemblages? *Ecological Indicators* **24**, 589–598.

Wagenhoff, A., Lange, K., Townsend, C.R. and Matthaei, C.D. 2013. Patterns of benthic algae and cyanobacteria along twin-stressor gradients of nutrients and fine sediment: a stream mesocosm experiment. *Freshwater Biology* **58**, 1849–1863.

Wall, G.J. and Wilding, L.P. 1976. Mineralogy and related parameters of fluvial suspended sediments in northwestern Ohio. *J. Environmental Quality* **5**, 168–173.

Wallbrink, P.J., Murray, A.S. and Olley, J.M. 1998. Determining sources and transit times of suspended sediment in the Murrumbidgee River, New South Wales, Australia, using fallout ^{137}Cs and ^{210}Pb. *Water Resources Research* **34**, 879–887.

Walling, D.E. 2006. Tracing versus monitoring: New challenges and opportunities in erosion and sediment delivery research. In: Owens, P.N. and Collins, A.J. (eds) *Soil Erosion and Sediment Redistribution in River Catchments*, CABI, Wallingford: pp. 13–27.

Walling, D.E. 2012. The Use of Radiochemical Measurements to Investigate Soil Erosion and Sedimentation. In R.A. Meyers (ed.) *Encyclopedia of Sustainability Science and Technology*. Springer, New York: pp. 3705–3768.

Walling, D.E. 2013. The evolution of sediment source fingerprinting techniques in fluvial systems. *Journal of Soils and Sediments DOI* 10.1007/s11368-013-0767-2.

Walling, D.E. and He, Q. 1997. Use of fallout ^{137}Cs in investigations of overbank deposition on floodplains. *Catena*, **29**, 263–282.

Walling, D.E. and He, Q. 1999a. Improved models for estimating soil erosion rates from cesium-137 measurements. *Journal of Environmental Quality* **28**, 611–622.

Walling, D.E. and He, Q. 1999b. Using fallout lead-210 measurements to estimate soil erosion on cultivated land. *Soil Science Society of America Journal* **63**, 1404–1412.

Walling, D.E. and Collins, A.L. 2008. The catchment sediment budget as a management tool. *Environmental Science and Policy* **11**, 136–143.

Walling, D.E., Peart, M.R., Oldfield, F. and Thompson, R. 1979. Suspended sediment sources identified by magnetic measurements. *Nature* **281**, 110–113.

Walling, D.E., Webb, B.W. and Woodward, J. C. 1992. Some sampling considerations in the design of effective strategies for monitoring sediment-associated transport. In: *Erosion and Sediment Transport Monitoring Programmes in River Basins (Proceedings of the Oslo Symposium, August 1992)*. IAHS Publ. no. 210, IAHS Press, Wallingford: pp. 279–288.

Walling, D.E., He, Q. and Blake, W. H. 1999. Use of ^{7}Be and ^{137}Cs measurements to document short- and medium-term rates of water-induced soil erosion on agricultural land. *Water Resources Research* **35**, 3865–3874.

Walling, D.E., Collins, A.L., Sichingabula, H.W. and Leeks, G.J.L. 2001. Integrated assessment of catchment sediment budgets. *Land Degradation and Development* **12**, 387–415.

Walling, D.E., Russell, M.A., Hodgkinson, E.A. and Zhang, Y. 2002. Establishing sediment budgets for two small lowland catchments. *Catena* **47**, 323–353.

Walling, D.E., Collins, A.L. and McMellin, G.K. 2003. A reconnaissance survey of the source of interstitial fine sediment recovered from salmonid spawning gravels in England and Wales. *Hydrobiologia* **497**, 91–108.

Walling, D.E., Collins, A.L., Jones, P.A., et al. 2006. Establishing finegrained sediment budgets for the Pang and Lambourn LOCAR catchments, UK. *Journal of Hydrology* **330**, 126–141.

Walling, D.E., Schuller, P., Zhang, Y and Iroumé, A. 2009. Extending the timescale for using beryllium-7 measurements to document soil redistribution by erosion. *Water Resources Research* **45**, W02418, doi:10.1029/2008WR007143.

Walling, D.E., Zhang, Y. and He, Q. 2011. Models for deriving estimates of erosion and deposition rates from fallout radionuclide (caesium-137, excess lead-210 and beryllium-7) measurements and the development of user-friendly software for model implementation. In *Impact of Soil Conservation Measures on Erosion Control and Soil Quality*, IAEA TECDOC-1655, IAEA Vienna: pp. 11–33.

Walling, D.E., Porto, P., Zhang, Y. and Du, P. 2014. Upscaling the use of fallout radionuclides in soil erosion and sediment budget investigations: Addressing the challenge. *International Soil and Water Conservation Research* **2**, 1–21.

Wicks, J.M. and Bathurst, J.C. 1996. SHESED: a physically based, distributed erosion and sediment yield component for the SHE hydrological modelling system. *Journal of Hydrology* **175**, 213–238.

Wilkinson, S.N., Henderson, A., Chen, Y. and Sherman, B. 2004 SedNet User Guide, Version 2. Client

Report. Canberra: CSIRO Land and Water. http://\ignorespaceswww.toolkit.net.au/sednet.

Wilkinson, S.N., Prosser, I.P., Rustomji, P. and Read, A.M. 2009. Modelling and testing spatially distributed sediment budgets to relate erosion processes to suspended sediment yields. *Environmental Modelling and Software* **24**, 489–501.

Wilkinson, S.N., Hancock, G.J., Bartley, B., et al. 2011. Using sediment tracing to identify sources of fine sediment in grazed rangelands draining to the Great Barrier Reef. *Agriculture, Ecosystems and Environment* **180**, 90–102.

Williams, N.D., Walling, D.E. and Leeks, G.J.L. 2007. High temporal resolution *in situ* measurement of the effective particle size characteristics of fluvial suspended sediment. *Water Research* **41**, 1081–1093.

Williams, N.D., Walling, D.E. and Leeks, G.J.L. 2008. An analysis of the factors contributing to the settling potential of fine fluvial sediment. *Hydrological Processes* **22**, 4153–4162.

Wood P.J. and Armitage, P.D. 1997. Biological effects of fine sediment in the lotic environment. *Environmental Management* **21**, 203–217.

Yang, C.T., Treviño, M.A. and Simes, F.J.M. 1998. *User's Manual for GSTARS 2.0 (Generalized Stream Tube model for Alluvial River Simulation Version 2.0)*. Technical Service Center, U.S. Bureau of Reclamation, Denver, Colorado.

Zapata, F. 2002. *Handbook for the Assessment of Soil Erosion and Sedimentation Using Environmental Radionuclides*. Kluwer, Dordrecht.

Zhang, Y., Collins, A.L., Murdoch, N., et al. 2014. Cross sector contributions to river pollution in England and Wales: updating waterbody scale information to support policy delivery for the Water Framework Directive. *Environmental Science and Policy* **42**, 16–32.

CHAPTER 4

Linking the past to the present: the use of palaeoenvironmental data for establishing reference conditions for the Water Framework Directive

Ian Foster[1,2] and Malcolm T. Greenwood[3]
[1] *Department of Environmental & Geographical Sciences, University of Northampton, Northampton, UK*
[2] *Department of Geography, Rhodes University, Grahamstown, South Africa*
[3] *Department of Geography, Loughborough University, Loughborough, UK*

Introduction

We live in a world of change, and the increasing manipulation of the atmosphere and land surface (deliberately or inadvertently) often produces a response in the fluvial system (physically, chemically and/or ecologically), which is rarely recognised until after the change has occurred. Typically, only after the event do we engage in monitoring in order to establish the nature and magnitude of the potential problem and often we have little or no baseline information regarding the conditions that existed prior to disturbance. Long-term monitoring of rivers globally is irregular, spatially variable and inconsistent in terms of parameters measured, and will only rarely provide a record of change spanning more than a few decades. We have therefore largely missed those periods of environmental change when the most dramatic natural changes or the growing impact of human activity occurred on our planet. Separating natural from anthropogenic impacts has pervaded many recent palaeoenvironmental research projects and remains an important focus for determining key drivers of environmental change (e.g., Currás *et al.*, 2012; Foster *et al.*, 2012). This separation is also important because current EU legislation requires the establishment of reference conditions against which to compare the current ecological status of European rivers and provide a target in terms of river restoration and catchment management.

Identifying reference conditions is argued to be important in any programme for ecological assessment (e.g., Smol, 2008; Moss, 2011) and provides a baseline against which to identify human-induced changes through time. The reference condition concept is enshrined in the Common Implementation Strategy for the Water Framework Directive (WFD-CIS) (European Commission, 2003). Reference conditions are required for rivers,

River Science: Research and Management for the 21st Century, First Edition.
Edited by David J. Gilvear, Malcolm T. Greenwood, Martin C. Thoms and Paul J. Wood.

canals, lakes and reservoirs and are formally defined as:

> For any surface water body type reference conditions or high ecological status is a state in the present or in the past where there are no, or only very minor, changes to the values of the hydromorphological, physico-chemical, and biological quality elements which would be found in the absence of anthropogenic disturbance. Reference conditions should be represented by values of the biological quality elements in calculation of ecological quality ratios and the subsequent classification of ecological status.
>
> *(European Commission, 2003: p. 79)*

Surface water bodies must be grouped into types and reference conditions estimated for each of the identified types prior to determining ecological status (Figure 4.1a). The approach recognises that water bodies can, and have, changed through time and started from very different initial states related to their geographical location and catchment features. Reference conditions will therefore be different across and within Member States to account for these regional differences and it will be necessary to estimate type-specific reference conditions for the relevant hydro-morphological, physico-chemical and biological quality elements (Figure 4.1a; European Commission, 2003).

The WFD-CIS Guidance (European Commission, 2003) allows reference conditions to cover periods of minor disturbance, which means that human impacts are permissible as long as there are no, or only limited, ecological effects. A key element of this statement is that the Member States have commonly agreed that reference conditions should accommodate a level of impact compatible with the extent of pre-intensification land-use pressures, and should therefore accommodate a level of direct morphological alteration compatible with ecosystem adaptation and recovery to a level of biodiversity and ecological functioning equivalent to unmodified natural water bodies (European Commission, 2003).

Assessment of ecological status is based on the calculation of an ecological quality ratio that compares current conditions with that of the reference condition (Figure 4.1b). Ecological quality ratios are defined as the:

> Ratio representing the relationship between the values of the biological parameters observed for a given body of surface water and values for these parameters in the reference conditions applicable to that body. The ratio shall be represented as a numerical value between zero and one, with high ecological status represented by values close to one and bad ecological status by values close to zero.
>
> *(European Commission, 2003: p. 78)*

Environmental Quality Standards (EQS) are included in the assessment of ecological status in Figure 4.1a (central box, 2nd row). There is therefore a clear distinction to be made in the WFD-CIS (European Commission, 2003) between the role of general physico-chemical quality elements and specific pollutants in the classification of ecological status. For good ecological status, physico-chemical quality elements must lie within the range necessary to ensure ecosystem functioning but must additionally meet EQS set in accordance with section 1.2.6 of the Water Framework Directive (European Parliament, 2000).

A range of methods is available to determine reference conditions. These include the use of survey data from existing reference sites, historical data, palaeoenvironmental data (including multi-proxy palaeoecological studies), hydro-chemical transfer functions (predominantly for lakes), modelling studies (e.g., land-use and critical load models) and expert judgement. Developments in palaeoecology have made significant progress in defining reference conditions and in evaluating ecological response to physico-chemical conditions in

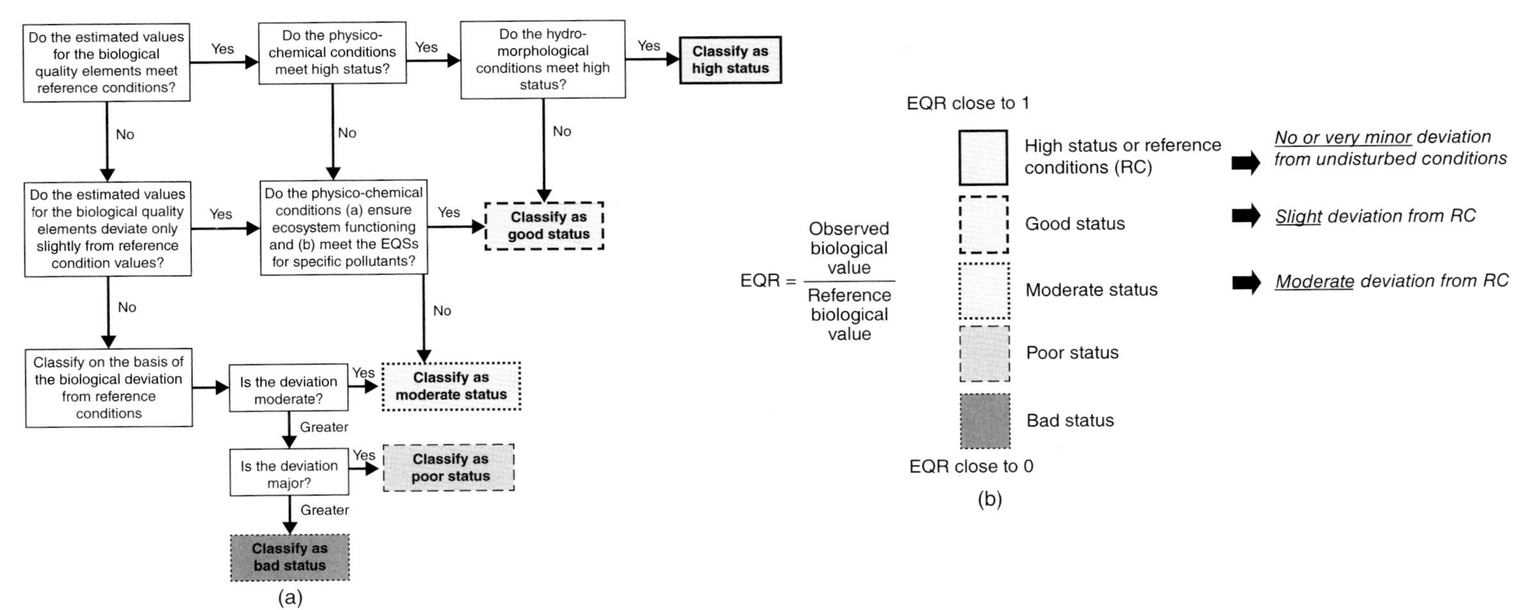

Figure 4.1 Indication of the relative roles of biological, hydromorphological and physico-chemical quality elements in ecological status classification (a); basic principles for classification of ecological status based on Ecological Quality Ratios (b) (after European Commission, 2003). Common Implementation Strategy for the Water Framework Directive (2000/60/EC), Guidance Document No 10: Rivers and Lakes – Typology, Reference Conditions and Classification Systems.

lakes for a range of relevant parameters (e.g., flood histories, sediment yields, nutrient status and acidity; Bennion and Battarbee, 2007; Battarbee and Bennion, 2011). Similar methods have only rarely been proposed for determining reference conditions for rivers (Collins *et al.*, 2010; Foster *et al.*, 2011), yet a range of sedimentary environments exist in river catchments which might provide relevant information at an appropriate timescale. (A key question here is 'what is an appropriate timescale?'). Direct human impacts on rivers have been documented as far back as the thirteenth century in Europe (Petts *et al.*, 1989; Petts, 1998) and to 3000 BCE in Egypt (Smith, 1971). A key issue in delivering the WFD objectives is to identify the most appropriate period for establishing reference conditions (Bennion *et al.*, 2011; Seddon *et al.*, 2012).

Various locations in catchments preserve evidence of past conditions. Active floodplains, river terraces, floodplain lakes and abandoned palaeochannels provide temporally and spatially incomplete records. They can provide information, for example, about channel morphology and substrate conditions, from which former velocity and discharge might be calculated. They may also tell us about sediment-associated contaminant status and the former ecology. By contrast, lakes and reservoirs usually preserve continuous records of past environmental conditions since the time of their formation or construction. However, records preserved in lakes require careful scrutiny and interpretation as the sediments (and biology) may more closely reflect the conditions existing within the lake rather than those of the contributing rivers and prove inadequate for establishing reference conditions for the river, or catchment, itself.

The aim of this chapter is to review the palaeoenvironmental evidence that might be obtained in order to establish reference conditions; we evaluate the barriers to their effective establishment and, finally, we explore the potentially uncomfortable relationship that appears to exist between policy, management and science in effectively delivering improvements in European river conditions. We explore the potential value of several depositional environments for determining reference conditions in the context of the WFD but have deliberately omitted a detailed review of evidence available from river terraces. Terraces preserve palaeoenvironmental information episodically over very long timescales, are often not well preserved in the landscape, and are unlikely to be of significant value in determining reference conditions for contemporary fluvial systems.

The fluvial landscape: floodplains, palaeochannels and connectivity

The fluvial landscape reflects the interaction of hydrological, geomorphological and biological processes (Petts and Amoros, 1997) and offers a rich tapestry of habitats, from those reflecting the contemporary position of the main channel to the remnants of the historic record hidden in deposited palaeochannel sediments. The spatial pattern of existing biotopes also reflects stages in the evolution of the fluvial landscape and, in doing so, offers an insight into the degree of connection and of the temporal sequence each stage represents; from main channel, (Parapotamic environments), to abandoned channels (Plesiopotamic – ox-bow lakes) and to the final terrestrialised stage, (Palaeopotamic – field ponds).

Recognising the importance of the historic archive, Amoros *et al.* (1987) describe both the spatial and temporal nature of the ecosystem by comparing the palaeoecological assemblages from deposits (diachronic analysis) with the community composition of the present-day biotopes occurring at different stages of the succession (synchronic analysis) and in doing so suggest a methodology for the study of floodplain evolution. The organic-rich sediments, either exposed from sites of gravel extraction or obtained by a variety of coring devices from sites on the floodplain, are processed in the laboratory, either by flotation methods and/or by using sieves of varying sizes. The total organic content of samples, a useful indicator of ecosystem maturity, is also determined by loss on ignition in a high-temperature oven, typically 550 °C; higher organic matter content suggesting a higher degree of ecosystem maturity (see below).

In order to describe the nature of the sedimentary record and the biological communities preserved in the organic deposits, a variety of descriptors have been used to track and provide an indication of environmental change. In lakes (closed ecosystems) many faunal and floral groups, such as non-biting midges and diatoms, have been used (see Smol *et al.*, 2001), but for open fluvial ecosystems, choice of taxonomic group is more restricted (Greenwood *et al.*, 2003; Howard *et al.*, 2010). A selection of palaeoenvironmental indicators and their potential applications are outlined in Table 4.1 and selected examples are illustrated further below. A major driver determining the structure and function of most aquatic communities and those of the river margin, is related to hydraulic condition and sediment supply and the patterns, relating to the temporal sequence in the development of the different water-bodies, can be identified across the floodplain. (See Table 4.1 and the three proxies (beetles, non-biting midges and caddisflies) reviewed in more detail below.)

In a study of the sedimentology of a section of the Rhône River floodplain, the distribution of sediment particle size found in the sequence of floodplain habitats, from the main channel to standing water habitats, indicated that each major biotope could be characterised by differentiated granulometric patterns associated with distinct flow conditions (Figure 4.2a adapted from Amoros *et al.*, 1987). Similarly, in a study of the upper Rhône and Ain floodplains (France), Castella *et al.* (1984), identified three assemblages of macroinvertebrate taxa that characterised the palaeochannels: (i) those weakly influenced by floods and characterised by lentic species; (ii) an intermediate group, mostly from former channels, mainly connected to the river at their downstream end and subject to variable flow conditions and (iii) a collection of channels that were highly influenced by floods and where lotic conditions prevailed for the majority of time. In a further study, Roux and Castella (1987) also illustrate the dominant groupings of palaeochannels on the upper Rhône River, based upon specific caddisfly assemblages (Insecta: Trichoptera), with each habitat type, and relates these to the degree of connectivity each has with the main channel (Figure 4.2b). Gandouin *et al.* (2006) recorded similar responses using assemblages of non-biting midges (Insecta: Chironomidae). A development of this idea is illustrated in Figure 4.2c where a macroinvertebrate index (PalaeoLIFE) has been used to hind-caste the flow conditions from a range of palaeodeposits from floodplain sites. This metric is based on the contemporary work of Extence *et al.* (1999) and the notion that the fluvial energy in the system is a key driver in determining the composition of each macroinvertebrate assemblage. The data used in Figure 4.2c is derived from the assemblages of sub-fossil

Table 4.1 Selected palaeoenvironmental indicators and their potential uses.

Proxy	For	Habitat scale of information Channel (C) Floodplain (F) Lake (L) Catchment (Ca)	Examples
(a) Physical/chemical			
Particle size	Rainfall/river discharge	Ca	Lapointe *et al.*, 2012 Partridge *et al.*, 2004
Sediment yield/sediment accumulation Rates	Catchment disturbance/change (natural/anthropogenic)	Ca	Boyle *et al.*, 2011 Foster *et al.*, 2012 Andreev *et al.*, 2004
Low and high temperature loss on ignition	Organic matter/carbonate	Ca L	Heiri *et al.*, 2001 Shuman, 2003
Environmental magnetism	Sediment sources, pollution history	Ca	
Geochemical analysis	Sediment sources Pollution history	Ca	Foster *et al.*, 2007 Hollins *et al.*, 2011
Radionuclides	Dating/sediment source tracing	Ca	Currás *et al.*, 2012 Foster 2006
Stable isotopes	Pollution sources (natural/anthropogenic)	Ca	Renburg *et al.*, 2002 Woodward *et al.*, 2012
Fly-ash	Atmospheric pollution/acidification/dating	Ca	Rose and Appleby 2005 Pittam *et al.*, 2009
Organic geochemistry (e.g., C:N ratios and isotopes)	Organic matter sources, vegetation and climate change	Ca L	Lane *et al.*, 2011 Rodysill *et al.*, 2012
(b) Biological			
Pollen and Spores	Vegetation/vegetation change/cultivation history/deforestation.	Ca	Pittam *et al.*, 2006 Bennett and Willis, 2001
Non-pollen palynomorphs	Grazing intensity	Ca	Davis and Shafer, 2006 Mighall *et al.*, 2012
Plant macrofossils	Local ecology	C L	Birks, 2001 Bjune, 2005
Charcoal	Fire history/industrial archaeology	Ca	Mighall *et al.*, 2012 Whitlock and Larsen, 2001
Diatoms, chrysophyte scales./cysts	Acidity nutrient status/ salinity	C L	Battarbee *et al.*, 2011a & b
Cladocera and Ostracods	Temperature/water quality	F C	Nazarova *et al.*, 2011 Irvine *et al.*, 2012
Caddisflies (Insecta: Trichoptera)	Methodology. Environmental reconstruction	F C L	Williams, 1989 Solem and Birks, 2000
Non-biting midges (Insecta: Chironomidae)	Multiproxy environmental reconstruction. Methodology	F C L	Bedford *et al.*, 2004. Gandouin *et al.*, 2006 Brooks *et al.*, 2007
Beetles (Insecta: Coleoptera)	Application and environmental reconstruction	F C L	Coope, 1994 Ponel *et al.*, 2007

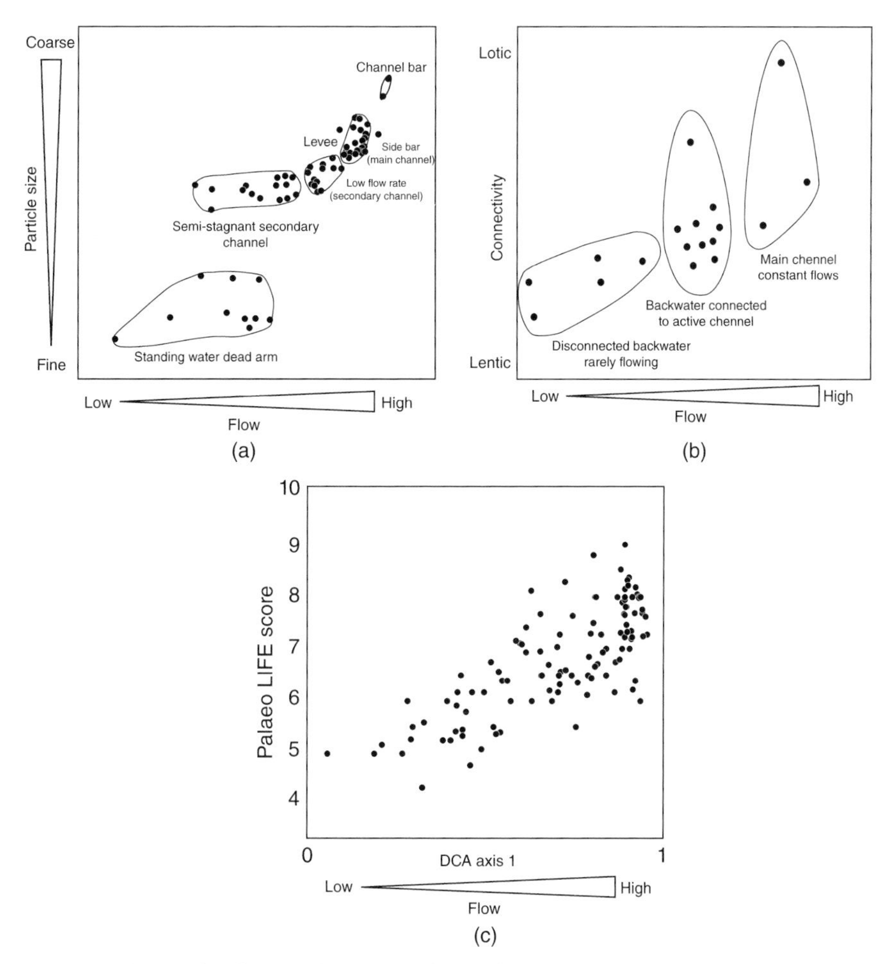

Figure 4.2 (a) Particle-size distribution of sediments from different geomorphic units of a section of the River Rhône floodplain and association with speed of flow (adapted from Amoros *et al.,* 1987). (b) Distribution and characterisation of 18 former channels of the rivers Rhône and Ain, based upon the contemporary caddisfly (Trichoptera) assemblages (adapted from Roux and Castella, 1987). (c) Relationship between DCA axis 1 and the PaleoLIFE score based upon sub-fossil caddisfly (Trichoptera) data from 17 palaeochannels from the River Trent floodplain; dated Late-Glacial to recent historic past (adapted from Greenwood *et al.,* 2006).

larval caddisflies across a timescale from the Late-Glacial period (ca. 13,000–10,000 years BP) to the recent historic past from the River Trent (UK): a high PalaeoLIFE score representing high flows in the main channel and a low score attributed to rarely flowing or disconnected backwater habitats (Greenwood *et al.,* 2006). The similarity in pattern of all three figures illustrates a link between the present and the past.

Floodplains as archives of change

Fragments of river channels, often in the form of meander cutoffs, preserve important information about river ecology as organisms are preserved in sediments accumulating at these locations. These sites could provide information to allow the reconstruction of historical reference conditions.

However, increased floodplain occupancy since Medieval times and the increase in sediment delivery from hillslopes to rivers due to forest clearance and agriculture have combined to modify river morphology to such an extent that in Europe floodplains are now almost entirely an artefact of human manipulation (Petts *et al.*, 1989; Bailey *et al.*, 1998; Brown, 2002). In a recent review, Lewin (2010) suggested that early river habitats would have contained a range of channel forms, including braided, meandering and anastomosing, while Brown (2002) stresses the significance of woody debris and beavers in pre-Medieval river systems prior to major disturbance. However, floodplain sites offered significant opportunities for the establishment of defensive sites, settlement, transport pathways and the exploitation of valuable natural resources. As a result, many of the pre-Medieval river types, especially anastomosing channels and floodplain wetlands, have all but disappeared from the landscape and are unlikely to be re-instated despite evidence that for many European rivers, these are the most likely reference conditions. The connectivity between various components of the channel and floodplain, exemplified by the data shown in Figure 4.2, is disrupted as a result of human intervention on the floodplain. Construction of flood embankments, levees and communication routes disrupts the lateral connectivity between the channel and floodplain (sensu Ward, 1989) and provide buffers (sensu Fryirs *et al.*, 2007) between hillslopes and channels. These changes reduce the transfer of water and essential nutrients, including carbon (see Langhans *et al.*, 2013), to aquatic communities occupying various habitats in the channel and on the floodplain. In combination with the extensive use of floodplains for agriculture, it is unlikely that reference conditions could ever be re-established in such locations.

Changes in river types are not only caused by direct human occupation and/or by changes in the delivery of sediment from the hill-slope to the valley. Late Holocene sea level rise and isostatic adjustment to variable ice loading have naturally changed the gradients of many low-lying coastal rivers. With sea level rise, palaeoenvironmental evidence often demonstrates a change from steep gradient braided or anastomosing styles to single thread meandering rivers (Trimble, 2010). Similar changes are likely to continue if rates of sea level rise increase as a consequence of future global warming. Isostatic adjustments may produce a spatially more complex response. Shennan *et al.* (2009), for example, showed that relative sea levels are rising, falling or stable in different parts of the British Isles at the present time.

There is a well-documented correlation between soil erosion and floodplain sedimentation over the Holocene and more recent timescales. Long term sedimentation rates are usually dated using ^{14}C, while rates over the last century can be dated using a combination of ^{137}Cs and ^{210}Pb (Walling and Foster, 2015). Lewin *et al.* (2005), for example, suggest that on major floodplains, sedimentation began in the early Holocene while Macklin *et al.* (2010) and others have shown that the bulk of sedimentation happened much later. Widespread and rapid sedimentation in the Medieval period (eleventh to fourteenth centuries) occurred at rates approaching ten times those of the Early Holocene. Contemporary sedimentation rate surveys and twentieth- to twenty-first-century reconstructions of sediment accumulation using ^{137}Cs and ^{210}Pb chronologies demonstrate that floodplains continue to act as important sediment stores in the sediment cascade (e.g., Rumsby, 2000; Foster, 2001; Trimble, 2010).

Floodplain depressions and floodplain lakes (defined by Foster (2006) as 'on-line' lakes if they remained in permanent contact with the river or 'off-line' lakes if they were only connected to the river at times of over-bank discharge) often preserve a temporally discontinuous palaeoenvironmental record. Techniques similar to those described above, and in a later section dealing with lake sediment records, can be used to provide information on palaeoflood frequency and on contaminant concentrations (e.g., Winter *et al.*, 2001; Paine *et al.*, 2002; Wolfe *et al.*, 2006) that could support the establishment of reference conditions for selected parameters for rivers.

A major limitation to the use of floodplains (and associated palaeochannels) as palaeoenvironmental archives is that floodplains do not develop in steep headwater catchments. They are scale-dependent emergent features of the landscape and are best developed in higher order streams. As a result floodplains, and/or associated palaeochannels, are generally not available for defining reference conditions in small headwater catchments.

Lake sediment-based archives

The analysis of lake sediment archives provides an opportunity to explore past catchment conditions and determine what some aspects of the catchment system were like before disturbance (Smol, 2008). In this context, the preservation of a continuous sedimentary record in the deepwater zone of lakes may help to provide realistic reference conditions. This approach raises two fundamental questions: How far do we need to go back in the record to identify what the reference conditions should be and what are the best indicators of change? A plethora of palaeoenvironmental proxy indicators are available (Table 4.1) and the choice of proxy will be determined by the research question posed. However, there are limitations in reconstructing river conditions from an analysis of the sedimentary archive contained in lakes and reservoirs. Palaeochannels contain ecological assemblages that reflect flow conditions, water quality and substrate type and can be interpreted at the patch and river reach scale rather than the catchment scale. These can provide information on almost all of the requirements set out in Figure 4.1 and the only major limitation is the spatial and temporal continuity of available reference locations. Lake and reservoir sediments preserve organisms that more closely reflect the water quality of the lake itself, which may or may not be dominated by processes operating in the upstream catchment. Each lake will differ in terms of the relative dominance in the sedimentary record of catchment inputs and/or atmospheric inputs and will also reflect the ability of the catchment itself to buffer these inputs before the river water reaches the lake. Studying a particular issue therefore requires careful selection of appropriate sites and selection of appropriate techniques from the list provided in Table 4.1. Bennion and Battarbee (2007) provided an overview of the potential for palaeolimnology to give information on reference conditions for the European Water Framework Directive and Battarbee and Bennion (2011) provided detailed case studies of how such reference conditions might be identified. Most of these case studies suggest that human impacts on nutrient status, atmospheric and industrial pollution and lake acidification were detectable from the early to the mid nineteenth century onwards and that reference conditions could be determined from a variety of proxy information preserved in the lake sediment column.

Palaeoenvironmental reconstructions using sediments accumulating at the point of inflow to small lakes and reservoirs (deltas) are rare and generally focus on palaeoflood conditions (see Foster, 2010) reconstructed from changes in the particle size distributions of accumulating coarse (largely sand and gravel sized) sediments. A number of studies have focused on the palaeoecology of major river deltas (e.g., Andreev *et al.*, 2004; Flessa, 2009), but the lack of small catchment-scale reconstruction suggests that these high-energy sedimentary environments have either been overlooked or that preservation of ecological evidence is poor.

It is unlikely that any assessment of river ecology (Column 1, Figure 4.1a) can be made directly using the palaeoecological record contained in lake and reservoir sediments alone. Site-specific information of the type preserved in floodplains and palaeochannels will be the only direct source of information for determining ecological reference conditions and hydromorphological status for lotic environments. However, there is significant potential for the lake sediment archive to provide information about physico-chemical conditions within the catchment (Bennion and Battarbee, 2007).

The evidence base for establishing reference conditions

In the following section we explore the value of selected physico-chemical and biological proxies that have been used for palaeoenvironmental reconstruction and that might form the basis for the establishment of reference conditions. They are selected to demonstrate the different types of information that might be obtained at different habitat scales, as summarised in Table 4.1.

Physico-chemical proxies

Sediment accumulation rates and reconstructed sediment yields in lakes and reservoirs

Changes in sediment accumulation rates (SARs) in European lakes were compiled by Rose *et al.* (2011) and were argued to reflect two potential contributions – an increase in the delivery of fine sediment from the contributing catchment (increased erosion) and/or increasing productivity and deposition of organic matter as reflected in trends in loss on ignition (see below). While SARs are not specified as reference conditions under WFD, Rose *et al.* (2011) analysed 207 dated lake sediment profiles to assess how SARs changed through time (in 25 year classes) for lakes of different types. Seventy-one percent of the sites showed near-surface SARs higher than 'basal' (mainly nineteenth century) rates. Eleven per cent showed no change and 18% showed a decline. Little change in SAR was observed prior to 1900 and most increases were observed in the twentieth and twenty-first centuries, in particular 1950–75 and post-1975. This indicates a general acceleration in SAR in European lakes during the second half of the twentieth century. Reference SARs were estimated for six lake-types. Contemporary SARs in all lake types exceeded reference conditions and greatest SARs were found in shallow low altitude lakes.

Foster *et al.* (2011) and Collins *et al.* (2012) focused on the potential for reconstructed sediment yields to determine reference conditions for catchment suspended sediment yields. Existing legislation with regard to fine sediment was set by the EU Freshwater Fish Directive with a guideline (mean annual suspended sediment concentration) of 25 mg l^{-1}. Foster *et al.* (2011) argued that the application of a single national standard is inappropriate for a pollutant that is strongly controlled by spatial variations in key catchment drivers and that sediment yield reconstruction offered an approach for assessing background sediment pressures on watercourses, enabling determination of values for periods pre-dating recent agricultural intensification (taken as ~ pre-1945). They proposed that Modern Background Sediment Delivery to Rivers (MBSDR) across England and Wales could be determined to quantify a feasible maximum sediment reduction and used 19 existing lake/reservoir sediment-based yield reconstructions to map the spatial variability in MSBDR for England and Wales. They proposed that the MBSDR could be taken to represent ecological demand for sediment inputs into watercourses required to support healthy aquatic habitats. Foster *et al.* (2011) also attempted to relate SARs in the 19 lakes analysed to the sediment yields for the same catchments. While a reasonable correlation was obtained for lowland agricultural catchments, the entire database showed a poor correlation between SAR and sediment yield, suggesting that the database used by Rose *et al.* (2011) would be unlikely to provide sufficiently robust data from which to estimate sediment pressures directly from the measurement of SAR alone. The lack of a strong correlation between SAR and sediment yield is unsurprising as rates of sedimentation at a single point in a lake will be partly controlled by temporal variations in sediment focusing (Dearing and Foster, 1993).

Sediment source fingerprinting

Sediment source fingerprinting methods for identifying the origin of fine (<63 μm) particulate sediment transported by rivers or deposited on floodplains and in lakes can be traced back to the 1970s and the work of researchers such as Klages and Hsieh (1975), Wall and Wilding (1976) and Walling *et al.* (1979) In these early studies, potential fine sediment sources were not clearly defined and the assessment of their relative contribution was essentially qualitative. Subsequent studies have focused on discriminating source types (e.g., urban street dust, cultivated land, channel banks, road verges), rather than spatial sources, and have begun to use a multi-tracer approach in order to obtain the best discrimination of sources (see Foster, 2000; Walling and Foster, 2015). To date, this has included the use of sediment geochemical signatures, mineral magnetism and gamma-emitting radionuclides for determining the origin of the inorganic fraction of the fine sediment load.

Space does not permit a detailed discussion of these issues (see Koiter *et al.*, 2013; Walling and Collins, Chapter 3), but in the context of establishing reference conditions with respect to WFD, the ability to use fingerprinting in an historical context (using dated floodplain and/or lake sediments) can provide a unique source of information that, in combination with sediment yield reconstruction, has the potential to inform managers about how much sediment is moving, where it is coming from and whether sources have changed through time. To date, major refinements of the fingerprinting methodology have focused on actively transported sediments (e.g., Collins *et al.*, 2010) and remains to be fully evaluated in a palaeoenvironmental context.

Loss on ignition

Most palaeolimnological studies make routine measurements of basic physical properties of the sediment including wet and dry bulk density and loss on ignition. Dry density is required for SAR and sediment yield calculation and many UK studies report sediment yields on a minerogenic basis so that the relative proportion of internally and externally produced organic matter is excluded from the calculation (Foster, 2006; Foster *et al.*, 2011). Loss on ignition is also used to exclude organic matter concentrations from the calculation of magnetic susceptibility and remanence concentration parameters that are often used for sediment source tracing (Walling and Foster, 2015). The established procedure requires the ignition of a pre-weighed and oven-dried sample in a muffle furnace at ~ 550 °C and combusting it for a period of time after which no further weight loss is recorded (Heiri *et al.*, 2001). Once weight has stabilised, all organic matter should have been removed from the sample and the remaining sediment is 'minerogenic'.

Biological proxies

Many early studies reconstructing palaeoenvironments have been based upon the analysis of a single faunal group, such as Coleoptera (Coope and Angus, 1975), but more recently the trend has been to adopt a multi-proxy approach where groups of experts pool knowledge and techniques to give complementary accounts for a particular field site. The advantages are the clear reinforcement of lines of evidence from different environmental signals, so as to provide a wider perspective (Howard, 2007; Schreve *et al.*, 2013). In fluvial deposits the sub-fossil remains of the larval stages of non-biting midges (Diptera: Chironomidae) and caddisflies (Trichoptera) provide a proxy for the aquatic condition, whereas the remains of adult beetle (Coleoptera) assemblages broaden the interpretation by offering both aquatic and riparian/terrestrial signals.

Beetles (Insecta: Coleoptera)

Beetle fragments of the head, thorax and wing cases (elytra) are prevalent in the waterlogged palaeodeposits; the material preserves well and identification of taxa to species level is often possible (Figure 4.3). From the initial purely descriptive studies and the accumulation of data from many sites, further advances have been made in interpreting characteristics associated with each faunal assemblage, for example the character of the thermal climate at the time of burial. The first such method was developed using beetle assemblages that had been radiocarbon dated from a wide range of geographic locations. Using details of each sub-fossil assemblage the Mutual Climatic Range methodology (MCR) was developed and based on the principle of establishing the range of climates occupied at the present-day by each beetle taxon represented in the fossil assemblage: hence climatic space being derived from geographic space (Atkinson *et al.*, 1986). Using this technique, climatic reconstructions have been made (e.g., Atkinson *et al.*, 1987; Maddy *et al.*, 1998) and verification of the temperature oscillations, especially across the Late-Glacial to Holocene transition period, has come from the Greenland Ice Core Project (GRIP) data (Coope *et al.*, 1999).

A useful recent advance has been the development of a database, Bugs Coleopteran Ecology Package (BugsCEP), incorporating the palaeorecord from many sites with

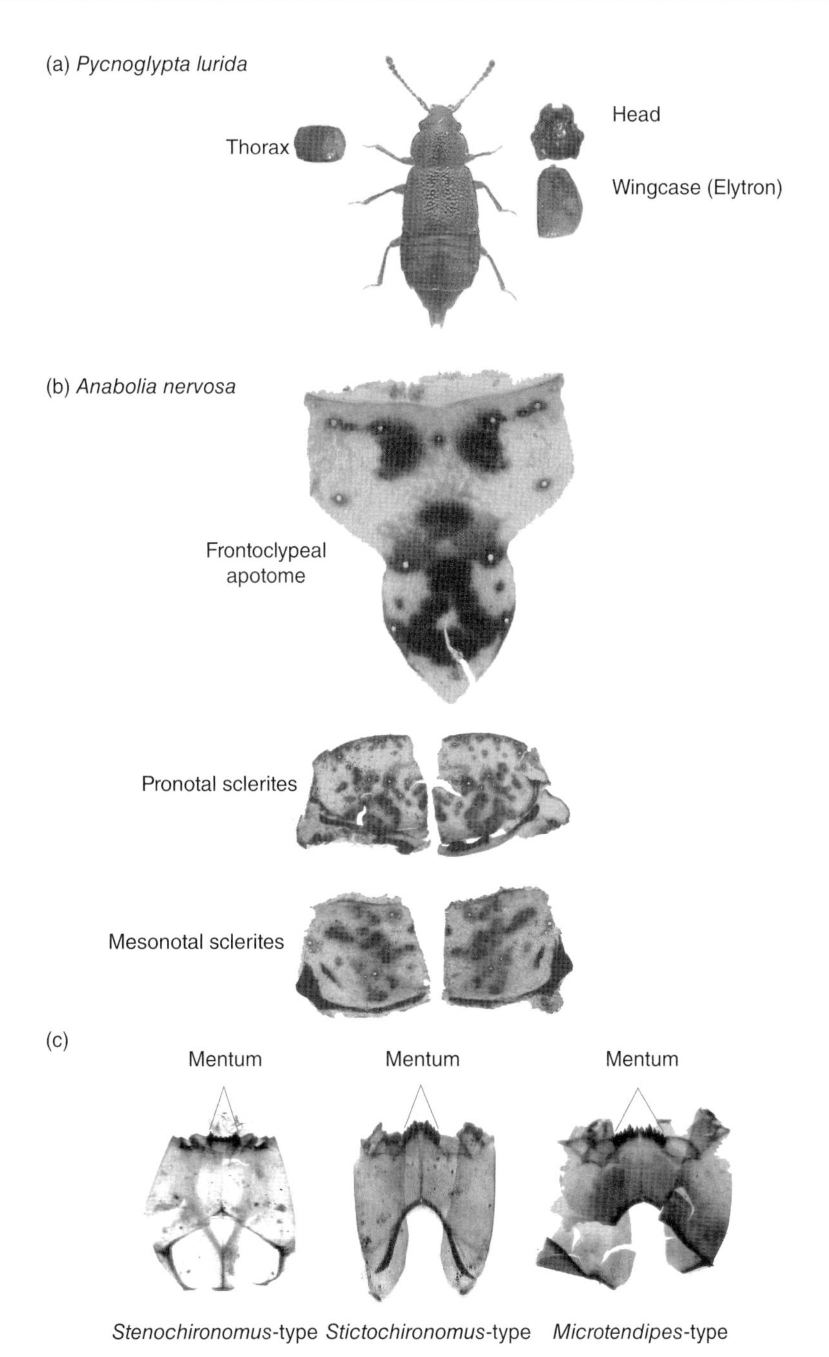

Figure 4.3 Sub-fossil fragments of beetle, caddisfly and non-biting midges collected from palaeochannel deposits from the River Trent floodplain. (a) *Pycnoglypta lurida* (Coleoptera: Staphylinidae), a rove beetle found in Late-Glacial deposits but no longer present in UK. (b) *Anabolia nervosa* (Trichoptera: Limnephilidae), a cased-caddisfly of rivers, lakes and ponds but not of temporary waterbodies. (c) Head capsules of non-biting midges *Stenochironomus*-type (a wood miner of lentic habitats), *Stictochironomus*-type (associated with macrophytes in littoral zone) and *Microctendipes*-type (common in lentic water-bodies) (Diptera: Chironomidae). Note the differing pattern of the toothed mentum. (*See colour plate section for colour figure*).

both ecological and distributional data of the modern fauna, so enabling rapid and more precise reconstructions of past environments to be made (http://www.bugscep.com, Buckland and Buckland, 2006).

Non-biting midges (Insecta: Diptera: Chironomidae)

Similar progress has been made using the non-biting midges (Chironomidae), a diverse group of two-winged flies with an aquatic larval stage. The head capsules of the aquatic larvae are numerous in silty deposits and detailed characters, such as the toothed mentum, allows for taxonomic identification (Figure 4.3). As for beetles and caddisflies, assemblages of the non-biting midges also characterise the types of flood-plain habitat (lentic taxa, e.g., *Gylptotendipes* sp., lotic taxa, e.g., *Cricotopus* sp.) and many species are stenothermic (e.g., *Pseudodiamesa* cf *arctica* – see Gandouin *et al.*, 2006). This adaptation by a range of taxa to the thermal properties of the environment has been used in the reconstruction of past climates. One approach taken differs from that for MCR methodology in that it is based upon surface sediment samples taken from a latitudinal transect of lakes in western Norway (Brooks and Birks, 2000). Having identified each assemblage a Chironomidae dataset, calibrated to July air temperatures, has produced a palaeotemperature inference model or transfer function that allows a temperature reconstruction to be made (Brooks and Birks, 2000). In a recent reconstruction of faunal remains from Mid Devensian sedimentary deposits around the skeleton of a woolly rhinoceros (*Coelodonta antiquitatis*) from the River Tame (UK), comparative calculations of beetle and chironomid palaeotemperature estimations made on the same samples, yielded strong agreement between the two methodologies (see Schreve *et al.*, 2013).

Caddisflies (Insecta: Trichoptera)

Of the many macroinvertebrate groups that use riverine habitats within their lifecycle, the Trichoptera (caddisflies) are particularly useful. Taxa are associated with almost all types of water-body, from ephemeral pools to large lowland rivers and the larval sclerites (from head and thorax) are numerous and well preserved in the silt and waterlogged peats. From field samples, the chitinous fragments are extracted by kerosene flotation, cleaned and mounted on microscope slides. Identification is based primarily on the shape, size, colour pattern and microsculpture of the fragments and identified by matching against reference material and figures from standard texts (Figure 4.3).

As for the preceding groups, the ecological detail for many taxa allows for a reconstruction of the habitat and biological traits of the community. For example, taxa can be categorised into their functional feeding groups, scraper/grazers (*Tinodes* spp.), deposit feeders (*Molanna* spp.), filter feeders (*Hydropsyche* spp.), shredders (*Limnephilus* spp.) and predators (*Agrypnia* spp.) and also each assigned to a preferential range of flows (or flow group), both strong determinants in the distribution of larval caddisfly taxa across the range of floodplain habitats. Taxa related to a range of flow groups (*sensu* Extence *et al.*, 1999) are, for fast-flowing waters, *Rhyacophila* spp. (Flow group I) and *Hydropsyche* spp (Flow group II), to those taxa inhabiting still water habitats, *Limnephilus flavicornis* (Flow group V) and *Trichostegia minor* (Flow group VI). Ecological complexities arise when using multi-proxies as each taxon adapts to both physical (e.g., temperature;

flow) and biological (e.g., competition for food resources and available space), but a multi-proxy approach using, for example, the three major components of the fauna illustrated here, on the same sample, can refine the environmental interpretation (see Howard *et al.* 2009, 2010).

Some debate may arise regarding the naming of taxa, as the sub-fossil material available is in this case made up of the robust parts of the exoskeleton, with many of the more delicate structures being lost. However, by using a multi-proxy approach from a range of sites and with a supporting ecological database, a picture can emerge. For this approach to develop, more integration between limnologists and fluvial scientists is to be encouraged.

Discussion and conclusion

Not all of the elements of Figure 4.1 can be determined using palaeoenvironmental information and Table 4.2 provides a summary of those reference conditions that might be determined using the range of sedimentary evidence and techniques considered above. The choice of sedimentary environment will ultimately be determined by the timescale over which reference conditions need to be established, but this also depends on matching the right timescale and spatial scale with the availability of evidence. While there have been increases in the number of published studies in all environments identified in Table 4.2, there are a limited number of case studies for UK rivers, yet implementation of the WFD requires the establishment of reference conditions for all rivers. There are five issues that emerge from the evidence presented above. Each of these is discussed briefly below.

Identifying change and human impact

The use of palaeoenvironmental data to identify reference conditions depends on the extent to which past human impacts can be identified (quantified) and whether a specific human impact can be related to a specific change in the proxy signal. In the context of palaeolimnology, Battarbee *et al.* (2011b) suggest that this may vary from site to site and from pressure to pressure and will also depend on whether the change can be ascribed to a climatic or a human forcing factor or will simply lie within the bounds of natural variability.

Late Quaternary and Early Holocene environmental change is unlikely to have been driven by human activities, yet many of the studies reported here have demonstrated that change occurs gradually and naturally. In the British Isles, thin soil and little vegetation would have existed in the Late Glacial. Geomorphological processes operating on a deglaciated or permafrost impacted landscape would have increased chemical weathering to produce soils that were slowly colonised by vegetation emerging from refugia. The availability of weathered parent material and soil would have given rise to fine sediment being transported by streams and rivers leading to the creation or further development of well-defined river channels, floodplains and a range of other floodplain habitats that would likely have attracted early human settlers. Inevitably, reference conditions are not static but change through time due to natural drivers (climate/sea level). Human disturbance and/or natural evolution of catchment soils and sediments may also mean that the past reference condition no longer exists and cannot be recreated. A similar issue arises with the absence of anastomosing rivers and floodplain wetlands that were 'obliterated'

Table 4.2 What can we know about reference conditions in rivers using sedimentary archives?

Source of information	Channel characteristics	Flow conditions	Sediment calibre/yield	Sediment quality	River ecology	Catchment climate/land use characteristics
Floodplain		√	√	√		√
Palaeochannel	√	√	√	√	√	√
Open water lake sediment		√	√	√		√

from Medieval landscapes. Using history to establish reference conditions may therefore be flawed in the absence of relevant comparable environments today. Although the WFD requires river managers to prevent deterioration of ecological status, to date there has been no formal provision for the effects of future climate change and the impact that this might have on the trajectories of river (including floodplain) morphology or river and lake ecology (Battarbee *et al.*, 2011b)

Relevance of the proxy variable and quantifying the amount of change

While some palaeoenvironmental reconstructions can provide direct evidence of the biological assemblages present in rivers, many cannot, especially those reconstructions based on palaeolimnological evidence. Figure 4.4, for example, plots the same variables as those plotted in Figures 4.2a and b and maps the flow, sediment size and connectivity conditions in relation to the palaeoenvironments reviewed in this chapter. Apart from abandoned palaeochannels, the range of flow and substrate conditions are not represented in the remaining palaeoenvironments and therefore significant amounts of information about the river itself cannot be reconstructed. While we may be able to utilise fossil assemblages to establish what likely nutrient, pH and salinity conditions

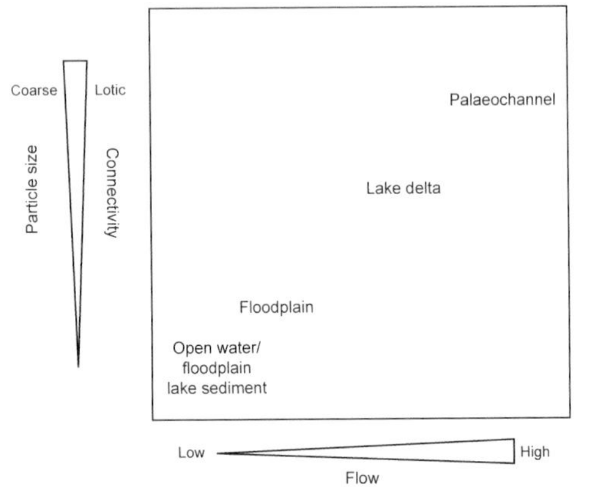

Figure 4.4 Flow, particle size and connectivity in relation to palaeoenvironments (note only palaeochannels will represent most of the conditions in contemporary river and floodplain habitats shown in Figures 4.2a and b).

prevailed and what the broad landscape ecology looked like, we cannot directly reconstruct the morphology and ecology of the inflowing rivers as the lake sediment record tells us little or nothing about these landscape elements.

Bennion *et al.* (2011) argued that the use of chemical criteria might enable a reasonably precise assessment of the degree of change from reference conditions based on an analysis of lake sediment record, they also note that:

> Quantifying the degree of change using biological measures is less easy, especially as in Europe under the WFD it is not only the degree of change between reference and present day status that is important; the degree of change needs to be assessed to classify sites into 'high', 'good',

> 'moderate', 'poor' and 'bad' status based on the degree to which present day conditions deviate from reference conditions. The 'good/moderate' distinction is especially important, as restoration to "at least good" status is the specific objective of the legislation.
>
> *(Bennion* et al., *2011: p. 538)*

Similar problems arise in directly using biological information from palaeochannels as not all organisms are represented in deposits remaining at such sites.

Availability of reference sites

In rivers, significant floodplain development generally does not occur in streams smaller than first or second order, especially in steep upland catchments, which means that floodplains and meander cutoffs are rare or non-existent. Natural lakes exist in some of these environments, largely as a legacy of Quaternary glaciations over the last ~ 2 million years, but lakes outside of ice limits are poorly represented (excepting coastal freshwater lagoons) although some artificial lakes and reservoirs might have existed for long enough for reference conditions to be established (Foster, 2006; Foster *et al.*, 2011). Development of floodplains, or their drainage and subsequent use for agriculture, means that many potential reference sites have also been lost from lowland floodplain systems.

Connectivity, complex response and uncertainty

While these concepts are familiar to all students of fluvial geomorphology, river managers may not fully understand why geomorphologists are nervous when asked the question: 'What will happen to rates of sediment transport in rivers if we rehabilitate part or all of a catchment in an attempt to reduce sediment yields?' Intuitively we might expect sediment transport rates to decline rapidly but the classic paper of Meade (1982) demonstrated that this was not necessarily the case as eastern seaboard US rivers simply tapped an alternative source of sediment after rehabilitation and maintained sediment yields at the pre-rehabilitation level. Similarly, Foster *et al.* (2012) have shown from a series of palaeolimnological reconstructions in the Eastern Cape, South Africa, that despite several decades of catchment rehabilitation (e.g., de-stocking, eliminating fire as a management tool) sediment yields have remained stubbornly high, as vegetation recovery in the catchment has been slow.

Interpretation of change in the fluvial system is fraught with difficulties. The drivers and subsequent catchment responses are complex, and include internal and external triggers that can be natural or human-induced. For example, increased agricultural production in Medieval times increased erosion on hill-slopes but simultaneously increased deposition in valley bottoms (complex response). It is essential that these basic concepts are understood as recovery of a river system to some form of catchment rehabilitation is not immediate and may take as long as decades to centuries to achieve, especially if the initial damage removed much of the soil or nutrient resource from the catchment. For centuries humans have altered the structure and function of entire catchment systems, disrupting horizontal, lateral and vertical connectivity in river channels and between river channels and their adjacent floodplains (Ward, 1989) and have altered the coupling between hillslopes and channels (Fryirs *et al.*, 2007).

Conceptual developments in geomorphology also lead to questions concerning our ability to predict river behaviour. Church (2010), for example, noted a major paradigm shift since the 1990s from one that encompassed ideas of equilibrium, linearity and predictability to a paradigm that treats fluvial landscapes as complex systems characterised by disorder, irregularity, instability, unpredictability and non-linearity.

Are reference conditions appropriate?

Like many Directives, the EU WFD sets broad aims and leaves the definition of objectives, and their implementation, to Member States. However, the WFD-CIS Guidance (European Commission, 2003) reviewed in the introduction sets out detailed guidelines that have reference conditions at the core of river evaluation in relation to ecology, habitat and water quality. What the guidance does not do is to specify how those reference conditions might be established. Our discussion has focused on the possible use of palaeoenvironmental methods, and their limitations, for establishing what those conditions should be. Our analysis suggests that we will only be able to define a limited range of conditions at the catchment and reach scale and that our ability to fully define reference conditions in the sense made explicit by WFD is essentially unachievable, simply based on the palaeoenvironmental record alone.

The validity of non-palaeoenvironmental methods identified in the introduction, namely modelling and expert judgement, for achieving the same goal is equally questionable because the concept of a reference condition is arguably flawed, both scientifically and practically, in the context of river management, leaving managers striving to achieve the unachievable. The apparent gap between what policy-makers want to implement and what river managers and scientists can deliver appears to have widened as a result of the debate surrounding the establishment of reference conditions. Debates surrounding water resource issues highlight similar problems of integrating science and management into policy well beyond the boundaries of the European Union. South Africa, for example, has often been identified as a leader in water policy (see Rowntree and du Preez, 2008, for a detailed review). In 1998, the post-apartheid South African Government defined the 'ecological reserve' for South African Rivers as the amount of water required to protect the aquatic ecosystems of the water resource in addition to a basic human needs reserve of ~ 25 litres per day of potable water per person. Despite over a decade of research effort, Rowntree and Du Preez (2008) note that high levels of uncertainty cloud decisions on flow requirements. Schreiner and Hassan (2011) and, more recently, Hering and Ingold (2012) discuss these issues further and suggest that less ambition, in terms of managing water resource requirements, may result in better delivery of objectives. Such a comment seems highly relevant in the light of the difficulty in establishing reference conditions for rivers.

Acknowledgements

Our thanks to Mark Szegner (Loughborough University) and Paul Stroud (University of Northampton) for compiling and drawing the figures. We are also grateful to Dr Paul Wood and Dr Helen Proffitt for their advice and helpful suggestions in writing this chapter.

References

Andreev A, Tarasov PE, Schwamborn G, *et al.* (2004) Holocene paleoenvironmental records from Nikolay Lake, Lena River Delta, Arctic Russia. *Palaeogeography, Palaeoclimatology, Palaeoecology* **209**, 197–217.

Amoros CA, Roux AL, Reygrobellet JL, *et al.* (1987) A method for applied ecological studies of fluvial hydrosystems. *Regulated Rivers* **1**, 17–36.

Atkinson TC, Briffa KR, Coope GR, *et al.* (1986) Climatic calibration of coleopteran data. In: Berglund BE (ed.). *Handbook of Holocene Palaeoecology and Palaeohydrology*. Chichester, John Wiley & Sons Ltd. 851–858.

Atkinson TC, Briffa KR and Coope GR (1987) Seasonal temperatures in Britain during the last 22,000 years. *Nature* **325**, 587–592.

Bailey, RG, José, PV and Sherwood, BR (eds) (1998) *United Kingdom Floodplains*. Otley, Westbury.

Battarbee RW and Bennion H (2011) Palaeolimnology and its developing role in assessing the history and extent of human impact on lake ecosystems. *Journal of Paleolimnology* **45**(4), 399–404.

Battarbee RW, Simpson GL, Bennion H and Curtis C (2011a) A reference typology of low alkalinity lakes in the UK based on pre-acidification diatom assemblages from lake sediment cores. *Journal of Paleolimnology* **45**(4), 489–505.

Battarbee RW, Morley D, Bennion H, *et al.* (2011b) A palaeolimnological meta-database for assessing the ecological status of lakes. *Journal of Paleolimnology* **45**(4), 405–414.

Bedford A, Jones, RT, Lang, B, *et al.* (2004) A Late-glacial Chironomid record from Hawes Water, northwest England. *Journal of Quaternary Science* **19**, 281–290.

Bennett, KD and Willis KJ (2001) Pollen. In: Smol JP, Birks HJB and Last WM (eds) *Tracking Environmental Change Using Lake Sediments*. Volume 3, Terrestrial, Algal and Siliceous Indicators. Dordrecht, Kluwer. 5–32.

Bennion H, and Battarbee RW (2007) The European Union Water Framework Directive: opportunities for palaeolimnology. *Journal of Paleolimnology* **38**, 285–295.

Bennion H, Battarbee RW, Sayer CD, *et al.* (2011) Defining reference conditions and restoration targets for lake ecosystems using palaeolimnology: a synthesis. *Journal of Paleolimnology* **45**(4), 533–544.

Birks HH (2001) Plant macrofossils. In: Smol JP, Birks HJB and Last WM (eds) *Tracking Environmental Change Using Lake Sediments*. Volume 3, Terrestrial, Algal and Siliceous Indicators. Dordrecht, *Kluwer*. 49–74.

Bjune, AE (2005) Holocene vegetation history and tree-line changes on a north–south transect crossing major climate gradients in southern Norway – evidence from pollen and plant macrofossils in lake sediments. *Review of Palaeobotany and Palynology* **133**, 249–275.

Boyle JF, Plater AJ, Mayers C, *et al.* (2011) Land use, soil erosion, and sediment yield at Pinto Lake, California: comparison of a simplified USLE model with the lake sediment record. *Journal of Paleolimnology* **45**, 199–212.

Brooks SJ and Birks HJB (2000). Chironomid-inferred late-glacial and early Holocene mean July air temperatures for Kråkenes Lake, Western Norway. *Journal of Paleolimnology* **23**, 77–89.

Brooks, SJ, Langdon, PG and Heiri, O (2007). The identification and use of Palaearctic Chironomidae larvae in palaeoecology. QRA Technical Guide No. 10, Quaternary Research Association, London.

Brown AG (2002) Learning from the past: palaeohydrology and palaeoecology. *Freshwater Biology* **47**, 817–829.

Buckland PI and Buckland PC (2006). Bugs Coleopteran Ecology Package (Versions: BugsCEP v7.63; Bugsdata v72; BugsMCR v2.02; BugStats v1.22).

Castella E, Richardt-Coulet M, Roux C and Richoux P (1984) Macroinvertebrates as 'describers' of morphological and hydrological types of aquatic ecosystems abandoned by the Rhône River. *Hydrobiologia* **119**, 219–225.

Church M (2010) The trajectory of geomorphology. *Progress in Physical Geography* **34**, 265–286.

Collins AL, Walling DE, Webb L and King P (2010) Apportioning catchment scale sediment sources using a modified composite fingerprinting technique incorporating property weightings and prior information. *Geoderma* **155**, 249–261.

Collins, AL, Foster, I, Zhang, Y, *et al.* (2012). Assessing 'modern background sediment delivery to rivers' across England and Wales and its use for catchment management. In: Collins, AL, Golosov, V, Horowitz, AJ, *et al.* (eds) *Erosion and Sediment Yields in the Changing Environment*. International Association of Hydrological Sciences (IAHS) Publication No **356**, Wallingford, UK. 125–131.

Coope GR (1994) The response of insect faunas to glacial-interglacial climatic fluctuations. *Philosophical Transactions of the Royal Society London B* **344**, 19–26.

Coope GR and Angus RB (1975) An ecological study of a temperate interlude in the middle of the last glaciation, based on fossil coleoptera from Isleworth, Middlesex. *Journal of Animal Ecology* **44**, 365–391.

Coope GR, Elias S, Miller, RF and Lowe JJ (1999) Late-glacial GRIP Event stratigraphy compared with Mutual Climatic Range temperature profiles from Coleoptera in Maritime Canada and Great Britain. Abstracts XV INAUA Congress, Durban, South Africa, 45.

Currás A, Zamora L, Reed JM, *et al.* (2012) Climate change and human impact in central Spain during Roman times: High-resolution multi-proxy analysis of a tufa lake record (Somolinos, 1280 m asl). *Catena* **89**, 31–53.

Davis OK and Shafer DS (2006). Sporormiella fungal spores, a palynological means of detecting herbivore density. *Palaeogeography, Palaeoclimatology, Palaeoecology* **237**, 40–50.

Dearing JA and Foster IDL (1993) Lake sediments and geomorphological processes; some thoughts. In: McManus J and Duck R (eds) *The Geomorphology and Sedimentology of Lakes and Reservoirs*. Chichester, Wiley. 5–14.

European Commission (2003) Common Implementation Strategy for the Water Framework Directive (2000/60/EC). Guidance on establishing reference conditions and ecological status class boundaries for inland surface waters. Available at: http://forum.europa.eu.int/Public/irc/env/wfd/library. Accessed September 2012.

European Parliament (2000) Parliament and Council Directive 2000/60/EC. Establishing a framework for community action in the field of water policy. Official Journal PE-CONS 3639/1/00 REV 1.European Union, Brussels, Belgium.

Extence CA, Balbi, DM and Chadd RP (1999) River flow indexing using British benthic macroinvertebrates: a framework for setting hydroecological objectives. *Regulated Rivers: Research and Management* **15**, 543–574.

Flessa KW (2009) Putting the dead to work: Translational paleoecology. In: Dietl GP and Flessa KW (eds) *Conservation Paleobiology. Using the Past to Manage for the Future*. The Paleontological Society Papers, Vol. 15. New Haven, CT, The Paleontological Society.

Foster IDL (ed.) (2000) *Tracers in Geomorphology*. Chichester, Wiley.

Foster IDL (2001) Fine particulate sediment transfers in lowland rural environments. In: Higgitt DL and Lee M (eds) *Geomorphological Processes and Landscape Change: Britain in the last 1000 Years*. Oxford, Blackwell. 215–236.

Foster IDL (2006) Lakes in the sediment delivery system. In: Owens P and Collins AJ (eds) *Soil Erosion and Sediment Redistribution in River Catchments*. Wallingford, CAB International. 128–142.

Foster IDL (2010) Lakes and reservoirs in the sediment cascade. In: Burt TP and Allison RJ, *Sediment Cascades: An Integrated Approach*. Chichester, Wiley. 345–376.

Foster, IDL, Boardman, J and Keay-Bright, J (2007) The contribution of sediment tracing to an investigation of the environmental history of two small catchments in the uplands of the Karoo, South Africa. *Geomorphology* **90**, 126–143.

Foster IDL, Collins AL, Naden PS, *et al.* (2011) The potential for paleolimnology to determine historic sediment delivery to rivers. *Journal of Paleolimnology* **45**(2), 287–306.

Foster IDL, Rowntree KM, Boardman J, and Mighall TM (2012) Changing sediment yield and sediment dynamics in the Karoo uplands, South Africa; post-European impacts. *Land Degradation & Development*. **23**: 508–522. DOI: 10.1002/ldr.2180

Fryirs K, Brierley G, Preston N and Kasai, M (2007) Buffers, barriers and blankets: The (dis)connectivity of catchment scale sediment cascades. *Catena* **70**, 49–67.

Gandouin E, Maasri A, Van Vliet-Lanoe B and Franquet E (2006). Chironomid (insecta: Diptera) assemblages from a gradient of lotic and lentic waterbodies in river floodplains of France: a methodological tool for paleoecological applications. *Journal of Paleolimnology* **35**,149–166.

Greenwood MT, Agnew MD and Wood PJ (2003) The use of caddisfly fauna (Insecta: Trichoptera) to characterise the Late-glacial River Trent, England. *Journal of Quaternary Science* **18**, 645–661.

Greenwood MT, Wood PJ and Monk WA (2006) The use of fossil caddisfly assemblages in the reconstruction of flow environments from floodplain paleochannels of the River Trent, England. *Journal of Paleolimnology* **35**,747–761.

Heiri O, Lotter AF and Lemcke G. (2001) Loss on ignition as a method for estimating organic and carbonate content in sediments: reproducibility and comparability of results. *Journal of Paleolimnology* **25**, 101–110

Hering, JG and Ingold KM (2012) Water Resources Management: What should be integrated? *Science* **336**, 1234–1235.

Hollins SE, Harrison JJ, Jones BG, *et al.* (2011) Reconstructing recent sedimentation in two urbanised coastal lagoons (NSW, Australia) using radioisotopes and geochemistry. *Journal of Paleolimnology* **46**, 579–596

Howard L (2007) The Reconstruction of River Flow and habitats within the River Trent Catchment Based upon Sub-Fossil Insect Remains: A Multi-Proxy Approach. Doctoral Thesis, Loughborough University.

Howard LC, Wood PJ, Greenwood MT and Rendell HM (2009) Reconstructing riverine paleo-flow regimes using subfossil insects (Coleoptera and Trichoptera): the application of the LIFE methodology to paleochannel sediments. *Journal of Paleolimnology* **42**, 453–466.

Howard LC, Wood PJ, Greenwood MT *et al.* (2010) Sub-fossil Chironomidae as indicators of palaeoflow regimes: integration into the PalaeoLIFE flow index. *Journal Quaternary Science* **25**, 1270–1283.

Irvine F, Cwynar LC, Vermaire JC and Rees ABH (2012) Midge-inferred temperature reconstructions and vegetation change over the last ~15,000 years from Trout Lake, northern Yukon Territory, eastern Beringia. *Journal of Paleolimnology* **48**,133–146.

Klages MG and Hsieh Y (1975) Suspended solids carried by the Galatin River of Southwestern Montana: II. *Using mineralogy for inferring sources. Journal of Environmental Quality* **4**, 68–73.

Koiter AJ, Owens PN, Petticrew EL and Lobb DA (2013) The behavioural characteristics of sediment properties and their implications for sediment fingerprinting as an approach for identifying sediment sources in river basins. *Earth Science Reviews* **125**, 24–42.

Lane CS, Horn SP, Mora CI *et al.* (2011) Sedimentary stable carbon isotope evidence of late Quaternary vegetation and climate change in highland Costa Rica. *Journal of Paleolimnology* **45**, 323–338.

Langhans SD, Richard U, Rueegg J, *et al.* (2013) Environmental heterogeneity affects input, storage, and transformation of coarse particulate organic matter in a floodplain mosaic. *Aquatic Sciences* **75**, 335–348.

Lapointe F, Francus P, Lamoureux SF, *et al.* (2012) 1750 years of large rainfall events inferred from particle size at East Lake, Cape Bounty, Melville Island, Canada. *Journal of Paleolimnology* **48**,159–173.

Lewin J (2010) Medieval environmental impacts and feedbacks: the lowland floodplains of England and Wales. *Geoarchaeology* **25**(3), 267–311.

Lewin J, Macklin MG and Johnstone, E (2005) Interpreting alluvial archives: Sedimentological factors in the British Holocene record. *Quaternary Science Reviews* **24**, 1873–1889.

Macklin MG, Jones AF and Lewin J (2010) River response to rapid Holocene environmental change; Evidence and explanation in British catchments. *Quaternary Science Reviews* **29**, 1555–1576.

Maddy D, Lewis SG, Scaife, RG, *et al.* (1998) The Upper Pleistocene deposits at Cassington, near Oxford, England. *Journal of Quaternary Science* **13**, 205–231

Meade RH (1982) Sources, sink, and storage of river sediments in the Atlantic drainage of the United States. *Journal of Geology* **90**, 235–252.

Mighall TM, Foster IDL, Rowntree KM and Boardman J (2012) Reconstructing recent land degradation in the semi-arid karoo of South Africa: a palaeoecological study at Compassberg, Eastern Cape. *Land Degradation & Development* (Special Issue on Land Degradation in South Africa). DOI: 10.1002/ldr.2176

Moss B (2011) *Ecology of Fresh Waters*, 4th edn. Chichester, Wiley Blackwell.

Nazarova L, Herzschuh U, Wetterich S, *et al.* (2011) Chironomid-based inference models for estimating mean July air temperature and water depth from lakes in Yakutia, northeastern Russia. *Journal of Paleolimnology* **45**, 57–71.

Paine JL, Rowan JS and Werritty A (2002) Reconstructing historic floods from floodplain sediments: Evidence from the River Tay, Scotland. In: Dyer FJ, Thoms MC and Olley JM (eds) *The Structure Function and Management Implications of Fluvial Sedimentary Systems*, IAHS Publication No **276**. 211–218.

Partridge TC, Scott L and Schneider RR (2004) Between Agulhas and Benguela: response of Southern African climates of the Late Pleistocene to current fluxes, orbital precession and the extent of the Circum-Antarctic vortex. In: Battarbee RW, Gasse F and Stickley CE (eds) *Past Climate Variability through Europe and Africa*. Dordrecht, Springer. 45–68.

Petts, GE (1998) Floodplain rivers and their restoration: A European perspective. In: Bailey, RG, José, PV and Sherwood, BR (eds) *United Kingdom Floodplains*. Otley, Westbury. 29–41.

Petts GE and Amoros C (eds) (1997) *Fluvial Hydrosystems*. Chapman and Hall, London.

Petts GE, Möeller H and Roux AL (eds) (1989) *Historical Change of Large Alluvial Rivers*. Chichester, Wiley.

Pittam NJ, Foster IDL and Mighall TM (2009) An integrated lake-catchment approach for determining sediment source changes at Aqualate Mere, Central England. *Journal of Paleolimnology* **42**, 215–232.

Pittam N, Mighall TM and Foster IDL (2006) Determination of the effect of sediment source changes upon preserved pollen assemblages in lake sediments. *Water Air and Soil Pollution* **6**, 677–683.

Ponel P, Gandouin E, Coope GR, *et al.* (2007) Insect evidence for environmental and climate changes from Younger Dryas to Sub- Boreal in a river floodplain at St-Momelinn (St-Omer basin, northern France), Coleoptera and Trichoptera. *Palaeogeography, Palaeoclimatology, Palaeoecology* **245**, 483–504.

Renberg I, Brannvall M-L, Bindler R and Emteryd O (2002) Stable lead isotopes and lake sediments-a useful combination for the study of atmospheric lead pollution history. *Science of the Total Environment* **292**, 45–54.

Rodysill JR, Russell JM, Bijaksana S, *et al.* (2012) A paleolimnological record of rainfall and drought from East Java, Indonesia during the last 1,400 years. *Journal of Paleolimnology* **47**, 125–139.

Rose NL and Appleby PG (2005) Regional applications of lake sediment dating by spheroidal carbonaceous particle analysis 1: United Kingdom. *Journal of Paleolimnology* **34**, 349–361.

Rose NL, Morley D, Appleby PG, *et al.* (2011) Sediment accumulation rates in European lakes since AD 1850: trends, reference conditions and exceedence. *Journal of Paleolimnology* **45**(4), 447–468.

Roux C and Castella E (1987) Les peuplements larvaires de trichopteres des anceins lits fluviaux dans trois secteurs de la plaine alluviale Du Haut-Rhône Francais. In: *Proceedings of the 5th International Symposium on Trichoptera.* Dr W Junk, Publishers, Dordecht. 305–312.

Rowntree KM and du Preez, L (2008) Application of integrative science in the management of South African rivers. In: Brierley GJ and Fryirs KA (eds) *River Futures: An Integrative Scientific Approach to River Repair.* Washington, Island Press. 237–254.

Rumsby BA (2000) Vertical accretion rates in fluvial systems: A comparison of volumetric and depth-based estimates. *Earth Surface Processes and Landforms* **25**, 617–631.

Schreiner B and Hassan R (2011) Lessons and conclusions. In: Schreiner B and Hassan R (eds) *Transforming Water Management in South Africa.* Designing and Implementing a New Policy Framework. Dordrecht, Springer. 271–276.

Schreve D, Howard AJ, Currant A, *et al.* (2013) A Middle Devensian woolly rhinoceros from Whitemoor Haye Quarry, Staffordshire (UK): palaeoenvironmental context and significance. *Journal of Quaternary Science* **28**, 118–130.

Seddon, EL, Wood PJ, Mainstone, CP *et al.* (2012). The use of palaeoecological techniques to identify reference conditions for river conservation management. In: Boon PJ and Raven PJ (eds) *River Conservation and Management.* Chichester, John Wiley and Sons Ltd. 217–221.

Shennan I, Milne G and Bradley SL (2009) Late Holocene relative land- and sea-level changes: providing information for stakeholders. *GSA Today* **19**, 52–53.

Shuman, B (2003) Controls on loss-on-ignition variation in cores from two shallow lakes in the northeastern United States. *Journal of Paleolimnology* **30**(4), 371–385

Smith, N (1971) *A History of Dams.* London, Peter Davies.

Smol J (2008) *Pollution of Lakes and Rivers: A Paleoenvironmental Perspective.* 2nd edn. Malden, Blackwell.

Smol JP, Birks HJB and Last WM (2001). *Tracking Environmental Change Using Lake Sediments.* Volume 4: Zoological Indicators. Dordrecht, Kluwer Academic Publishers.

Solem, JO and Birks, HH (2000). Late-glacial and early Holocene Trichoptera (Insecta) from Kråkenes Lake, western Norway. *Journal of Paleolimnology* **23**, 49–56.

Trimble SW (2010) Streams, valleys and floodplains in the sediment cascade. In: Burt TP and Allison RJ (eds) *Sediment Cascades: An Integrated Approach.* Chichester, Wiley. 307–343.

Wall, GJ and Wilding LP (1976). Mineralogy and related parameters of fluvial suspended sediments in northwestern Ohio. *Journal of Environmental Quality* **5**, 168–73.

Walling DE, Peart MR, Oldfield F, and Thompson R (1979). Suspended sediment sources identified by magnetic measurements. *Nature* **281**, 110–113.

Walling DE and Foster IDL (2015) Environmental radionuclides, mineral magnetism and sediment geochemistry. In: Kondolf M and Piegay H (eds) *Tools in Fluvial Geomorphology.* 2nd edn. Chichester, Wiley.

Ward JV (1989) The four-dimensional nature of the lotic ecosystem. *Journal of the Benthological Society* **8**, 2–8.

Whitlock, C and Larsen CPS (2001) Charcoal as a fire proxy. In: Smol JP, Birks HJB and Last WM (eds) *Tracking Environmental Change Using Lake Sediments.* Volume 3, Terrestrial, Algal and Siliceous Indicators. Dordrecht, Kluwer. 75–97.

Williams, NE (1989) Factors affecting the interpretation of caddisfly assemblages from Quaternary sediments. *Journal of Paleolimnology* **1**, 241–248.

Winter LT, Foster IDL, Charlesworth SM and Lees JA (2001) Floodplain lakes as sinks for sediment-associated contaminants – a new source of proxy hydrological data? *Science of the Total Environment* **266**, 187–194.

Wolfe BB, Hall RI, Last WM, *et al.* (2006) Reconstruction of multi-century flood histories from oxbow lake sediments, Peace-Athabasca Delta, Canada. *Hydrological Processes* **20**, 4131–4153.

Woodward CA, Potito AP and Beilman DW (2012) Carbon and nitrogen stable isotope ratios in surface sediments from lakes of western Ireland: implications for inferring past lake productivity and nitrogen loading. *Journal of Paleolimnology* **47**, 167–184.

CHAPTER 5

Achieving the aquatic ecosystem perspective: integrating interdisciplinary approaches to describe instream ecohydraulic processes

John M. Nestler[1], Claudio Baigún[2] and Ian Maddock[3]

[1] *IIHR Hydroscience and Engineering, University of Iowa, Iowa City, USA*

[2] *Instituto de Investigaciones e Ingeniería Ambiental, Universidad de San Martín, Argentina*

[3] *Institute of Science and the Environment, University of Worcester, Worcester, UK*

Introduction

The need to identify impacts of altered flow regimes and channel morphologies on the physical and biotic components of river systems has been heightened by recognition of the effects of previous management, ecological decline, increased environmental awareness amongst the general public, new legislation and a rise in the use of river restoration and rehabilitation techniques. This has led workers from traditional subjects associated with river science, such as hydrology, fluvial geomorphology, aquatic ecology and engineering hydraulics, to create *multidisciplinary* collaborations in which they apply and pool their individual knowledge, approaches and skills, More importantly new *interdisciplinary* approaches have developed at the interface of these traditional subjects, for example hydromorphology (Orr *et al.*, 2008), hydroecology (Wood *et al.*, 2007), ecohydraulics (Nestler *et al.*, 2007a), ecogeomorphology (Thoms and Parsons, 2002), ecohydrology (Zalewski *et al.*, 1997) and ecohydromorphology (Clarke *et al.*, 2003; Vaughan *et al.*, 2009) in response to the technology needs of river managers. From a historical perspective, the long-term pattern in the evolution of river science is clear – traditional disciplines will be increasingly integrated into interdisciplinary teams in acknowledgement that rivers function as a system of tightly integrated components (Nestler *et al.*, 2012).

The importance of an interdisciplinary approach is illustrated by conceptual models (Gentile *et al.*, 2001) describing the fundamental dynamics of flowing water systems. For example, the Upper Mississippi River conceptual model used to guide management and restoration planning (Lubinski and Barko, 2003) simplifies river condition into five broad categories of variables called essential ecosystem characteristics (EECs): (i) hydrology and hydraulics; (ii) biogeochemical cycling; (iii) geomorphology; (iv) habitat and (v) population dynamics. The

River Science: Research and Management for the 21st Century, First Edition.
Edited by David J. Gilvear, Malcolm T. Greenwood, Martin C. Thoms and Paul J. Wood.

conceptual model further describes the dynamics of the river as interactions among these five EECs. Note that each EEC identified by the conceptual model is associated with a major discipline of River Science and that each interaction among the EECs is mirrored by an interdisciplinary approach listed in the first paragraph (e.g., the interaction of hydrology and geomorphology is captured in hydromorphology). The holistic study of river systems, either for scientific discovery or wise management, requires an interdisciplinary approach because efforts by any single discipline neither effectively produces benefits to society through wise river management nor creates opportunities for scientific discovery.

The connections among the disciplines that comprise river science as depicted in conceptual models are based on the very nature of water as a relatively dense and nearly incompressible fluid. The movement of this fluid over a landscape or through a channel exerts a force that can significantly erode, transport, or deposit sediments, nutrients and biota. From a fluvial geomorphology perspective, the present and future shapes of natural river channels ultimately depend on landscape characteristics, rainfall and flow regimes, and sediment types and loads. From a biogeochemical cycling perspective, rivers are the destination of materials from surrounding landscapes and the distribution and processing of these materials are partially dependent on the work performed by moving water (Nestler *et al.*, 2012). From an ecohydraulics perspective, many aquatic species have evolved specialised structures, body shapes, behaviours and life histories to live in flowing or standing water (Vogel, 1996). Clearly, scientifically valid descriptions of instream ecohydraulics processes, particularly from a system perspective, require multidisciplinary integration.

Conceptually, inter-relationships among disciplines that comprise river science, as described above, are relatively clear and simple. However, effective integration among these disciplines is difficult because different scientific paradigms often underpin the separate disciplines that each contribute to an interdisciplinary approach. Individual disciplines may have different jargons, conventions, and traditions as they each grow and evolve at their own rates in their own directions. The inherent divergence of disciplines as they mature and adapt to new findings makes technology transfer and integration among them difficult. Moreover, growth of each discipline can be comparatively irregular as scientific controversies are addressed, new core concepts emerge and innovative technologies break down barriers to research (Kuhn, 1962). These irregularities are magnified at the paradigm boundary because they create even more profound differences that must be overcome for successful interdisciplinary collaboration. Development of conceptual frameworks to lessen the barriers separating disciplines becomes critical to achieve efficient and effective integration of technologies.

Perhaps the most difficult two disciplines to integrate into an interdisciplinary approach are hydraulic engineering and fluvial ecology, because their foundational concepts of determinism and empiricism, respectively, are so different. Empiricists believe knowledge is gained primarily through observation and experience. Empiricism is the foundation of the scientific method in which an investigator observes a part of the natural world of interest, carefully crafts a hypothesis which when tested provides insight into the workings of nature, collects data within an experimental framework, and then infers a truth based on analysis (usually statistical) of collected data. This is a typical approach associated with aquatic ecology. Alternatively, determinists believe that strong cause and effect relationships structure the natural world to the point that nature appears to obey certain equations. Determinism is a prevalent perspective of those that study fluid dynamics (either mathematically, or experimentally), and believe that knowledge is gained primarily by mathematical derivation using principles such as Newton's Laws. Therefore, integrating principles found in ecohydraulics may provide useful insight into other interdisciplinary challenges.

One integrating concept in ecohydraulics is the idea of the governing equation. This concept is usually narrowly defined by computational fluid dynamicists as a deterministic equation that explains fluid motion pattern with such accuracy and resolution that it appears to 'govern' fluid flow. In a sense, empiricists using statistical inference are also searching for a quantitative description of natural processes. That is, they are searching for their own version of a governing equation, but with a different concept of causality and, therefore, different expectations about realism, generality, resolution, precision and accuracy. For example, statistical inference of data typically exhibits relatively loose cause–effect relationships. Many ecological processes are less connected to one another or exhibit more diffuse cause and effect relationships because numerous other chemical and biotic (some of which are difficult to measure) variables are also important. In contrast, deterministic models are mathematically derived so that expectations for both accuracy and resolution are relatively greater. Consequently, for any instream process, in a perfect Newtonian world and in a setting where data availability is not limiting, members of both paradigms should ultimately converge on the same governing equation, each using their own approach, because there is only a single physical reality. It is important for all river scientists interested in quantitative descriptions of instream processes to understand that many different disciplines have the same goal of scientific discovery, but achieve their goals using different approaches.

The example described above for ecohydraulics suggests that the field of river science should generally also exhibit organising principles, even if these principles may not be immediately evident because of apparently divergent perspectives, approaches, tools, and methods of the member disciplines. To continue to expand the interdisciplinary perspective of river science requires a broad context so that scientists in one discipline of river science can communicate their work to scientists working in another. Without this context, the complex issues facing river managers cannot be adequately addressed because single-discipline findings are a poor substitute for holistic, multidisciplinary solutions. An evaluation of the breadth of studies that together comprise river science identifies patterns that can be used by multidisciplinary researchers to give their work the broader context needed to further the development of river science. We propose that each of the many multidisciplinary studies of instream processes in the field of river science can be broadly categorised using two principles (Figure 5.1):

- Scale principle: classification into similar ranges of temporal and spatial scales (scale constant) versus different ranges of time and space scales (scale divergent).
- Causality principle: classification into determinism (high causality alternatively expressed as low uncertainty) versus

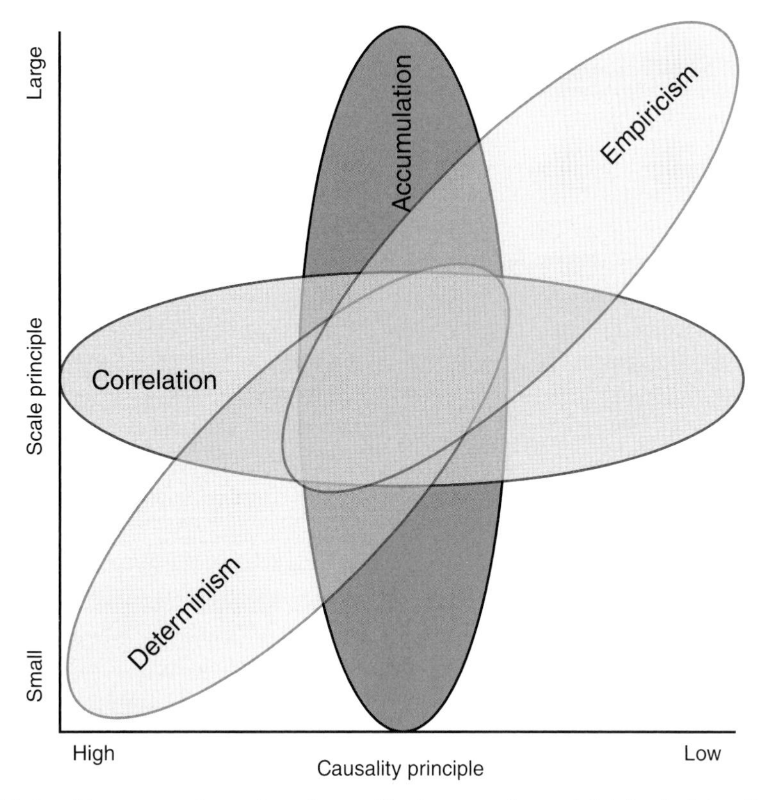

Figure 5.1 Relationships among the scale and causality (inverse of uncertainty) principles. Together, these principles explain the general trends in ecohydrology and ecohydraulics and are, therefore, key to classifying a study into either a deterministic or empirical framework. Note the presence of considerable overlap among approaches that can lead to disagreements among an interdisciplinary team.

empiricism (reduced causality alternatively expressed as high uncertainty).

The scale principle recognises that each instream process is classifiable by the range of scales over which it inherently occurs and, therefore, the scale at which it must be measured (Nestler *et al.*, 2005). For example, an estimate of water velocity in a dynamic river using a current meter may have a spatial scale of several metres and a temporal scale measured in seconds to hours. In contrast, an estimate of sediment transport through a reach will have a lateral and vertical spatial scale of the river cross section and a temporal scale of days or weeks, depending upon the characteristics of the hydrograph. Processes from different disciplines that occur over similar ranges of scales can be studied and analysed more easily using a correlative approach than processes that differ greatly in scale (Figure 5.2). Often, these studies are organised at a particular scale stratum by a specific issue or question such as habitat requirements of a target species. However, these studies must assume that ignoring the causality principle will not substantially affect results.

River science studies that address processes that occur across a broad range of scales are usually intent on accumulating the effects of smaller-scale processes to a system-level description of effect using a discipline such as hydrology that is particularly

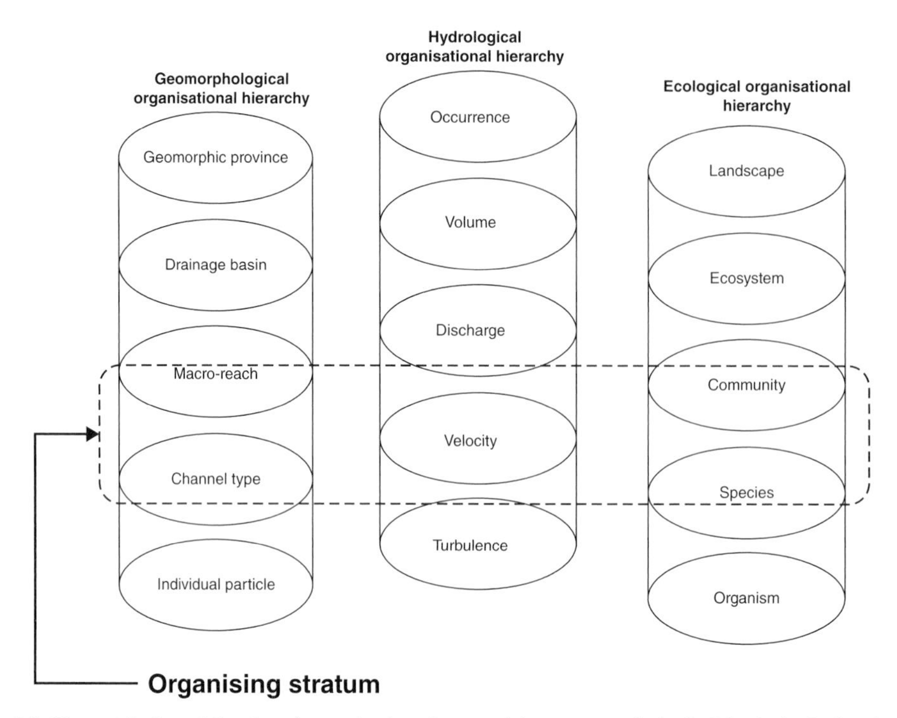

Figure 5.2 Hierarchical spatial scales of organisation characterising geomorphological, hydrological and ecological subsystems of a river. Note organising substratum (often habitat) to focus each discipline to an optimum contribution to an interdisciplinary approach (modified from Dollar *et al.*, 2007, p 150). Reproduced with permission from Elsevier.

useful for system-level integration and analysis. In the process of this integration, causality must be assumed to be relatively constant (Figure 5.1) but scale must, by necessity, be assumed to be unimportant. For example, runoff of nutrients from a large spatial domain may occur over a mosaic of landscape types and scales, but effects of the runoff may accumulate as eutrophication at a river-wide scale.

Together, the two principles provide a useful template to classify optimally a study of multiple instream process into either a deterministic or empiricist framework. For example, descriptions of fish habitat will likely best be addressed using an empiricist approach because characterisation of habitat rarely produces clear and concise results, particularly in large rivers. This lack of clarity arises for a variety of causes ranging from sampling and measurement challenges to uncertain conceptual foundations for the measurements because the scales at which different fishes react to their environment is not completely known. In contrast, description of phosphorous cycling will likely be addressed using a deterministic approach because chemical kinetics of the phosphorous cycle and their dependency on advection and dispersion in aquatic systems are relatively well known.

The great challenge of river science is not an advocacy for one interdisciplinary approach over another, but rather to determine how disparate approaches can best be melded to maximise scientific understanding of instream processes. Scientifically credible integration among two or more disciplines requires that the foundation principles underpinning each separate discipline must be identified and reconciled. In addition, principles must be used to guide the integration of individual disciplines into a greater construct to address the many challenges of river management. These integrating principles must themselves be as robust as the foundational principles that underpin each separate discipline. Classification of the large number of diverse instream processes of interest to different disciplines using the scale and causality principles is a useful way to optimally organise interdisciplinary studies so that they can contribute to river science. For clarity and brevity, we focus on classifying interdisciplinary studies into either a deterministic or empirical framework, although hybrid approaches are also possible.

Empiricism, classification and the scale principle

The classification of physical and biological processes at a range of temporal and spatial scales provides an example of how the scale principle has been applied to understand aquatic ecosystem structure and function. Mosley (1985) provided an early review of the intrinsic relationship between river channel morphology, instream flow requirements, and habitat. Maddock (1999) stressed the importance of physical habitat, illustrating the role of fluvial geomorphology in determining the physical template of a river system, which when overlaid by an associated flow pattern determines the spatial and temporal pattern of hydraulic variables such as water depths, velocities and turbulence. To help simplify and understand this complexity, river scientists have developed a variety of morphological and habitat classification systems to cluster features and controlling factors and thereby describe patterns among different studies, places or times. In accordance with the scale principle, features that occur at similar scales (spatial and temporal) are grouped together (*scale constant*) and separated from features or processes that operate at different scales (*scale divergent*). Fluvial geomorphologists have engaged in classification of river systems for decades in order to try and understand the form of river channels, their pattern, morphology, dimensions, bed features and the factors that control them (Schumm, 1963; Rosgen, 1994; Brierley and Fryirs, 2005). Each classification system has unique characteristics which in part depend on the region where it was developed and hence the type of river system where it is primarily applicable.

Scale divergent classifications

River channel classification systems also vary in terms of the individual or range of spatial scales that are addressed (e.g., they may examine a single or multiple scales, and if multiple, the number and the terms associated with them can vary), the specific characteristics of the channel that are assessed (e.g., gradient, width, depth, planform shape and channel pattern), and the application of the classification system (e.g., for research into the processes that determine river geomorphology or for river management purposes). Schumm (1985) highlighted five spatial scales in his fluvial geomorphological classification of river systems (network, reach, meander bend, bedform and individual grains).

Frissell *et al.* (1986) distinguished spatially nested hierarchies of scale with five similar levels of identification to that of Schumm's approach, that is stream system, segment, reach, pool/riffle and microhabitat. The key difference is that Schumm (1985) was focusing on the geomorphology of the river's planform as a means to understanding the variety in form, function and process operating at these different scales. The Frissell *et al.* (1986) approach was aimed at understanding the structure and function of the river system in both physical and biological terms through the role of habitat. This early example provides a good illustration of how fluvial geomorphologists and biologists have often been applying similar approaches to tackle the challenges associated with river science research and the management of river systems. Both approaches have the merit of perceiving the classification of river systems at different scales, but have done this separately within their own fields of interest. This has not only led to a simple duplication of effort, but to a variety of scale divergent classification systems, approaches, techniques and nomenclature that reflects their own sub-discipline's paradigms, terminology, scientific approaches and priorities.

Thoms and Parsons (2002) defined the interdisciplinary study of river at the interface of geomorphology, hydrology and ecology as 'ecogeomorphology'. A series of papers that arose from the Binghamton conference on Geomorphology and Ecosystems highlighted the difficulties facing the integration of these two subject areas at their interface (Renschler *et al.*, 2007). Dollar *et al.* (2007) presented a framework to integrate the different scales of assessment associated with the traditional disciplines of geomorphology, hydrology and ecology.

This approach not only helps provide a greater understanding of the links in spatial scales used between the disciplines, but makes the case for interdisciplinary river science that examines the interaction of geomorphology, hydrology and ecology to evaluate these interactions at multiple spatial scales. To date the use of multiple spatial scales that attempt to link features and processes that are scale divergent and that also integrate these three disciplines has not been widespread, but has been growing in number (Maddock *et al.*, 1995; Klaar *et al.*, 2009).

Scale constant approaches: the meso-scale example

Although the importance of recognising that river systems can be described across a nested hierarchy of spatial scales has been known since the mid 1980s, most studies in river science, whether associated primarily with fluvial geomorphology or ecology or both, have tended to focus on only one or two of these scales, and have usually focused on the extremes of scale, such as the large scale associated with the river catchment or the smaller scale associated with individual habitat units. Therefore, most studies have focused on the scale constant approach. For example, geomorphologists have often studied the factors determining large-scale drainage networks, causes of channel patterns such as braided, meandering, straight or anastomosing, or controls on morphological units such as pools and riffles. Concepts and studies associated with assessing aquatic communities and population dynamics often involve models of spatial distributions at the larger catchment scale (e.g., the River Continuum Concept: Vannote *et al.*, 1980) or sampling at the small morphological unit scale (Pedersen and Friberg, 2006). Fausch *et al.* (2002) stressed the importance of assessing river morphology, habitat and ecological interactions at the intermediate scale because this bridges the gap between the two extremes. For some types of biota, such as fish, this scale is equally if not more important than the two extremes, and demonstrates the need to examine river systems as continuous entities rather than features made up of individual sites, units or reaches that are separate and isolated from one another.

The differences in approaches between traditional disciplines when tackling similar issues is neatly illustrated by the desire of the different disciplines using a scale constant approach to classify features at the intermediate or 'meso'-scale of interest in river channels. Understanding the form and process of river morphology is central to the study of fluvial geomorphology, and the two features most commonly studied by geomorphologists at this scale are the pool and riffle. Richards (1976) published a seminal paper assessing their morphology, and since then pool–riffle sequences have been studied with respect to numerous aspects including their morphology (Carling and Orr, 2000), lateral spacing (Keller and Melhorn, 1978), hydraulics (MacWilliams *et al.*, 2006) and influence on channel pattern (Thompson 1986). However, it was a group of fisheries biologists who developed a more detailed classification system of morphological units in order to understand salmonid fish distributions (Bisson *et al.*, 1982). The units were referred to as habitat units, but their designation is primarily a morphological one. The Bisson *et al.* (1982) system was subsequently revised and adapted and has led to a proliferation of classification systems that identify units at this particular meso-spatial scale. Hawkins *et al.* (1993) subdivided the geomorphologists pool and riffle classification into finer levels of resolution to explain ecological differences at a smaller spatial resolution. The sub-categories include riffles, and other morphological features or 'channel geomorphic units' that are of a similar scale (e.g., rapid, chute, run), but were sometimes difficult to differentiate because they depend on subjective field observations.

The importance of this spatial scale can be recognised by the proliferation of classification systems that have been developed. Some utilise the channel morphology to distinguish units, such as the channel geomorphic units outlined by Hawkins *et al.* (1993), and similar systems have been used by geomorphologists to assess the determining factors influencing the distribution of 'channel units' (Halwas and Church, 2002). Similar approaches have been used to assess features described as morphological units (Moir and Pasternack, 2008), geomorphic units (Howard and Cuffey, 2003), hydromorphic units (Gilvear *et al.*, 2004), hydromorphological units (Hauer *et al.*, 2009), physical biotopes (Wadeson, 1994) and hydraulic biotopes (Padmore, 1997). Others have referred to them as habitats, that is mesohabitats, which implies an ecological association with the features of interest, but in fact the terms for features are similar to the others named above, and the definition used to distinguish the units is a physical one (Pardo and Armitage, 1997).

Clearly, there is a range of classification systems based on the definition of physical units at the mesoscale. Some use common terms (e.g., pools or riffles) but define them differently, and most have different numbers and types of units associated with the classification system, raising difficulties in comparing methods and conclusions between studies. The confusion in the literature over the use of the terminology and definition of mesoscale features, combined with the difficulty in objectively identifying them in the field depending on which system or protocol is being used (Whitacre *et al.*, 2007) and observer variability (Roper *et al.*, 2008) provide additional challenges.

Further difficulties are highlighted when comparing classifications at the mesoscale that focus on units defined by their physical structure with other classification systems, usually created, modified and adapted by biologists, to distinguish the ecological functions that are associated with different parts of the channel. For example, Schwartz and Herricks (2008) used a mesohabitat scale approach, with units designated based on a combination of their hydraulics, geomorphology and the biological resource needs of fish. Newson and Newson (2000) show how two different approaches to classifying a channel reach based on the physical biotopes present (i.e., the physical features), or the functional habitats (i.e., the ecological function) of the reach may look. Landscape processes that each occur over a large range of scales are more difficult to integrate into a synthetic whole. For example, habitats in low-order streams are usually easily defined by visible boundaries (e.g., Bisson *et al.*, 1982; Hawkins *et al.*, 1993) that are much more difficult to define in large rivers.

Causality principle at small and large scales

Over decades, the scale principle implemented using empirical approaches has contributed to a number of important advances in aquatic ecology, particularly for small and moderate size streams exhibiting relatively hard bottoms whose shape persists over a range of hydrologic variability (e.g., Figure 5.3). However, the successes of empiricism for smaller systems have not been duplicated for large rivers, particularly for floodplain rivers with alluvial beds. Such systems exhibit both dynamic hydrographs and highly erodible, moveable beds changing over time and space. Consequently, the interconnections among fish location, flow pattern and solid channel features were not explained until computational fluid dynamics (CFD) modelling, a tool from the realm of determinism, was integrated with traditional approaches into a new multidisciplinary collaboration (Goodwin *et al.*, 2006). Key to this integration was the

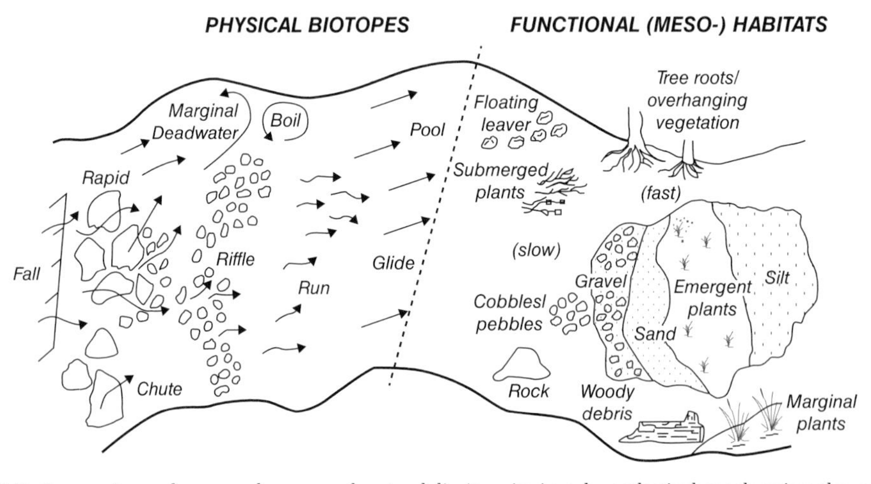

Figure 5.3 Comparison of mesoscale approaches to delimit units in a hypothetical reach using the geomorphic characteristics (physical biotopes) and habitat units distinguishing their ecological function (functional mesohabitats) (Newson and Newson, 2000, p. 200). Reproduced with permission from SAGE.

development of the Integrated Reference Frameworks Concept that laid out the basic principles for coupling together models and approaches that were traditionally focused on ecohydraulics processes occurring over vastly different scales (Nestler *et al.*, 2007a).

Small-scale, well-defined geomorphic units useful for habitat mapping in smaller systems are overwhelmed by the high-energy, dynamic flow patterns and spatial complexity of large, alluvial river systems. As a consequence, optimum characterisation of these systems shifts from classification of solid structures (e.g., micro to mesoscale 'patches' such as stumps, rock outcrops and small scale bedforms) to understanding how instream processes are dynamically distributed over large time and space scales (Nestler *et al.*, 2007b). In relatively unimpaired large rivers, the dynamic balance between channel morphology and the erosive force of water creates ever-changing, complex flow patterns within which large river fishes navigate in three dimensions over a range of lateral and longitudinal scales. From a moving fish's perspective, a large river is best represented as a waterscape of solid and fluid features (and associated instream processes) which gradually shift over a range of time and space scales.

Understanding the fluid environment from a fish's perspective is important for river restoration and management of the impacts of dams and other structures that alter river flow fields. The notion that fish respond to gradients as cues to select movement paths or locate habitats in highly dynamic systems is in contrast to the habitat mosaic concept described earlier for low-order rivers. By responding to such gradients, fish are able to move in the flow field within geomorphologic complexity searching for areas where they can optimally feed, avoid predation, thermally regulate and successfully reproduce. Unlike the habitat patch approach where fish position in a specific habitat is envisioned as a space limited by discrete depth, current or temperature values, in a gradient perspective habitat limits and fish spatial movements are defined by velocity gradients. For example, a detritivorous fish following a streamline of decreasing velocity gradient (hydraulic strain) and increasing pressure (depth) will be able to locate a low-velocity deposition zone where organic matter may be deposited (Figure 5.4). By continuing to follow decreasing velocity gradient, this fish can relocate the deposition zone as it shifts in response to changes in flow or geomorphology.

The recognition that migrating fish respond to velocity magnitude and velocity magnitude gradients leads to a more complete and holistic understanding of how fish are linked to ecohydraulics processes in relatively unimpaired systems and how these linkages can be disrupted in degraded systems. In the following section we describe two examples of the explanatory power achieved by coupling fluid dynamics and fish movement behaviour.

Causality at small scales: understanding fish movement in highly modified systems

Tailrace sections of rivers are initially altered by dam construction and further modified over time by disruptions in hydrology and sediment transport caused by reservoir operation. These alterations, usually created by the configuration of the dam or downstream energy dissipation structures, create unique hydrogeological conditions and hydrodynamic patterns not typically found in natural rivers. Installation of hardened, angular in-channel structures that are not the product of natural erosion

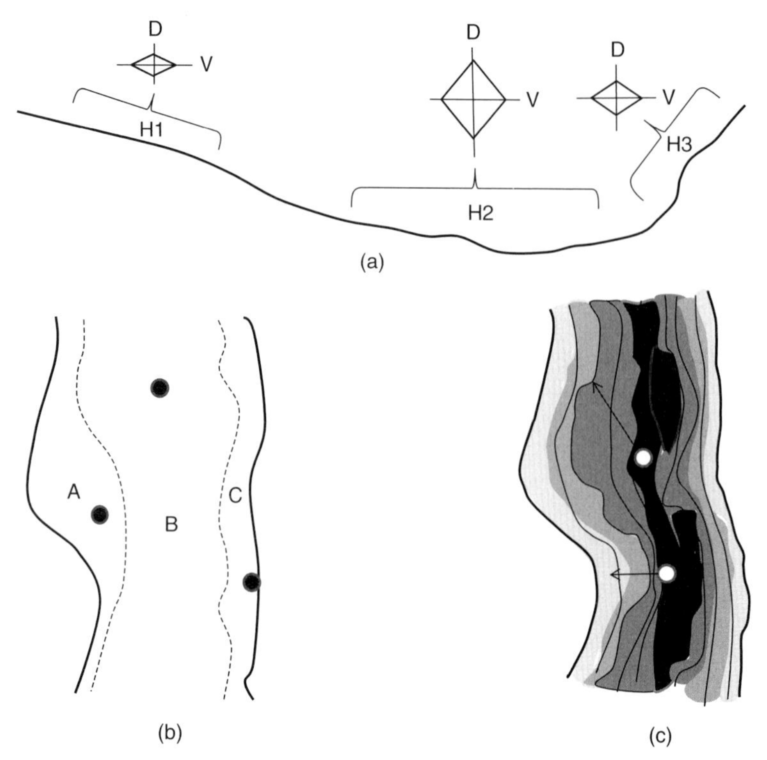

Figure 5.4 (a) Comparative framework of habitat concept based on a patchy perspective where three habitat types are recognised (H1, H2, H3) and a gradient perspective based on depth (black lines) and water velocity fields (shadow areas). Diamonds describe depth (D) and water velocity (V) magnitude at each habitat. (b) Dotted lines in the figure represent isobaths. Circles indicate fish position. (c) Arrows indicate the possible movement directions according to both depth and water velocity gradient analysis.

and depositional processes may alter the relationship between flow field pattern, channel geomorphology and other eco-hydraulics processes of natural channels. These alterations may affect the relationship between the near-field cues used by fish to make movement decisions and the habitats to which these cues should lead. Therefore, the signals produced by these highly modified structures are not part of the evolutionary training set of fishes and, consequently, appear to confuse or disorient them (Nestler *et al.*, 2008).

As a consequence, fish often delay their migration even at dams with fishways or concentrate in areas of the tailrace that seem odd or ineffectual for an actively migrating fish. For example, at Yacyreta Dam on the Paraná River, Argentina, Oldani and Baigún (2002) and Oldani *et al.* (2007) suggested that low fish elevator passage efficiency for large migratory catfish species (e.g., *Pseudoplatystoma* spp. and *Zungaro zungaro*) could be related to the highly turbulent discharge plumes downstream of the turbine draft tubes. The high-energy flows in the dam tailrace are in contrast to the natural pattern of the low-gradient middle Paraná River. The discharge plumes, as they expand downstream of the powerhouse as the channel widens, will obliterate the small, relatively uniform attracting flows released at the mouth of each fishway and hence mask the entrance of the fishway to upstream migrating fishes. At a finer scale, in the pre-impoundment condition, thalweg-oriented large migratory fish use the centre of the alluvial channel as their migratory corridor towards their spawning grounds or other types of habitats. Typically few angular, rigid, discrete features occur in this corridor so that the response of migrating benthic fish to the sharp velocity gradients associated with a constructed structure such as a fishways entrance is unknown. The first step in the challenge of increasing fishway efficiency can be reduced to the problem of designing and operating the fishway, powerhouse and spillway in such a way that a continuous path to the entrance of the fishway is maintained using hydrodynamic cues that mimic natural rivers (Nestler *et al.*, 2008). Therefore, unravelling the relationship among fish movement behaviour, flow pattern and fluvial geomorphology and forecasting the response of fishes to altered flows and manmade instream structures requires close collaboration among aquatic ecologists, fluid dynamicists and hydrogeomorphologists.

Causality at large scales: linking local fish behaviour to watershed processes

In general, South American rivers are less impaired than North American and European rivers so that natural connections between fish abundance and landscape processes can still be studied and understood. In these rivers hydrologic pattern and related variables are the primary factors that appear to govern the evolution of fish life history traits. Periodic floods in large rivers modify erosion and deposition patterns that affect habitat structure and thereby create spatial heterogeneity and temporal fluctuations enhancing the persistence of ecological communities (Bayley, 1995). Moreover, flood pulses also shape the evolutionary history of the aquatic biota and ecological processes (Naiman *et al.*, 2002) and regulate community assembly patterns and regional diversity levels (Poully and Rodriguez, 2004) influencing beta (i.e., among different sites on the same river) and gamma diversity (i.e., among different rivers) in river floodplains (Arrington and Winemiller, 2004).

The importance of hydrology to river dynamics has led to a number of studies to identify variables that describe hydrologic pattern. The flood hydrograph has been described using variables such as timing, amplitude, duration, rapidity of change and smoothness (Welcomme, 2001; Petts, 2007). Most studies have emphasised the importance of the peak flow of the hydrograph (e.g., Junk *et al.*, 1989; Lake 2007), but did not stress important processes occurring during the low-water period which are important to fish life cycles. A more complete description of a flood pulse hydrograph that is more useful to link with life histories of floodplain river fishes has been proposed by Neiff (1990, 1999). He pointed out the equal importance of the flood period (potamophase) when water exceeds bankfull and floods the alluvial valley and the dry period (limnophase) when the floodplain becomes isolated from main channel. This separation of the hydrograph into two periods of equal importance overcomes the problem of underestimating the importance of the dry phase (Neiff et *al.*, 1994). This more comprehensive approach differentiates the spatial and temporal components of the pulse dynamic into hydrological variables of Frequency, **I**ntensity, **T**ension, **R**ecurrence, **A**mplitude and **S**easonality (the FITRAS function (Table 5.1) (Neiff, 1990; Casco *et al.*, 2005) which is similar to the Indicators of Hydrologic Alteration

Table 5.1 Definitions of flood pulse temporal (frequency, recurrence and seasonality) and spatial (amplitude, intensity and tension) attributes (FITRAS function adapted from Neiff, 1990).

Variable	Definition
Frequency	Number of times that a selected flow takes place during a standard time period (usually 100 years)
Intensity	Greatest or lowest flow value of a flood or drought during a time period
Tension	Standard deviation between maximum and minimum flows in a multi-year hydrograph
Recurrence	Probability of a flood or drought occurrence over a standard period (usually 100 years)
Amplitude (Duration)	Time duration a river remains in a flood or low flow phase
Seasonality	Seasonal frequency of floods or droughts
Elasticity	Ratio between areas inundated during period of greatest flooding (potamophase) lowest flow (limnophase).

(IHA) (Richter *et al.*, 1996). The hydrologic variables of FITRAS also appear relevant to the study of behavioural guilds or life history patterns.

The mechanistic connection between landscape processes and instream hydraulic processes is important to river ecology (e.g., Vannote *et al.*, 1980: Junk *et al.*, 1989; Poff *et al.*, 1997). Temporal habitat use and life history stages of migratory fishes in large rivers can only be understood by integrating terrestrial and ecohydraulic processes over a range of different spatial scales. These processes represent a chain of events linking terrestrial and aquatic organic production ultimately manifested in fish biomass. For example, an incremental increase in discharge represents rainfall runoff within the basin that will transport sediments, organic matter, and nutrients from the watershed into the river. In addition, the local increase in depth or velocity appears to trigger reproductive migrations that range from hundreds to thousands of kilometres in length. These migrations allow different life stages of fishes to utilise seasonally available habitats or conditions at large scales.

The connection between watershed processes and fish ecology is exemplified by the ubiquitous, important neotropical genera *Prochilodus* and *Semaprochilodus* that appear to adjust their migratory and reproductive cycles to flood pulse intensity in the Upper Parana River (Gomes and Agostinho, 1997). They also exhibit variable migratory movements of different distance and directions depending upon basin characteristics (Lucas *et al.*, 2001). Like many migratory species, maturation and spawning appears synchronised with water level increases so that their semi-buoyant eggs and small but numerous larvae drift rapidly downstream until they enter floodplain lagoons where they remain for at least one year (Winemiller and Rose, 1992; Winemiller, 2005). The distance that eggs and larvae drift before entering floodplain lagoons as well as the surface area and persistence of these lagoons depends on flood pulse amplitude and intensity (Table 5.1 and Figure 5.5). Post-spawning adults either migrate downstream or laterally to feed in the alluvial valley (Agostinho *et al.*, 1993). As the water recedes, adults leave the floodplain but juveniles remain in lagoons and channels of the floodplain for two years until they recruit to the main channel (Figure 5.5).

Migratory species like *Prochilodus* and *Semaprochilodus* have evolved complex life histories and sophisticated movement

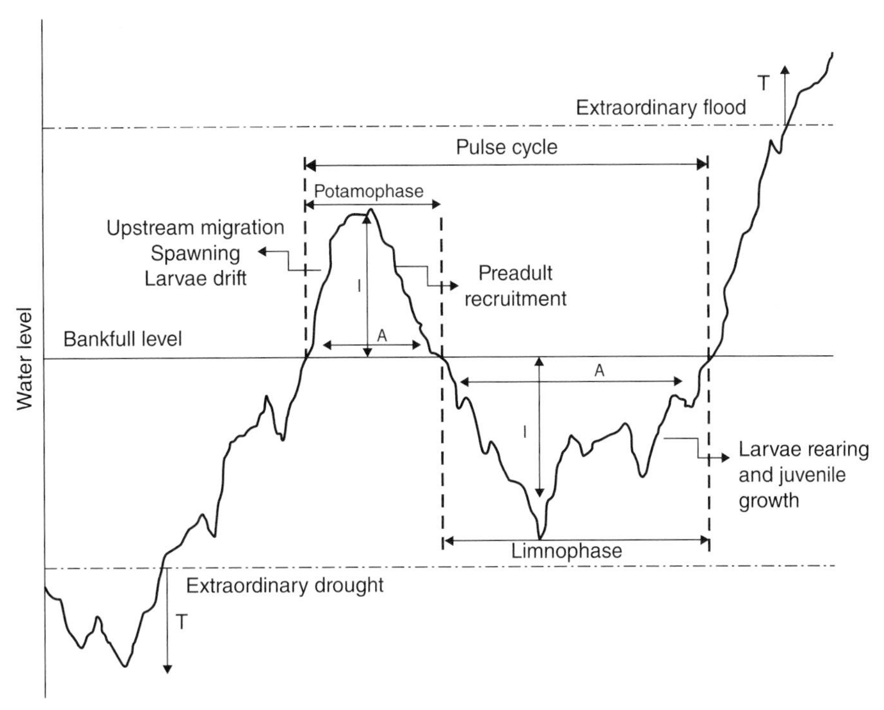

Figure 5.5 Representation of the pulse concept based on several attributes coupled to migratory fish that use the main channel and floodplain areas for completing their life cycles. A: Amplitude: I: Intensity: T: tension.

behaviours to take advantage of the spatially complex patterns of increased primary and secondary production created by flood pulses (Pringle, 2001). Seasonal flood pulses trigger and promote complex biological processes (Junk *et al.*, 1989; Neiff, 1990; Junk and Wantzen, 2004) and the attributes of the flood pulse are key components of floodplain river integrity (Poff *et al.*, 1997). These attributes drive the exchange and storage of organic matter between the floodplain and main channel, and influence biodiversity and biotic abundance by producing seasonal disturbances. For example, the flood pulse appears to govern wetland production in the Amazon (Klinge *et al.*, 1990) and lower Paraguay and Paraná rivers (Neiff *et al.*, 2001). In this context detritivorous species contribute to nutrient recycling and the regulation of carbon transport in rivers (Winemiller *et al.*, 2006; Taylor *et al.*, 2006) and support part of the energy cycle by feeding on organic-rich sediment (Bowen, 1983; Bowen *et al.*, 1984; Jepsen and Winemiller, 2002).

Discussion

River science has a rich and abundant history in which many guiding principles have been proposed to describe patterns in instream processes and river functions across continents, geologic provinces and stream sizes. However, integrating these principles and approaches into a holistic synthesis to understand how river abiotic and biotic features interact is a relatively young paradigm. In some respects, studies on smaller river systems have benefitetd from a more complete interdisciplinary integration than has occurred for large rivers. Habitat analyses conducted in smaller, wadeable systems can be coupled to hydraulic models allowing river scientists to forecast the effects of incremental flow changes on fish habitat. Unlike in smaller systems, the multidisciplinary integration of CFD, hydrogeology and aquatic ecology is still at a relatively early stage in large river studies. As a consequence, there is opportunity for enhanced interdisciplinary integration for river scientists working in the world's large rivers across the range of traditional disciplines to better understand the linkages among hydrogeology, aquatic ecology and fluid dynamics. It is only very recently that river scientists have proposed frameworks for integrating disciplines using a combination of scale constant and scale divergent approaches, that is, between similar and across varying temporal and spatial scales. Frameworks such as the Integrated Reference Frameworks Concept (Nestler *et al.*, 2007a) and the River Machine Conceptual Model (Nestler *et al.*, 2012) are available to support improved interdisciplinary integration. The next stage is to apply these frameworks for the application of interdisciplinary approaches to advance river science research and ensure sustainable river management.

Historically, some of the most important advances in river science have been made by interdisciplinary teams. For example, the Instream Flow Incremental Methodology (IFIM) (Bovee and Millhous, 1978) was developed by an interdisciplinary team at the US Fish and Wildlife Service driven by the need to develop water management strategies to mitigate impacts of altered hydrographs on instream habitat. It became one of the most widely used techniques for defining environmental flows during the 1980s and 1990s (Gore *et al.*, 2001). Similarly, one of the more recent conceptual models in river science, the Riverine Ecosystem Synthesis, is an interdisciplinary approach proposed by a multidisciplinary team (Thorp *et al.*, 2006). Clearly, an interdisciplinary approach will continue to be important if river scientists are going to make relevant environmentally sustainable management recommendations.

Advances in computer technology, numerical methods and mesh typologies have made the integration of computational fluid dynamics into river science studies more feasible. A more complete integration of CFD modelling into river science will help bridge the range of disparate scales often encountered in the subject. A mesh or grid is used to approximate the physical domain (tessellation) in a typical CFD modelling application and a governing equation is solved at each node of the mesh (discretisation of the governing equation). The use of CFD modelling as part of an interdisciplinary approach is important, particularly for large rivers, because the computational mesh can be used to address the scale issue that often limits large river studies. For example, understanding fish movement and habitat selection requires a description of the local conditions that can be represented at a single node. In contrast, understanding how carbon cycle dynamics relates to fish movement and habitat selection requires landscape scales that can be approximated by boundary conditions within reaches or segments of the model mesh. We believe the increased use of CFD modelling and its ability to serve as a template to integrate river processes across scales is critical for the continued advancement of river science.

Achieving a broader synthesis requires not only a multidisciplinary team, but also a concept of how a multidisciplinary team can optimally interface. Each discipline within a multidisciplinary team is guided by their individual principles. Creation of a fully integrated interdisciplinary team also requires guiding principles that can be used to merge the perspectives of each discipline into a greater synthesis. We hope that the classification of existing river science studies using the scale and causality principles will help river scientists forge the necessary collaborative partnerships to further guide and develop the discipline.

Acknowledgements

We thank our colleagues who suffered through multiple versions of this chapter for their patience and helpful suggestions. We also thank the anonymous reviewers for their helpful suggestions.

References

Agostinho AA, Vazzoler AE, Gomes LC, Okada EK. 1993. Estratificación espacial de *Prochilodus scrofa* en distintas fases del ciclo de vida en la planicie de inundación del alto Río Paraná y el Embalse Itaipu, Paraná, Brasil. *Revue Hidrolbiologie Tropicale* **26**: 79–90

Arrington DA, Winemiller KO. 2004. Organization and maintenance of fish diversity in shallow waters of tropical floodplains. In: Welcomme R, Petr T (eds) *Proceedings of the Second International Symposium of the Management of Large Rivers for Fisheries. Volume* **II**. FAO Regional Office for Asia and the Pacific, Bangkok, Thailand, RAP Publication 2004/17.

Bayley PB. 1995. Understanding large river-floodplain ecosystems. *BioScience* **45**: 153–158.

Bisson PA, Nielsen JL, Palmason RA, Grove LE. 1982. A system of naming habitat types in small streams, with examples of habitat utilization by salmonids during low stream flow. In: Armantrout NB (ed.) *Acquisition and Utilization of Aquatic Habitat Inventory Information*. Proceedings of a symposium, 28–30 October 1981, Portland Oregon. American Fisheries Society.

Bovee KD, Milhous RT. 1978. *Hydraulic Simulation in Instream Flow Studies: Theory and Techniques*. U.S. Fish and Wildlife Service Biological Services Program FWS/OBS-78/33.

Bowen SH. 1983. Detritivory in neotropical fish communities. *Environmental Biology of Fishes* **9**: 137–144.

Bowen SH, Bonetto A, Ahlgren MO. 1984. Microorganisms and detritus in the diet of a typical neotropical riverine detritivore, *Prochilodus platensis* (Pisces, Prochilodontidae). *Limnology and Oceanography* **29**: 1120–1122.

Brierley GJ, Fryirs KA. 2005. *Geomorphology and River Management: Applications of the River Styles Framework*. Blackwell Publishing; Oxford, UK.

Carling PA, Orr HG. 2000. Morphology of riffle-pool sequences in the River Severn, England. *Earth Surface Processes and Landforms* **25**: 369–384.

Casco S, Neiff M, Neiff JJ. 2005. Biodiversidad en ríos del litoral fluvial. Utilidad del software PULSO. Temas de la biodiversidad del litoral Argentino II. *INSUGEO, Miscelánea* **14**: 419–434.

Clarke SJ, Bruce-Burgess L, Wharton G. 2003. Linking form and function: towards an eco-hydromorphic approach to sustainable river restoration. *Aquatic Conservation: Marine and Freshwater Ecosystems* **13**: 439–450.

Dollar ESJ, James CS, Rogers KH, Thoms MC. 2007. A framework for interdisciplinary understanding of rivers as ecosystems. *Geomorphology* **89**: 147–162.

Fausch KD, Torgersen CE, Baxter CV, Li HW. 2002. Landscapes to riverscapes: bridging the gap between research and conservation of stream fishes. *BioScience* **52**: 483–498.

Frissell CA, Liss WJ, Warren CE, Hurley MD. 1986. A hierarchical framework for stream habitat classification: viewing streams in a watershed context. *Environmental Management* **10**: 199–214.

Gentile JH, Harwell MA, Cropper Jr, W, et al. 2001. Ecological conceptual models: a framework and case study on ecosystem management for South Florida sustainability. *The Science of the Total Environment* **274**: 231–253.

Gilvear DJ, Davids C, Tyler AN. 2004. The use of remotely sensed data to detect hydromorphology; River Tummel, Scotland. *River Research and Applications* **20**: 795–811.

Gomes LC, Agostinho AA. 1997. Influence of the flooding regime on the nutritional state and juvenile recruitment of the curimba, *Prochilodus scrofa (*Steindachner) in the Upper Paraná River, Brazil. *Fisheries Management Ecology* **4**: 263–274.

Goodwin RA, Nestler JM, Anderson JJ, et al. 2006. Forecasting 3-D fish movement behavior using an Eulerian–Lagrangian-agent method (ELAM). *Ecological Modelling* **192**: 197–223.

Gore JA, Layzer JB, Mead J. 2001. Macroinvertebrate instream flow studies after 20 years: A role in stream management and restoration. *Regulated Rivers: Research and Management* **17**: 527–542.

Halwas KL, Church M. 2002. Channel units in small, high gradient streams on Vancouver Island, British Columbia. *Geomorphology* **43**: 243–256.

Hawkins CP, Kershner JL, Bisson PA, et al. 1993. A hierarchical approach to classifying stream habitat features. *Fisheries* **18**: 3–10.

Hauer C, Mandlburger G, Habersack H. 2009. Hydraulically related hydro-morphological units: description based on a new conceptual mesohabitat evaluation model (MEM) using LIDAR data as geometric input. *River Research and Applications* **25**: 29–47.

Howard JK, Cuffey KM. 2003. Freshwater mussels in a California North Coast Range river: occurrence, distribution, and controls. *Journal of the North American Benthological Society* **22**: 63–77.

Jepsen DB, Winemiller KO. 2002. Structure of tropical food webs revealed by stable isotope. *Oikos* **96**: 46-55

Junk WJ, Wantzen KM. 2004. The flood pulse concept: new aspects, approaches and applications. An update. In: Welcomme R, Petr T (eds), *Proceedings of the Second International Symposium of the Management of Large Rivers for Fisheries. Volume* **II**. FAO Regional Office for Asia and the Pacific, Bangkok, Thailand, RAP Publication 2004/17:117–140.

Junk WJ, Bayley PB, Sparks RE. 1989. The flood pulse concept in river floodplain systems. In: Dodge DP (ed.), *Proceedings of the International Large River Symposium* **106**: 110–127.

Keller EA, Melhorn WN. 1978. Rhythmic spacing and origin of pools and riffles. *Geological Society of America Bulletin* **89**: 723–730.

Klaar MJ, Maddock IP, Milner AM. 2009. The development of hydraulic and geomorphic complexity in recently formed streams in Glacier Bay National Park, Alaska. *River Research and Applications* **25**: 1331–1338.

Klinge H, Junk WJ, Revilla CJ. 1990. Status and distribution of forested wetlands in tropical South America. *Forest Ecology and Management* **33**/34: 81–101.

Kuhn T. 1962. *The Structure of Scientific Revolutions*, (1st edn). University of Chicago Press, Chicago.

Lake PS. 2007. Flow-generated disturbances and ecological responses: Flood and droughts. In: Wood PJ, Hannah DM, Sadler JP (eds) *Hydroecology and Ecohydrology: Present, Past and Future*. John Wiley and Sons Ltd, Chichester: 225–252.

Lubinski K, Barko J. 2003. Upper Mississippi River – Illinois Waterway System navigation feasibility study: environmental science panel report. Prepared for U.S. Army Engineer Districts: Rock Island, Rock Island, IL; St. Louis, St. Louis, MO, and St. Paul, St Paul, MN.

Lucas MC, Baras E, Hom TJ, et al. 2001. *Migration of Freshwater Fishes*. Blackwell Science, Oxford.

MacWilliams ML, Jr., Wheaton JM, Pasternack GB, et al. 2006. Flow convergence routing hypothesis for pool-riffle maintenance in alluvial rivers. *Water Resources Research* **42**: W10427.

Maddock IP, Petts GE, Bickerton MA. 1995. River channel assessment – a method for defining channel sectors on the River Glen, Lincolnshire, UK. *International Association of Hydrological Sciences* **230**: 219–226.

Maddock I. 1999. The importance of physical habitat assessment for evaluating river health. *Freshwater Biology* **41**: 373–391.

Moir HJ, Pasternack GB. 2008. Relationships between mesoscale morphological units, stream hydraulics and Chinook salmon (*Onchorhynchus tshawytscha*) spawning habitat on the Lower Yuba River, California. *Geomorphology* **100**: 527–548.

Mosley MP. 1985. River channel inventory, habitat and instream flow assessment. *Progress in Physical Geography* **9**: 494–523.

Naiman RJ, Bunn SE, Nilsson C, et al. 2002. Legitimizing fluvial ecosystems as users of water. *Environmental Management* **30**: 455–467.

Neiff JJ. 1990. Ideas para la interpretación ecológica del Paraná. *Interciencia* **15**: 424–441.

Neiff JJ. 1999. El régimen de pulsos en ríos y grandes humedales de Sudamérica. In: Malvarez AI (ed.), *Tópicos sobre humedales subtropicales y templados de Sudamérica*. Universidad de Buenos Aires, Buenos Aires: 97–146.

Neiff JJ, Poi de Neiff AS, Casco SA. 2001. The effect of prolonged floods on *Eichhornia crassipes* growth in Paraná River floodplain lakes. *Acta Limnologica Brasiliensia* **13**: 51–60.

Neiff JJ, Iriondo MH, Carignan R. 1994. Large tropical South American wetlands: an overview. In: Link GL, Naiman RJ (eds): *The Ecology and Management of Aquatic-terrestrial Ecotones*. Proceedings Book, University of Washington, Washington: 156–165.

Nestler JM, Goodwin RA, Loucks DP. 2005. Coupling of biological and engineering models for ecosystem analysis. *ASCE Journal of Water Resources Planning and Management* **131**(2): 101–109.

Nestler JM, Goodwin RA, Smith DL, Anderson JJ. 2007a. A mathematical and conceptual framework for hydraulics. In: Wood PJ, Hannah DM, Sadler JP (eds), *Hydroecology and Ecohydrology: Past, Present and Future*, John Wiley & Sons, Ltd, Chichester: 205–224.

Nestler, JM, Baigún CR, Oldani NO, Weber LJ. 2007b. Contrasting the Middle Paraná and Mississippi Rivers to develop a template for restoring large floodplain river ecosystems. *Journal River Basin Management* **5**(4): 305–319.

Nestler JM, Goodwin RA, Smith DL, et al. 2008. Optimum fish passage and guidance designs are based in the hydrogeomorphology of natural rivers. *River Research and Applications* **24**: 148–168.

Nestler JM, Pompeu P, Goodwin RA, et al. 2012. The River Machine: A template for fish movement and habitat, fluvial geomorphology, fluid dynamics, and biogeochemical cycling. *River Research and Application* **28**(4): 490–503 doi:10.1002/rra.1567.

Newson MD, Newson CL. 2000. Geomorphology, ecology and river channel habitat: mesoscale approaches to basin-scale challenges. *Progress in Physical Geography* **24**: 195–217.

Oldani NO, Baigún CRM. 2002. Performance of a fishway system in a major South American dam on the Parana River (Argentina-Paraguay). *River Research and Applications* **18**: 171–183

Oldani NO, Baigún CRM, Nestler JM, Goodwin RA. 2007. Is fish passage technology saving fish resources in the lower La Plata River basin? *Neotropical Ichthyology* **5**: 89–102.

Orr HG, Large ARG, Newson MD, Walsh CL. 2008. A predictive typology for characterising hydromorphology. *Geomorphology* **100**: 32–40.

Padmore CL. 1997. Biotopes and their hydraulics: a method for determining the physical component of freshwater habitat quality. In: Boon PJ, Howell DL (eds) *Freshwater Quality: Defining the Indefinable*, HMSO, Edinburgh: 251–257.

Pardo I, Armitage PD. 1997. Species assemblages as descriptors of mesohabitats. *Hydrobiologia* **344**: 111–128.

Pedersen ML, Friberg N. 2006. Two lowland stream riffles – linkages between physical habitats and macroinvertebrates across multiple spatial scales. *Aquatic Ecology* **41**: 475–490.

Petts G. 2007. Hydroecology: the scientific basis for water resource management and water regulation. In: Wood PJ, Hannah DM, Sadler JP (eds) *Hydroecology and Ecohydrology: Present, Past and Future*. John Wiley and Sons Ltd, Chichester: 225–252.

Poff NL, Allan JD, Bain MB, et al. 1997. The natural flow regime: a paradigm for river conservation and restoration. *BioScience* **47**: 769–784.

Poully M, Rodriguez MA. 2004. Determinism of fish assemblage structure in neotropical floodplain lakes: Influence of internal and landscape lakes conditions. In: Welcomme R, Petr T (eds) *Proceedings of the Second International Symposium of the Management of Large Rivers for Fisheries. Volume* **II**. FAO Regional Office for Asia and the Pacific, Bangkok, Thailand, RAP Publication 2004/17:

Pringle CM. 2001. Hydrologic connectivity and the management of biological reserves: A global perspective. *Ecological Applications* **11**: 981–998.

Renschler CS, Doyle MW, Thoms M. 2007. Geomorphology and ecosystems: challenges and keys for success in bridging disciplines. *Geomorphology* **89**: 1–8.

Richards KS. 1976. The morphology of riffle-pool sequences. *Earth Surface Processes and Landforms* **1**: 71–88.

Richter BD, Baumgartner JV, Powell J, Braun DP. 1996. A method for assessing hydrologic alteration within ecosystems. *Conservation Biology* **10**: 1163–1174.

Roper BB, Buffington JM, Archer E, et al. 2008. The role of observer variation in determining Rosgen stream types in northeastern Oregon mountain streams. *Journal of the American Water Resources Association* **44**: 417–427.

Rosgen DL. 1994. A classification of natural rivers. *Catena* **22**: 169–199.

Schumm SA. 1963. A tentative classification of alluvial river channels. *United States Geological Survey Circular* **477**: 10.

Schumm SA. 1985. Patterns of alluvial rivers. *Annual Review of Earth and Planetary Sciences* **13**: 5–27.

Schwartz JS, Herricks EE. 2008. Fish use of ecohydraulics-based mesohabitat units in a low-gradient Illinois stream: implications for stream restoration. *Aquatic Conservation: Marine and Freshwater Ecosystems* **18**: 852–866.

Taylor B, Flecker AS, Hall RO. 2006. Loss of a harvested fish species disrupts carbon flow in a diverse tropical river. *Science* **313**: 833–836.

Thompson A. 1986. Secondary flows and the pool-riffle unit: A case study of the processes of meander development. *Earth Surface Processes and Landforms* **11**: 631–641.

Thoms MC, Parsons M. 2002. Eco-geomorphology: an interdisciplinary approach to river science. *International Association of Hydrological Sciences* **276**: 113–119.

Thorp JH, Thoms MC, Delong MD. 2006. The riverine ecosystem synthesis: biocomplexity in river networks across space and time. *River Research and Applications* **22**: 123–147.

Vannote RL, Minshall GW, Cummins KW, et al. 1980. The river continuum concept. *Canadian Journal of Fisheries and Aquatic Science* **37**: 130–137.

Vaughan IP, Diamond M, Gurnell AM, et al. 2009. Integrating ecology with hydromorphology: a priority for river science and management. *Aquatic Conservation: Marine and Freshwater Ecosystems* **19**(1): 113–125. DOI: 10.1002/aqc.895

Vogel S. 1996. *Life in Moving Fluids: The Physical Biology of Flow* (2nd edn). Princeton University Press, Princeton.

Wadeson LA. 1994. A geomorphological approach to the identification and classification of instream flow environments. *South African Journal of Aquatic Sciences* **20**: 1–24.

Welcomme RL. 2001. *Inland Fisheries: Ecology and Management*. FAO, Fishing News Books, Blackwell Science Ltd, Oxford.

Whitacre HW, Roper BB, Kershner JL. 2007. A comparison of protocols and observer precision for measuring physical stream attributes. *Journal of the American Water Resources Association* **43**: 923–937.

Winemiller KO. 2005. Life history strategies, population regulations and implications for fisheries management. *Canadian Journal of Aquatic Sciences* **62**: 877–885.

Winemiller KO, Rose KA. 1992. Patterns of life-history diversification in North American fishes: implications for population regulation. *Canadian Journal of Fisheries and Aquatic Sciences* **49**: 2196–2218.

Winemiller KO, Montoya JV, Roelke DL, et al. 2006. Seasonally varying impact of detritivorous fishes on the benthic ecology of a tropical floodplain river. *Journal of North American Benthological Society* **25**: 250–262

Wood PJ, Hannah DM, Sadler JP. 2007. *Hydroecology and Ecohydrology, Past, Present and Future*. John Wiley & Sons, Chichester.

Zalewski M, Janauer GA, Jolankai G. 1997. *Ecohydrology: A New Paradigm for the Sustainable Use of Aquatic Resources*. Unesco, Paris. IHP-V Technical Documents in Hydrology no. 7.

CHAPTER 6

Measuring spatial patterns in floodplains: A step towards understanding the complexity of floodplain ecosystems

Murray Scown[1], Martin C. Thoms[1] and Nathan R. De Jager[2]
[1] *Riverine Landscapes Research Laboratory, Geography and Planning, University of New England, Australia*
[2] *US Geological Survey, Upper Midwest Environmental Sciences Center, Wisconsin, USA*

Introduction

Floodplains can be viewed as complex adaptive systems (Levin, 1998) because they are comprised of many different biophysical components, such as morphological features, soil groups, and vegetation communities, which interact and adapt over time (Stanford *et al.*, 2005). Interactions and feedbacks among the biophysical components often result in emergent phenomena occuring over a range of scales, often in the absence of any controlling factors (sensu Hallet, 1990). The emergence of new biophysical features and rates of processing feeds back into floodplain adaptive cycles and can lead to alternative stable states of floodplain structure and function which are dynamic over multiple scales (cf. Hughes, 1997; Stanford *et al.*, 2005). Interactions between different biophysical components, feedbacks, self emergence, and scale are all key properties of complex adaptive systems (Levin, 1998; Phillips, 2003; Murray *et al.*, 2014) and therefore will influence the manner in which we study and view floodplain spatial patterns.

Measuring the spatial patterns of floodplain biophysical components is a prerequisite to examining and understanding these ecosystems as complex adaptive systems. Elucidating relationships between pattern and process, which are intrinsically linked within floodplains (Ward *et al.*, 2002), is dependent upon an understanding of spatial pattern. This knowledge can help river scientists determine the major drivers, controllers, and responses of floodplain structure and function, as well as the consequences of altering those drivers and controllers (Hughes and Cass, 1997; Whited *et al.*, 2007). Interactions and feedbacks between physical, chemical, and biological components of floodplain ecosystems create and maintain a structurally diverse and dynamic template (Stanford *et al.*, 2005). This template influences subsequent interactions between components that consequently affects system trajectories within floodplains (sensu Bak *et al.*, 1988).

River Science: Research and Management for the 21st Century, First Edition.
Edited by David J. Gilvear, Malcolm T. Greenwood, Martin C. Thoms and Paul J. Wood.

Constructing and evaluating models used to predict floodplain ecosystem responses to natural and anthropogenic disturbances therefore requires quantification of spatial pattern (Asselman and Middelkoop, 1995; Walling and He, 1998). Quantifying these patterns also provides insights into the spatial and temporal domains of structuring processes as well as enabling the detection of self-emergent phenomena, environmental constraints or anthropogenic interference (Turner *et al.*, 1990; Holling, 1992; De Jager and Rohweder, 2012). Thus, quantifying spatial pattern is an important building block on which to examine floodplains as complex adaptive systems (sensu Levin, 1998).

Approaches to measuring spatial pattern in floodplains must be cognisant of scale, self-emergent phenomena, spatial organisation, and location. Fundamental problems may arise when patterns observed at a site or transect scale are scaled-up to infer processes and patterns over entire floodplain surfaces (Wiens, 2002; Thorp *et al.*, 2008). Likewise, patterns observed over the entire spatial extent of a landscape can mask important variation and detail at finer scales (Riitters *et al.*, 2002). Indeed, different patterns often emerge at different scales (Turner *et al.*, 1990) because of hierarchical structuring processes (O'Neill *et al.*, 1991). Categorising data into discrete, homogeneous and predefined spatial units at a particular scale (e.g., polygons) causes limitations and errors associated with scale and subjective classification (McGarigal *et al.*, 2009; Cushman *et al.*, 2010). These include loss of information within classified 'patches', as well as the ability to detect the emergence of new features that do not fit the original classification scheme. Many of these issues arise because floodplains are highly heterogeneous and have complex spatial organisations (Carbonneau *et al.*, 2012; Legleiter, 2013). As a result, the scale and location at which measurements are made can influence the observed spatial patterns, and patterns may not be scale independent or applicable in different geomorphic settings (Thoms and Parsons, 2011). We argue that it is more appropriate to allow patterns to 'self-emerge' from quantitative data obtained at scales appropriate for the questions being asked (Dollar *et al.*, 2007), or across multiple scales of space and time.

Established research paradigms often display positive feedback loops, which reinforce popular study designs often to the detriment of scientific advancement (Schumm, 1998; Delong and Thoms, Chapter 2). Existing perceptions of spatial patterns in floodplains have dictated how and where data are collected, the type of data collected, the scale of observations, and what is measured and analysed. For example, floodplains have been perceived as linear ecotones with distinct spatial gradients from the main channel to distal parts of the floodplain. This has led to data being collected in one direction along these perceived gradients, with results reinforcing the perception of floodplains as gradients (e.g., Chafiq *et al.*, 1992; Glavac *et al.*, 1992; Désilets and Houle, 2005). Similarly, when spatial pattern in floodplains is perceived under the patch-mosaic paradigm, data are categorised into discrete units before the patterns are measured (e.g., Kalliola and Puhakka, 1988; Arscott *et al.*, 2000; Whited *et al.*, 2007). Any subsequent observations of spatial pattern made are therefore determined by the categorisation imposed. Many problems surround this circular approach to examining spatial pattern in floodplains, and these problems are reinforced by lack of appreciation for patterns in multiple directions or beyond the scale of investigation (Robertson and Gross, 1994; Cooper *et al.*, 1997; Thoms and Parsons, 2011). Quantifying spatial patterns in floodplains must be based on the concept of self-emergence of patterns or structures rather than the measurement of data collected within, or categorised based on, preconceived structures. It is argued here that the former has largely been neglected in favour of the latter in conventional approaches to measuring spatial pattern in floodplains.

The development of new technologies, especially those associated with remotely-sensed data capture, increases the ability to quantitatively measure the spatial complexity of floodplain surfaces (Scown *et al.*, 2015). Satellite imagery, aerial photography and airborne laser scanning (LiDAR) now provide quantitative numerical data on many physical and biological attributes of floodplain ecosystems, at increasingly fine resolutions and over vast spatial extents (Mertes, 2002; Notebaert *et al.*, 2009; Legleiter and Overstreet, 2013). Quantitative, self-emergent characterisation protocols, along with increased computer processing power, have enabled robust and meaningful analyses of spatial pattern from such datasets in many terrestrial and aquatic ecosystems (Thorp *et al.*, 2008; Fonstad and Marcus, 2010; Carbonneau *et al.*, 2012). However, many of these techniques and approaches have not been widely applied in floodplain research. There is a tendency to categorise and simplify new, fine resolution, quantitative data in order to adhere to past convention or perception of spatial pattern in floodplains.

These issues outlined above are addressed in this chapter. First, a review of the development of studies of spatial pattern in floodplains is undertaken, including a meta-analysis of the literature from 1934–2013 in which spatial pattern in floodplains has been investigated. Trends in floodplain ecosystem research, types of data used, conceptual constructs of spatial pattern, measurement approaches, and the scales of observation are outlined. Second, relationships between data used and the development of floodplain conceptual models are highlighted to illustrate feedback loops in the study of floodplain ecosystems, in which data type and study design reinforce the ruling conceptual model of the day and vice versa. The implications of this reinforcement are discussed along with other limitations, arguing that the approach to measuring spatial patterns in floodplains and the type of data used should not dictate, or be dictated by, a certain floodplain conceptual model. Third, a case study of the Upper Mississippi River floodplain is presented to demonstrate the importance of enabling patterns, and hence perceptions of floodplains, to quantitatively emerge from the data, whilst being particularly cognisant of scale and location. Finally a discussion of the implications of the meta-analysis and findings of the case study is presented in the context of future floodplain research, as well as the potential of using complex adaptive systems as a framework for directing floodplain science and management in the twenty-first century.

A history of spatial pattern in floodplain research

Research on spatial pattern in floodplains has a rich history spanning many decades. Understanding trends in this research enables important concepts, dominant paradigms, limitations, and knowledge gaps to be identified. To facilitate this review, a meta-analysis of floodplain spatial pattern research from 1934–2013 was undertaken.

Table 6.1 List of attributes recorded from each of the publications reviewed.

Component(s) of interest	• Hydrology • Geomorphology • Vegetation	• Elevation • Surface cover • *Other*
Type(s) of data	• Field surveys/sampling • Maps • Aerial photography	• Satellite/aerial multispectral imagery • Digital elevation model • Literature review
Data collection and representation	• Qualitative observation • Sites • Transects	• Patches • 3-dimensional surfaces
Conceptual paradigm	• Gradient • Patch mosaic	• Hybrid • *Other*
Measure of spatial pattern	• Description/mapping • Composition and/or variability of floodplain features/properties • Diversity of floodplain features or habitats • Composition/variability/diversity and spatial organisation of floodplain features/properties	
Scale(s) of investigation	• Single/multiple scales	• 10^1, 10^2, 10^3, 10^4+ metres

This meta-analysis was limited to publications describing or measuring spatial patterns in floodplains, and thus does not encompass all literature on the subject. The search included publications listed in Google Scholar, Science Direct, ProQuest, JSTOR, and EBSCO using the keywords: floodplain, spatial pattern, heterogeneity, diversity, complexity, patch mosaic, gradient, ecotone, landscape ecology. Relevant publications contained in references obtained from this search were also included and these were mainly authored books, Ph.D. theses and pre-1990 journal articles. Attributes of each publication were recorded in regards to the floodplain component(s) of interest, type(s) of data used, how the data were collected and represented, the conceptual paradigm(s) used, and if/how and at which scale(s) spatial pattern was measured (Table 6.1).

The number of publications investigating spatial pattern in floodplains increased dramatically in the 1980s (Figure 6.1). This coincided with the emergence and development of landscape ecology as a discipline; a discipline that focused specifically on pattern and process (Turner, 1989). Around the same time, concepts of landscape ecology were being applied to rivers and floodplains (e.g., Décamps, 1984). The period 1994–2003 was associated with three special scientific journal issues with specific relevance to pattern and process

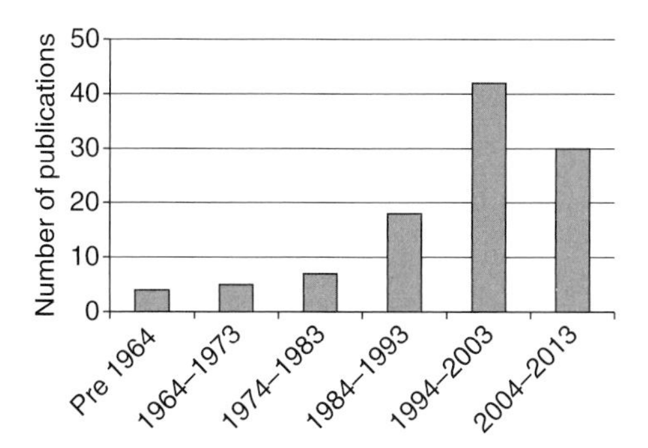

Figure 6.1 The number of floodplain landscape research publications found in this literature review grouped per decade after 1964.

in floodplains (Statzner *et al.*, 1994; Brown *et al.*, 1997; Tockner *et al.*, 2002).

Research on spatial pattern in floodplains prior to 1964 tended to be dominated by studies on floodplain geomorphology, vegetation, topography, and sediment character (Figure 6.2). Relationships between flood disturbance, vegetation succession and floodplain age were also a common theme of studies during this period (e.g., Shelford, 1954; Everitt, 1968). Spatial pattern was generally described and mapped using both qualitative and quantitative approaches in many of these earlier floodplain studies. It was not until the 1980s that spatial pattern in floodplain hydrology became prevalent (Figure 6.2). Increasingly, a spatial perspective of various hydrological properties such as inundation frequency and duration, water depth, velocity, temperature, turbidity, and other physico-chemical parameters of floodplains were presented (e.g., Hughes, 1990; Malard *et al.*, 2000). The reporting of the spatial pattern of floodplain fauna including macroinvertebrates, fish, reptiles, and mammals has also become a focus of floodplain research in recent decades (e.g., Chafiq *et al.*, 1992; Townsend and Butler, 1996).

The relative contribution of different scientific disciplines to published floodplain research has changed over time. Studies reporting on floodplain geomorphology from a spatial perspective peaked between 1974 and 1983 (Figure 6.2). Since this period there has been a decline in this focus with a corresponding increase in floodplain hydrology studies. Studies reporting on

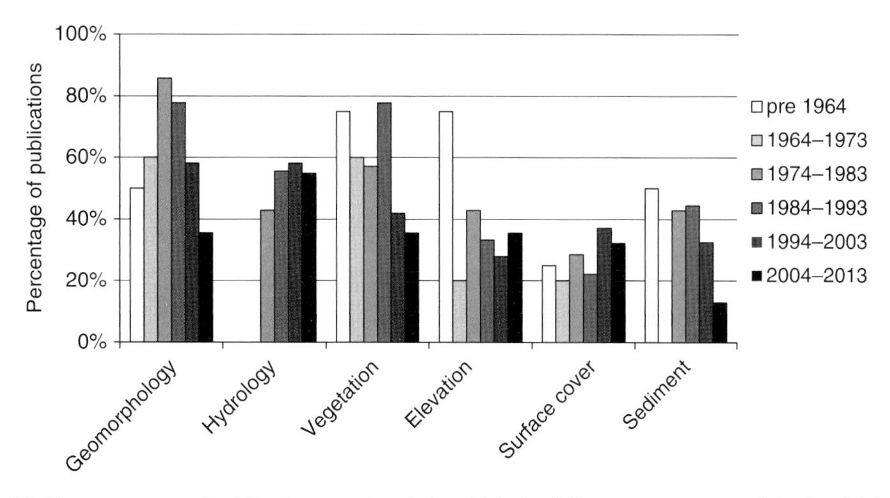

Figure 6.2 The percentage of publications per decade in which the different components of the floodplain were investigated. Note: percentages total more than 100 per decade because most publications investigated multiple components, hence could be counted more than once within each decade.

spatial pattern in floodplain vegetation and topography have fluctuated over time but there has been a decline since the period 1984–93. A similar decline has also occurred for floodplain sediment research. The declines observed in most of these floodplain components do not indicate that the total number of publications reporting floodplain spatial pattern has decreased, rather that their relative contribution decreased. This is a result of a change in the nature of floodplain research from that which had a focus on multiple floodplain components in any one study to a more singular floodplain component focus. However, 79 % of the publications reviewed report on multiple aspects of floodplain spatial pattern, reflecting the importance of the many components of floodplain systems.

There has also been a change in how floodplain data has been collected, from solely field-based studies of spatial pattern to a significant increase in the use of remotely-sensed data gathering technologies. The proportion of studies that collected field data decreased over time with a corresponding increase in the use of multispectral imagery, especially in the last decade (Figure 6.3). The production of three-dimensional floodplain surfaces has also increased in prominence, over the last two decades, as a tool for measuring spatial pattern in floodplains. Such surfaces have been used to describe and measure patterns in elevation, sedimentation rates, soil properties, nutrient concentrations, suspended sediment load, and flood water depth (e.g., Walling and He, 1998; Alsdorf *et al.*, 2007; Legleiter, 2013). The use of aerial photography for describing and measuring spatial patterns in floodplains has remained fairly constant over time (Figure 6.3).

Paradigms of spatial pattern in floodplains

Spatial pattern in floodplains has been investigated under two common paradigms since the 1950s; the *gradient* and *patch mosaic* paradigms. The gradient paradigm suggests that ecosystem variables change continuously in space across the landscape (Gustafson, 1998; Manning *et al.*, 2004) typically from near channel to distal floodplain regions. Many floodplain properties

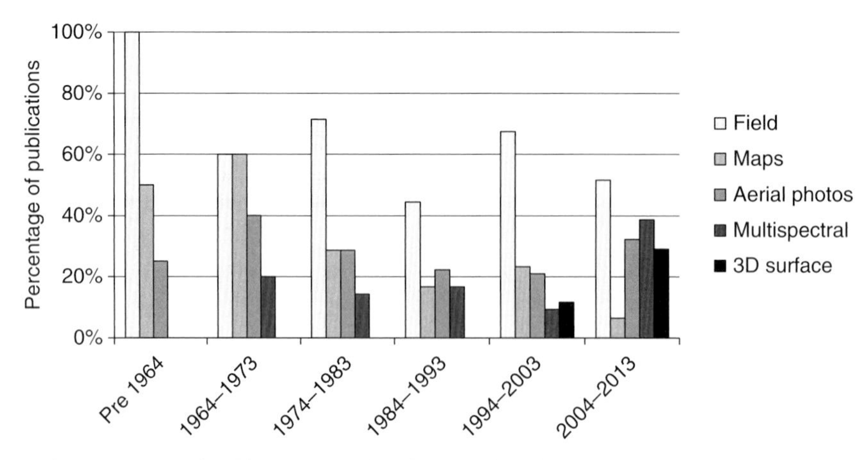

Figure 6.3 The percentage of publications per decade (excluding literature reviews) which used each type of data. Note: percentages total more than 100 per decade because most publications used multiple types of data, hence could be counted more than once within each decade.

and processes have been conceptualised and modelled as gradients in this fashion, including vegetation, soil character, suspended sediment load, sedimentation rates, inundation frequency, and surface–groundwater exchanges. In contrast, under the patch mosaic paradigm, floodplain landscapes are perceived as an assemblage of discrete categorical units, which are considered internally homogeneous, significantly different from neighbouring units and have distinct boundaries (Forman and Godron, 1981). Floodplain patches have been categorised based on vegetation type, surface cover, geomorphic features, and aquatic habitat character. An early investigation of floodplain spatial gradients was that of Turner (1934). This study described the gradual change in plant community assemblage along an elevation gradient from the main channel to distal floodplain regions. The results of which have been supported many times since, illustrating the dominance of the gradient or ecotone paradigm prior to 1964 (Figure 6.4). The earliest description of floodplain patches was by Shelford (1954), who mapped and described distinct biotic communities in the Lower Mississippi Valley that were spatially organised as a mosaic. The patch mosaic paradigm has gained popularity since the 1980s, being adopted in an increasing proportion of the publications reviewed. The number of publications in which the gradient paradigm was adopted has remained relatively constant; however, these publications constitute an ever-declining proportion of the literature (Figure 6.4). Other, process-based, paradigms within which spatial pattern in floodplains has been considered have included the intermediate disturbance hypothesis (Connell, 1978), hydrological connectivity (Amoros and Roux, 1988) and ecological succession (Clements, 1916).

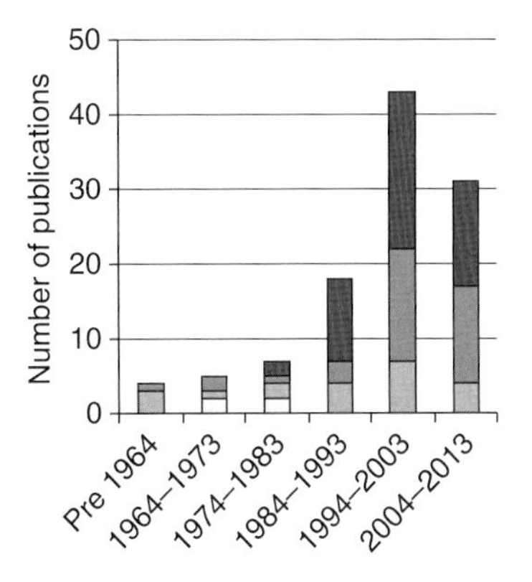

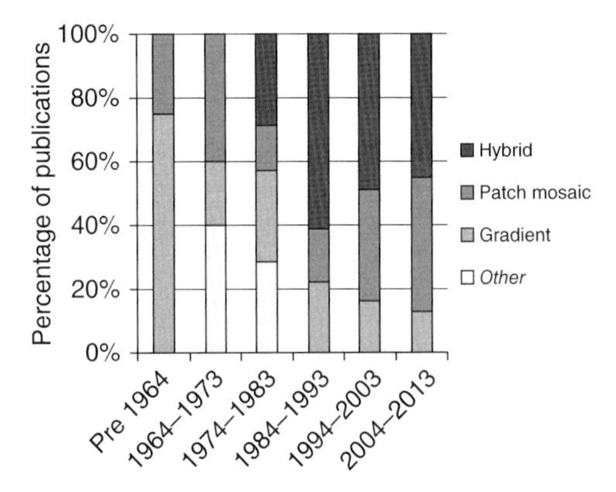

Figure 6.4 Trends in the conceptual framework within which spatial pattern in floodplains has been viewed.

The contrasting gradient and patch mosaic paradigms dominated how spatial pattern in floodplains was perceived until the 1980s (Figure 6.4). However, these two paradigms are not mutually exclusive, especially when spatial pattern is considered within a scalar context. Pattern can appear as a gradient at one scale and as a patch mosaic at another (Turner *et al.*, 1990). Thus, an increasing number of '*hybrid*' paradigms have emerged in which both patch and gradient patterns are considered in floodplains over various spatial scales (van Coller *et al.*, 2000; Wiens, 2002). Hybrid paradigms of floodplain spatial pattern suggest that gradients or ecotones may occur as areas of rapid change in an ecosystem variable within boundary regions (Naiman *et al.*, 1988) and that discrete patches can be superimposed onto broader-scale gradients or trends in a particular floodplain landscape character (Hughes, 1997; van Coller *et al.*, 2000; Wiens, 2002). One of the earliest studies of discrete zones or patches of vegetation along an elevation gradient was by Hawk and Zobel (1974), thus representing an initial hybrid floodplain pattern paradigm. Since the 1980s, there has been a rapid increase in the number of publications investigating spatial pattern in floodplains from either a patch mosaic or hybrid perspective (Figure 6.4).

Approaches to quantifying spatial pattern in floodplains

Approaches to investigating spatial pattern in floodplains have varied over time (Table 6.2), changing from being a descriptive exercise to one that is more data driven and analytical. Traditionally, contour, planform, geomorphic, and vegetation mapping were common but mainly for descriptive purposes. Floodplain cross-sections and sediment cores were also used to characterise spatial pattern in floodplains. These approaches often involved quantitative field measurements; however, further analysis was generally qualitative and the spatial patterns described were rarely measured from a landscape perspective.

More recently, floodplain spatial pattern has been quantitatively measured using spatial and non-spatial statistics (Table 6.2). Non-spatial statistics, when applied to categorical data, measure the composition, heterogeneity or diversity of floodplain features. In the publications reviewed, features of interest were most often based on vegetation, geomorphology, and/or aquatic habitat type. This approach has also been applied to numerical data such as water temperature, turbidity, and nutrient concentrations. Spatial statistics, on the other hand, measure the spatial organisation of landscape structure using either categorical or numerical data. These analyses provide a measurement of spatial variability and organisation such as the variogram. Spatial statistics have been applied to such floodplain properties as vegetation patches, soil characteristics, and elevation. These statistics generally provide a single reach- or floodplain-averaged value that does not account for location (Gustafson, 1998; McGarigal *et al.*, 2009).

Table 6.2 Examples of approaches which have been used to describe and measure spatial pattern in floodplains.

Descriptive	Analytical	
Mapping	***Non-spatial statistics***	***Spatial statistics***
• Contour	*Categorical data*	*Categorical data*
• Geomorphic	Number of patches	Patch shape
• Vegetation	Patch richness	Patch density
Cross-sections	Shannon diversity	Proximity index
Vegetation transects	Simpson's evenness	Interspersion
Sediment profiles		Aggregation
	Numerical data	*Numerical data*
	Range	Variogram
	Coefficient of variation	Fractal dimension
	Standard deviation	Autocorrelation

However, such values can obscure complexity at different scales and locations within floodplains (Thoms and Parson, 2011). There are techniques to account for scale and location using such metrics, which are discussed later in this chapter.

Basic non-spatial measures of floodplain landscapes emerged in the 1980s, although there are some earlier exceptions (Figure 6.5). Non-spatial analyses of floodplain patterns have used the number and relative proportions of different floodplain features or habitats, changes in floodplain width downstream, channel and shoreline length at different discharges, number of channel nodes, and the coefficient of variation of factors such as flood frequency, turbidity, water temperature, and nutrient concentrations measured at sites throughout the floodplain. Such measures provide useful information on the area of habitat available to organisms and the spatial variability of physico-chemical conditions in floodplains.

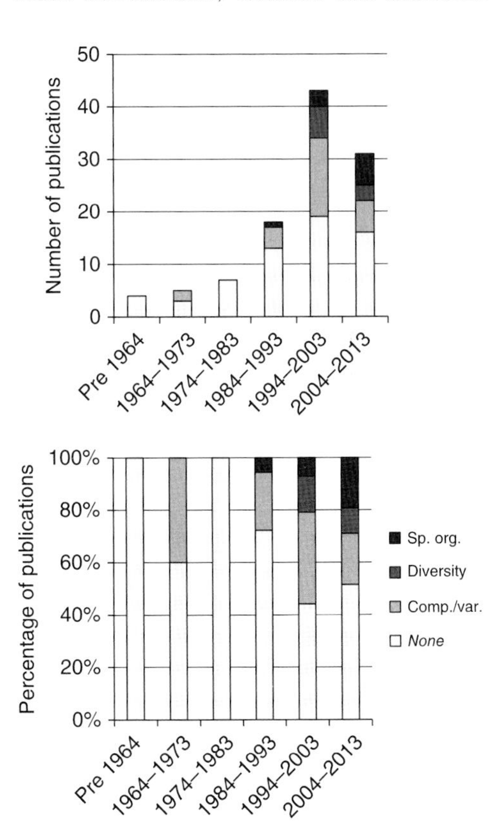

Figure 6.5 Trends in the approach used to measure spatial pattern in floodplains. Comp./var. = composition and/or variability of floodplain features/properties; Diversity = diversity of floodplain features or habitats; Sp. org. = composition/variability/diversity and spatial organisation of floodplain features/properties.

Diversity indices have also been applied to measuring spatial pattern in floodplains. These too are non-spatial; however, they provide a more robust measure of spatial pattern than basic compositional measures (Pielou, 1975). Often it is not just the number, area or relative proportions of landscape features that are important for ecosystem complexity, but an interaction of the three, which can be measured using diversity indices. In a landscape context, diversity indices have been applied to floodplain features or habitat types as '*species*' and their area as '*abundance*' (Arscott *et al.*, 2000). Habitat diversity indices were first applied to measuring spatial pattern in floodplain landscapes in the late 1990s (Figure 6.5), with the Shannon diversity and Simpson's evenness indices the most common. Habitat diversity is important because it interacts with the diversity of species within habitats (α-diversity) and the turnover of species between habitats (β-diversity) to determine overall biodiversity (γ-diversity) in floodplain landscapes (Ward *et al.*, 1999). Increased habitat diversity in floodplains due to their biogeomorphic complexity is thought to contribute to higher biodiversity than surrounding river and hillslope environments (Naiman *et al.*, 1988; Ward *et al.*, 1999).

Spatial organisation and location are also important components of spatial pattern in floodplains. The spatial organisation of habitats in floodplain landscapes can influence their hydrological connectivity, exchanges of materials, organisms and energy throughout the floodplain, and overall biodiversity and productivity (Thoms, 2003). Such attributes are measured using spatial indices of landscape pattern (Gustafson, 1998), as opposed non-spatial measures already described (Table 6.2). The number of publications in which spatial organisation was measured along with non-spatial statistics has increased in the last two decades (Figure 6.5). Spatial metrics used in floodplain research have included patch shape, juxtaposition and topology indices, as well as geostatistical tools such as variograms (Table 6.2). Variograms are useful tools indicating both variability and spatial organisation while accounting for scale, which are all important in quantifying spatial pattern (Cushman *et al.*, 2010; Legleiter, 2013). Non-spatial statistics have also been calculated using moving window analyses to account for location and multiple scales simultaneously in floodplain research (De Jager and Rohweder, 2012). Such approaches appear promising for better understanding of spatial pattern in floodplains in the future.

Limitations to traditional approaches

The increase in the number of publications measuring floodplain spatial pattern (Figure 6.5) is intrinsically linked to the type of data available and that used to measure spatial pattern (Figure 6.3). However, many of the approaches used have been influenced and therefore limited by a number of factors. These include: (i) preconceptions of spatial pattern in floodplains; (ii) the types of data used to measure spatial pattern and (iii) the scales at which observations are made. A feedback loop has existed between these three factors thereby reinforcing the approach to investigating floodplain spatial pattern. Thus, studies of floodplain spatial pattern have suffered from circularity. Our literature review indicates that the community of riverine landscape ecologists has arrived at a time when it is possible to reflect on how new data-capture technologies and spatial-analytical tools can be used to quantify spatial patterns in floodplains. Here we outline some of the issues that became apparent from the meta-analysis.

Almost half of the published studies on gradients in floodplain ecosystems have been undertaken with an initial study design based at-a-site or along transects (Figure 6.6). These studies have been designed to sample within a gradient; that is, based on a preconceived gradient. They are often undertaken at small scales and reinforce the perception of the gradient because they have not sampled areas of the floodplain beyond that gradient. There is often little evidence to suggest that the patterns observed occur across large scales or in multiple directions (Thoms and Parsons, 2011). Additionally, spatial pattern in floodplains may not be continuous, as is implied by the gradient paradigm (Southwell and Thoms, 2006), and important boundaries in floodplains may be overlooked due to scale and resolution limitations.

Similarly, the majority of research on spatial pattern in floodplains from the patch mosaic perspective has been based upon existing categorical maps (Figure 6.6). Rarely is the presence or reliability of the patches themselves questioned before their assemblage or organisation is measured (Robertson and Gross, 1994; Cooper *et al.*, 1997). Patches are often delineated qualitatively and subjectively, and may not reflect what is perceived as a patch by an organism (Wiens and Milne, 1989; Manning

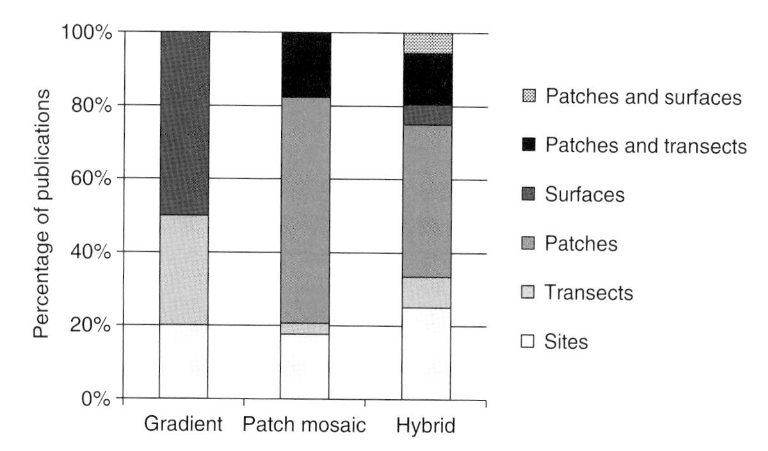

Figure 6.6 Percentage contribution of each data type to the publications under each paradigm (excluding literature reviews).

et al., 2004), nor account for any substantial amount of variability in ecosystem processes (Cushman *et al.*, 2010). Once delineated, all variation within patches is lost, and the scale of any attempt to quantify spatial pattern is limited to the scale of the initial categorical map (McGarigal *et al.*, 2009). By imposing qualitative categorisations of floodplain patches and not allowing patterns to emerge from the data over multiple scales, fundamental properties of complex adaptive systems are undermined.

Hybrid paradigms have been proposed to overcome many of the limitations associated with the gradient and patch mosaic paradigms. However, many hybrids are still of a conceptual nature; recognising the importance of both continuous and discrete spatial patterns at various scales but often not quantifying those patterns. Rudimentary descriptions of spatial pattern have occurred within a hybrid framework (e.g., van Coller *et al.*, 2000; Tockner *et al.*, 2003); however, it is often concluded that spatial patterns in floodplains are *'complex'*, with no definition or quantification of what *'complex'* is. Most attempts at quantifying spatial pattern under hybrid paradigms revert back to patch- or site-based approaches (Figure 6.6), mainly due to the availability of data and conventionality of measuring spatial pattern using patches. Consequently, hybrid models suffer from the same limitations as the original paradigms.

The metrics traditionally used to quantify spatial pattern in floodplains also have limitations. Until recently, non-spatial statistics dominated the quantification of spatial pattern (Figure 6.5). However, these provide no indication of spatial organisation, which is another important factor contributing to the complexity of floodplain ecosystems. Compositional and diversity statistics provide highly valuable information in many cases but they fail to capture the important spatial component of complexity. Further, most of these statistics provide a single floodplain-averaged value, which is useful for comparing floodplains of different reaches or rivers; however, they fail to account for location within the floodplain, which can be important when considering particular species' habitats or ecosystem processes (De Jager and Rohweder, 2012). Spatially-aware (contextual) approaches to measuring spatial pattern in floodplains are

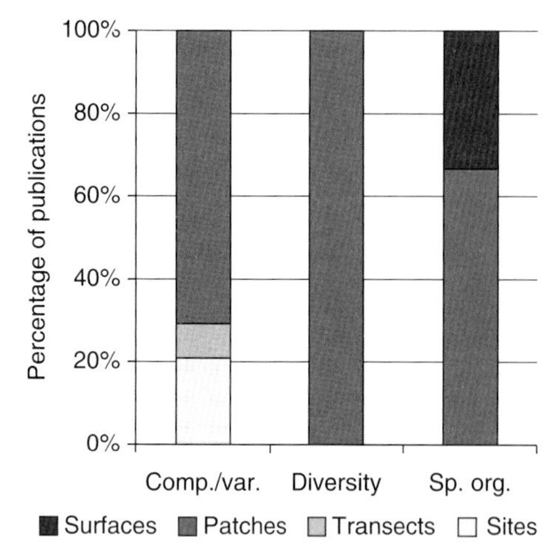

Figure 6.7 Percentage contribution of each data type used to measure spatial pattern under each approach (see Figure 6.5 for abbreviations).

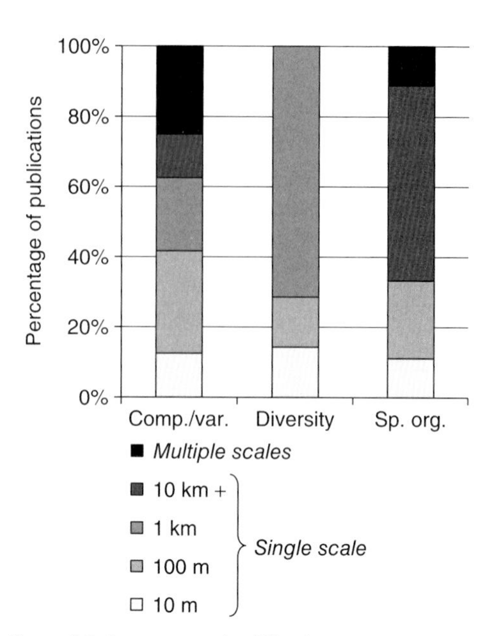

Figure 6.8 Percentage of publications measuring spatial pattern using each approach at different scales (see Figure 6.5 for abbreviations).

becoming more common (Figure 6.5); however, the majority of applications so far have been based on patch data (Figure 6.7), with a few notable exceptions (e.g., Gallardo, 2003; Legleiter, 2013). In fact, categorical patch data has dominated all approaches to measuring spatial pattern in floodplains (Figure 6.7). Hence, most of these studies are subject to the limitations of patch data already outlined.

The scale of investigation of spatial pattern in floodplain research has also been limited. More than 80% of the publications reviewed in which spatial pattern was quantified focused on a single scale (Figure 6.8). Although the scales of investigation ranged from tens of metres to hundreds of kilometres, multiple scales were rarely considered simultaneously (Figure 6.8). Scale is a fundamental consideration when investigating spatial pattern (Turner *et al.*, 1989; Levin, 1992). Pattern can appear as a gradient at one scale and as a patch mosaic at another (Turner *et al.*, 1990). Different ecosystem processes can operate over vastly different scales, and different organisms perceive and respond to the landscape at vastly different scales (Wiens and Milne, 1989; Holling, 1992). Therefore if research is to be focused at a particular scale, it is important to be explicit about which process or organism(s) that scale is relevant to (Parsons *et al.*, 2004). More lucrative in pattern analysis, however, is the consideration of multiple scales simultaneously (Bar Massada and Radeloff, 2010; De Jager and Rohweder, 2012). This enables the emergence of patterns at different scales to occur, rather than pattern at one scale being imposed, and may be useful in identifying scales over which dominant structuring processes occur, the relative importance of top-down versus bottom-up influences, or scales at which anthropogenic interference has occurred (Robertson and Gross, 1994; Thorp *et al.*, 2008). Future research into spatial pattern in floodplains should therefore consider multiple scales whenever possible and/or relevant.

A new approach for measuring spatial pattern in floodplains

As new, continuous, high-resolution data for floodplains are collected, it is possible to apply new techniques for quantifying spatial pattern. In this section the importance of scale, self-emergence, spatial organisation, location, and metric type when measuring spatial pattern in floodplains is demonstrated. We used surface metrics and moving window analyses to measure spatial pattern from quantitative topographic data in the form of a gridded digital elevation model (DEM) for a portion of the Upper Mississippi River floodplain. The DEM was derived from airborne laser scanning (LiDAR) and is a continuous numerical representation of floodplain surface elevation. Surface metrics can be used to measure spatial pattern from this type of data without the need to delineate patches (McGarigal *et al.*, 2009). They are based on *'pixels'* rather than *'patches'* as the basic structural elements of a landscape (Cushman *et al.*, 2010). As such, there are no issues associated with arbitrary categorisation of data into discrete patches or with boundary delineation (Cushman *et al.*, 2010). Using a moving window analysis, each surface metric can be calculated for a neighbourhood around each pixel (cell) in the DEM in order to account for location. Increasing the size of the neighbourhoods enables floodplain structure to be characterised across multiple scales and patterns to self-emerge (Bar Massada and Radeloff, 2010; De Jager and Rohweder, 2012). Such an approach accounts for the spatial organisation and scale of landscape structure and pattern, and is useful not only for DEM data but any gridded ecological data.

Case study

Study area

This case study was conducted in Pool 9 of the Upper Mississippi River (UMR), which lies between the states of Minnesota, Iowa and Wisconsin in the USA (Figure 6.9). The northern part of the UMR is divided into 29 navigation pools by a series of locks and dams mostly constructed during the 1930s. The river valley bottom in Pool 9 is approximately 50 river kilometres in length and generally between five and six kilometres wide (Figure 6.9). The floodplain in much of the lower half of the pool is permanently inundated due to water impoundment behind the lock and dam. The impounded water, river channels, and floodplain together occupy over 210 square kilometres. Major geomorphic floodplain features observed in the upper half of the pool include the main and anastomosing channels, islands, natural levees, crevasse splays, backwaters, and swamps. This portion of the UMR is constrained between dolostone-capped sandstone bluffs on either side, which can rise up to 200 metres above the river valley bottom. The river valley was carved by glacial meltwater following the last glacial maxim. Climate in the UMR is continental, with marked seasonal temperature and precipitation patterns. This results in a seasonal hydrograph, with average monthly discharge for Pool 10 (directly downstream of Pool 9) peaking around 2600 m^3s^{-1} during April and falling to around 670 m^3s^{-1} during January, for the period from 1982 to 2012 (USGS, 2013).

Methods

A gridded bare-earth DEM with 1×1 m^2 cell size was used as the base dataset for this case study. The DEM was derived from airborne laser scanning (LiDAR), which was obtained in 2007 by the US Army Corps of Engineers'

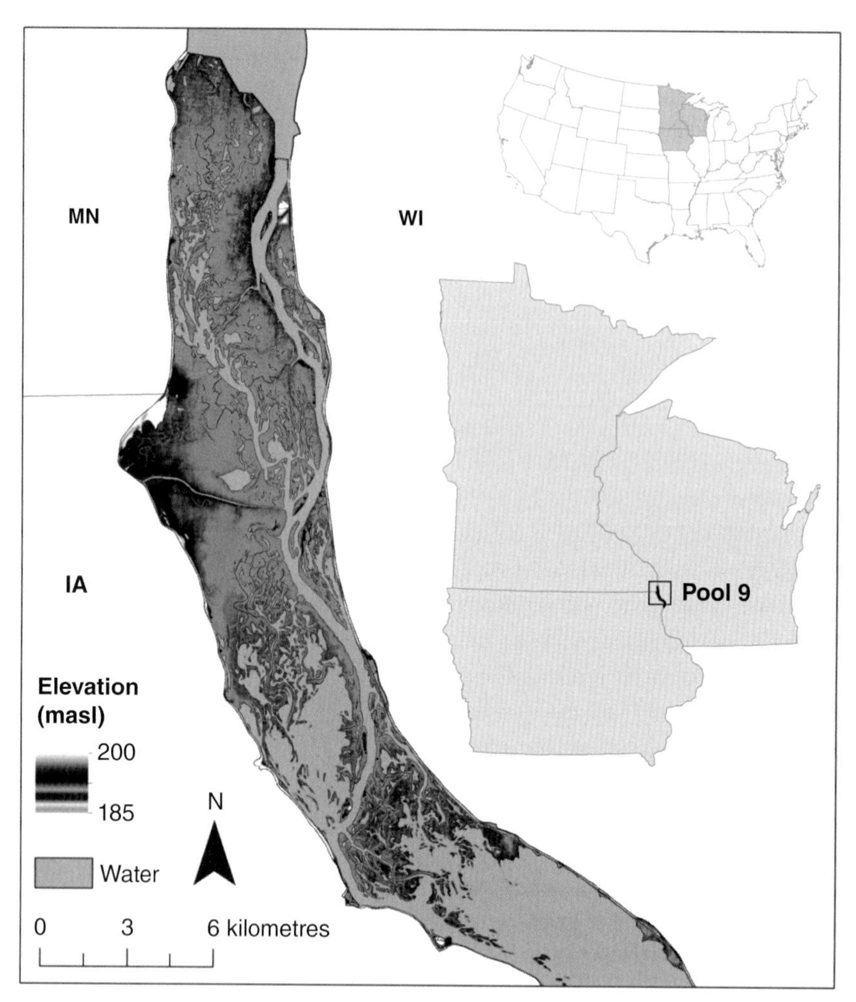

Figure 6.9 Regional location of Pool 9 of the Upper Mississippi River and detail of floodplain topography in its upper reaches. Scown *et al.*, 2015. Reproduced with permission from Elsevier. (*See colour plate section for colour figure*).

Upper Mississippi River Restoration Program (UMRR). LiDAR typically provides spot heights with a horizontal accuracy within one metre and a vertical accuracy within 15 centimetres, although they can be much better. From the LiDAR spot heights, the US Geological Survey's Upper Midwest Environmental Sciences Center (UMESC) created a *Tier 2* DEM which has undergone rigorous quality assurance testing (UMESC, 2013). The *Tier 2* DEM for Pool 9 was used in this case study.

The DEM was clipped to the extent classified in the Upper Mississippi River Restoration Environmental Management Program's Long-Term Resource Monitoring (LTRM) floodplain land cover/use data polygons for Pool 9 (UMESC, 2012). A 50 m buffer within the outer edge of this extent was removed to prevent valley side effects. All areas classified as 'open water', 'agriculture', 'developed', 'levee', 'pasture', 'plantation', and 'roadside grass/forbes' in these polygons were also removed to minimise the effects of significant man-made structures and errors in the LiDAR data associated with water. These areas constituted only a small proportion of the DEM. Man-made sand piles on the banks of the main channel where dredged sediment has been deposited were included as these now present significant physical features of the floodplain. The lower part of Pool 9 where most of the floodplain is now permanently submerged was also removed as there were only a few small islands recorded in the DEM in this area.

The clipped DEM was then de-trended relative to the 30-year mean low water level to remove the overall downstream slope. A surface with the same resolution and extent as the clipped DEM was interpolated between the river mile contours that contained a 30-year mean low water level value. This surface represented an estimate of the mean low water height above mean sea level in Pool 9, incorporating the overall downstream slope. This surface was subtracted from the clipped DEM to produce a de-trended DEM, which contained a height value above the 30-year mean low water level at that location downstream for every $1 \times 1\ m^2$ cell. Any negative values (i.e., below mean low water level) were removed from the de-trended DEM.

Five surface metrics were chosen to measure spatial patterns in the topography of the floodplain (Table 6.3). *Range* determines the range of surface elevations within an area. This is an important structural property of floodplains due to the influence of elevation on inundation and vegetation patterns. A higher range of surface elevations within an area corresponds to a greater range of flood frequencies, and potentially greater habitat diversity for various floodplain plant species (Hamilton *et al.*, 2007). *SD* measures the variability in surface topography around the mean elevation in an area, while *CV* measures the variability relative to the mean. Higher *SD* and *CV* reflect increased topographic variability in an area. Variability in surface heights increases the spatial variation of flood frequencies and soil saturation in floodplains, both of which contribute to higher biodiversity in floodplains (Pollock *et al.*, 1998). SD_{CURV} and *Rugosity* relate to surface roughness (Mark, 1975). Surface roughness is a particularly important structural property of floodplains since it can create a diverse array of hydraulic and geomorphic conditions across the surface (Nicholas and Mitchell, 2003). SD_{CURV} measures the standard deviation of total curvature within an area. Curvature varies depending on the type of local landform: high point, depression, ridge, valley, saddle, spur, shoulder, and so on (Nogami, 1995; Iwahashi and Pike, 2007). It also determines whether material will diverge or converge at that location, or whether flowing substances will accelerate or decelerate across that location (Evans, 1972). *Rugosity* is calculated as the ratio between the actual surface area and that of a flat plane occupying the same x, y extent. It reflects the convolutedness of the surface within an area as well as the actual surface area available as habitat. Actual habitat area is important in many ecosystems when competition for space is a key structuring process (Hoechstetter *et al.*, 2008), this may be particularly relevant in densely vegetated floodplains such as that of the Amazon (Salo *et al.*, 1986; Hamilton *et al.*, 2007). Higher values in both SD_{CURV} and *Rugosity* indicate a more convoluted and topographically complex surface.

These five surface metrics were calculated from three input grids both globally (for the entire grid) and locally (within a neighbourhood around every cell in the grid). The input grid from which *Range*, *SD*

Table 6.3 Description of the five surface metrics calculated.

Metric	Description	Indicates	References
Range	The difference between the lowest and highest points in the DEM or within a neighbourhood	Magnitude of topographic relief within an area	Nogami (1995) Wilson *et al.* (2007) Walker *et al.* (2009)
Standard deviation (SD)	The standard deviation of all surface height values in the DEM or within a neighbourhood	Variability of the surface about the mean height within an area	Evans (1972) Mark (1975) Hoechstetter *et al.* (2008) McGarigal *et al.* (2009)
Coefficient of variation (CV)	The coefficient of variation of all surface height values in the DEM or within a neighbourhood	The magnitude of surface height variability relative to the mean height within an area	McCormick (1994) Pollock *et al.* (1998)
SD_{CURV}	The standard deviation of *total curvature* (Jenness, 2012) of each cell in the DEM or within a neighbourhood	Variability of the shape of the surface within an area	Tarolli *et al.* (2012)
Rugosity	The ratio of the true surface area of the DEM or within a neighbourhood to that of a flat plane occupying the same (x, y) extent	Convolutedness of the surface within an area	Hobson (1972) Jenness (2004) Kuffner *et al.* (2007) Wilson *et al.* (2007) Walker *et al.* (2009)

and *CV* were calculated was the de-trended DEM, while two other input grids derived from the de-trended DEM were used to calculate SD_{CURV} and *Rugosity*. These input grids were the *total curvature* and *surface–area ratio* of each cell in the de-trended DEM, respectively, and were created using the *DEM Surface Tools* toolbox in ArcGIS 10.0 (Jenness, 2012). Global surface metrics were calculated using the value of every cell in the input grid and provide a single floodplain-averaged metric value. Local surface metrics were calculated for each cell in the grid based on all values within a neighbourhood (window) around that cell using a moving window analysis. The global median for each surface metric measured at each window size was also estimated from the raster quartiles in ArcGIS 10.0.

The moving window analyses were conducted using the *FocalStatistics* tool in ArcGIS 10.0. Windows were circular and centred over the target cell. Fourteen window sizes were used with radius = 10, 20, 30, 40, 50, 100, 150, 200, 250, 300, 400, 500, 750, 1000 metres. The output of the moving window analysis is the metric value measured within the specific sized window around that cell. Hence, cells in the output grid contain metric values rather than height values as in the de-trended DEM. Due to the scarcity of input grid values in some areas of the floodplain, some windows contained only a very small proportion of data. Whenever a window around a target cell did not contain more than 60% data, this cell was removed from the output grid. This percentage was chosen based on examination of (i) the relationship curves between minimum proportion of window containing data and percentage of all cells from the DEM retained, and (ii) maps of the proportion of window containing data for each window size. Removing the cells, which did not contain at least 60% data in a particular window size, minimised spurious values due to insufficient sample sizes while maximising the number and spatial distribution of cells in the output grid.

The scaling characteristics of the surface metrics were quantified using linear regression of log-transformed data for both (i) the global median of each metric measured for each window size, and (ii) the local metric value at 18 random sample cells for each window size. The base 10 logarithms of window radius (in metres) and metric value were plotted and linearly regressed in Microsoft Excel. Relationships were significant at $p < 0.05$. A straight line on the log–log plot is the same as a power function fit to the untransformed data.

Differences in scaling characteristics between metrics were investigated using the range standardised slopes of the regression lines. Global median values were range standardised to between 1 and 2 for each metric, and the base 10 logarithm of the range standardised metric score was regressed against that of window radius to give a range standardised slope. Differences in the scaling characteristics of each metric at different spatial locations were investigated in a 3 × 2.5 km sample section of the floodplain. This area contained most of the major physical features present in the floodplain including the main and side channels, natural levees, crevasse splays, backswamps, and islands. Eighteen sample cells were randomly chosen within this area and the base 10 logarithm of each surface metric was regressed against that of window radius to determine the scaling characteristics of each metric for each cell. The coefficient of variation of the absolute slope of all significant regression lines was calculated for each metric from the sample cells and used as an indicator of the spatial variability of the scaling characteristics of each metric.

Results

Global metric values greatly overestimated local metric values in the majority of the floodplain, particularly when local metrics were measured at small scales (Figure 6.10). The global medians of each local metric were highly dependent upon scale, with each increasing with window size (Figure 6.10). This indicates that greater variability in the topographic structure of the floodplain is observed with increasing measurement scale. There was a significant log–log relationship between global median and window radius for all surface metrics (Table 6.4). For *Range*, *SD*, *CV* and SD_{CURV}, the straight line fitted to this relationship had an r^2 value greater than 0.98, while for *Rugosity* it had an r^2 of 0.832 (Table 6.4). Although a straight line was a good fit for *Rugosity*, there also appeared to be a scale (around window radius = 100 m) at which *Rugosity* values reached an asymptote for the scale range investigated (Figure 6.10e). The global *Rugosity* value was still higher than this apparent asymptote. The slopes of the lines when each metric was range standardised were similar, although slightly higher for SD_{CURV} (Table 6.4), suggesting that on average all five metrics are similarly sensitive to scale.

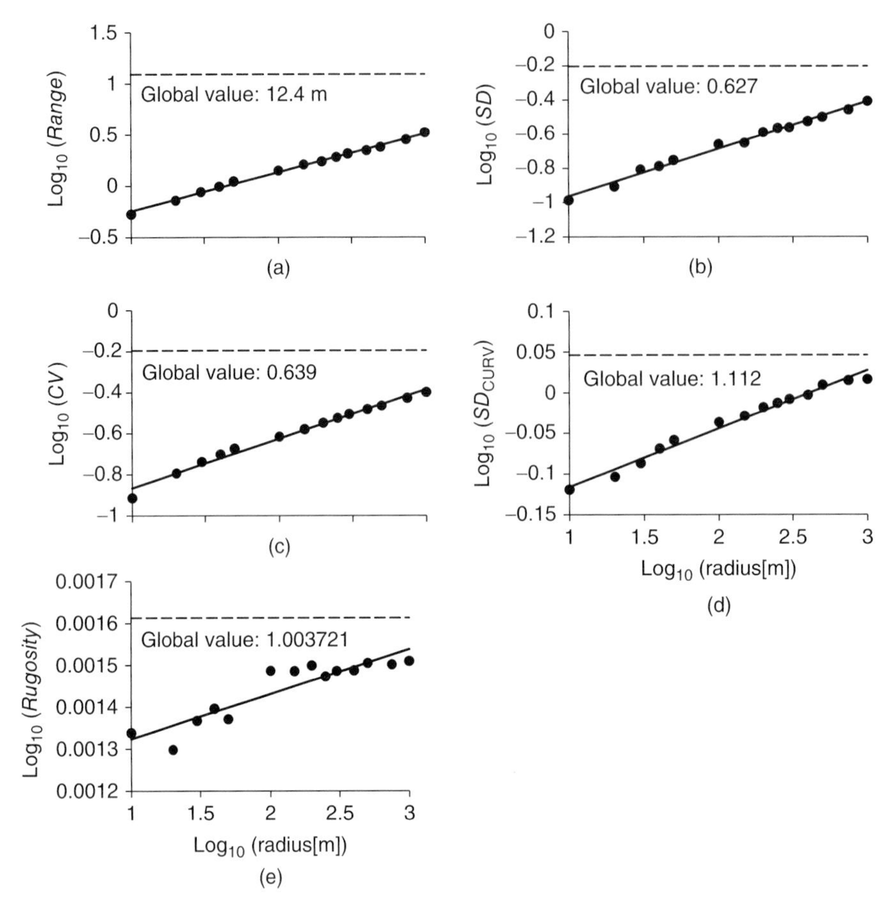

Figure 6.10 The linear relationships between Log_{10}(window radius) and Log_{10}(global median metric value) for (a) *Range*, (b) *SD*, (c) *CV*, (d) SD_{CURV} and (e) *Rugosity*. Global metric value also noted (dashed line indicates base 10 logarithm). Note different *y*-axis scales.

Table 6.4 Linear regression results of Log_{10}(window radius) against Log_{10}(metric value) for global medians. Slope and y-intercept given to four significant figures, *RS* = range standardised.

Metric	F	*p*	*d.f.*	r^2	Slope	*y*-intercept	*RS slope*
Range	3081.9	0.000	1,12	0.996	0.3791	−0.6241	6.270×10^{-2}
SD	1255.3	0.000	1,12	0.991	0.2777	−1.240	6.332×10^{-2}
CV	892.2	0.000	1,12	0.987	0.2407	−1.106	6.248×10^{-2}
SD_{CURV}	695.1	0.000	1,12	0.983	7.168×10^{-2}	−0.1874	6.766×10^{-2}
Rugosity	59.5	0.000	1,12	0.832	1.073×10^{-4}	1.216×10^{-3}	6.248×10^{-2}

Local surface metric values were highly dependent upon scale, but also on location. Results of the five surface metrics measured locally at three window sizes (radius = 10, 100 and 1000 m) are shown in Figure 6.11a for the 3 × 2.5 km sample section of the floodplain. For *Range*, *SD*, *CV* and *Rugosity*, most of this area was dominated by low (dark blue) metric values at the 10 m and 100 m window sizes, whereas SD_{CURV} had relatively higher values (greens and yellows). *Range* and *SD* appeared to have similar spatial patterns when measured at the 10 m and 100 m window sizes, but quite different ones at the largest size. High values for both these metrics were visually evident at the 100 m window size, but were absent at the 1000 m window size for *SD* while dominant at that scale for *Range*. *CV* appeared to distinguish relatively higher values in the western (low-lying) quarter of the frame than any of the other metrics. Areas of high SD_{CURV} and *Rugosity* were evident at the 10 m and 100 m window sizes fringing many of the smaller channels and gullies, but were generalised or lost at the largest window size. High SD_{CURV} values were particularly prominent at the edges of the main channel and some smaller channels at the 100 m window size, while these areas were not distinguished by the other metrics at that scale.

The scaling characteristics of local surface metrics in the sample cells were not always consistent with the scaling characteristics of the global median for each metric. On average, r^2 values for local surface metric regressions were lower than those of the global medians (Tables 6.4, 6.5 and 6.6). The results of log–log regressions for the three example cells (A – circle, B – triangle and C – square) are shown in Figure 6.11b and Table 6.5. Locations of each of the cells are shown on the maps in Figure 6.11a. *Range* increased with window size at all three cells, but at different rates (slopes) to that of the global median. Cell A (circle) and cell C (square) had lower slopes than that of the global median while the slope of cell B (triangle) was higher. *SD* increased with window size at cells A and B, but decreased at cell C. Again, the slopes of these lines differed from that of the global median. *CV* was not significantly related to window size in a log–log way at cell C ($p = 0.269$), nor was the relationship well described by a straight line at cell A ($r^2 = 0.291$). The direction of change also depended upon location for SD_{CURV} and *Rugosity* values. SD_{CURV} and *Rugosity* both increased with window radius at cells A and B, but decreased at cell C. The absolute slope of the lines for both of these metrics at cell C was also much higher relative to that of the global medians.

The influence of scale on local surface metric values was also highly dependent upon location within the sample section of the floodplain. The scaling characteristics of *Rugosity* were by far the most variable in space within this section of the floodplain (Table 6.6). The coefficient of variation of absolute slope of the regression lines among the 14 significant sample sites for *Rugosity* was more than three times that of any of the other four metrics (Table 6.6). *Range* was significantly related to window radius in a log–log way at all 18 of the sample sites and this relationship was the least sensitive to location of any of the metrics, with the lowest coefficient of variation of absolute slope (Table 6.6). The slopes of the lines for *SD*, *CV* and SD_{CURV} were similarly variable in space in the sample floodplain section (Table 6.6).

Discussion

This case study highlights differences between metrics measured globally versus locally, which reflects the importance of scale and location for measuring spatial pattern or structural complexity in floodplains using surface metrics. The results suggest that one floodplain- or reach-averaged metric value may be insufficient for characterising the complexity of spatial pattern within a floodplain. Surface metrics measured locally can provide analyses of

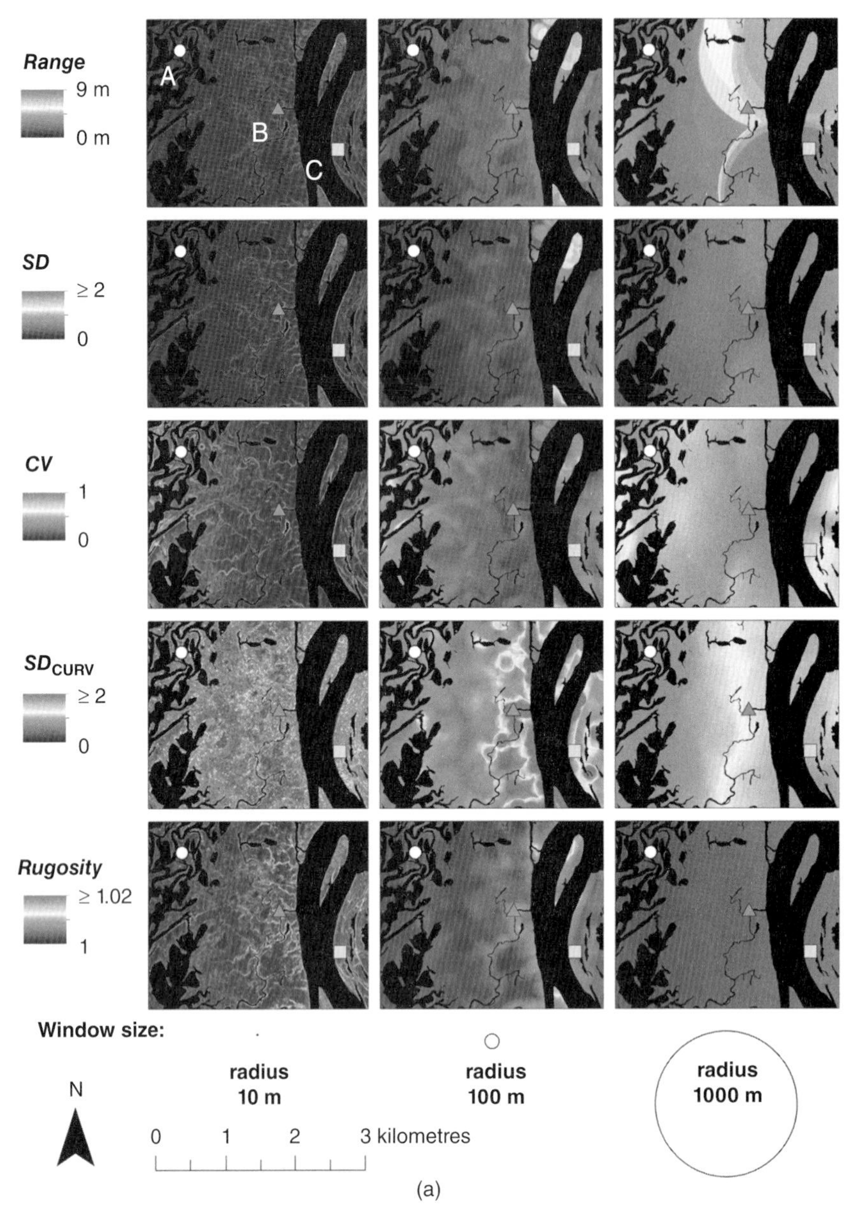

Figure 6.11 (a) Local surface metric results for three window sizes in the 3 × 2.5 km sample section of floodplain and locations of the three example cells, and (b) linear relationships between Log_{10} (window radius) and Log_{10} (global median metric value) for each surface metric at each example cell. (*See colour plate section for colour figure*).

floodplain structure tailored to specific scales and locations within the floodplain, which are relevant to the questions being asked. Such approaches have already been applied to measuring spatial pattern in many ecosystems (e.g., Riitters *et al.*, 2002; Bar Massada and Radeloff, 2010). On the other hand, global metric values may be appropriate when considering ecosystem processes or organisms that respond to

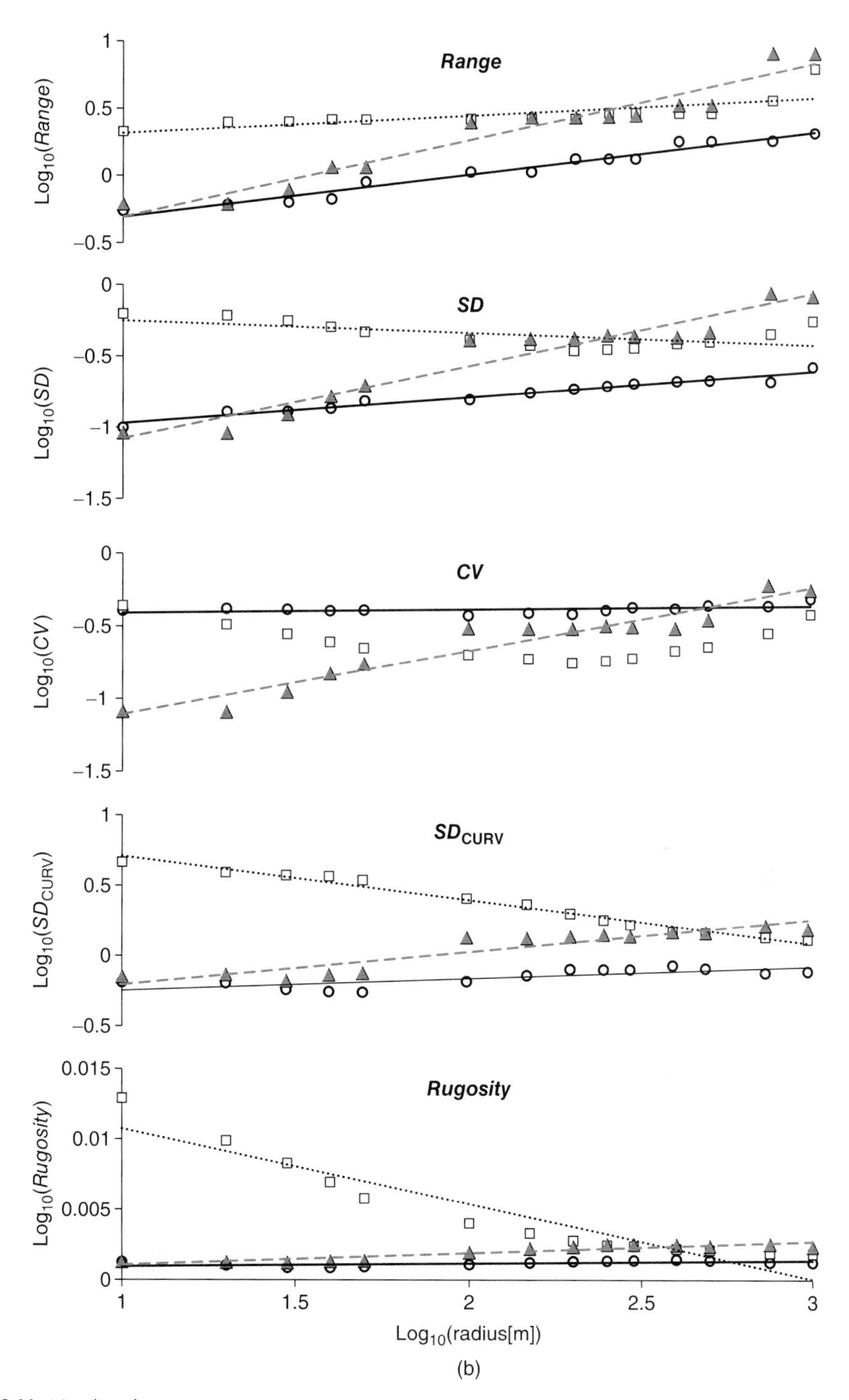

Figure 6.11 *(Continued)*

Table 6.5 Linear regression results of Log_{10}(window radius) against Log_{10}(metric value) for three local cells. Slope and y-intercept given to four significant figures, * indicates non-significant result, ** indicates r^2 less than 0.8.

Metric	Cell	F	*p*	*d.f.*	r^2	Slope	*y*-intercept
Range	○	359.8	0.000	1,12	0.968	0.3177	−0.6253
	▲	181.8	0.000	1,12	0.938	0.5751	−0.8830
	□	15.6	0.002	1,12	0.565**	0.1351	0.1782
SD	○	317.5	0.000	1,12	0.964	0.1849	−1.157
	▲	159.7	0.000	1,12	0.930	0.5111	−1.589
	□	5.9	0.032	1,12	0.329**	-8.258×10^{-2}	−0.1698
CV	○	4.9	0.047	1,12	0.291**	2.547×10^{-2}	−0.4348
	▲	148.3	0.000	1,12	0.925	0.4374	−1.544
	□	1.3	0.269*	1,12			
SD_{CURV}	○	19.5	0.001	1,12	0.619**	8.472×10^{-2}	−0.3320
	▲	72.0	0.000	1,12	0.857	0.2308	−0.4361
	□	534.2	0.000	1,12	0.978	−0.3098	1.019
Rugosity	○	8.8	0.012	1,12	0.423**	2.090×10^{-4}	7.480×10^{-4}
	▲	65.7	0.000	1,12	0.846	8.364×10^{-4}	2.209×10^{-4}
	□	99.7	0.000	1,12	0.893	-5.341×10^{-3}	1.607×10^{-2}

Table 6.6 Summary of linear regression results of Log_{10}(window radius) against Log_{10}(metric value) for the 18 randomly sampled cells, CV (abs. slope) is used as an indicator of the spatial variability of the scaling characteristics of each metric.

Metric	No. significant	Average r^2	CV (abs. slope)
Range	18	0.876	0.383
SD	16	0.830	0.422
CV	15	0.816	0.433
SD_{CURV}	15	0.790	0.439
Rugosity	14	0.725	1.475

spatial pattern at the scale of the entire floodplain. In these cases, the detail provided by local metrics may be irrelevant. Global metric values may also be useful for making general comparisons between floodplains or between reaches. This is likely the case not only for surface metrics, but also for any of the metrics described in the meta-analysis. Arscott *et al.* (2000) conducted such a comparison of the composition, Shannon Diversity, and Simpson's Index of Dominance of aquatic habitat types between six geomorphically distinct reach types in the Fiume Tagliamento, an alpine floodplain river in northern Italy. Others have compared measures of the diversity of hydraulic patches within river reaches between different discharges (Thoms *et al.*,

2006; Wallis *et al.*, 2012). To the best of our knowledge, no similar analysis of floodplain topography using surface metrics has been undertaken, so comparison of these metric values with other studies will require further research. It is worth noting, however, that river channel morphology has recently been investigated using variograms (Legleiter, 2013) and such an approach could be highly relevant for floodplain morphological research.

The value of different surface metrics for measuring floodplain complexity, as shown in this case study, is dependent upon the scale or extent over which the metric is measured. This has implications for comparing results between studies, but also for investigating pattern–process relationships as the scale(s) of measurement of spatial pattern must be relevant to the scale(s) over which the process of interest occurs (Turner *et al.*, 1989; Wiens and Milne, 1989). However, the scale dependence of surface metrics is not necessarily a bad thing. Although we must be mindful of scale when measuring metrics individually, the scaling characteristics (scalograms) of metrics can be used to ask questions about major structuring processes and the scales over which they operate or dominate (Bar Massada and Radeloff, 2010; De Jager and Rohweder, 2012). Distinct breaks in scaling characteristics, or departures from scaling regression lines (Figure 6.10), may indicate scales at which structuring processes and surface attributes change, or scales at which anthropogenic disturbance has overridden natural processes (De Jager and Rohweder, 2012). *Rugosity* best demonstrates this here. Although there was a significant log–log relationship, there also appeared to be little change in the global median between windows with radius = 100 m and 1000 m (Figure 6.10e). This may suggest that changes in *Rugosity* and its drivers are minimal between these scales. Below this scale range *Rugosity* values increased substantially more, while above they must increase again in order to reach the global value. Further investigation into this is required. Straight lines on a log–log plot such as those in this case study represent a power function in the untransformed data. There is a rich literature on the significance of power functions in many ecological domains (Holling, 1992; Levin, 1992; Milne *et al.*, 1992), which parallels with many principles of complex adaptive systems theory such as hierarchy, fractals, nonlinearity, chaos, and $1/f$ noise which have been investigated in geomorphic systems (Goodchild and Mark, 1987; Bak *et al.*, 1988; Turcotte, 2007). Future research should investigate the scaling characteristics of particular metrics throughout the floodplain within the context of such principles.

The importance of location for measuring spatial pattern in floodplains is also highlighted in this case study of the Mississippi floodplain. Moving window analyses, although computationally intensive, can provide spatially-contextual approaches for measuring spatial pattern, which account for location. Moving window analyses can be applied to any gridded data. This includes DEMs, as evidenced in this case study, but also raster datasets containing Normalised Difference Vegetation Index (NDVI), habitat type, soil conditions, sedimentation rates, water depth and the like. Such data are becoming increasingly available over larger areas and also for multiple snapshots in time thanks to advances in field and remote data capture techniques (Fuller and Basher, 2013; Legleiter, 2013). Further, moving windows can be used to not only calculate surface metrics, but also for almost any of the metrics described in the meta-analysis, particularly with advances in computer processing power and automated scripts. Regardless of whether moving window analyses are used, location should still be accounted for when measuring spatial pattern in floodplains. This can be achieved by maximising replicate samples with good spatial coverage, or with the aid of geostatistical tools such as variograms, which determine the degree of spatial autocorrelation within a dataset.

Allowing spatial patterns to self-emerge from quantitative data, rather than imposing subjective patches onto a floodplain at one particular scale is also highlighted in this case study. Emergent patterns reflect scales and locations at which particular characteristics of the floodplain surface are detected by the relevant surface metric. Take SD_{CURV}, for example. At the 100 m window size (Figure 6.11a) areas or *'patches'* of high SD_{CURV} were qualitatively apparent along several of the smaller channels and the shorelines of the main channel and islands. These areas were distinguished by this particular metric at this scale. However, at the 1000 m window size (Figure 6.11a), SD_{CURV} divided the frame roughly in half; higher values in the eastern half, lower values in the western half; while smaller channels and shorelines were not distinguished. The trend from high to low is much more gradual in space at the larger scale, but two different zones could qualitatively be drawn if that was the objective. This suggests that the main channel influences SD_{CURV} at large scales in the same way that smaller channels and shorelines influence this metric at mid scales. Put another way, zonation or *'patchiness'* based on SD_{CURV} is different, and is influenced by different physical features, at different scales.

In terms of interpreting the results of this case study and directions for future research, two more topics deserve brief discussion. First, any robust quantification of spatial pattern must be cognisant of scale, spatial organisation, self-emergence, and location before pattern–process linkages and complexity in floodplain ecosystems are examined. Further investigation into the implications of the spatial patterning and scaling characteristics of these metrics, and others, on floodplain ecosystem processes should be a focus of future research. Second, interpreting spatial pattern requires the consideration of multiple dimensions. It goes beyond just the metric value dimension (z), which is the focus of a plot of dependent against independent variables (Figures 6.10 and 6.11b). The spatial location (x, y) of each z is also important (Figure 6.11a). This poses many problems for displaying, interpreting, and communicating results, which should be addressed in future research. Geographic Information Systems (GIS) and geostatistics will likely provide valuable tools in such endeavours in future floodplain research, as they have already in geography and landscape ecology as well as in some previous floodplain research (Carbonneau *et al.*, 2012; Legleiter, 2013).

Synopsis and future directions

This chapter has focused on measuring spatial pattern in floodplains. We reviewed 108 publications from 1934–2013 to determine trends, dominant paradigms, and approaches to measuring spatial pattern in floodplains. Many advances in knowledge and techniques have emerged from the rich literature; however, conventional approaches have also had their limitations. Feedback loops have developed in which preconceptions about floodplain patterns dictate how and where data are collected, and these data then reinforce the preconceptions. This limits the possibility of exploring alternative perspectives of spatial pattern in floodplains, and their implications. The limited scale and location of observations in many studies, as well as the analyses which are employed, have also limited our understanding of spatial pattern in floodplains. The case study highlighted the importance of considering scale, self-emergence, spatial organisation, and location when measuring spatial pattern in floodplains. Quantitative investigations of spatial pattern cognisant of these are required in the future. These studies should be undertaken across a variety of scales, disciplines, geographic settings, and datasets. Such endeavours are becoming easier thanks to advances in data-capture technologies, computer-processing power, and interdisciplinary perspectives and approaches to measuring spatial pattern.

One possible and promising future direction for floodplain research lies within the context of complex adaptive systems theory. Floodplains, and river systems in general, have many properties typical of complex adaptive systems: a high diversity of system components, interactions, and feedbacks; self-emergent phenomena that can occur in the absence of any global controller; hierarchical structuring; and multiple stable states far from equilibrium. These properties promote and maintain the biodiversity and productivity of these valuable ecosystems. However, they also have important consequences for floodplain science and management. In the presence of thresholds (Church, 2002; Phillips, 2003), perturbations in complex adaptive systems may result in a significant restructuring of components and interactions (Bak and Paczuski, 1995; Rietkerk *et al.*, 2004). In floodplains, this means that anthropogenic interferences may have disproportionately large, unpredictable, and delayed ecological effects (Sparks *et al.*, 1990), and untangling the interactions between system components in order to predict a response can be difficult. Measuring spatial pattern is one of many steps towards understanding how floodplain ecosystems will respond to increasing pressures, identifying thresholds between multiple stable states, and maintaining the diversity of components, interactions, and feedbacks. These are all important tasks for the future and it is hoped that this chapter is useful in such endeavours.

Acknowledgements

The authors would like to thank the book editors and an anonymous reviewer for their comments, which greatly improved the chapter. We also thank the University of New England and the USGS Upper Midwest Environmental Sciences Center for their support. Any use of trade, product or firm names is for descriptive purposes only and does not imply endorsement by the US Government.

References

Alsdorf, D., Bates, P., Melack, J., *et al.* (2007). Spatial and temporal complexity of the Amazon flood measured from space. *Geophysical Research Letters*, 34: L08402.

Amoros, C. and Roux, A. L. (1988). 'Interaction between water bodies within the floodplains of large rivers: function and development of connectivity'. In: Connectivity in Landscape Ecology, Proceedings of the 2nd International Seminar of the International Association for Landscape Ecology, Müstersche Geographische Arbeiten 29. K. L. Schreiber (ed.), pp. 125–130.

Arscott, D. B., Tockner, K. and Ward, J. V. (2000). Aquatic habitat diversity along the corridor of an alpine floodplain river (Fiume Tagliamento, Italy). *Archiv für Hydrobiologie*, 149: 679–704.

Asselman, N. E. M. and Middelkoop, H. (1995). Floodplain sedimentation: Quantities, patterns and processes. *Earth Surface Processes and Landforms*, 20: 481–499.

Bak, P. and Paczuski, M. (1995). Complexity, contingency, and criticality. *Proceedings of the National Academy of Sciences of the United States of America*, **92**: 6689–6696.

Bak, P., Tang, C. and Wiesenfeld, K. (1988). Self-organized criticality. *Physical Review A*, 38: 364–374.

Bar Massada, A. and Radeloff, V. C. (2010). Two multi-scale contextual approaches for mapping spatial pattern. *Landscape Ecology*, 25: 711–725.

Brown, A. G., Harper, D. and Peterken, G. F. (1997). European floodplain forests: structure, functioning and management. *Global Ecology and Biogeography Letters*, 6: 169–178.

Carbonneau, P., Fonstad, M. A., Marcus, W. A. and Dugdale, S. J. (2012). Making riverscapes real. *Geomorphology*, 137: 74–86.

Chafiq, M., Gibert, J., Marmonier, P., *et al.* (1992). Spring ecotone and gradient study of interstitial fauna along two floodplain tributaries of the River Rhne, France. *Regulated Rivers: Research & Management*, 7: 103–115.

Church, M. (2002). Geomorphic thresholds in riverine landscapes. *Freshwater Biology*, 47: 541–557.

Clements, F. E. (1916). *Plant Succession: An Analysis of the Development of Vegetation*. Carnegie Institute, Washington, DC.

Connell, J. H. (1978). Diversity in tropical rain forests and coral reefs. *Science*, 199: 1302–1310.

Cooper, S. D., Leon, B., Sarnelle, O., *et al.* (1997). Quantifying spatial heterogeneity in streams. *Journal of the North American Benthological Society*, 16: 174–188.

Cushman, S. A., Gutzweiler, K., Evans, J. S. and McGarigal, K. (2010). 'The Gradient paradigm: a conceptual and analytical framework for landscape ecology'. In: *Spatial Complexity, Informatics, and Wildlife Conservation*. S. Cushman and F. Huettmann (eds). Springer, Japan: pp. 83–108.

Delong, M. D. and Thoms, M. C. (in press). An ecosystem framework for river sceicne and management.

De Jager, N. R. and Rohweder, J. J. (2012). Spatial patterns of aquatic habitat richness in the Upper Mississippi River floodplain, USA. *Ecological Indicators*, 13: 275–283.

Décamps, H. (1984). 'Towards a landscape ecology of river valleys'. In: *Trends in Ecological Research for the 1980's*. J. H. Cooley and F. B. Golley (eds). Plenum, New York: pp. 163–178.

Désilets, P. and Houle, G. (2005). Effects of resource availability and heterogeneity on the slope of the species-area curve along a floodplain-upland gradient. *Journal of Vegetation Science*, 16: 487–496.

Dollar, E. S. J., James, C. S., Rogers, K. H. and Thoms, M. C. (2007). A framework for interdisciplinary understanding of rivers as ecosystems. *Geomorphology*, 89: 147–162.

Evans, I. S. (1972). 'General geomorphometry, derivatives of altitude, and descriptive statistics'. In: *Spatial Analysis in Geomorphology*. R. J. Chorley (ed.). Methuen & Co., London: pp. 17–90.

Everitt, B. L. (1968). Use of the cottonwood in an investigation of the recent history of a flood plain. *American Journal of Science*, 266: 417–439.

Fonstad, M. A. and Marcus, W. A. (2010). High resolution, basin extent observations and implications for understanding river form and process. *Earth Surface Processes and Landforms*, 35: 680–698.

Forman, R. T. T. and Godron, M. (1981). Patches and structural components for a landscape ecology. *BioScience*, 31: 733–740.

Fuller, I. C. and Basher, L. R. (2013). Riverbed digital elevation models as a tool for holistic river management: Motueka River, Nelson, New Zealand. *River Research and Applications*, 29: 619–633.

Gallardo, A. (2003). Spatial variability of soil properties in a floodplain forest in northwest Spain. *Ecosystems*, 6: 564–576.

Glavac, V., Grillenberger, C., Hakes, W. and Ziezold, H. (1992). On the nature of vegetation boundaries, undisturbed flood plain forest communities as an example - a contribution to the continuum/discontinuum controversy. *Vegetatio*, 101: 123–144.

Goodchild, M. F. and Mark, D. M. (1987). The fractal nature of geographic phenomena. *Annals of the Association of American Geographers*, 77: 265–278.

Gustafson, E. J. (1998). Quantifying landscape spatial pattern: What is the state of the art? *Ecosystems*, 1: 143–156.

Hallet, B. (1990). Spatial self-organization in geomorphology: From periodic bedforms and patterned ground to scale-invariant topography. *Earth-Science Reviews*, 29: 57–75.

Hamilton, S. K., Kellndorfer, J., Lehner, B. and Tobler, M. (2007). Remote sensing of floodplain geomorphology as a surrogate for biodiversity in a tropical river system (Madre de Dios, Peru). *Geomorphology*, 89: 23–38.

Hawk, G. M. and Zobel, D. B. (1974). Forest succession on alluvial landforms of the McKenzie River Valley, Oregon. *Northwest Science*, 48: 245–265.

Hobson, R. D. (1972). 'Surface roughness in topography: a quantitative approach'. In: *Spatial Analysis in Geomorphology*. R. J. Chorley (ed.). Methuen & Co., London: pp. 221–246.

Hoechstetter, S., Walz, U., Dang, L. H. and Thinh, N. X. (2008). Effects of topography and surface roughness in analyses of landscape structure - A proposal to modify the existing set of landscape metrics. *Landscape Online*, 3: 1–14.

Holling, C. S. (1992). Cross-scale morphology, geometry, and dynamics of ecosystems. *Ecological Monographs*, 62: 447–502.

Hughes, F. M. R. (1990). The influence of flooding regimes on forest distribution and composition in the Tana River floodplain, Kenya. *Journal of Applied Ecology*, 27: 475–491.

Hughes, F. M. R. (1997). Floodplain biogeomorphology. *Progress in Physical Geography*, 21: 501–529.

Hughes, J. W. and Cass, W. B. (1997). Pattern and process of a floodplain forest, Vermont, USA: predicted response of vegetation to perturbation. *Journal of Applied Ecology*, 34: 594–612.

Iwahashi, J. and Pike, R. J. (2007). Automated classifications of topography from DEMs by an unsupervised nested-means algorithm and a three-part geometric signature. *Geomorphology*, 86: 409–440.

Jenness, J. S. (2004). Calculating landscape surface area from digital elevation models. *Wildlife Society Bulletin*, 32: 829–839.

Jenness, J. S. (2012). DEM Surface Tools. Jenness Enterprises, [Online] http://www.jennessent.com/arcgis/surface&uscore;area.htm

Kalliola, R. and Puhakka, M. (1988). River dynamics and vegetation mosaicism: a case study of the River Kamajohka, northernmost Finland. *Journal of Biogeography*, 15: 703–719.

Kuffner, I. B., Brock, J. C., Grober-Dunsmore, R., *et al.* (2007). Relationships between reef fish communities and remotely sensed rugosity measurements in Biscayne National Park, Florida, USA. *Environmental Biology of Fishes*, 78: 71–82.

Legleiter, C. J. (2013). A geostatistical framework for quantifying the reach-scale spatial structure of river morphology: 1. Variogram models, related metrics, and relation to channel form. *Geomorphology*, 205: 65–84.

Legleiter, C. J. and Overstreet, B. T. (2013). Retreiving river attributes from remoteley sensed data: an experimental evaluation based on field spectroscopy at the outdoor stream lab. *River Research and Applications* (Online).

Levin, S. A. (1992). The problem of pattern and scale in ecology. *Ecology*, 73: 1943–1943.

Levin, S. A. (1998). Ecosystems and the biosphere as complex adaptive systems. *Ecosystems*, 1: 431–436.

Malard, F., Tockner, K. and Ward, J. V. (2000). Physico-chemical heterogeneity in a glacial riverscape. *Landscape Ecology*, 15: 679–695.

Manning, A. D., Lindenmayer, D. B. and Nix, H. A. (2004). Continua and umwelt: Novel perspectives on viewing landscapes. *Oikos*, 104: 621–628.

Mark, D. M. (1975). Geomorphometric parameters: a review and evaluation. *Geografiska Annaler. Series A, Physical Geography*, 57: 165–177.

McCormick, M. I. (1994). Comparison of field methods for measuring surface topography and their associations with a tropical reef fish assemblage. *Marine Ecology Progress Series*, 112: 87–96.

McGarigal, K., Tagil, S. and Cushman, S. (2009). Surface metrics: an alternative to patch metrics for the quantification of landscape structure. *Landscape Ecology*, 24: 433–450.

Mertes, L. A. K. (2002). Remote sensing of riverine landscapes. *Freshwater Biology*, 47: 799–816.

Milne, B. T., Turner, M. G., Wiens, J. A. and Johnson, A. R. (1992). Interactions between the fractal geometry of landscapes and allometric herbivory. *Theoretical Population Biology*, 41: 337–353.

Murray, A. B., Coco, G. and Goldstein, E. B. (2014). Cause and effect in geomorphic systems: Complex systems perspectives. *Geomorphology*, 214: 1–9.

Naiman, R. J., Décamps, H., Pastor, J. and Johnston, C. A. (1988). The potential importance of boundaries of fluvial ecosystems. *Journal of the North American Benthological Society*, 7: 289–306.

Nicholas, A. P. and Mitchell, C. A. (2003). Numerical simulation of overbank processes in topographically complex floodplain environments. *Hydrological Processes*, 17: 727–746.

Nogami, M. (1995). Geomorphometric measures for digital elevation models. *Zeitschrift für Geomorphologie, Supplementband* 101: 53–67.

Notebaert, B., Verstraeten, G., Govers, G. and Poesen, J. (2009). Qualitative and quantitative applications of LiDAR imagery in fluvial geomorphology. *Earth Surface Processes and Landforms*, 34: 217–231.

O'Neill, R. V., Gardner, R. H., Milne, B. T., *et al.* (1991). 'Heterogeneity and Spatial Hierarchies'. In: *Ecological Heterogeneity*. J. Kolasa and S. T. A.

Pickett (eds). Springer-Verlag, New York: pp. 85–96.

Parsons, M., Thoms, M. C. and Norris, R. H. (2004). Using hierarchy to select scales of measurement in multiscale studies of stream macroinvertebrate assemblages. *Journal of the North American Benthological Society*, 23: 157–170.

Phillips, J. D. (2003). Sources of nonlinearity and complexity in geomorphic systems. *Progress in Physical Geography*, 27: 1–23.

Pielou, E. C. (1975). *Ecological Diversity*. John Wiley & Sons, New York.

Pollock, M. M., Naiman, R. J. and Hanley, T. A. (1998). Plant species richness in riparian wetlands - a test of biodiversity theory. *Ecology*, 79: 94–105.

Rietkerk, M., Dekker, S. C., de Ruiter, P. C. and van de Koppel, J. (2004). Self-organized patchiness and catastrophic shifts in ecosystems. *Science*, 305: 1926–1929.

Riitters, K. H., Wickham, J. D., O'Neill, R. V., *et al.* (2002). Fragmentation of continental United States forests. *Ecosystems*, 5: 0815–0822.

Robertson, G. P. and Gross, K. L. (1994). 'Assessing the heterogeneity of belowground resources: quantifying pattern and scale'. In: *Exploitation of Environmental Heterogeneity by Plants: Ecophysiological Processes Above- and Belowground*. M. M. Caldwell and R. W. Pearcy (eds). Academic Press, San Diego, CA: pp. 237–253.

Salo, J., Kalliola, R., Häkkinen, I., *et al.* (1986). River dynamics and the diversity of Amazon lowland forest. *Nature*, 322: 254–258.

Schumm, S. A. (1998). *To Interpret the Earth: Ten Ways to Be Wrong*. Cambridge University Press, Cambridge.

Scown, M. W., Thoms, M. C. and De Jager, N. R. (2015). An index of floodplain surface complexity. *Hydrology and Earth System Sciences Discussions*, 12: 4507–4540.

Shelford, V. E. (1954). Some lower Mississippi valley flood plain biotic communities: their age and elevation. *Ecology*, 35: 126–142.

Southwell, M. and Thoms, M. (2006). 'A gradient or mosaic of patches? the textural character of inset-flood plain surfaces along a dryland river system'. In: *Sediment Dynamics and the Hydromorphology of Fluvial Systems*. IAHS Press, Wallingford, UK, pp. 487–495.

Sparks, R., Bayley, P., Kohler, S. and Osborne, L. (1990). Disturbance and recovery of large floodplain rivers. *Environmental Management*, 14: 699–709.

Stanford, J. A., Lorang, M. S. and Hauer, F. R. (2005). The shifting habitat mosaic of river ecosystems. *Verhandlungen des Internationalen Verein Limnologie*, 29: 123–136.

Statzner, B., Resh, V. H. and Dolédec, S. (eds) (1994). Ecology of the Upper Rhône River: a test of habitat templet theories. *Freshwater Biology*, 31: 253–556.

Tarolli, P., Sofia, G. and Dalla Fontana, G. (2012). Geomorphic features extraction from high-resolution topography: landslide crowns and bank erosion. *Natural Hazards*, 61: 65–83.

Thoms, M. C. (2003). Floodplain–river ecosystems: lateral connections and the implications of human interference. *Geomorphology*, 56: 335–349.

Thoms, M. and Parsons, M. (2011). Patterns of vegetation community distribution in a large, semi-arid floodplain landscape. *River Systems*, 19: 271–282.

Thoms, M. C., Reid, M., Christianson, K. and Munro, F. (2006). 'Variety is the spice of river life: recognizing hydraulic diversity as a tool for managing flows in regulated rivers'. In: *Sediment Dynamics and the Hydromorphology of Fluvial Systems*. IAHS Press, Wallingford, UK: pp. 169–178.

Thorp, J. H., Thoms, M. C. and Delong, M. D. (2008). *The Riverine Ecosystem Synthesis: Towards Conceptual Cohesiveness in River Science*. Elsevier, Amsterdam.

Tockner, K., Ward, J. V., Arscott, D. B., *et al.* (2003). The Tagliamento River: a model ecosystem of European importance. *Aquatic Sciences - Research Across Boundaries*, 65: 239–253.

Tockner, K., Ward, J. V., Edwards, P. J. and Kollmann, J. (2002). Riverine landscapes: an introduction. *Freshwater Biology*, 47: 497–500.

Townsend, P. A. and Butler, D. R. (1996). Patterns of landscape use by beaver on the lower Roanoke River floodplain, North Carolina. *Physical Geography*, 17: 253–269.

Turcotte, D. L. (2007). Self-organized complexity in geomorphology: observations and models. *Geomorphology*, 91: 302–310.

Turner, L. M. (1934). Grassland in the floodplain of Illinois rivers. *American Midland Naturalist*, 15: 770–780.

Turner, M. G. (1989). Landscape ecology: the effect of pattern on process. *Annual Review of Ecology and Systematics*, 20: 171–197.

Turner, M. G., O'Neill, R. V., Gardner, R. H. and Milne, B. T. (1989). Effects of changing spatial scale on the analysis of landscape pattern. *Landscape Ecology*, 3: 153–162.

Turner, S. J., O'Neill, R. V., Conley, W., *et al.* (1990). 'Pattern and scale: statistics for landscape ecology'. In: *Quantitative Methods in Landscape Ecology: The Analysis and Interpretation of Landscape Heterogeneity*. M. G. Turner and R. H. Gardner (eds). Springer-Verlag, New York, pp. 17–49.

UMESC (2012). *The Long Term Resource Monitoring Program (LTRMP) Land Cover/Use Data*. Upper Midwest Environmental Sciences Center, La Crosse, WI. [Online] http://www.umesc.usgs.gov/data&uscore;library/land&uscore;cover&uscore;use/land&uscore;cover&uscore;use&uscore;data.html

UMESC (2013). *Light Detection And Ranging (LiDAR) Elevation Data*. Upper Midwest Environmental Sciences Center, La Crosse, WI. [Online] http://www.umesc.usgs.gov/mapping/resource&uscore;mapping&uscore;ltrmp&uscore;lidar.html

USGS (2013). *USGS Surface-Water Monthly Statistics for Iowa: USGS 05389500 Mississippi River at McGregor, IA*. U.S. Geological Survey WaterWatch, [Online] http://waterdata.usgs.gov/ia/nwis/monthly?referred&uscore;module=sw&search&uscore;site&uscore;no=05389500&format=sites&uscore;selection&uscore;links

van Coller, A. L., Rogers, K. H. and Heritage, G. L. (2000). Riparian vegetation-environment relationships: complimentarity of gradients versus patch hierarchy approaches. *Journal of Vegetation Science*, 11: 337–350.

Walker, B. K., Jordan, L. K. B. and Spieler, R. E. (2009). Relationship of reef fish assemblages and topographic complexity on southeastern Florida coral reef habitats. *Journal of Coastal Research*, 25: 39–48.

Walling, D. E. and He, Q. (1998). The spatial variability of overbank sedimentation on river floodplains. *Geomorphology*, 24: 209–223.

Wallis, C., Maddock, I., Visser, F. and Acreman, M. (2012). A framework for evaluating the spatial configuration and temporal dynamics of hydraulic patches. *River Research and Applications*, 28: 585–593.

Ward, J. V., Malard, F. and Tockner, K. (2002). Landscape ecology: a framework for integrating pattern and process in river corridors. *Landscape Ecology*, 17: 35–45.

Ward, J. V., Tockner, K. and Schiemer, F. (1999). Biodiversity of floodplain river ecosystems: Ecotones and connectivity. *Regulated Rivers: Research & Management*, 15: 125–139.

Whited, D. C., Lorang, M. S., Harner, M. J., *et al.* (2007). Climate, hydrologic disturbance, and succession: drivers of floodplain pattern. *Ecology*, 88: 940–953.

Wiens, J. A. (2002). Riverine landscapes: taking landscape ecology into the water. *Freshwater Biology*, 47: 501–515.

Wiens, J. A., Milne, B. T. (1989). Scaling of 'landscapes' in landscape ecology, or, landscape ecology from a beetle's perspective. *Landscape Ecology*, 3: 87–96.

Wilson, M. F. J., O'Connell, B., Brown, C., *et al.* (2007). Multiscale terrain analysis of multibeam bathymetry data for habitat mapping on the continental slope. *Marine Geodesy*, 30: 3–35.

CHAPTER 7

Trees, wood and river morphodynamics: results from 15 years research on the Tagliamento River, Italy

Angela M. Gurnell
School of Geography, Queen Mary, University of London, London, UK

Introduction

Trees, wood and river morphodynamics: a context

The physical character of rivers and their margins depends upon the processes of fluvial sediment transfer from headwaters to mouth and between the main river channel and its river corridor. These transfers influence the morphodynamics of the transitional zone between the low flow channel and the surrounding hillslopes. They result in a wide range of river channel and floodplain styles that have been related to the valley gradient, properties of the river's flow regime, and the calibre and quantity of sediment transported by the river (e.g., Leopold and Wolman, 1957; Schumm, 1977, 1985; Church,1992, 2002; Nanson and Croke, 1992).

Since the early 1980s, researchers have presented increasing evidence that vegetation is also important for river and floodplain morphodynamics. Initially, associations were recognised between the frequency and duration of inundation of a suite of river corridor landforms and the riparian plant communities that grew on them (e.g., Hupp and Osterkamp, 1985). Such associations reflect the fact that many riparian plant species depend on fluvial processes for seed and vegetative propagule dispersal (e.g., Mahoney and Rood, 1998; Merritt and Wohl, 2002; Gurnell *et al.*, 2008; Greet *et al.*, 2011) and for moist, wet or waterlogged soils to support their germination and growth (e.g., Lite and Stromberg, 2005; Williams and Cooper, 2005; Pezeshki and Shields, 2006; Gonzalez *et al.*, 2012). They also reflect the fact that riparian species are able to cope with the flow shear stresses, erosion and sedimentation disturbances that are found within river margins.

Riparian plants both affect and respond to fluvial processes. They affect the flow field (e.g., Liu *et al.*, 2010; Bennett *et al.*, 2008) and thus sediment retention and transfer (e.g., Prosser *et al.*, 1995; Ishikawa *et al.*, 2003). Plant roots and rhizomes influence mechanical and hydraulic soil properties (e.g., Docker and Hubble, 2008; Pollen-Bankhead and Simon, 2010), and thus the stability, erosion resistance and soil moisture regime of river margin landforms.

River Science: Research and Management for the 21st Century, First Edition.
Edited by David J. Gilvear, Malcolm T. Greenwood, Martin C. Thoms and Paul J. Wood.

Large wood produced by riparian forests, can also protect, reinforce and stabilise landforms (e.g., Abbe and Montgomery, 2003; Gurnell *et al.*, 2005, 2012; Collins *et al.*, 2012).

In recent years, these important interactions and feedbacks between riparian vegetation and fluvial processes, and their effect on the character and dynamics of the riparian habitat mosaic, have been explored in several major reviews (e.g., Corenblit *et al.*, 2007, 2009; Gurnell and Petts, 2011; Gurnell, 2012; Gurnell *et al.*, 2012; Osterkamp *et al.*, 2012; Camporeale *et al.*, 2013), underpinning the emerging field of fluvial biogeomorphology.

Trees, wood and river morphodynamics: an early conceptual model

This chapter synthesises research conducted since the late 1990s on the middle and lower reaches of the Tagliamento River (68 to 127 km from the river's source). The synthesis focuses on interactions between trees, wood and fluvial processes, and their consequences for river morphodynamics, using a conceptual model of island development as a framework for the synthesis (Figure 7.1, Gurnell *et al.*, 2001).

The conceptual model proposes that three broad categories of tree-related roughness elements contribute to the initiation of island development (seedlings, dead wood, and living (regenerating) wood). These are incorporated in three trajectories of vegetation growth (Figure 7.1a) on open bar surfaces. Trajectory (a) is initiated by dispersed seed germination across open gravel bar surfaces. Trajectory (b) is initiated by seed germination and regeneration from small pieces of living wood that accumulate with finer sediments in the lee of large (dead) wood accumulations. Trajectory (c) is initiated by regeneration from large living pieces of wood (often entire uprooted trees). All three trajectories involve interaction between the establishing woody vegetation and fluvial processes of erosion and deposition. Trajectory (c) involves the most rapid rates of vegetation growth, retention/aggradation of finer sediment, and development of root-reinforced, erosion resistant, vegetated landforms. Trajectory (a) shows the slowest rates of vegetation growth, sediment retention and landform development.

When these trajectories are set within the context of flood disturbances (Figure 7.1b), the model proposes that trajectory (a) is very unlikely to lead to the development of islands because the relatively slow growing dispersed seedlings are easily uprooted or buried by fluvial processes before they are able to develop into sizeable plants. Trajectory (c), which supports the most rapid vegetation growth, is most likely to resist flood disturbance and trap sediments to support rapid pioneer island development and coalescence to form building islands and, eventually, established islands (Figure 7.1a). Trajectory (b) has an intermediate chance of contributing to established island development rather than succumbing to removal of the vegetated patches and landforms by fluvial processes (Figure 7.1a and b). The relative success of the three trajectories in contributing to island development reflects their different rates of initial above- and below-ground vegetation growth, ability to trap and stabilise finer sediment, and to resist erosion/removal by the sequence of fluvial disturbances to which they are subjected. The same trajectories contribute to the expansion of building and established islands, and also islands dissected from the floodplain by avulsions, leading to the production of complex islands (Figure 7.1a).

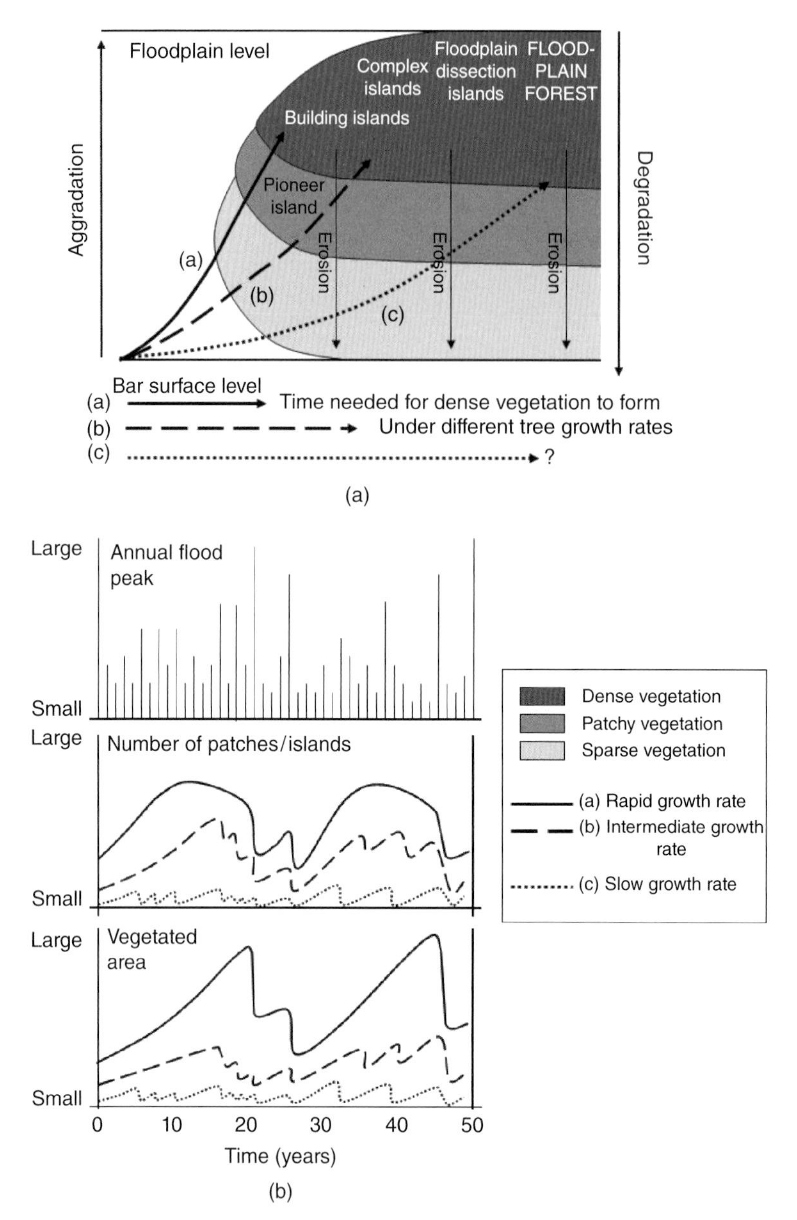

Figure 7.1 A conceptual model of island development (after Gurnell *et al.*, 2001). (a) Different rates of aggradation and island development (from bare bar surface through pioneer, building and established island development) according to different growth trajectories a, b and c (for explanation see text). (b) Changes in the number and area of islands under each of the three vegetation growth trajectories (a, b, c) in response to the same sequence of annual floods.

The Tagliamento River

The main stem of the Tagliamento rises close to the Passo della Mauria (1298 m.a.s.l.) in the southern fringe of the European Alps and drains approximately 170 km to the Adriatic Sea. The climate changes from alpine to mediterranean along the river's course. The river has a flashy flow regime: floods can occur at any time, but are concentrated in spring and autumn as a result of, respectively, snow-melt and thunderstorms.

Apart from its most downstream section, the river is not closely confined by flood embankments. It shows strong downstream changes in valley slope, discharge and bed sediment calibre (Figure 7.2, Gurnell *et al.*, 2000a; Petts *et al.*, 2000), and is bordered by riparian woodland, with distinct downstream changes in dominant tree species (Figure 7.3, Karrenberg *et al.*, 2003).

Karrenberg *et al.* (2003) surveyed samples of five 50 m^2 vegetated patches located within the active tract and spaced every 10 km along the main stem. They found a downstream reduction in woody species richness and average patch age (Figure 7.3a), with distinct variations in the basal area of the woody species along the river (Figure 7.3b). *Alnus incana* and *Salix eleagnos* dominated the headwaters, whereas *Populus nigra* was found along the middle and lower reaches (Figure 7.3c).

Populus nigra and several willow species that are present along the Tagliamento (*Salix alba, S. daphnoides, S. elaeagnos, S. purpurea, S. triandra*) are all members of the Salicaceae family and regenerate freely from deposited uprooted trees and wood fragments, whereas *Alnus incana* (Betulaceae) regenerates less readily in this way. This partly explains the transition from predominantly dead-wood deposits in the headwater reaches to widespread regeneration from deposited wood in the middle and lower reaches (Gurnell *et al.*, 2000b, Figure 7.4).

By focusing on reaches between 68 and 127 km from the river's source, this synthesis refers to reaches dominated by *Populus nigra* (Figure 7.3), where a large proportion of wood sprouts following deposition within the active tract (Figure 7.4), and the river is unconfined by embankments. In this chapter, the section on growth of riparian trees in disturbed riparian environments reviews elements of the lifecycle of the alluvial Salicaceae (poplars and willows) that are significant for the performance of this family as river ecosystem engineers that influence river morphodynamics, and then presents observations of the growth performance of *Populus nigra* along the Tagliamento. The section entitled 'Flow disturbance and vegetation cover' considers the impact of fluvial processes on vegetation dynamics and the following section presents evidence to support the role of vegetation in influencing river morphology. The final section considers the impact of changes in fine sediment supply, wood supply and flow regime on the performance of the conceptual model (Figure 7.1).

Growth of riparian trees in disturbed riparian environments

The alluvial Salicaceae

Riparian zones within the Northern Hemisphere are dominated by species from the Salicaceae (willow and poplar) family, and this is certainly true for the Tagliamento River. Alluvial Salicaceae species are pioneer woody species that have morphological, biomechanical and reproductive characteristics that make them particularly suited to disturbed riparian environments (Karrenberg *et al.*, 2002).

Alluvial Salicaceae species reproduce freely both sexually and vegetatively. In spring, enormous numbers of small, light seeds are produced (Braatne *et al.*, 1996; Imbert and Lefèvre, 2003; Karrenberg and Suter, 2003), which have a very short period

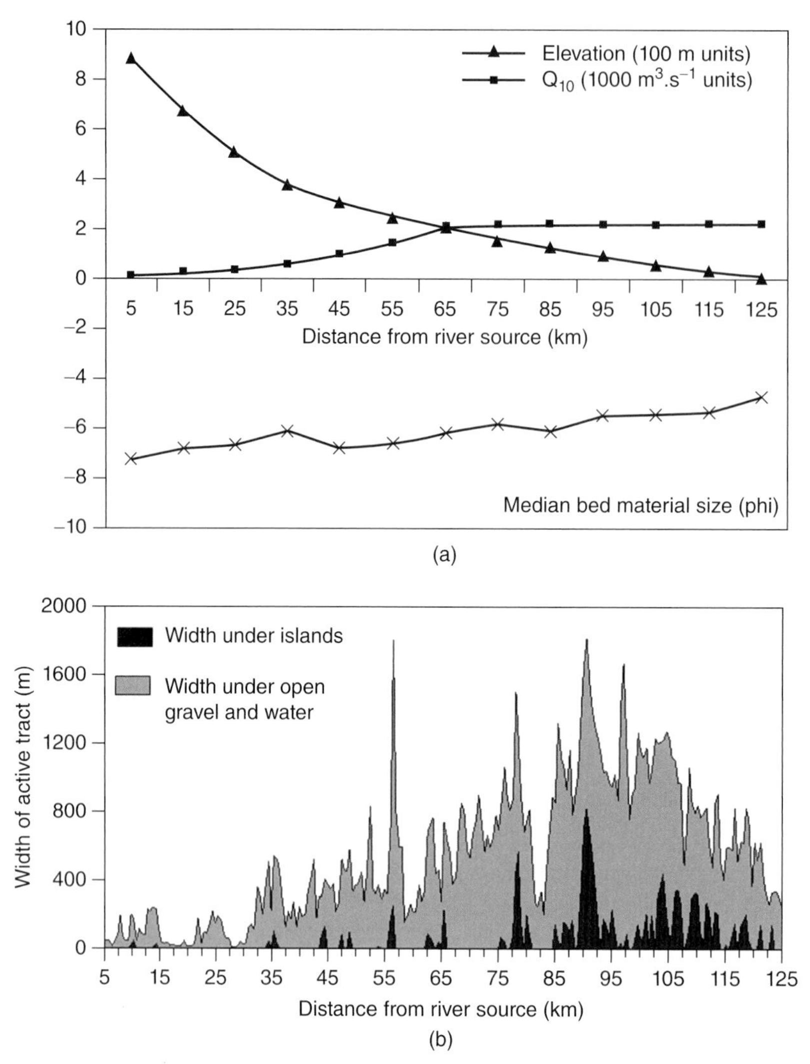

Figure 7.2 Downstream variations in river corridor properties along the Tagliamento main stem. (a) elevation, Q_{10} (method described in Gurnell *et al.*, 2000a), and median particle size of the coarsest patch of main channel edge sediment (method described in Petts *et al.*, 2000; values interpolated to 10 km intervals using partly unpublished data). (b) Width of the active tract at 0.5 km downstream intervals, subdivided into width under islands and width under open gravel and water (method described in Gurnell *et al.*, 2000a).

of viability. Karrenberg and Suter (2003) observed loss of viability in 75% seeds retained in dry storage after approximately 30 days for *P. nigra*, 16 to 18 days for *S. daphnoides* and *S. eleagnos*, and 9 to 12 days for *S. alba*, *S. purpurea* and *S. triandra*. Loss of viability in the field is likely to be faster. Each species produces seed within a brief time window. In 2000, Karrenberg and Suter (2003) observed a sequence of overlapping seeding periods between mid-April and early-June for *S. daphnoides*, *S.purpurea*, *S. eleagnos*, *S. triandra*, *S. alba* and *P. nigra* at their study site on the middle Tagliamento. Seed production by each species lasted two to three weeks.

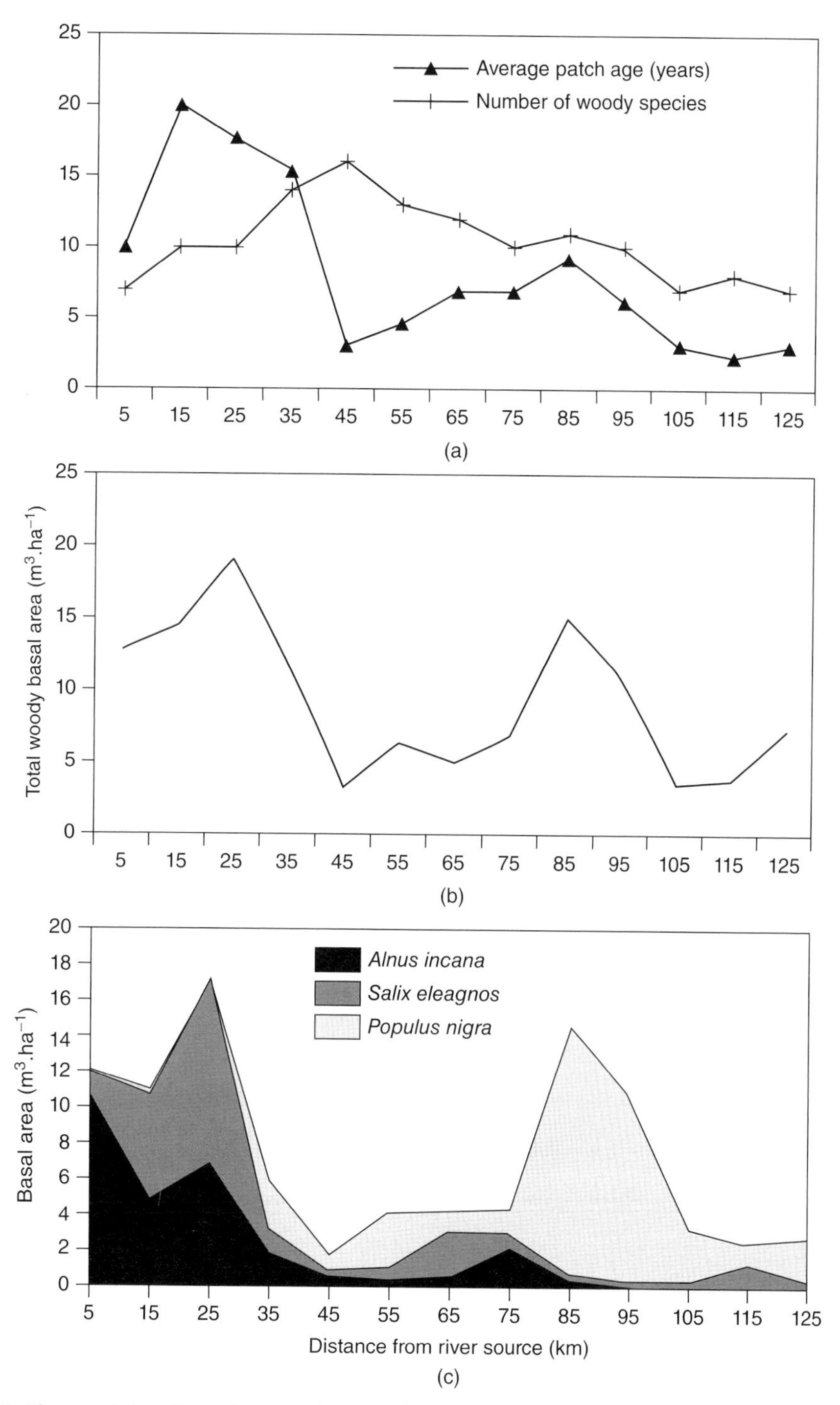

Figure 7.3 Characteristics of woody vegetation at 10 km intervals along the Tagliamento main stem. At each site, measurements were obtained within 5 × 50 m^2 plots located on vegetated patches within the active tract (data from Karrenberg *et al.*, 2003). (a) average age of oldest tree within each of the 5 plots and number of woody species present; (b) basal area of all woody species; (c) basal area of *A. incana*, *S. eleagnos* and *P. nigra*.

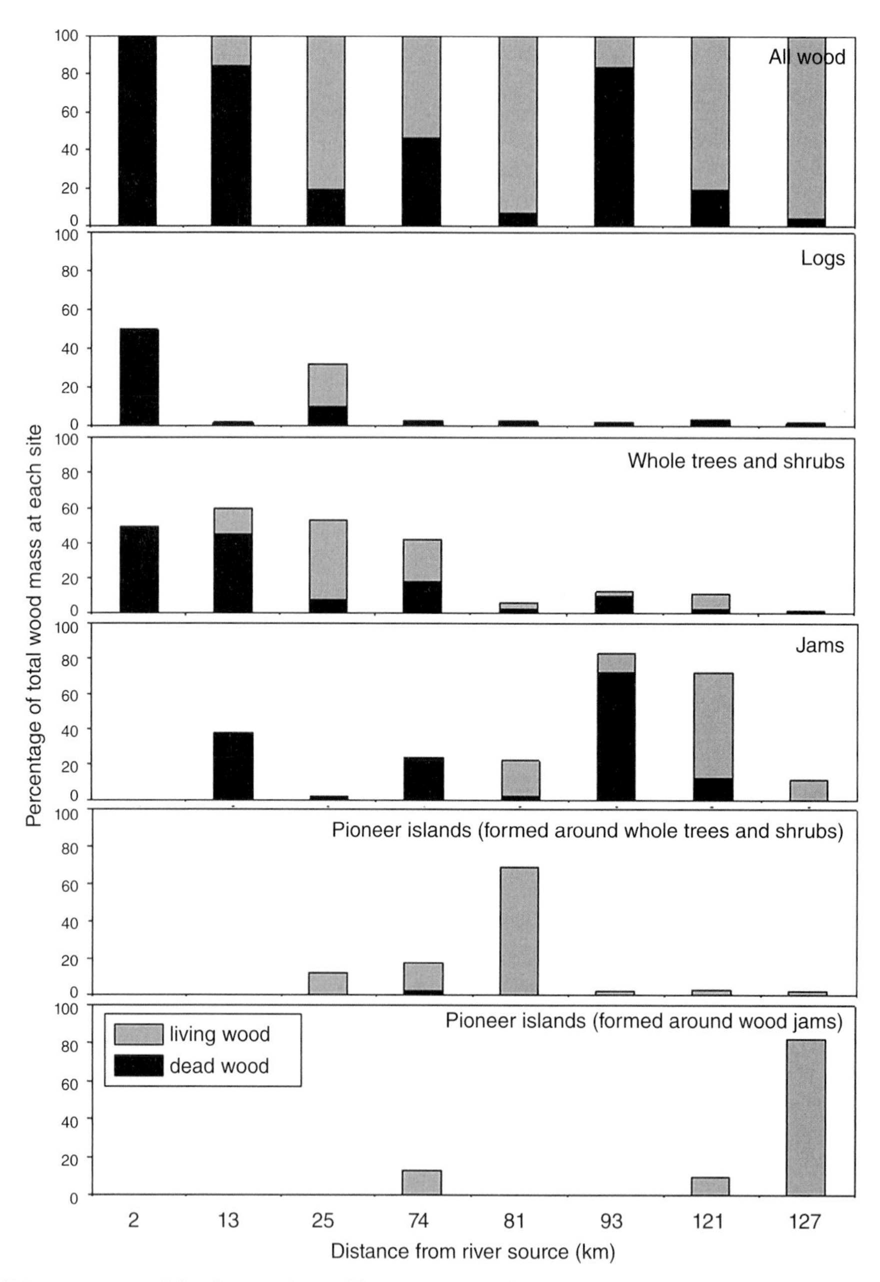

Figure 7.4 Percentage of the deposited wood biomass exposed on the surface of the active tract that is dead (black) or sprouting/alive (grey) at eight sites along the Tagliamento main stem (data from Gurnell *et al.*, 2000b). Data are presented for the total exposed wood biomass (top) and for different components of the biomass, illustrating a downstream trend in the proportions of the wood according to type and whether dead or sprouting.

The tiny, light seeds are widely dispersed by wind and water. If they are deposited on moist, bare sediment, they germinate almost immediately, resulting in spatially discrete areas of seedlings that reflect river levels at the time of seed dispersal for each species. Seedlings grow quickly following germination if the alluvial water table does not fall too rapidly and if the young plants are not disturbed by flooding or extreme drought during the first year or two of their development. The ideal soil moisture regime varies between species.

Due to the short period of seed viability, specific germination and early growth requirements, and high sensitivity of seedlings to flood or drought stress, few seedlings grow to maturity. Mahoney and Rood (1998) proposed a 'recruitment box' model that defined the river levels and rates of river level decline required at the time of seed dispersal for successful recruitment of the Salicaceae. Understanding of interrelationships between topographic position, alluvial sediment calibre and flow regime properties in relation to Salicaceae recruitment has advanced greatly in the last decade (e.g., Amlin and Rood, 2002; Guilloy-Froget *et al.*, 2002; Lytle and Merritt, 2004; Ahna *et al.*, 2007; Merritt *et al.*, 2010), allowing river flow regimes to be designed to promote recruitment of particular species in regulated systems (e.g., Hughes and Rood, 2003; Rood *et al.*, 2005).

The Salicaceae also reproduce asexually. This can occur at any time during the growing season, with regeneration observed from small or large vegetative fragments or entire uprooted trees. From a geomorphological perspective, these fragments can be viewed as 'living wood' (Gurnell *et al.*, 2001), since they can take on the geomorphological role and dynamics of dead wood, but can then gain additional anchorage and above-ground biomass by sprouting roots and shoots. An ability to produce adventitious roots is particularly important in dynamic riparian environments, since this gives these species a high tolerance of burial and resistance to uprooting, and has important implications for alluvial sediment retention, reinforcement and stabilisation. Young plants have relatively flexible canopies, allowing them to bend and reduce their flow resistance during floods. Therefore, asexual reproduction is most commonly initiated from established trees, whose canopy is relatively rigid and more susceptible to breakage, and whose substantial internal resources can support the early stages of regeneration in a range of environmental conditions.

In suitable soil moisture conditions, growth rates of seedlings and sprouting rates from vegetative fragments can be extremely high, allowing plants to establish quickly and gain root anchorage. Up to 3 mm day^{-1} main shoot growth in *Populus nigra*, *Salix alba* and *Salix eleagnos* seedlings, 10 mm day^{-1} main shoot growth in cuttings, and 15 mm day^{-1} shoot growth from uprooted deposited trees have been observed on the Tagliamento River (Francis and Gurnell, 2006; Moggridge and Gurnell, 2009). Root growth is also rapid, with average daily increments in vertical root penetration of sand and gravel substrates of 27 and 20 mm, respectively, for *Salix eleagnos*, and 15 and 10 mm, respectively, for *Populus nigra*, observed in experiments where the water table was manipulated to decline at a rate of 3 cm day^{-1} (Francis *et al.*, 2005).

Laboratory and field experiments on seedlings and cuttings and field observations of established shrubs and trees have also demonstrated that different species of the Salicaceae show varying tolerances to hydrological conditions such as inundation and flood disturbance (Amlin and Rood, 2001; Glenz *et al.*, 2006), and depth to water table and drought (Amlin and Rood, 2002, 2003). This sensitivity to hydrological processes is expressed in the distribution and growth performance of different riparian tree species within the riparian zone (e.g., Dixon *et al.*, 2002; Cooper *et al.*, 2006; Turner *et al.*, 2004; Friedman *et al.*, 2006; Robertson, 2006).

Riparian tree growth: observations from the Tagliamento River

Riparian tree growth performance is a key factor in the conceptual model (Figure 7.1). Many experiments and observations of riparian tree growth performance have been conducted along the Tagliamento, providing information on the growth rate of seedlings, cuttings and deposited trees of different species under varying (temporal and spatial) environmental conditions. In this section, relevant data are presented to characterise variations in growth performance along the middle and lower Tagliamento between 71 and 127 km from the source, focusing on the dominant riparian tree species in this section of the river: *P. nigra*.

Initial growth of *P. nigra* varies greatly according to propagule type (Figure 7.5). The figure provides box plots of the average daily growth achieved during the first growing season at a site 79 km from the source by the longest shoot of seedlings; 40 cm long 5–6 mm diameter cuttings; and entire uprooted, deposited trees (average length = 14.2 m, average diameter at 1 m above the root wad = 17 cm). In all cases the measurements were obtained from bar top locations representing sand to coarse gravel surface sediments, and were collected during 2003 and 2004. Initial growth is stronger from vegetative propagules than from seeds, with uprooted trees showing

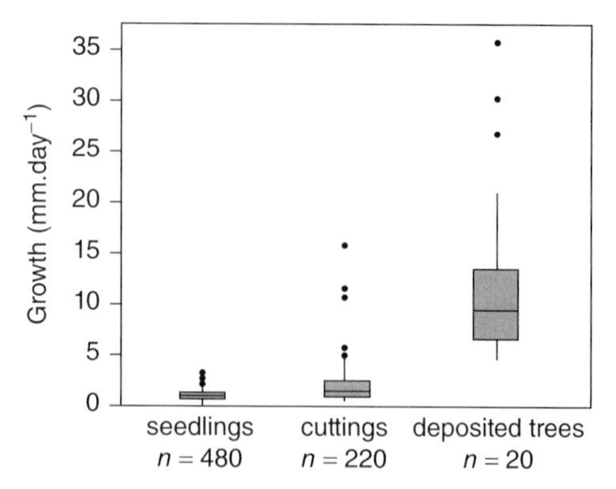

Figure 7.5 Average daily growth rates of *P. nigra* during the first growing season, based on the length of the main shoot from seedlings and cuttings and the longest of 10 shoots spaced evenly along the trunk of uprooted trees. All sampled individuals were growing on open bar tops on surface sediments ranging from silty-sand to coarse gravel. All measurements were taken at a site 79 km downstream from the source of the Tagliamento during the 2003 and 2004 growing seasons.

an order of magnitude larger shoot growth than small cuttings.

Growth rates of *P. nigra* at the same site after the first growing season are illustrated in Figure 7.6. The age of each plant was estimated by counting annual growth segments of the main shoot (younger trees) and annual growth rings at 1 m above-ground level (older trees). There is considerable scatter in annual growth increments between trees of the same age, but there appears to be an increasing average annual growth rate for older trees (up to the maximum 24 years of the sampled trees).

Figures 7.5 and 7.6 provide an insight into growth rates from different propagule types and at different growth stages (ages), but they give no indication of growth variability across space and time in response to changing environmental conditions, particularly moisture availability. In order to explore spatiotemporal variations in growth rate, measurements of 20 × 3 m tall trees

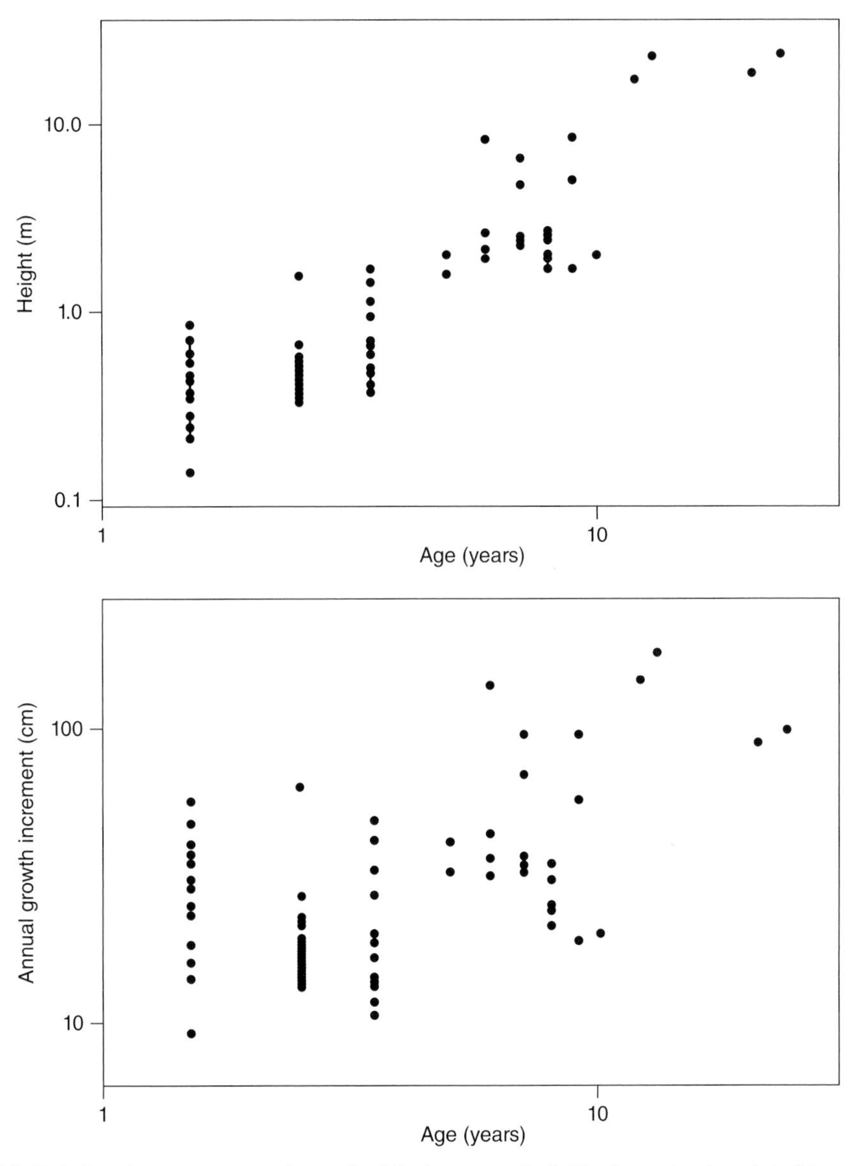

Figure 7.6 Variations in average annual growth of *P. nigra* across individuals ranging from 2 to 24 years old at a site 79 km from the source of the Tagliamento (observations collected in 2004 and 2005).

(+/- ca. 40 cm) were obtained from 15 sites along the Tagliamento River between 71 and 127 km from the source. Small (3 m) trees were selected to ensure that growth performance reflected recent growing conditions (the previous few years). The sampled trees were located on high bar tops (to control for topographic position) and each tree was isolated from surrounding trees (to control for light receipt and competition for water); 12 of the 15 sites were sampled in 2005 and 14 in 2010. Tree age was determined from growth rings at 1 m above the ground surface and the average annual growth rate

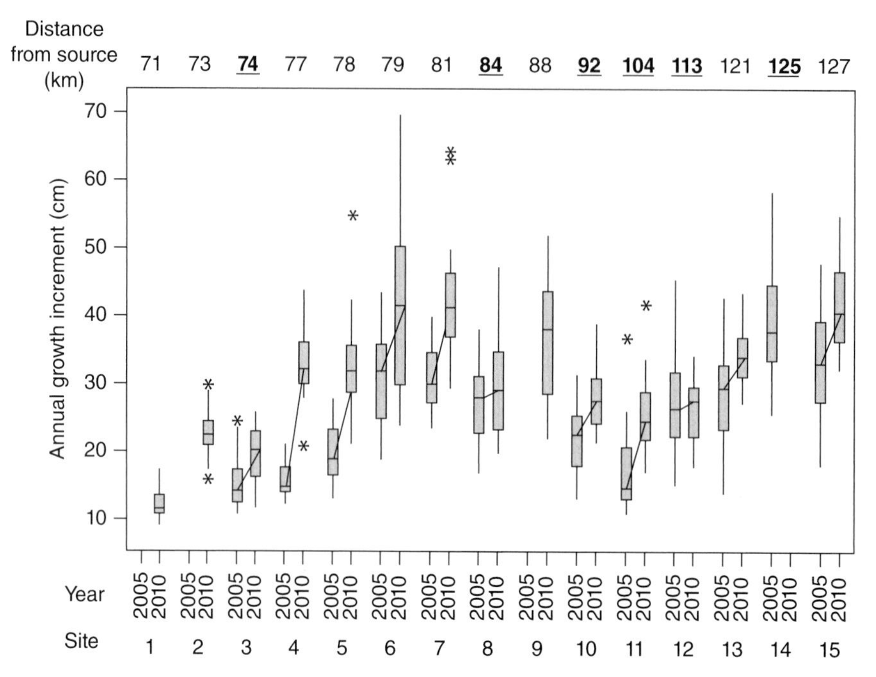

Figure 7.7 Box and whisker plots of the annual growth increments, measured in 2005 and 2010, of samples of 20 3-m tall *P. nigra* located at 15 sites along the Tagliamento between 71 and 127 km from the river's source.

was determined by dividing tree height in excess of 1 m by the number of growth rings. These measurements revealed a clear downstream pattern in annual growth rates (Figure 7.7 and 7.8), rising from very low rates at site 1 (71 km) to a maximum at sites 6 (79 km), where the data for Figures 7.5 and 7.6 were obtained) and 7 (81 km), then decreasing to site 11 (104 km) and increasing again to site 15 (127 km). These spatial changes in growth rates of the same tree species correspond to hydrological changes between 71 and 127 km from the source. Low flow discharges rise between sites 1 (71 km) and 7 (81 km) as three tributary streams join the main stem and water is funnelled by the surrounding mountains into a narrow gorge (130 m wide) at 83 km, inducing groundwater upwelling from the alluvial aquifer. Downstream between sites 7 (81 km) and 11 (104 km), there are no significant tributary confluences and the river loses water to a vast alluvial aquifer, leading to a downstream decrease in low flows with the river frequently drying out around site 11 (104 km) during the summer. Between sites 11 and 12 (104 to 113 km), in association with a regional spring line, low flows increase (Doering *et al.*, 2007). The downstream trends in *P. nigra* growth rates correlate closely with these downstream hydrological changes.

At the 11 sites where measurements were made in 2005 and 2010, the median of the average growth rates observed in 2010 was always greater than in 2005 (Figure 7.7). Two-way analysis of variance (ANOVA) applied to the observations from these 11 sites explained over 60% of the variation in growth rates, with significant differences in growth rates between sites ($p < 0.001$), and years ($p < 0.001$) and with a significant interaction between years and sites ($p < 0.001$), indicating spatial variations in the

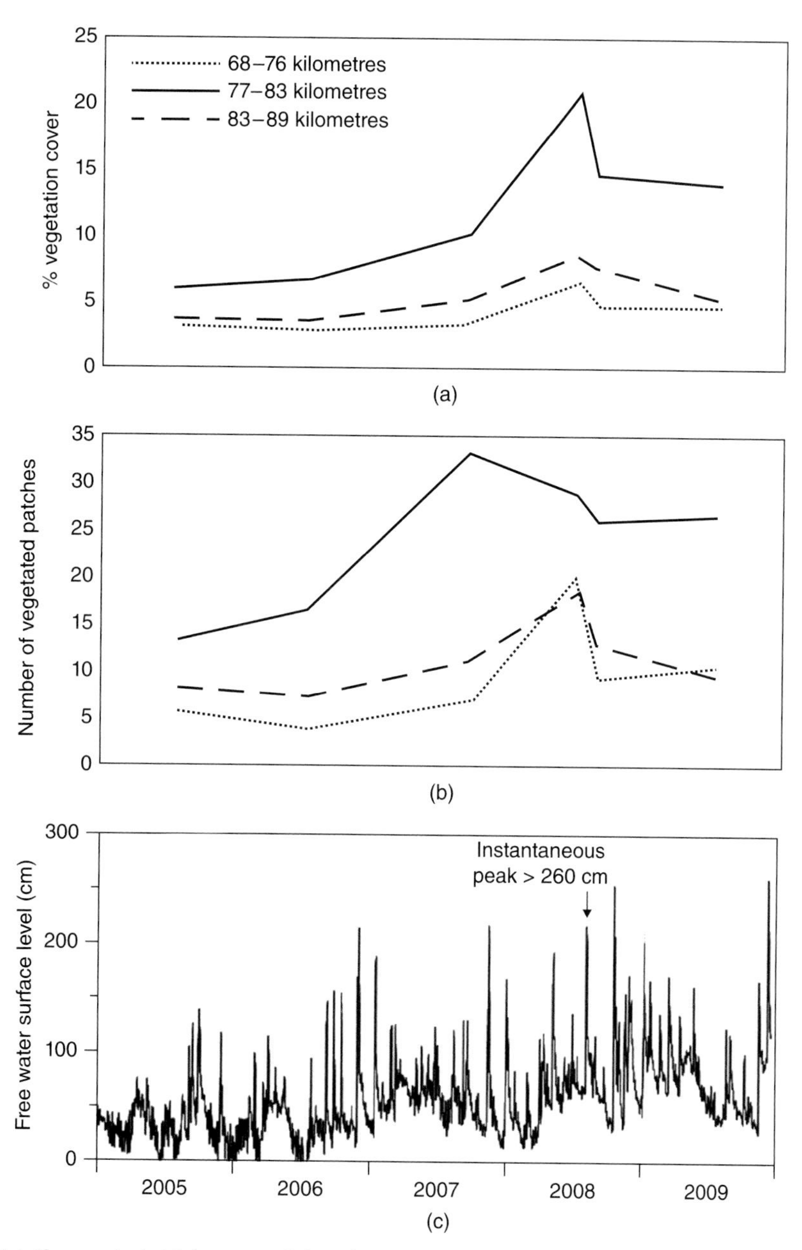

Figure 7.8 (a) Changes in total (sparse and dense) vegetation cover estimated between 68–76 km, 77–83 km and 83–89 km from six Aster images (August 2005, July 2006, September 2007, July 2008, August 2008, July 2009). (b) Changes in the number of discrete patches of vegetation (sparse and dense) estimated between river kilometres 68–76 km, 77–83 km and 83–89 km from the same six Aster images as (a). (c) Average hourly water surface levels relative to a local datum at a gauge located at 83 km from the river's source, 1 January 2005 to 31 December 2009 (data from Bertoldi *et al.*, 2011a).

change in growth rates between 2005 and 2010. A gauge located at 83 km from the river's source recorded numerous flow pulses in the range 100 to 200 cm stage in the period between 1 January 2006 and 31 December 2010 (Figure 7.8c), whereas flow pulses in this range were relatively rare between 1 January 2001 and 31 December 2005 (not illustrated). A river stage of 200 cm inundates the braid bars to the level above which large vegetation patches start to appear within the active tract in the reach between 74 and 83 km from the river's source (Bertoldi *et al.*, 2009; Figure 7.9), suggesting that groundwater levels within the alluvial aquifer were largely within the root zone of these large vegetated patches on frequent occasions between 2006 and 2010. This was not the case between 2001 and 2005, during which the river stage remained well below 100 cm for the majority of the summer growing season in every year (not illustrated). These data support a hydrological cause for the observed differences in average annual growth increment in 2005 and 2010 between river kilometres 74 and 83, since the upper 2 m growth of the sampled 3 m trees was largely achieved in the previous five years (Figure 7.7). Since all of the remaining 11 sites are downstream of the gauge at 83 km, they are subject to similar relative water level fluctuations (although related to different local base levels because of spatial variations in surface–groundwater exchange), further supporting a hydrological control on changing growth rates between 2005 and 2010 at all 11 sites.

Tree growth rates are one of the two major controls on island development in the conceptual model (Figure 7.1), and so these data suggest that spatial and temporal differences in vegetation growth and island dynamics might be expected within different reaches of the Tagliamento main stem.

Flow disturbance and vegetation cover

Flow disturbance sufficient to erode vegetation and fine sediment is another key factor in the conceptual model (Figure 7.1). The impact of flood events on vegetation extent is assessed through analysis of satellite imagery, river stage records and photographs, focusing on the section of the river between 68 and 79 km from the source.

An analysis of the changing percentage of the active tract that is vegetated was conducted using Thematic Mapper (TM) data for the period 1984 to 2001 (Henshaw *et al.*, 2013). Eighteen TM scenes captured between June and September (when riparian trees have a fully developed leaf cover) during low river flow and cloud-free sky conditions were analysed. The 30 m resolution pixels were separated into vegetated and unvegetated classes by applying a threshold value of 0.2 to the Normalised Difference Vegetation Index (NDVI, Rouse *et al.*, 1973) and were then accumulated over the area of the active tract to estimate the percentage vegetation cover at each of the 18 dates. Figure 7.10a compares the estimates of percentage vegetation cover with the occurrence of river flows exceeding an average 200 cm stage over one hour (the level above which flows start to interact with large vegetated patches along this section of the river, Figure 7.9, Bertoldi *et al.*, 2009). Between the August 1984 and September 1994 images, there was only one flow event that clearly exceeded the bankfull level of 300 cm. During this period there was a gradual increase in vegetation cover, with minor reductions following the highest flow levels. Three floods exceeded the 300 cm level in the period between 1995 and 2001, during which there were no suitable TM scenes for analysis. Indeed,

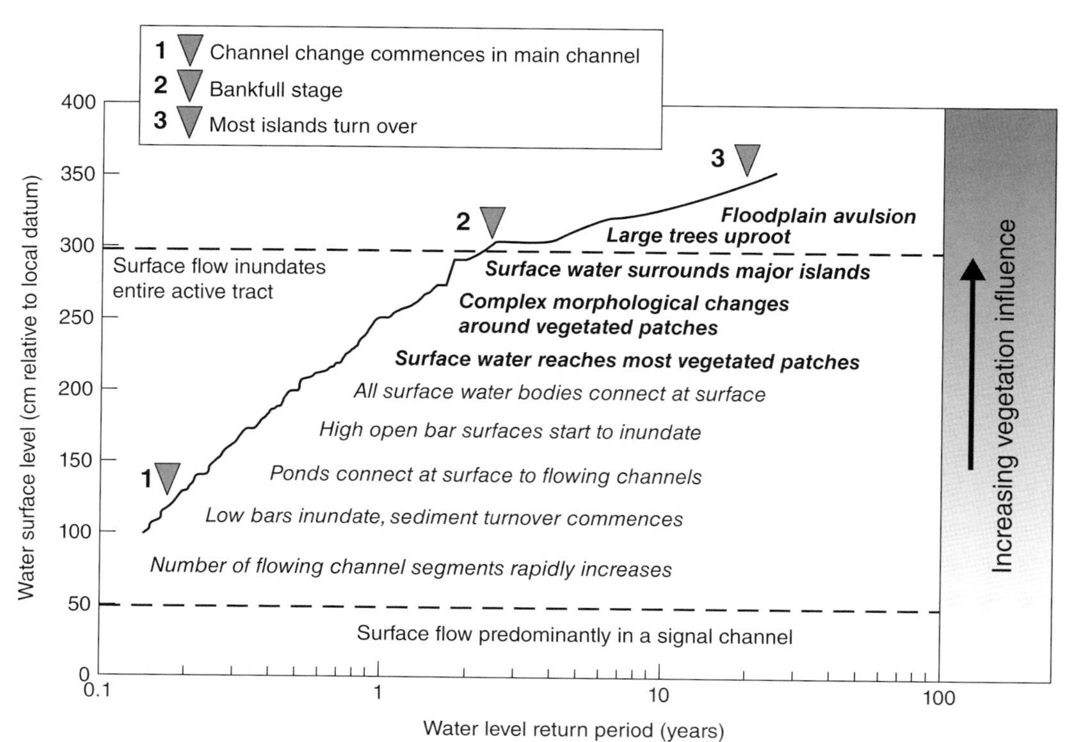

Figure 7.9 Significance of river stages (monitored at 83 km) for inundation of biogeomorphological features within the active tract between 74 and 83 km. The associations are interpreted from repeat photographs of the active tract obtained from 2003 to present. Small, young vegetated features start to appear at around 100 cm elevation, but large vegetated patches are generally confined to areas above 200 cm elevation (developed from Bertoldi *et al.*, 2009).

the largest flood in the period from 1982 to 2011 occurred in November 2000, shortly before the September 2001 image. The 2001 image shows a dramatic reduction in vegetation cover in comparison with the 1994 image. Following a bankfull event in 2004, vegetation cover gradually increased in the last five images that were analysed.

Estimates of vegetation cover between 68 and 89 km were also extracted from higher resolution Aster data (Bertoldi *et al.*, 2011a), allowing a more sensitive spatial analysis of vegetated area. Vegetation cover estimates were based on classification of the 15 m pixels into heavily vegetated, sparsely vegetated and unvegetated, using threshold NDVI values of 0.2 and 0.1. Figure 7.10b presents the total (sparse plus heavy) vegetated area of the active tract, expressed as a percentage, on three occasions (August 2005, July 2008, July 2009). Dense vegetation cover followed a similar downstream pattern on all three dates, but with highest overall cover in 2008 and lowest in 2005. Figure 7.10b illustrates a rapid expansion in vegetated area between 2005 and 2008, a period during which there were no flow events exceeding 200 cm river stage by more than 20 cm. However, hourly flood stages of 218 cm (instantaneous peak stage > 260 cm) in August 2008 and 255 cm in October 2008 resulted in a significant reduction in vegetation cover in the July 2009 image.

In reaches with the highest initial vegetation cover, vegetation expansion was proportionally largest between 2005 and 2008 and vegetation removal was proportionally lowest between 2008 and 2009 (Figure 7.10b), illustrating that in reaches where vegetation growth is strongest, young vegetation is also most resistant to removal by flooding, presumably because it grows at a very rapid rate. A more detailed temporal picture of changing vegetation cover is provided in Figures 7.8a and b, by including analyses of a further three Aster images (July 2006, September 2009, August 2008 – 14 days after the August flood), and separating the cover estimates into three sub-reaches of contrasting vegetation cover and rate of expansion (68–76 km, 78–83 km, 83–89 km). Figure 7.8 verifies the temporal aspect of the conceptual model (Figure 7.1b) by illustrating how the relatively modest vegetation expansion in the upstream (68–76 km) and downstream (83–89 km) sub-reaches between 2005 and 2008 was reflected in a steady increase in the number of vegetated patches, whilst the rapid vegetation expansion in the central sub-reach (78–83 km) was accompanied by an initial increase in the number of patches but then some coalescence between 2007 and 2008. Moreover, whereas the two high river stages in 2008 led to a reduction in vegetated area and number of vegetated patches almost back to 2005 levels in the upstream and downstream sub-reaches, only a relatively minor reduction in area and number of patches occurred in the middle sub-reach. Thus, where growth of the dominant riparian tree species, *P. nigra*, is most vigorous (between 78 and 83 km, Figure 7.7), vegetation cover in the newly vegetated area is sufficiently developed to resist erosion by the two flood events, whereas this is not the case in the other two sub-reaches, where *P. nigra* growth rates are significantly slower (68–76 km, 83–89 km, Figure 7.7).

Photographs of a section of the river between 80 and 81 km at times relevant to the changes illustrated in Figures 7.8 and 7.10 are presented in Figure 7.11. In September 2001 the vegetation cover was low almost a year after the very large flood in November 2000. Trees deposited by the flood on the bar top between the two islands in the top right of the photograph and on the

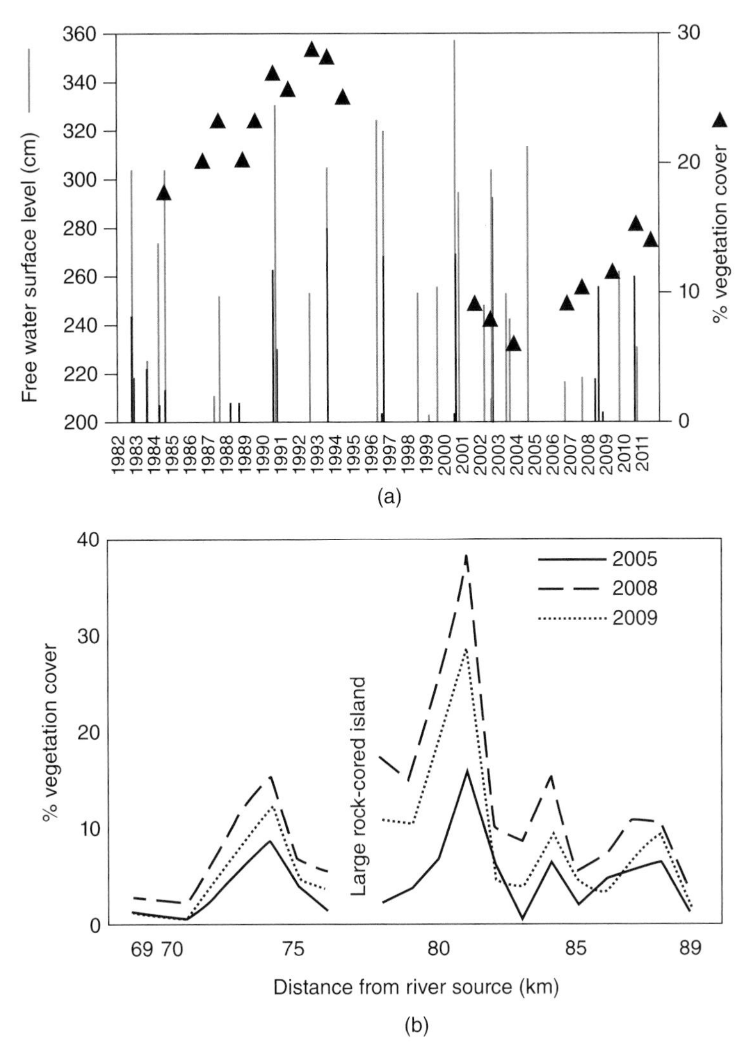

Figure 7.10 (a) Changing vegetation cover during the summer months (June to September) on 19 occasions between 1984 and 2011. The percentage vegetation cover (black triangles) was estimated for the entire active tract between 77 and 83 km following classification of the 30 m pixels of Thematic Mapper scenes into vegetated or unvegetated using a threshold NDVI value of 0.2 (for detailed methodology, see Henshaw *et al.*, 2013). (b) Changing vegetation cover in 1 km reaches of the Tagliamento active tract between 68 and 89 km from the river's source in summers 2005, 2008 and 2009 (a reach including a large vegetated, bedrock cored island is excluded). Cover was estimated from Aster data following classification of the 15 m pixels into heavily vegetated, sparsely vegetated and unvegetated using threshold NDVI values of 0.2 and 0.1 (for detailed methodology, see Bertoldi *et al.*, 2011a).

bar top towards the left centre of the photograph had sprouted to form pioneer islands, but elsewhere in the active tract, vegetation cover was negligible. By July 2005, the areas between the pioneer islands had become vegetated as the islands coalesced, and there were some new pioneer islands on the bar at the top of the photograph that had developed from trees deposited in a bankfull flood in October 2004. However,

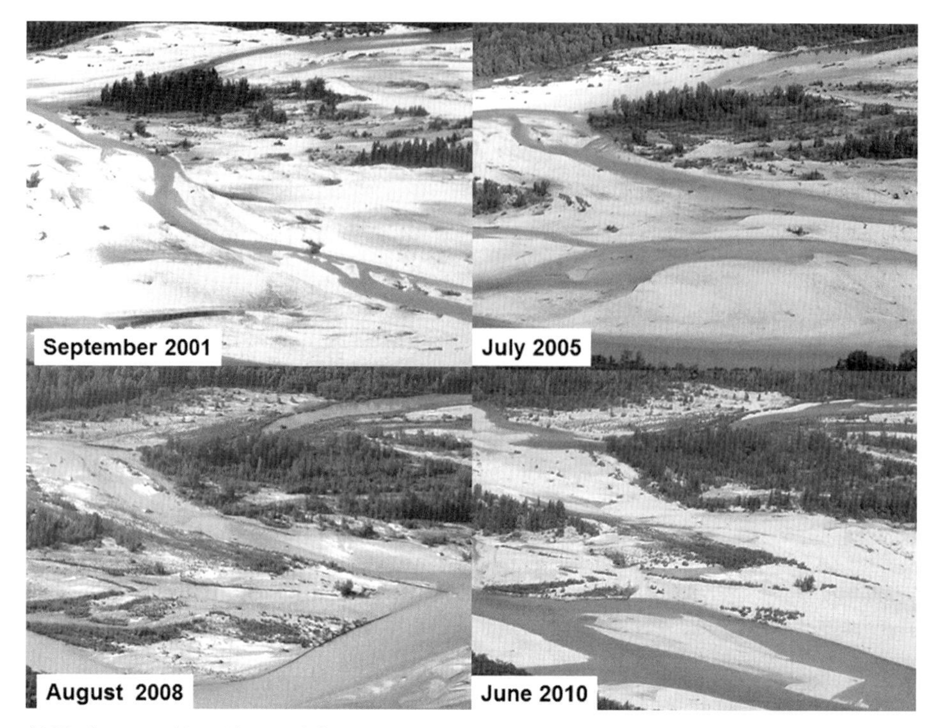

Figure 7.11 Photographic evidence of changing vegetation cover at a site between 80 and 81 km from the river's source in the summers of 2001, 2005, 2008 and 2010 (photographs by A.M. Gurnell).

other areas remained bare of vegetation, probably as a result of erosion by the same flood. By 2008, the vegetated areas in the 2005 photograph had coalesced to formed established islands and newly vegetated areas had developed across much of the remainder of the bar surfaces. In the 2010 photograph, the established islands present in 2008 remained, but approximately a half of the newly vegetated areas of 2005 had been eroded, presumably during the 250–260 cm stage events that occurred in the intervening period (Figure 7.10a).

The above evidence illustrates how rapidly the vegetated area can expand during periods with few floods, and also how quickly new vegetated areas can develop a tall vegetation cover forming established islands. It also shows that once the vegetation is established, it is able to resist erosion by all but the largest floods, and that even relatively young, short vegetation cover can survive moderate flood events.

Vegetation and fine sediment retention

The final key factor in the conceptual model (Figure 7.1) is the retention of sand and finer sediment by vegetation to construct root-reinforced island landforms. There is much field evidence for this process, with established island surfaces elevated over 1.5 m above the surfaces of the highest gravel bar tops within the active tract. However, a spatially comprehensive analysis is needed to ascertain whether this process has any significant effect on the morphology of the entire active tract. Such an analysis was applied to 1 km segments of the active tract between river kilometres 68 and 89 (Bertoldi *et al.*, 2011b).

Vegetation height/biomass and braid plane topography were both analysed using airborne Lidar data collected by the UK Natural Environment Research Council in May 2005. The elevation of the ground surface was estimated from the Lidar data to construct a Digital Terrain Model (DTM). Vegetation height was then calculated as the difference between the interpolated ground surface and the Lidar point cloud across a 5 m grid. A grid cell was deemed bare of vegetation if none of the points lay more than 1 m above the surface. Vegetation density was estimated as the proportion of points that lay more than 1 m above the ground surface in a grid cell. In the same way, the density of vegetation greater than 5, 10 and 20 m tall was estimated as the proportion of points that were higher than these elevations above the ground surface. Downstream slope was estimated from the moving average of bed elevation estimated for all active tract grid cells within an 800 m square window. In this way, the downstream gradient could be subtracted from the DTM, allowing the frequency distribution of bed elevation to be compared among 1 km segments of the active tract.

Nineteen 1 km segments of the river were analysed (segments containing a rock cored island and a bedrock confined gorge were excluded). Figure 7.12a and b display frequency distributions of bed elevation within the active tract for the most heavily vegetated (Figure 7.12a) and least heavily vegetated (Figure 7.12b) 1 km segments. Each bar in the frequency distributions is subdivided to show the proportion of the grid cells at that elevation which are unvegetated or support vegetation over 1, 5, 10 or 20 m tall. The shape of the two bed elevation frequency distributions in Figures 7.12a and b are completely different, with the least vegetated segment (Figure 7.12a) showing a more peaked and negatively skewed frequency distribution, whereas the most heavily vegetated segment shows a wide, symmetrical frequency distribution (Figure 7.12b).

When the skewness (Figure 7.12c) and kurtosis (Figure 7.12d) of the bed elevation frequency distributions for all 19 1-km segments are plotted against the average tree canopy height within each segment, it is apparent that the morphology of the active tract changes as the average canopy height changes. Bertoldi *et al.* (2011b) found statistically significant correlations between the skewness and kurtosis of the bed elevation frequency distribution and several other measures of vegetation cover and height (e.g., proportion of the active tract covered by vegetation taller than 5 m, median elevation of vegetated grid cells, tree growth rate (interpolated from Figure 7.7)). Thus, the bed elevation frequency distribution becomes more symmetrical and wider as tree growth rate, vegetation cover and height, and the median elevation of vegetated grid cells increases, illustrating a clear topographic signature of vegetation within the entire active tract.

Changing the controlling factors

This chapter has explored a range of supporting evidence to validate the conceptual model (Figure 7.1), at least for the middle and lower reaches of the Tagliamento. In particular, it has presented evidence to support the crucial role of riparian tree growth rate, river flow regime, and the retention of fine sediment by vegetation in

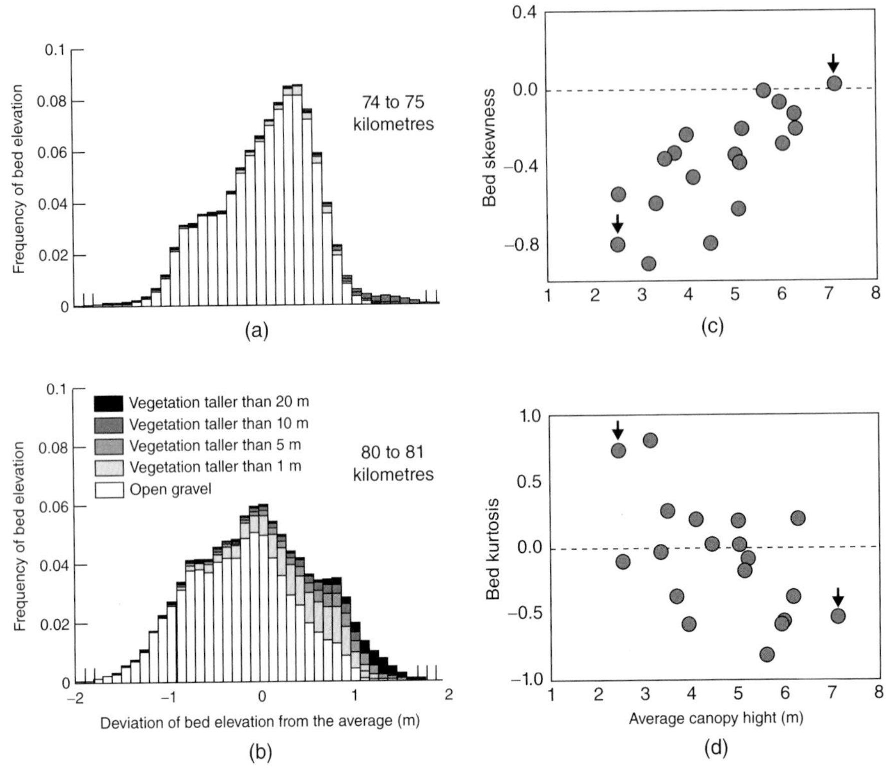

Figure 7.12 Frequency distributions of river bed elevation within 1 km segments of the active tract of the Tagliamento River between 68 and 89 km, in the most heavily (a) and least heavily (b) vegetated segments. The bars are subdivided according to the proportion of grid cells at that elevation that are bare gravel (vegetation shorter then 1 m), or under vegetation taller than 1, 5, 10 and 20 m. Relationships between average vegetation canopy height in 19 1-km segments of the Tagliamento River between 68 and 89 km and the skewness (c) and kurtosis (d) of the frequency distribution of river bed elevation (data from Bertoldi *et al.*, 2011b).

influencing island formation and morphology of the active tract. In this section, other evidence is presented to explore the consequences when these controls are changed or removed, and to provide information to generalise the functioning of the model across a wider range of environmental conditions.

Fine sediment supply

The supply of fine sediment increases downstream along the Tagliamento as the bed material also fines (Petts *et al.*, 2000). In the middle reaches, the bed material is typically pebble–cobble size (median particle size of lag deposits = −6 phi, average size of largest particles in lag deposits = −7.5 phi; Petts *et al.*, 2000) and there is a good supply of sand and finer size sediment. However, in the headwaters the bed material is much coarser, typically cobble–boulder size (median particle size of lag deposits = −7 phi, average size of largest particles in lag deposits = −8 phi; Petts *et al.*, 2000) and finer sediment supply is relatively limited. In the lower reaches, the bed material is typically pebble size (median particle size of lag deposits = −5 phi, average size of largest particles in lag deposits = −6.5 phi; Petts *et al.*, 2000) and there is a plentiful supply of sand and finer size sediment. However, other factors change along the active tract.

In the headwaters, dead wood dominates (Figure 7.4), and woody vegetation develops through predominantly sexual reproduction with seeds germinating in the shelter of boulders and dead wood accumulations. Much coarse sediment is delivered to the river corridor by hillslope (e.g., landslide) processes and islands develop opportunistically, often where coarse bed sediment and wood accumulations are not readily mobilised by the river. Vegetation establishment enhances the stability of sediment patches, but island topography results mainly from flow splitting and bed incision around wood and vegetation rather than by island surface aggradation (Gurnell *et al.*, 2001). This process of island development differs from the aggradation model proposed in Figure 7.1.

In the lower reaches, pioneer islands develop around accumulations or jams of living wood pieces that are often aligned parallel to the main flow direction and located towards the margins of the active tract. Here, the river adopts a wandering to meandering planform as bank sediments fine and the bed gradient reduces. Pioneer islands trap large quantities of finer sediment and frequently join to form living wood cored scroll bars. The buried wood sprouts rapidly, supported by the moist fine sediment, to form lines of shrubs and trees that aggrade quickly to floodplain level, inducing erosion of the opposite bank and meander migration (Gurnell *et al.*, 2001).

Over all, plentiful fine sediment and a moist growing environment supports the development of aggrading islands, but these islands have a different morphology in reaches of different planform. Furthermore, fine sediment is not essential to island development where sections of the active tract are sufficiently stable to allow vegetation to develop and establish more slowly.

River flow regime

Some insights into how the island model (Figure 7.1) might operate in lower energy conditions than are typical of the Tagliamento can be extracted from Figure 7.11. The model emphasises the importance of asexual reproduction in supporting island development. However, during the period of low flow disturbance from 2005 to 2008, the potential for sexual reproduction to support landform development in lower energy river systems became apparent.

Between the 2005 and 2008 photographs (Figure 7.11), numerous seeds germinated along the margins of the main low flow braid channels in reaches with the highest tree growth rates (Figure 7.7). The seedlings grew rapidly and, by 2008, channel margin strips of vegetation had a cover exceeding 80%, and canopy height exceeding 1.5 m. Fine sediment, trapped by the vegetation strips during minor flow pulses, formed levee-like structures up to 45 cm deep. This process appears to be the first stage of bank construction along the low-flow channel margins. These vegetated strip landforms, like those centred on living and dead wood, can also be viewed as (elongated) pioneer islands, which may develop, coalesce, aggrade and eventually merge with the floodplain, but may also be removed or dissected during disturbance events, as illustrated by the 2010 photograph (Figure 7.11). This evidence indicates that any of the three trajectories shown in Figure 7.1 may have importance for river morphodynamics depending on the river energy/flow disturbance regime (Gurnell *et al.*, 2012).

Wood supply

The crucial role of wood in driving island development on the high energy Tagliamento River has been revealed by an analysis of historical air photographs covering the main stem between 67 and 83 km (Zanoni *et al.*, 2008). Although numerous other controlling factors may have changed between the 1940s and the present, widespread removal of riparian trees on the floodplain and adjacent hillslopes could be clearly observed in air photographs dated 1944 and 1946. In the same photographs the active tract was much wider than at present and wood accumulations and islands were virtually absent. Over the following decades, the riparian trees recovered, and wood and islands started to spread across the active tract until the mid 1980s. During the 1980s, attachment of islands to the floodplain resulted in active tract narrowing to the current width coupled with the maintenance of an island-braided planform. These processes were particularly extensive in the river segments that support the highest rates of riparian tree growth.

Conclusions

Over the last 15 years, a great deal has been learnt about trees, wood and the morphodynamics of the Tagliamento River, some of which has been summarised in this chapter. Although focusing on one river system, this research contributes to the rapidly expanding field of fluvial biogeomorphology, and has benefited from 15 years during which hydrological conditions have varied greatly. As a result, it has been possible to investigate the biogeomorphic consequences of periods of low or high flood disturbance and to consider their relevance to a broad spectrum of humid temperate river types.

In a recent review, Gurnell *et al.* (2012) linked observations across several river systems to identify a range of vegetation-dominated pioneer landforms that characterise humid temperate rivers of different energy. To progress this research, more biogeomorphological datasets are needed. At the same time, further research on the Tagliamento will characterise the growth performance of other riparian tree species, and their contributions to the fluvial biogeomorphological functioning of this system, particularly in the headwaters where dead wood, sexual reproduction and other riparian tree species than *P. nigra*, dominate. This information is of scientific interest and is also crucial to developing sustainable approaches to river management that incorporate the key natural processes that drive river morphodynamics.

Acknowledgements

Many people have contributed to biogeomorphological research on the Tagliamento River. Research was initiated through a collaboration with Peter Edwards, Geoffrey Petts, Klement Tockner and James Ward. Over the ensuing 15 years, numerous others have contributed, with Walter Bertoldi, Robert Francis, Alex Henshaw and Helen Moggridge being involved for sustained periods. In the early days, the research was supported by two small grants from the UK Natural Environment Research Council (Refs: GR9/03249, NER/B/S/2000/00298). More recently the research on the Tagliamento has been supported by a major grant from the Leverhulme Trust (Ref: F/07 040/AP) and broader research on interactions between vegetation and fluvial geomorphology has been supported by the REFORM (REstoring rivers FOR effective catchment Management) collaborative project funded by the European Union Seventh Framework Programme under grant agreement 282656. Geoff Petts was involved in the Tagliamento research from the beginning and he has maintained an active interest, contributing many good ideas despite his 'limited understanding of green things'.

References

Abbe, T.B., Montgomery, D.R., 2003. Patterns and processes of wood debris accumulation in the Queets river basin, Washington. *Geomorphology* 51: 81–107.

Ahna, C., Mosera, K.F., Sparks, R.E., White, D.C., 2007. Developing a dynamic model to predict the recruitment and early survival of black willow (Salix nigra) in response to different hydrologic conditions. *Ecological Modelling* 204: 315–325.

Amlin, N.A., Rood, S.B., 2001. Inundation tolerances of riparian willows and cottonwoods. *Journal of the American Water Resources Association* 37(6): 1709–1720.

Amlin, N.M., Rood, S.B., 2002. Comparative tolerances of riparian willows and cottonwoods to water-table decline. *Wetlands* 22(2): 338–346.

Amlin, N.M., Rood, S.B., 2003. Drought stress and recovery of riparian cottonwoods due to water table alteration along Willow Creek, Alberta. *Trees* 17: 351–358.

Bennett, S.J., Wu, W., Alonso, C.V., Wang, S.S.Y., 2008. Modeling fluvial response to in-stream woody vegetation: implications for stream corridor restoration. *Earth Surface Processes and Landforms* 33: 890–909.

Bertoldi, W., Gurnell, A.M., Surian, N., et al. 2009. Understanding reference processes: linkages between river flows, sediment dynamics and vegetated landfroms along the Tagliamento River, Italy. *River Research and Applications* 25: 501–516.

Bertoldi, W., Drake, N., Gurnell, A.M., 2011a. Interactions between river flows and colonising vegetation on a braided river: exploring spatial and temporal dynamics in riparian vegetation cover using satellite data. *Earth Surface Processes and Landforms*, doi: 10.1002/esp.2166, available on-line.

Bertoldi, W., Gurnell, A.M., Drake, N., 2011b. The topographic signature of vegetation development along a braided river: results of a combined analysis of airborne lidar, colour air photographs and ground measurements. *Water Resources Research*, doi:10.1029/2010WR010319, available on-line.

Braatne J.H., Rood S.B., Heilman P.E. 1996. Life history, ecology, and conservation of riparian cottonwoods in North America. In: R.F. Stettler, H.D. Bradshaw, P.E. Heilman and T.M. Hinckley (eds) *Biology of Populus and its Implications for Management and Conservation*, NRC Research Press, Ottawa, pp. 57–86.

Camporeale, C., Perucca, E., Ridolfi, L., Gurnell, A.M., 2013. Modeling the interactions between river morphodynamics and riparian vegetation. *Reviews of Geophysics* 51: 2012RG000407.

Church, M., 1992. Channel morphology and typology. In: C. Callow and G. Petts (eds), *The Rivers Handbook: Hydrological and Ecological Principles*. Blackwell, Oxford, pp. 126–143.

Church, M., 2002. Geomorphic thresholds in riverine landscapes. *Freshwater Biology* 47: 541–557.

Collins, B.D., Montgomery, D.R., Fetherston, K.L., Abbe, T.M. 2012. The floodplain large-wood cycle hypothesis: A mechanism for the physical and biotic structuring of temperate forested alluvial valleys in the North Pacific coastal ecoregion. *Geomorphology* 139–140: 460–470.

Cooper, D.J., Dickens, J., Hobbs , *et al.* 2006. Hydrologic, geomorphic and climatic processes controlling willow establishment in a montane ecosystem. *Hydrological Processes* 20(8): 1845–1864.

Corenblit, D., Tabacchi, E., Steiger, J., Gurnell, A.M., 2007. Reciprocal interactions and adjustments between fluvial landforms and vegetation dynamics in river corridors: A review of complementary approaches. *Earth-Science Reviews* 84(1–2): 56–86.

Corenblit, D., Steiger, J., Gurnell, A.M., et al. 2009. Control of sediment dynamics by vegetation as a key function driving biogeomorphic succession within fluvial corridors. *Earth Surface Processes and Landforms* 34(13): 1790–1810.

Dixon, M.D., Turner, M.G., Jin, C.F., 2002. Riparian tree seedling distribution on Wisconsin River sandbars: Controls at different spatial scales. *Ecological Monographs* 72(4): 465–485.

Docker, B.B., Hubble, T.C.T., 2008. Quantifying root-reinforcement of river bank soils by four Australian tree species. *Geomorphology* 100(3–4): 401–418.

Doering, M., Uehlinger, U., Rotach, A., et al. 2007. Ecosystem expansion and contraction dynamics along a large Alpine alluvial corridor (Tagliamento

River, Northeast Italy). *Earth Surface Processes and Landforms* 32, 16931704.

Francis, R.A., Gurnell, A.M., Petts, G.E., Edwards, P.J. 2005. Survival and growth responses of Populus nigra, Salix elaeagnos and Alnus incana cuttings to varying levels of hydric stress. *Forest Ecology and Management* 210: 291–301.

Francis, R.A., Gurnell, A.M., 2006. Initial establishment of vegetative fragments within the active zone of a braided gravel-bed river (River Tagliamento, NE Italy). *Wetlands* 26(3): 641–648.

Friedman, J.M., Auble, G.T., Andrews, E.D., et al. 2006. Transverse and longitudinal variation in woody riparian vegetation along a montane river. *Western North American Naturalist* 66(1): 78–91.

Glenz, C., Schlaepfer, R., Iorgulescu, I., Kienast, F., 2006. Flooding tolerance of Central European tree and shrub species. *Forest Ecology and Management* 235: 1–13.

González, E., González-Sanchis, M., Comín, F.A., Muller, E., 2012. Hydrologic thresholds for riparian forest conservation in a regulated large Mediterranean river. *River Research and Applications* 28(1): 71–80.

Greet, J.O.E., Angus Webb, J., Cousens, R.D., 2011. The importance of seasonal flow timing for riparian vegetation dynamics: a systematic review using causal criteria analysis. *Freshwater Biology*, DOI: 10.1111/j.1365-2427.2011.02564.x, available on-line.

Guilloy-Froget, H., Muller, E., Barsoum, N., Hughes, F.M.R., 2002. Dispersal, germination and survival of *Populus nigra L.* (Saliacaceae) in changing hydrologic conditions. *Wetlands* 22(3): 478–488.

Gurnell, A.M. 2012. Wood in fluvial systems. In: J. Shroder and E. Wohl (eds) *Treatise on Geomorphology*. San Diego, CA, Academic Press.

Gurnell, A.M., Petts, G.E., 2011. Hydrology and ecology of river systems. In: P. Wildere (ed.), *Treatise on Water Science*. Academic Press, Oxford, pp. 237–269.

Gurnell, A.M., Petts, G.E., Harris, N., et al. 2000a. Large wood retention in river channels: The case of the Fiume Tagliamento, Italy. *Earth Surface Processes and Landforms* 25(3): 255275.

Gurnell, A.M., Petts, G.E., Hannah, D.M., et al. 2000b. Wood storage within the active zone of a large European gravel-bed river. *Geomorphology* 34(1–2): 55–72.

Gurnell, A.M., Petts, G.E., Hannah, D.M., et al. 2001. Riparian vegetation and island formation along the gravel-bed Fiume Tagliamento, Italy. *Earth Surface Processes and Landforms* 26(1): 31–62.

Gurnell, A.M., Tockner, K., Edwards, P.J., Petts, G.E., 2005. Effects of deposited wood on biocomplexity of river corridors. *Frontiers in Ecology and Environment* 3(7): 377–382.

Gurnell, A.M., Thompson, K., Goodson, J., Moggridge, H., 2008. Propagule deposition along river margins: linking hydrology and ecology. *Journal of Ecology*, 96: 553–565.

Gurnell, A.M., Bertoldi, W., Corenbit, D. 2012. Changing river channels: the roles of hydrological processes, plants and pioneer landforms in humid temperate, mixed load, gravel bed rivers. *Earth Science Reviews* 111: 129–141.

Henshaw, A.J., Gurnell, A.M., Bertoldi, W., Drake, N.A., 2013. An assessment of the degree to which Landsat TM data can support the assessment of fluvial dynamics, as revealed by changes in vegetation extent and channel position, along a large river. *Geomorphology*, early view.

Hughes, F.M., Rood, S.B., 2003. Allocation of River Flows for Restoration of Floodplain Forest Ecosystems: A Review of Approaches and Their Applicability in Europe. *Environmental Management* 32(1): 12–33.

Hupp, C.R., Osterkamp, W.R., 1985. Bottomland vegetation distribution along Passage Creek, Virginia, in relation to fluvial landforms. *Ecology*, 66: 670–681.

Imbert, E., Lefevre, F., 2003. Dispersal and gene flow of *Populus nigra* (Salicaceae) along a dynamic river system. *Journal of Ecology* 91: 447–456.

Ishikawa, Y., Sakamoto, T., Mizuhara, K., 2003. Effect of density of riparian vegetation on effective tractive force. *Journal of Forest Research* 8: 235–246.

Karrenberg, S., Edwards, P.J., Kollmann, J., 2002. The life history of Salicaceae living in the active zone of floodplains. *Freshwater Biology* 47: 733–748.

Karrenberg, S., Kollmann, J., Edwards, P.J., et al. 2003. Patterns in woody vegetation along the active zone of a near-natural Alpine river. *Basic and Applied Ecology* 4: 157–166.

Karrenberg, S., Suter, M. 2003. Phenotypic trade-offs in the sexual reproduction of Salicaceae from flood plains. *American Journal of Botany* 90(5): 749–754.

Leopold, L.B., Wolman, M.G., 1957. River channel patterns-braided, meandering and straight. U.S. *Geological Survey Professional Paper*, 282B 39–85.

Lite, S.J., Stromberg, J.C., 2005. Surface water and ground-water thresholds for maintaining

Populus–Salix forests, San Pedro River, Arizona. *Biological Conservation* 125: 153–167.

Liu, D., Diplas, P., Hodges, C.C., Fairbanks, J.D., 2010. Hydrodynamics of flow through double layer rigid vegetation. *Geomorphology* 116(3–4): 286–296.

Lytle, D.A., Merritt, D.M., 2004. Hydrologic regimes and riparian forests: A structured population model for cottonwood. *Ecology* 85(9): 2493–2503.

Mahoney, J.M., Rood, S.B., 1998. Streamflow requirements for cottonwood seedling recruitment: an integrative model. *Wetlands* 18: 634–645.

Merritt, D.M., Wohl, E.E. 2002. Processes governing hydrochory along rivers: hydraulics, hydrology and dispersal phenology. *Ecological Applications* 12: 1071–1087.

Merritt, D.M., Scott, M.L., LeRoy Poff, N., et al. 2010. Theory, methods and tools for determining environmental flows for riparian vegetation: riparian vegetation-flow response guilds. *Freshwater Biology* 55(1): 206–225.

Moggridge, H.L., Gurnell, A.M., 2009. Controls on the sexual and asexual regeneration of Salicaceae along a highly dynamic, braided river system. *Aquatic Sciences* 71: 305–317.

Nanson, G.C., Croke, J.C. 1992. A genetic classification of floodplains. *Geomorphology* 4(6): 459–486.

Osterkamp, W.R., Hupp, C.R., Stoffel, M., 2012. The interactions between vegetation and erosion: new directions for research at the interface of ecology and geomorphology. *Earth Surface Processes and Landforms* 37(1): 23–36.

Petts, G.E., Gurnell, A.M., Gerrard, A.J., et al. 2000. Longitudinal variations in exposed riverine sediments: a context for the development of vegetated islands along the Fiume Tagliamento, Italy. *Aquatic Conservation: Marine and Freshwater Ecosystems* 10: 249266.

Pezeshki, S. R., F.D. Shields 2006. Black willow cutting survival in streambank plantings, southeastern United States. *Journal of the American Water Resources Association* 42(1): 191–200.

Pollen-Bankhead, N., Simon, A., 2010. Hydrologic and hydraulic effects of riparian root networks on streambank stability: Is mechanical root-reinforcement the whole story? *Geomorphology* 116(3–4): 353–362.

Prosser, I.P., Dietrich, W.E., Stevenson, J., 1995. Flow resistance and sediment transport by concentrated overland flow in a grassland valley. *Geomorphology* 13: 71–86.

Robertson, K.M., 2006. Distributions of tree species along point bars of 10 rivers in the south-eastern US Coastal Plain. *Journal of Biogeography* 33: 121–132.

Rood, S.B., Samuelson, G.M., Braatne, J.H., et al. 2005. Managing river flows to restore floodplain forests. *Frontiers in Ecology and the Environment* 3(4): 193–201.

Rouse, J.W., Haas, R.H., Schell, J.A., Deering, D.W. 1973. Monitoring vegetation systems in the Great Plains with ERTS. *Third ERTS Symposium*, NASA SP-351 I, pp. 309–317.

Schumm, S.A., 1977. *The Fluvial System*. Wiley, New York.

Schumm, S.A., 1985. Patterns of alluvial rivers. *Annual reviews of Earth and Planetary Science* 13: 5–27.

Turner, M.G., Gergel, S.E., Dixon, M.D., Miller, J.R., 2004. Distribution and abundance of trees in floodplain forests of the Wisconsin River: Environmental influences at different scales. *Journal of Vegetation Science* 15: 729–738.

Williams, C.A., Cooper, D.J. 2005. Mechanism of riparian cottonwood decline along regulated rivers. *Ecosystems* 8: 382–395.

Zanoni, L., Gurnell, A.M., Drake, N., Surian, N., 2008. Island dynamics in a braided river from an analysis of historical maps and air photographs. *River Research and Applications* 24: 1141–1159.

CHAPTER 8

The Milner and Petts (1994) conceptual model of community structure within glacier-fed rivers: 20 years on

Alexander M. Milner
School of Geography, Earth and Environmental Sciences, University of Birmingham, Birmingham, UK

Introduction

Glaciers store about 75% of the world's freshwater, contribute significantly to river flow and water resources at high altitude and latitude (Fleming and Clarke, 2005), and maintain stream flow during the summer dry season when rivers in non-glacierised basins display low flow (Hannah *et al.*, 2007). Rivers with glacier meltwater inputs sustain important downstream ecosystems such as lakes, wetlands and meadows (Buytaert *et al.*, 2011) and provide important habitat for fisheries (Milner *et al.*, 2009; Stahl *et al.*, 2008) and a number of rare and endemic macroinvertebrate species (Snook and Milner, 2001; Brown *et al.*, 2007; Muhlfeld *et al.*, 2011). Their meltwaters also provide important ecosystem services as they provide predictable water storage and release during the summer when other water sources are low and thereby facilitate socioeconomic needs, including water resource provision, hydro-power production, agriculture (irrigation) and tourism (de Groot *et al.*, 2010). Glacierised environments are one of the most vulnerable systems to climate change due to connections between atmospheric forcing, snowpacks/glacier mass-balance, stream flow, water quality and hydrogeomorphology (physico-chemical habitat), and river ecology (Smith *et al.*, 2001; Hannah *et al.*, 2007). Until the turn of the century, biological communities in glacier-fed rivers were not extensively studied, although a number of classic studies had overviewed macroinvertebrates in these systems and highlighted the important role of water temperature (e.g., Stefan, 1971; Saether,1968).

Within this context, Geoff Petts gave a seminar at the University of Stirling in November of 1992 that stimulated subsequent discussion on the nature of glacier-fed rivers. Typical of post-seminar activities, we adjourned to a local public house. Towards the end of the evening Geoff and I used the back of an empty cigar-packet to sketch some ideas on the main characteristic features of glacier-fed rivers from our collective experiences in Arolla and Alaska. These doodles subsequently became the basis of a review paper that presented conceptual models of the structure of glacier-fed rivers.

River Science: Research and Management for the 21st Century, First Edition.
Edited by David J. Gilvear, Malcolm T. Greenwood, Martin C. Thoms and Paul J. Wood.

The review 'Glacial rivers: physical habitat and ecology' (Milner and Petts, 1994) was presented at a special session of the North American Benthological Society (NABS) Annual Meeting in Calgary, Alberta, Canada in May 1993 entitled 'Ecology of cold streams; running waters at high latitudes and elevations'. The review became one of 14 papers in a special issue of *Freshwater Biology* (1994; Volume 32) edited by Professor Mike Winterbourn, which included 11 papers presented at the meeting and an additional five solicited manuscripts.

Overview of the conceptual model

The Milner and Petts paper (1994) examined a physicochemical habitat template of glacial rivers with characteristic seasonal and diurnal flow and water temperature regimes, sediment fluxes and channel form and morphology. A generalised conceptual model of a glacial river was developed that examined the relationship between the zoobenthic community and river channel form and stability, water temperature and time (Figure 8.1). The paper highlighted the unique deterministic response of the biotic communities to this template, particularly with respect to macroinvertebrates. These were suggested to show distinct patterns according to distance downstream or time since deglaciation, principally as a function of changing water temperature and channel stability. Where maximum water temperature was < 2°C, communities were predicted to be dominated by *Diamesa* (Chironomidae). Downstream, as water temperature and channel stability increase, other Diamesinae, Orthocladiinae and Simuliidae colonise when T_{max} > 2 °C < 4°C.

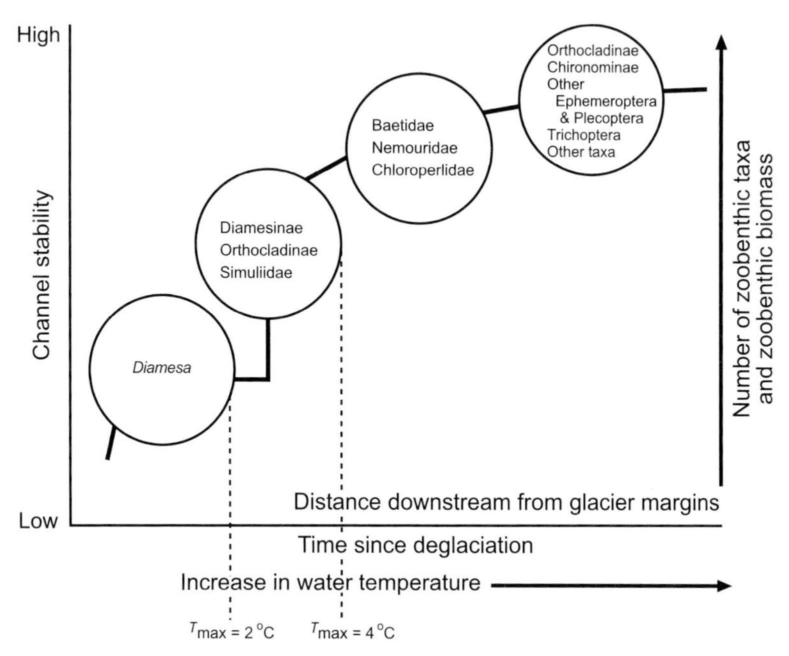

Figure 8.1 The original model for glacier-fed rivers predicting the macroinvertebrate community with relation to water temperature and channel stability. Milner and Petts, 1994. Reproduced with permission from John Wiley & Sons.

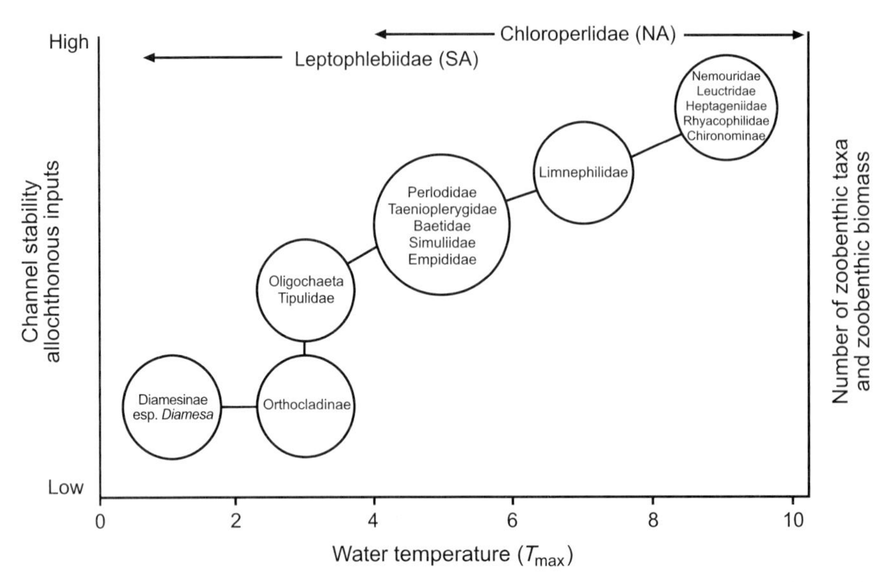

Figure 8.2 Modification of the original glacier-fed model following the AASER project and other studies revising the predicted macroinvertebrate community with relation to water temperature and channel stability and including allocthonous inputs. Milner *et al.*, 2001. Reproduced with permission from John Wiley & Sons.

Baetidae, Nemouridae and Chloroperlidae were suggested to become characteristic members of the community when $T_{max} > 4\,°C$ (Figure 8.1 – see Milner and Petts, 1994; Milner *et al.*, 2001a,b). However, even if water temperature is suitable, low channel stability may retard colonisation and permit cold-tolerant taxa (e.g., *Diamesa*) to remain dominant in the community. Groundwater tributaries, lakes and valley confinement were identified as modifiers of the proposed longitudinal patterns, mainly through effects on downstream water temperature and channel stability. The original model was developed principally on conceptual ideas and some primary data from Alaska.

AASER and the validation of the original model

Between 1996 and 1999, the conceptual models were tested using similar field protocols at a range of European glacier-fed river systems from the French Pyrenees to Svalbard within the framework of a European Union project entitled Arctic and Alpine Stream Ecosystem Research (AASER). The seven AASER sites selected represented a latitudinal gradient from 43 °N to 79 °N and a climatic gradient from oceanic to a more continental pattern. The findings were incorporated into a special issue of *Freshwater Biology* that included studies from other parts of Europe, Greenland and New Zealand entitled 'Glacial-fed rivers – unique lotic ecosystems' edited by Brittain and Milner (2001). For example, for Iceland, Gislasson *et al.* (2001) found that macroinvertebrate communities were in general agreement with the predictions of the Milner and Petts (1994) model for the upstream reaches. Assemblages consisted mainly of Orthocladiinae and Diamesinae (Chironomidae), although other taxa such as Simuliidae, Plecoptera and Trichoptera were also found in low numbers. Distance from the glacier, altitude, bryophyte biomass and the Pfankuch Index of channel stability were the most significant explanatory variables determining the structure of macroinvertebrate communities. However, distance from the glacier was likely to be correlated with T_{max}.

Maiolini and Lencioni (2001), investigating the longitudinal distribution of macroinvertebrate assemblages in a glacially influenced stream system in the Italian Alps, found some exceptions to the original model. The dipteran families Empididae and Limoniidae were more abundant at the upper stations than Simuliidae, while Nematoda were also numerous at some sites. Similar patterns were found by Lods-Crozet *et al.* (2001) for the glacier-fed Mutt in the Upper Rhone valley of Switzerland. Leuctridae, Taeniopteryidae and Nemouridae were the first Plecoptera to appear and Heptageniidae were more abundant than Baetidae at the glacial sites in the Italian Alps. In the French Pyrenees, however, discontinuity between sites due to steep gradient changes, indicated that a *Diamesa* dominated community could be sustained at T_{max} of 13°C because the channel stability was extremely low.

Using a set of 11 environmental variables generalised additive models (GAMs), adopted to predict macroinvertebrate taxa diversity across the seven AASER glacier-fed river sites, indicated maximum water temperature and channel stability (as estimated by the bottom component of the Pfankuch Index) accounted for the greatest deviance (measure of variance) in the models (Milner *et al.*, 2001a,b). These findings confirmed the original concept of the Milner and Petts (1994) model that water temperature and channel stability were the two principal variables driving zoobenthic communities in these river systems. Other variables incorporated in the models were tractive force, Froude number, water conductivity and suspended solids. Individual GAMs allowed a refinement of the proposed zoobenthic response to water temperature criteria from the original model describing their likely first appearance, particularly at T_{max} water temperatures above 4 °C. Additional groups were added to the model between a T_{max} of 6 to 8°C including Perlodidae, Taeniopterygidae and Empididae, and rather than Trichoptera and other Ephemeroptera and Plecoptera, specific T_{max} values were ascribed to Limnephilidae, Rhyacophilidae, Leuctridae and Heptageniidae. Simuliidae and Nemouridae were reassigned to higher T_{max} values than in the original model. For New Zealand and South America, the family Leptophlebiidae was incorporated into the model where water temperature was < 4°C, as the genus *Deleatidium* is found close to the glacier margin where T_{max} < 2°C, thereby replacing *Diamesa,* the chironomid typical of the Northern Hemisphere glacier-fed streams.

Further relevance of the model

In the glacier-fed streams of the southern Tibetan Plateau in China, Murakami *et al.* (2012) found that the coldest stream (4 km from the glacier), where T_{max} was < 6°C, the community was dominated by one nemourid stonefly, *Illiesonemoura* sp., and Oligochaetae. Where the water temperature increased to 8 °C, macroinvertebrate communities were more similar to the model, as *Baetis* mayflies and chironomids became dominant. No sites were included where T_{max} was < 2°C and the chironomid taxonomy was not undertaken beyond family level. In contrast, Hamerlik and Jacobsen (2012) focused solely on this family during an investigation of eight high mountain streams in southern Tibet with differing degrees of glacial influence. The proportion of Diamesinae (mostly the genus *Diamesa*) increased with greater glacial contribution to flow as predicted, but Orthocladiinae (subgenus *Euorthocladius*) dominated at all sites, probably due to T_{max} exceeding 2 °C. With decreasing glacial influence, Chironominae became more abundant at the expense of Orthocladiinae as predicted by the conceptual model. Interestingly, this study found that *Diamesa* was able to sustain high relative abundance (> 40%) at a water temperature of 6.3 °C.

Hieber *et al.* (2005) examined 10 sites in alpine headwater streams of the Swiss Alps, some of which were glacially influenced, and found channel stability and water temperature to be the main drivers of the benthic macroinvertebrate community. Glacially dominated sites were generally characterised by low taxon richness, dominated by the chironomid subfamily Diamesinae. Other taxa occurring in low numbers at these sites included Baetidae, Heptageniidae, Nemouridae, Leuctridae and Taeniopterygidae, and other dipterans such as Orthocladiinae and Limoniidae. Other taxa in the conceptual model were found only in low abundance at some of the kryal sites (e.g., Perlodidae, Simuliidae) or were absent (e.g., Rhyacophilidae, Chironominae, Empididae, Tipulidae). Interestingly, the invertebrate assemblages of kryal sites also showed no general differences between lake outlets and other streams. These findings are contrary to the model predictions of lakes enhancing downstream channel stability and water temperature and changing the macroinvertebrate community composition.

Finn *et al.* (2010) examined the newly created 482 m reach of a glacier-fed river downstream of a glacial snout in the Swiss Alps that had recently receded, and found that the colonising taxa were typical occupants of the uppermost reaches of proglacial streams (predominantly the Chironomidae subfamily Diamesinae), as predicted by the model. Although a greater temperature gradient was evident downstream of the glacier due to the addition of warmer lake tributary water, four other taxa migrated upstream. In a similar vein, the reduction in percent glacierisation of the Wolf Point Creek catchment (Glacier Bay National Park, southeast Alaska) from 78% in 1977 to 0% in 1992 is equivalent to the longitudinal zonation downstream from a glacial margin, as glacial influence becomes reduced. Interestingly some groups did not colonise until T_{max} water temperature was considerably higher than that proposed by the model, for example Simuliidae at 8 °C, Oligochaetae at 10 °C and Empididae at 13 °C. These taxa may have been limited by dispersal constraints in colonising across high mountain barriers and fjords.

Jacobsen *et al.* (2010) tested the conceptual model in the equatorial Ecuadorian Andes by sampling benthic macroinvertebrates and measuring environmental variables at nine sites between 4730 and 4225 m altitude, along a 4.3 km stretch of a glacier-fed stream. Taxon richness and overall density (from 4 individuals m^{-2} to 825 individuals m^{-2}) increased with distance from the glacier, similar to the pattern predicted. At the sites closest to the glacier, the subfamily Podonominae was abundant but became less important further downstream. Orthocladiinae were important, both in terms of abundance and species richness at all sites, whereas Diamesinae were numerous only in the middle reaches and were completely absent from the upper three sites, where water temperature was colder. The limited importance of Diamesinae, and its replacement by Podonominae, is different from the typical pattern observed in north-temperate glacier-fed streams, principally as the genus *Diamesa* is missing from the Neotropics. Stream temperature and channel stability were found to explain most of the variability in faunal composition and richness, thereby supporting the model for the Ecuadorian glacier-fed streams. These findings were later supported when Kuhn *et al.* (2011) expanded the study to include three neighbouring equatorial glacier-fed systems in the same region.

In New Zealand, Milner *et al.* (2001a,b) examined the longitudinal downstream zonation of macroinvertebrates along the glacier-fed Fox and Waiho rivers of the west coast of the South Island. Water temperature and macroinvertebrate richness increased downstream as predicted by the model, but channel stability and total macroinvertebrate abundance did not. Similar to the Neotropics, *Diamesa* is absent from New Zealand glacier-fed rivers and its role fulfilled by species of the ubiquitous New Zealand ephemeropteran genus *Deleatidium* typically found in all reaches, together with the chironomid genera *Eukiefferiella* and *Maoridiamesa*. Plecopterans and trichopterans were found only at sites > 6 km from the glacier terminus, where water temperature averaged 4–5 °C. Because of the large size of these two rivers and their source glaciers, water warmed only gradually downstream, thereby limiting macroinvertebrate community development closer to the glacier, unlike alpine glaciers. Results indicated a separation of fauna into two distinct zones (a low diversity upper zone and a richer lower zone), rather than a gradual species transition. Unlike the typical alpine setting of many European glacier-fed streams, these west coast glacier-fed rivers of New Zealand flow through beech forest with the glacier terminus below the treeline.

Subsequently, Cadbury *et al.* (2010) examined macroinvertebrate species assemblages of a glacier-fed river on the eastern side of the southern alps of New Zealand. The Rob Roy stream, a tributary of the Matukituki River, is located in Mt Aspiring National Park near Wanaka, with a glacier at a higher altitude (2644 m cascading down to 900 m) than either the Fox or the Waiho glaciers, which are near sea level. Although habitat stability increased downstream from the glacier terminus, disturbance remained a limiting factor to macroinvertebrate abundance and species richness at the lower sites, with channel stability remaining low due to the short steep nature of the channel. Average water temperature increased by 2 °C over a distance of 3.3 km downstream, with habitat heterogeneity and species richness highest at the downstream site before the confluence with the Matukituki. The ephemeropterans *Deleatidium cornutum* and *Nesameletus* dominated at the upper two sites with the chironomids *Eukiefferiella* and *Maoridiamesa*, but became co-dominant with *Deleatidium angustum* at the lowermost site, where other plecopteran and trichopteran taxa were collected. *Nesameletus* was only found at the lower sites in the Fox and Waiho rivers. The most remarkable finding, relating to the longitudinal zonation patterns within the mayfly genus *Deleatidium*, was the replacement of *Deleatidium cornutum* at the upper most sites by *D. angustum* as water temperature increased. *Deleatidium cornutum* was found to have high rates of productivity, potentially a strong adaptation to persisting at low water temperature (Winterbourn *et al.*, 2008). Further downstream in the Matukituki

River other *Deleatidium* species dominated as water temperature increased (Winterbourn, *unpublished data*). This unique record of a mayfly species at such low water temperatures close to the glacier has not been found elsewhere, nor has the distinct zonation of different species within the same genus as water temperature increased downstream. Stenothermic species of the orthoclad *Eukiefferiella* occur in both New Zealand and European streams. They have been found in European glacier-fed streams within 200 m of the glacier terminus, co-existing with *Diamesa* (Lods-Crozet *et al.*, 2001). The paucity of Plecoptera and Trichoptera at the lowest water temperature in the Rob Roy stream was also similar to European glacier-fed streams, except Trichoptera were found at lower channel stability than predicted by the model. Cadbury *et al.* (2010) used multivariate analysis of presence–absence data for the macroinvertebrate communities of the Rob Roy and the Fox and Waiho rivers to develop a conceptual model for New Zealand glacier-fed streams using the same key variables of channel stability and maximum water temperature as in the original model (see Figure 8.3).

As in New Zealand, biogeographical constraints are very important when considering glacier-fed rivers located in Arctic/sub-Arctic areas where the size of the colonising pool of macroinvertebrate taxa is low. Nowhere is this better illustrated than in the Svalbard Archipelago, where initial studies in the Ny Alesund area during the AASER campaign found only three taxa groups represented. No Empheroptera, Plecoptera or Trichoptera were collected (Castella *et al.*, 2001). The fauna was dominated by Chironomidae, of which 29 species were found, the majority from the genus

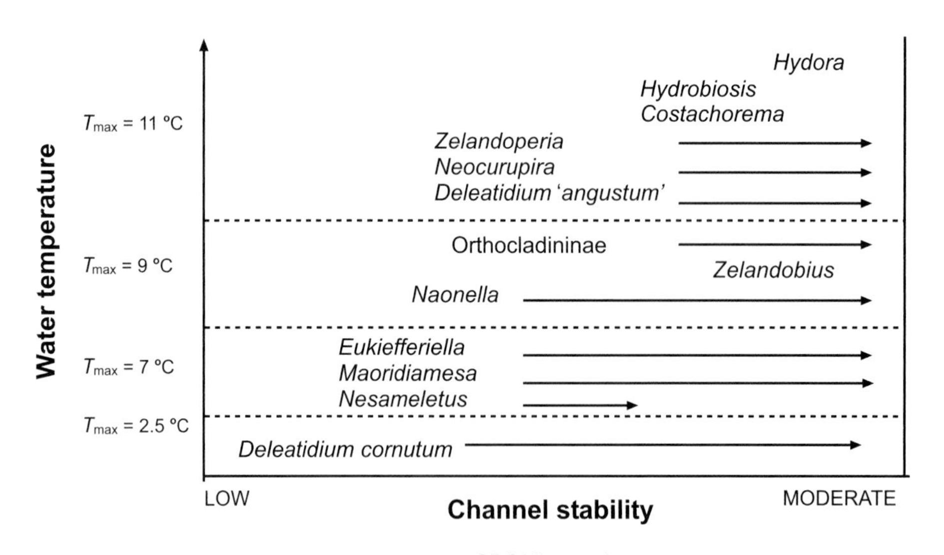

Figure 8.3 Modified conceptual model of macroinvertebrate communities for glacier-fed rivers in New Zealand with relation to water temperature and channel stability. Cadbury *et al.*, 2011. Reproduced with permission from John Wiley & Sons.

Diamesa. Diamesa aberrata and *D. bohemani* dominated sites close to the glacier while *D. artica* and *D. betrami* dominated sites fed primarily by snowmelt (Lods-Crozet, 2007). Blaen *et al.* (2014) undertook further extensive investigations of streams of varying water sources in this region and found a greater diversity of chironomids in non glacier-fed systems. These investigations led Blaen *et al.* (2014) to develop a conceptual model for northwest Svalbard rivers (see Figure 8.4), which differed from mainland Europe. Only three taxa groups were included, *Diamesa*, Orthocladiinae and Oligochaetae. With a T_{max} between 4 °C and 6 °C, Oligochaetae differed from the original model where T_{max} was 4 °C.

Similarly, Friberg *et al.* (2001) examined 16 streams in Greenland, a number of which were glacier-fed, and placed the findings in the framework of the conceptual model. Greenland was found to support six species of Trichoptera, one of Ephemeroptera, but no Plecoptera, similar to Svalbard. These authors found the only variables that were significant in influencing macroinvertebrate species diversity were channel stability, water temperature and suspended sediments and, for abundance, water temperature and conductivity. This finding was related to different water sources, as high suspended sediments arise from glacial runoff and are potentially correlated with low channel stability. In Greenland the most abundant species was *Orthocladius thienemanni*, which was found in the majority of streams, including those most strongly influenced by glacial runoff. *Eukiefferiella claripennis*, *Hydrobaenus* spp., *Cricotopus* sp. and *Micropsectra recurvata* were the next most abundant taxa. *Diamesa* sp. was not dominant or abundant in streams with glacial influence, except at one site, and is thus a significant departure from the conceptual model.

Studies have demonstrated the importance of groups that have not been extensively investigated in glacier-fed streams and which were not included in the conceptual model. For example, Maiolini and Lencioni (2001) and Lods-Crozet (2001) both highlighted the importance of

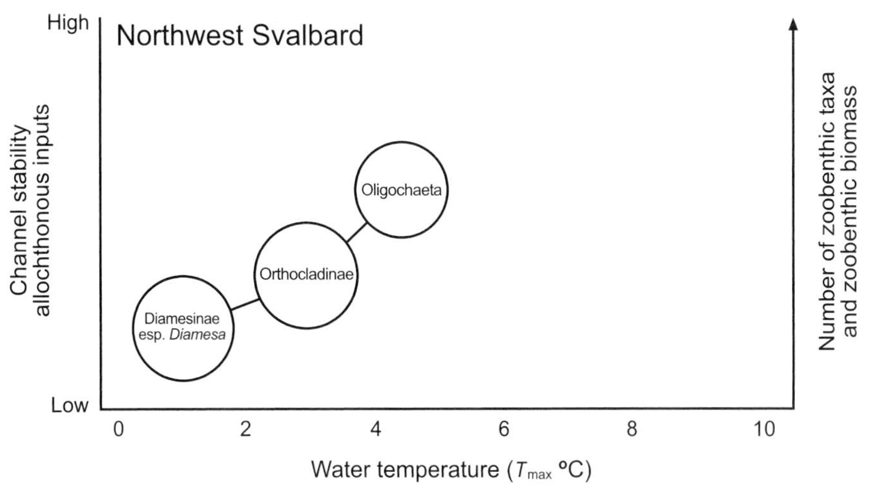

Figure 8.4 Modified conceptual model of macroinvertebrate communities for glacier-fed rivers in Svalbard. Blaen *et al.*, 2014. Reproduced with permission from John Wiley & Sons.

Nematoda at sites close to glacial sources in streams of the Italian and Swiss Alps. Eisendle (2008) specifically examined the role of free-living nematodes in glacier-fed river reaches of the Austrian Alps and found this group dominated the benthic fauna in each reach, with abundance higher than that of other invertebrates. Overall, nematode spatiotemporal distribution patterns agreed well with the model, increasing in abundance and diversity from the glacial source. Temporal differences were also evident as water sources changed, particularly with respect to recolonisation during the autumnal period at one site as glacial influence decreased. Among nematodes, cold tolerance and resistance is commonplace and thus may be advantageous to inhabitants of glacier-fed rivers for surviving winter conditions.

Exciting new directions have involved examining the role of the microbial biofilm communities in glacier-fed rivers and the major role they play in controlling numerous stream ecosystem processes (Freimann *et al.*, 2013) with potential implications for downstream biodiversity and biogeochemistry (Wilhelm *et al.*, 2013). The most dominant phyla detected in glacial habitats were Proteobacteria, Bacteroidetes, Actinobacteria and Cyanobacteria/chloroplasts.

Glacial Index and ARISE classification system

In a re-analysis of the AASER data, Jacobsen and Dangles (2012) extended the latitudinal and altitudinal gradient by including sites from Ecuador and New Zealand. Using GLM models, the strongest relationship with taxon richness was found to be the Glacial Index (GI). The GI combines glacier size with distance from the glacial terminus and was considered by Jacobsen and Dangles (2011) to perform better than T_{max} and the Pfankuch Index (channel stability) as it potentially integrates the effect of other important variables (e.g., suspended solids and substrate type). The authors propose the GI as a simple and useful measure of environmental harshness in glacier-fed streams, which is easier to measure than T_{max} and more robust than the subjective Pfankuch Index of channel stability. By extending the latitudinal and elevational range of the AASER sites to the tropics, and standardising environmental harshness through the GI, the relationship between taxon richness and latitude and elevation was hump shaped with a peak at mid-latitudes rather than linear as reported by Castella *et al.* (2001) for the AASER sites. However this hump-shaped response was absent for sites closest to the glacier, which varied only slightly along the latitudinal gradient due to strong deterministic filter of environmental harshness. This GI should not be confused with another GI Index, the Glaciality Index, of Ilg and Castella (2006), which is a measure of glacial runoff based on four environmental variables (water temperature, channel stability, conductivity and suspended sediment concentration).

A major shift in our understanding of glacier-fed rivers was the proposal of a new classification system by Brown *et al.* (2003) to better describe spatial and temporal variability in glacial, snowmelt and groundwater inputs to alpine streams, based upon the mix of proportions of water contributed from each of these sources. This was furthered by the validation of the Alpine RIver and Stream Ecosystem classification (ARISE) approach, using data collected from three streams in the French Pyrenees, which examined methods to quantify runoff from different water sources in alpine catchments (Brown *et al.*, 2009). Brown *et al.* (2010) compared the ARISE approach with the GI Index of Ilg and Castella (2006) and found both were significant predictors of macroinvertebrate taxonomic richness, beta diversity, the number of EPT genera and total abundance, although regression models were typically stronger for the ARISE meltwater contribution approach. However, at the species level, ARISE performed better for predicting the abundance of 13 of the 20 most common taxa.

Brown *et al.* (2007) suggested that as meltwater contributions decline (associated with lower suspended sediment concentrations, higher water temperature and electrical conductivity) alpha diversity at a glacier-fed river site will increase. However, beta diversity (between-sites) will be reduced, because the habitat heterogeneity associated with spatiotemporal variability of water source contributions will become lower. Brown *et al.* (2007) predicted that some endemic alpine aquatic species (such as the Pyrenean caddis fly *Rhyacophila angelieri*) and cold water stenothermic taxa would become extinct with decreasing glacial cover, leading to a decrease in gamma diversity (regional). Jacobsen and Dangles (2011) followed up with similar conclusions from analysis of the extended AASER dataset and found that environmental harshness increased beta diversity. Further analysis of datasets from Ecuadorian Alps, the Coast Range Mountains of south east Alaska and the Italian and Swiss Alps by Jacobsen *et al.* (2012) demonstrated that local (alpha) and regional (gamma) diversity, as well as turnover among reaches (beta-diversity), will be consistently reduced by the shrinkage of glaciers. The authors demonstrated that 11–38% of the regional species pools, including endemics, would be expected to be lost following a total disappearance of glaciers in a catchment. Also a steady shrinkage of the glacier is likely to reduce taxon turnover in proglacial river systems and local richness at downstream reaches, where glacial cover in the catchment is less than 5–30%. These sensitive macroinvertebrate taxa may be important biological indicators of environmental change in glacierised river basins and the reduction in glacial mass (Brown *et al.*, 2007).

Khamis (2015) assessed biodiversity patterns across a full spectrum of glacial influence, using end-member mixing models to quantify meltwater contributions, and identified a unimodal community level response (e.g., richness and abundance), which was consistent across basins of differing geology. These findings were similar to Jacobsen and Dangles (2011) and the hump-shaped response along a gradient of environmental harshness. However, it is important to note that the specific mechanisms for higher diversity at intermediate meltwater disturbance sites may be due to colonisation – competition trade off, due to the patchiness of refugia and resources at intermediate meltwater sites, rather than disturbance reducing competition/competitive exclusion.

Summary and future directions

Looking back to that night at the public house in Stirling 24 years ago and the back of the cigar packet doodles, we had no idea that the conceptual model for the structure of glacier-fed rivers and the communities that these systems potentially support would lead us on a path of research to test the model in various parts of the world and would inspire others to do likewise. Although the initial model was not accurate in certain regions and modifications were necessary, overall the conceptual framework has proven relatively robust due to the overriding dominance of the variables of water temperature and channel stability in creating deterministic patterns within the macroinvertebrate community.

Glacier-fed rivers are predominantly at high latitudes and altitudes, environments that are some of the most sensitive to climate change. In the current phase of global climate warming, many glaciers are shrinking (Barry, 2006) and thus it is essential we fully understand the effect on glacier-fed river communities. Loss of snow and ice-masses will alter spatial and temporal dynamics in runoff with important changes in the relative contributions of snowmelt, glacier-melt and groundwater to stream flow (Milner *et al.*, 2009). Using a dose–response relationship model (i.e., glacier cover – taxa abundance relationship), Khamis *et al.* (2014a) simulated anticipated changes in taxa abundance representing key groups (i.e., predator, shredder, grazer, filterer and a cold stenotherm) based on projected changes in glacier cover to 2080 from TOPKAPI modelling. The model predicted the complete loss of the cold stenotherm (*Diamesa latitarsis* grp.) from the Taillon–Gabiétous basin in the French Pyrenees by 2070 as glacier cover disappeared. The model also indicated increased abundance of more generalist shredding, grazing and filtering taxa (e.g., *Leuctra* spp., *Rhithrogena* spp. and *Simulium* spp.) at the expense of the cold stenotherms, as water temperature and channel stability increase in line with the conceptual model. This predicted replacement of the glacial stream specialist supports earlier suggestions that gamma and beta diversity will be reduced as glaciers recede (Brown *et al.*, 2007; Jacobsen *et al.*, 2012; Finn *et al.*, 2013).

With relation to beta diversity, Finn *et al.* (2013) investigated genetic beta diversity in a population of *Baetis alpinus* from the French Pyrenees by investigating mitochondrial DNA haplotypes for 1113 individuals. The greatest genetic diversity for this species was in headwater areas, with more homogenisation evident in downstream reaches. Finn *et al.* (2013) concluded that extreme conditions (e.g., low temperature, high instability, isolation) in high-glaciality streams probably enhance beta diversity at both the genetic and at the community level. Similarly, Wilhelm *et al.* (2013) found that the beta diversity of biofilms decreased with increasing streamwater temperature away from the glacier margin, suggesting that glacier retreat may contribute to the homogenisation of microbial communities among glacier-fed streams. Clearly genetic investigations would allow further insights into diversity responses to shrinking levels of glacierisation at different levels of the community.

The conceptual model of glacier-fed rivers is focused purely on the structure of the community and not how it functions. Indeed, it may be that even with a loss of beta and gamma diversity the functioning of the system remains similar or may even be more efficient with greater trait diversity and routes for carbon fluxes. It is important that other attributes of macroinvertebrate communities are examined with respect to changing water sources and reduced glacial influence to fully understand their effects. The examination of macroinvertebrate traits with relation to changing glacial influence (similar to the study by Brown and Milner 2010) would be of value and provide insights into ecosystem function.

For example, Clitherow *et al.* (2013) examined the food web structure in the harsh environment of a glacier-fed river in Austria and found low taxon richness, highly connected individuals and a short mean food chain length compared to other studies. These data suggest that changes at one node will rapidly spread through the network.

Another potentially interesting research avenue could be the identification of critical thresholds and tipping points of these systems as glacial runoff becomes reduced. For example, in the first study of this nature, Khamis *et al.* (2014b) used TITAN (Threshold Indicator Taxa ANalysis) to identify critical threshold changes in community composition of river taxa in the French Pyrenees as < 5.1% glacier cover and contributions of meltwater as < 66.6%. Below these thresholds the cold stenothermic benthic invertebrate taxa, *Diamesa* spp. and the Pyrenean endemic *Rhyacophila angelieri* were lost. Generalist taxa including *Protonemura* sp., *Perla grandis, Baetis alpinus, Rhithrogena loyolaea* and *Microspectra* sp. increased when glacier cover was < 2.7 % and meltwater was < 52 %. This kind of analysis would be interesting to carry out on a wider spatial scale, although it would probably have to be carried out using percent glacial cover, due to the lack of present data using percent meltwater.

Development of a monitoring network of studies of glacier-fed rivers: sentinels of climate change

Glaciers behave differently across different climatic zones, hence glacier-fed stream habitats and ecology also differ (Jacobsen *et al.*, 2010). Even within the same climatic zone, glaciers differ with respect to their distribution, size, slope and geology and this is expected to have profound effects on the physicochemical habitat template, biodiversity and function of downstream ecosystems. It is critical that we obtain a broader understanding of glacier-fed freshwater ecosystems across a wider spatial and temporal scale to improve our understanding of these systems as sentinels for climate change. It is thus essential that we develop a worldwide monitoring network of glacier-fed systems within different climatic zones where investigators follow the same protocols in both physico-chemical and biological investigations using the cutting edge methodologies that have recently been developed. The establishment of a network was furthered by a recent European Science Foundation workshop attended by specialists in the field of glacier-fed rivers in Birmingham in September 2013. The title of the workshop was GLACier-fed rivers, HYDRoECOlogy and climate change; current knowledge and future NETwork of monitoring sites (GLAC-HYDRECO-NET). Some potential sites are well established where previous research has been undertaken, others would have to be initiated. Regions that should be included are the: Arctic – Svalbard, Greenland; Subarctic – Alaska, Iceland, Norway; Temperate – Switzerland, Austria, Italy, Pyrenees, New Zealand; Subtropical – China, Tibet, Nepal; Tropical – Bolivia; and Equatorial – Ecuador. By following similar protocols, data collected from sites in glacier-fed watersheds would be comparable and driving variables and underlying trends would be easier to distinguish across a range of glacier types and climatic, elevational and latitudinal gradients. It would be important that well-designed experiments (e.g., the use of mesocosm channels) be a key component of this network to evaluate responses to changing conditions, such as the role of increased abundance of large-bodied predators on a benthic community as water sources change and environmental harshness ameliorates (see Khamis, 2015), and also the effect on ecosystem functioning.

Acknowledgements

I would like to thank the many colleagues that I have worked with on glacier-fed rivers since 1994 to increase our knowledge, including all the AASER group, David Hannah of the University of Birmingham and Lee Brown of the University of Leeds. I am grateful for the efforts of former and present graduate students that have worked on glacier-fed rivers in various parts of the world: Debbie Snook, Lee Brown, Sarah Hainie, Christopher Mellor, Philip Blaen, Kieran Khamis and Catherine Docherty, and who have significantly increased our understanding of these systems.

References

Barry R.G. (2006) The status of research on glaciers and global glacier recession: a review. *Progress in Physical Geography* 30: 285–306.

Blaen, P.J., Hannah, D.M., Brown, L.E. & Milner, A.M. (2014) Hydrological drivers of macroinvertebrate communities in High Arctic rivers (Svalbard). *Freshwater Biology* 59*:* 378–391.

Brittain, J.E. & Milner, A.M. (2001) Ecology of glacier-fed rivers: current status and concepts. *Freshwater Biology* 46: 1571–1578.

Brown, L.E., Hannah, D.M. & Milner, A.M. (2003) Alpine stream habitat classification: an alternative approach incorporating the role of dynamic water source contributions. *Arctic, Antarctic and Alpine Research* 35: 313–322.

Brown, L.E, Milner, A.M. & Hannah, D.M. (2007) Vulnerability of alpine stream biodiversity to shrinking snowpacks and glaciers. *Global Change Biology* 13: 958–966.

Brown, L.E, Hannah, D.M. & Milner, A.M. (2009) ARISE: A classification tool for Alpine RIver and Stream Ecosystems. *Freshwater Biology* 54: 1357–1369

Brown, L.E., Milner, A.M. & Hannah, D.M. (2010) Predicting river ecosystem response to glacial meltwater dynamics: a case study of quantitative water sourcing and glaciality index approaches. *Aquatic Sciences* 72: 325-334.

Buytaert, W., Cuesta-Camacho, F. & Tobon, C. (2011) Potential impacts of climate change on the environmental services of humid tropical alpine regions. *Global Ecology and Biogeography* 585: 19–33.

Cadbury, S.L., Milner, A.M. & Hannah, D.M. (2010) Hydroecology of a New Zealand glacier-fed river: linking longitudinal zonation of physical habitat and macroinvertebrate communities. *Ecohydrology* 4: 520–531.

Castella, E., Adalsteinsson, H., Brittain, J.E., *et al.* (2001) Macrobenthic invertebrate richness and composition along a latitudinal gradient of European glacier-fed streams. *Freshwater Biology* 46(12): 1811–1832.

Clitherow, L.R., Carrivick, J.L. & Brown, L.E. (2013) Food web structure in a harsh glacier-fed river. *PLOS ONE:* 8.

Eisendle, U. (2008) Spatiotemporal distribution of free-living nematodes in glacial-fed stream reaches (Hohe Tauern, Eastern Alps, Austria). *Arctic, Antarctic, and Alpine Research* 40(3): 470–480.

de Groot, R.S., Alkemade, R., Braat, L., *et al.* (2010) Challenges in integrating the concept of ecosystem services and values in landscape planning, management and decision making. *Ecological Complexity* 7: 260–272.

Freimann, R., Bu, H., Findlay, S.E.G. & Robinson, C.T. (2013). Response of lotic microbial communities to altered water source and nutritional state in a glaciated alpine floodplain. *Limnology and Oceanography* 58: 951–965.

Friberg, N., Milner, A.M., Svendsen, L.M. *et al.* (2001) Distribution of macroinvertebrates in streams of Greenland. *Freshwater Biology* 46: 1753–1764.

Fleming, S.W. & Clarke, G.K.C. (2005) Attenuation of high-frequency interannual streamflow variability by watershed glacial cover. *Journal of Hydraulic Engineering* 131: 615–618.

Finn, D.S., Rasanen, K. & Robinson, C.R. (2010) Physical and biological changes to a lengthening stream gradient following a decade of rapid glacial recession. *Global Change Biology* 16: 3314–3326.

Finn, D., Khamis, K. & Milner, A.M. (2013) Loss of small glaciers will diminish beta diversity in Pyrenean streams at two levels of organization. *Global Ecology and Biogeography* 22: 1466–8238.

Freimann, R., Bu, H., Findlay, S.E.G. & Robinson, C.T. (2013) Response of lotic microbial communities to altered water source and nutritional state in a glaciated alpine floodplain. *Limnology and Oceanography* 58: 951–965.

Gislason, G.M., Adalsteinsson, H., Hansen, I., *et al.* (2001) Longitudinal changes in macroinvertebrate assemblages along a glacial river system in central Iceland. *Freshwater Biology* 46: 1737–1752.

Hamerlik, L. & Jacobsen, D. (2012) Chironomid (Diptera) distribution and diversity in Tibetan streams with different glacial influence. *Insect Conservation and Diversity* 5: 319–326.

Ilg, C. & Castella, E. (2006) Patterns of macroinvertebrate traits along three glacial stream continuums. *Freshwater Biology* 51: 840–853.

Hannah, D.M., Brown, L.E., Milner, A.M., *et al.* (2007) Integrating climatology-hydrology-ecology for alpine river systems. *Aquatic Conservation: Marine & Freshwater Ecosystems* 17: 636–656

Hieber, M., Robinson, CT., Uehlinger, U., *et al.* (2005) A comparison of benthic macroinvertebrate assemblages among different alpine streams. *Freshwater Biology* 50: 2087–2100.

Jacobsen, D. & Dangles, O. (2012) Environmental harshness and global richness patterns in glacier-fed streams. *Global Ecology and Biogeography* 21: 647–656.

Jacobsen, D., Dangles, O., Andino, P., *et al.* (2010) Longitudinal zonation of macroinvertebrates in an Ecuadorian glacier-fed stream: do tropical glacial systems fit the temperate model? *Freshwater Biology* 55: 1234–1248.

Jacobsen, D., Milner, A.M., Brown, L.E. & Dangles, O. (2012) Biodiversity under threat in glacier-fed river systems. *Nature Climate Change* 2: 361–364.

Khamis, K., Hannah, D.M., Clarvis, M.H., *et al.* (2014a) The use of invertebrates as indicators of environmental change in alpine rivers and lakes. *Science of The Total Environment* 493: 1242–1254.

Khamis, K., Hannah, D.M., Brown, L.E. & Milner, A.M. Predation modifies macroinvertebrate community structure in alpine stream mesocosms. *Freshwater Science* 34: 66–80.

Khamis, K. (2014b) Climate change and glacier retreat in the French Pyrenees: implications for alpine river ecosystems. unpublished PhD thesis University of Birmingham.

Khamis, K., Hannah, D.M., Brown, L.E. & Milner, A.M. (2015) Glacier-groundwater stress gradients control alpine river biodiversity. Ecohydrology.

Kuhn, J., Andino, P., Calvez, R., *et al.* (2011) Spatial variability in macroinvertebrate assemblages along and among neighbouring equatorial glacier-fed streams *Freshwater Biology* 56: 2226–2244.

Lods-Crozet, B., Lencioni, V., Olafsson, J.S., *et al.* (2001) Chironomid (Diptera: Chironomidae) communities in six European glacier-fed streams. *Freshwater Biology* 46: 1791–1810.

Lods-Crozet, B., Lencioni, V., Brittain, J.E. & Marziali, L. (2007) Contrasting chironomid assemblages in two high Arctic streams on Svalbard. *Fundamental and Applied Limnology* 170:211–222.

Maiolini, B. & Lencioni, V. (2001) Longitudinal distribution of macroinvertebrate assemblages in a glacially influenced stream system in the Italian Alps. *Freshwater Biology* 46: 1625–1640.

Milner, A.M. & Petts, G.E. (1994) Glacial rivers: Physical habitat and ecology. *Freshwater Biology* 32: 295–307.

Milner, A.M., Brittain, J.E., Castella, E. & Petts, G.E. (2001a) Trends of macroinvertebrate community structure in glacier-fed streams in relation to environmental conditions: a synthesis. *Freshwater Biology* 46: 1833–1848.

Milner, A.M., Taylor, R.C. & Winterbourn, M.J. (2001b) Longitudinal distribution of macroinvertebrates in two glacier-fed New Zealand rivers. *Freshwater Biology* 46: 1765–1776.

Milner, A.M., Brown, L.E. & Hannah, D.M. (2009) Hydroecological effects of shrinking glaciers. *Hydrological Processes* 23: 62–77.

Murakami, T., Hayashi, Y., Minami, M., *et al.* (2012) Limnological features of glacier-fed rivers in the Southern Tibetan Plateau, China. *Limnology* 13: 301–307.

Muhlfeld, C.C., Giersch, J.J., Hauer, F.R., *et al.* (2011) Climate change links fate of glaciers and an endemic alpine invertebrate. *Climatic Change* 106: 337–345.

Saether, O.A. (1968) Chironomids of the Finse Area, Norway, with special reference to their distribution in a glacier brook. *Freshwater Biology* 64: 426–483.

Smith, B.P.G., Hannah, D.M., Gurnell, A.M., & Petts, G.E. (2001) A hydrogeomorphological framework for the ecology of alpine proglacial rivers. *Freshwater Biology* 46: 1579–1596.

Snook, D.L. & Milner, A.M. (2001) The influence of glacial runoff on stream macroinvertebrates in the Taillon catchment, French Pyrenees. *Freshwater Biology* 46: 1609–1624.

Stahl, K., Moore, R.D., Shea, J.M., *et al.* (2008) Coupled modelling of glacier and streamflow response to future climate scenarios. *Water Resources Research* 44: W02422.

Steffan, A.W. (1971) Chironomid (Diptera) biocoenoses in Scandinavian glacier brooks. *The Canadian Entomologist* 103: 477–486.

Wilhelm, L., Singer, G.A., Fasching, C. *et al.* (2013) Microbial diversity in glacier-fed streams. *ISME Journal* 7: 1651–1660

Winterbourn, M.J., Cadbury, S., Ilg, C. & Milner, A.M. (2008) Mayfly production in a New Zealand glacial stream and the potential effect of climate change. *Hydrobiologia* 603: 211–219.

CHAPTER 9

Remote sensing: mapping natural and managed river corridors from the micro to the network scale

David J. Gilvear[1], Peter Hunter[2] and Michael Stewardson[3]

[1] *School of Geography, Earth and Environmental Sciences, Plymouth, Devon, UK*

[2] *School of Natural Sciences, University of Stirling, Stirling, Scotland*

[3] *Department of Infrastructure Engineering, Melbourne School of Engineering, The University of Melbourne, Victoria, Australia*

Introduction

At the core of river science is quantifying the spatial and temporal dynamics of river flow, water quality, physical habitat, and plants and animals at multiple scales (Gilvear *et al.*, Chapter 1, this volume). The advent and development of remote sensing, with now near global coverage at medium scales of spatial resolution, has revolutionised our capacity to map and analyse spatial and temporal variability of the individual components of river ecosystems and aquatic processes in natural and managed systems (Marcus and Fonstad, 2010). This capacity of remote sensing can also provide output in terms of 2-dimensional vertical or horizontal or 3-dimensional perspectives and normally at all but the smallest scales (e.g., sub-centimetre), and the shortest (sub-hourly) and longest timescales (>50 years). In this chapter remote sensing is deemed to be earth observation from spaceborne, airborne and terrestrially based sensors that measure the electromagnetic radiation emitted or reflected from land and water surfaces within the river corridor network – namely in-channel, riparian and floodplain. Its application to river science and management will be examined in relation to hydrology, water quality, geomorphology, ecology and river engineering.

In essence the physical template of river ecosystems is made up of three principal land–water cover components, namely water and its chemical constituents, minerogenic matter (i.e., soils, sediment and bedrock) and living (e.g., algae, aquatic macrophytes and alluvial woodland) and dead plant material (e.g., woody debris) draped over the river corridor topography. Remote sensing has the ability to survey topography and the three cover components all have unique spectral signatures and can often be differentiated with appropriate sensors and use of discrete wavelengths within the electromagnetic spectrum. Hence, remote sensing can reveal the basic spatial and temporal patterns of the physical habitat of rivers and associated processes. The challenge for remote sensing is to interrogate specific wavelengths to discover the suite of

River Science: Research and Management for the 21st Century, First Edition.
Edited by David J. Gilvear, Malcolm T. Greenwood, Martin C. Thoms and Paul J. Wood.

attributes of each land–water component. Components include channel bathymetry, floodplain topography, above and below the water line bed material particle size, water quality and temperature, floating, submerged and emergent aquatic plant biomass and 3-dimensional terrestrial vegetation structure. In terms of managed river systems, there is the added complexity of detection of human modifications such as bridges, flood embankments, weirs and dams, and buildings. Detection of these parameters and features involves shape recognition, sometimes in association with other spectral properties, to differentiate largely similar exposed soil and sediment land cover types.

The potential of remotely sensing rivers is simply a function of resolving key issues surrounding the size and nature of the river, the river science phenomena of interest and the remote sensing platforms and sensors available. In relation to the nature of the river, whether the phenomena of interest is above or below water is a key issue that affects the approach taken. The extent to which any feature is obscured by overhead canopies or cast in shadow is another issue. If the phenomena of interest is below the water line, depth and water clarity are critical. The key remote sensing factors determining the outcome of a river science-focused remote sensing study is that of atmospheric conditions, imagery spatial coverage and resolution, length and temporal resolution of datasets (Figure 9.1), sensor spectral resolution and retrieval of information through image analysis. These factors will all be considered in this chapter but is worth highlighting at the start the key areas of the electromagnetic spectrum and their relevance to river science. At its simplest, with regard to the visible part of the spectrum, blue wavelengths are best for water penetration and bathymetry and substrate mapping. Green wavelengths are best for mapping water turbidity and vegetation types and plant vigour. In the near infra-red delineation of water bodies is most easily achieved and soil moisture and vegetation communities are well discriminated. Thermal data is good for water temperature determination, soil moisture variability and vegetation stress. Wavelengths outside of the visible spectrum are useful because they can penetrate cloud cover, measure variables such as surface temperature of land and water and ground elevation.

The challenge for river science, at the broadest level, is firstly to optimise river remote sensing within existing research frameworks. More importantly the challenge is to make use of the remote sensing capability in terms of developing new approaches towards river science that lead to better understanding and model capability of natural and human-modified river ecosystems. Meeting the challenge will require an interdisciplinary approach and one where the river scientists and remote sensing communities can appreciate each other's needs, approaches and goals. This chapter aims to highlight the development of river remote sensing through time, illustrate the opportunities and challenges of river remote sensing and highlight its future potential in terms of research and application.

A chronology of the science of remote sensing of river systems

Changing remote sensing technology and river science

The very earliest form of remote sensing was traditional panchromatic photography. Early photographs from the late Victorian period when compared to modern day

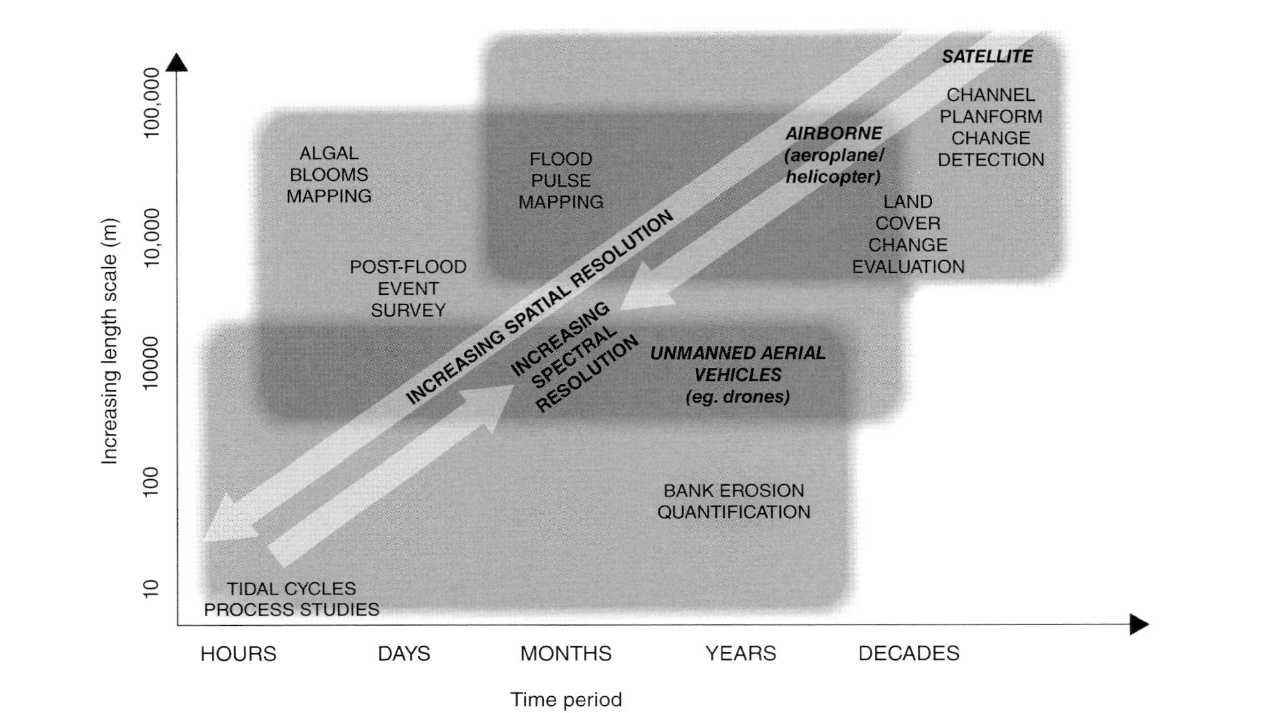

Figure 9.1 Remote sensing platforms (satellite, airborne and unmanned aerial vehicles) and their capability in terms of temporal and spatial components of river systems, spatial and spectral resolution and examples of river science application.

images can allow historical reconstruction of changes in channel morphology and riparian corridor habitat. The first landmark event for river science was aerial photography (initially black and white then colour and infra-red), which now provides coverage of some river corridors back to the 1940s, providing a 70-year historical archive (Figure 9.2). For example, using sequential sets of aerial photography, Gurnell *et al.* (1994) constructed a 60-year timeline of bank erosion on the River Dee, England. However, early coverage was sparse and patchy, rarely focused on river corridors and temporally highly intermittent, often with gaps of decades. Since then image analysis has been used to maximise the utility of old imagery, such as black and white aerial photography, beyond basic river network and channel planform mapping (e.g., Gilvear *et al.*, 1998; Westaway *et al.*, 2003) but the basic problems of spatial coverage, and lack of multi-spectral capability, will remain for river scientists trying to reconstruct mid-twentieth-century channel dynamics. Remotely sensed multi-spectral data with global coverage but at low spatial resolution followed the launch of the first Landsat satellite in 1972. Since 1972 there have been seven Landsat satellites and two are still in existence. These satellites maintain a 16–18 day cycle and thus can give high temporal resolution, but except for on the world's largest rivers the spectral resolution is limiting. Only in the case of the panchromatic mode of the current ETM satellite is a spatial resolution of better than 30 metres obtained, with pixel size being 15 metres. France *et al.* (1986) working in Wales concluded that Landsat TM data could record streams down to 3–5 metres width with acceptable accuracy. Thirty-three first-order streams were thus detected. However, scrutiny of 1:10,000 aerial photographs revealed 156 first-order streams, many of which were less than 1.0 m wide. As such the earliest sensors only provided detailed reach-scale information on the world's largest river systems. In low-order streams, where channel widths are smaller, spatial

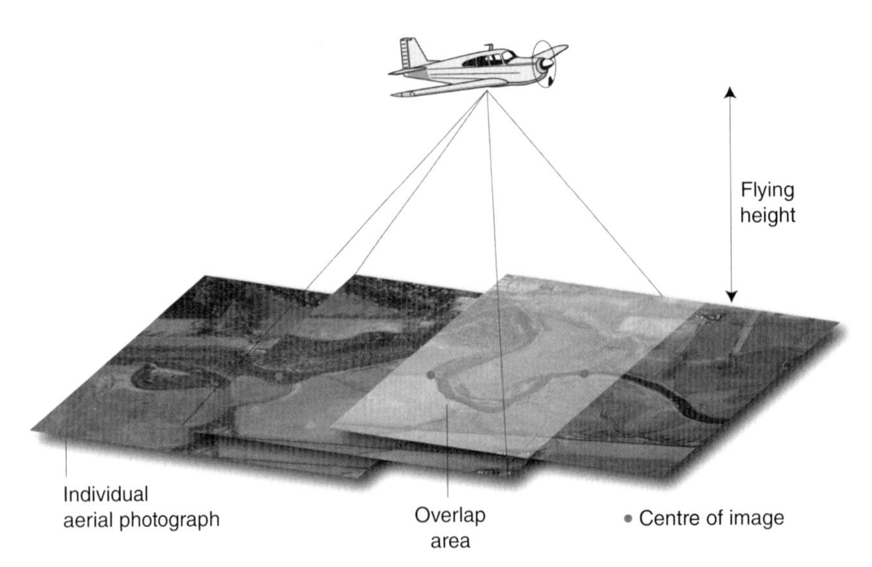

Figure 9.2 A mosaic of aerial photographs taken from a plane showing the river corridor of the River Tay, Scotland. (*See colour plate section for colour figure*).

resolution becomes especially critical. SPOT satellite imagery, coming online in 1986 and with a five-day off-vertical imaging capability, provided an improvement in spatial resolution (5–10 metres). A large number of spaceborne sensors have been launched in the last two decades each with either or both spatial and spectral resolution improved. This considerably advanced the potential for river scientists to use remotely sensed data for discriminating individual fluvial features on large rivers and extend remote sensing for basic mapping of small and medium sized rivers (Figure 9.1). For example, IKONOS launched in 1999 is 4-metre resolution in multi-spectral mode and 1 metre in panchromatic mode. Quickbird launched in 2002 has 2.5 metres in multi-spectral and 0.6 metres in panchromatic modes. Other sensors have high spectral capability, for example the 12-bit 36-band MODIS operating in the 0.4 to 14.4 μm wavelengths, but for most applications the 250 m+ resolution is too coarse – although it very much depends upon application.

Issues around the coarse spatial resolution provided by satellite sensors led to significant research using sensors mounted on airborne platforms, where specially commissioned aircraft flights followed river courses and were able to provide sub 2 m resolution data with either multi-spectral or hyper-spectral capability. Airborne thematic mapper data with 11 spectral bands including one in the thermal part of the spectrum was often the data of choice in the 1990s. Post 2000, the compact airborne imaging spectrometer with 256 spectral bands to choose from has been the sensor most often used in mapping channel morphology and physical habitat ,particularly on European and North American rivers (e.g., Legleiter *et al.*, 2004; Marcus *et al.*, 2003; Winterbottom and Gilvear, 1997). Airborne LiDAR has also become integral to the river scientist's toolkit for morphological mapping above the water line in terms of interrogating the morphology of exposed channel bars and floodplains (Charlton *et al.*, 2003). Helicopter-based videography also provides an effective method for gaining imagery of river morphology and habitat, but quantification of habitat from the imagery is problematic (Kleynhans, 1996).

For small streams and reach-scale studies sensors mounted on various low-altitude unmanned aircraft (UAVs) and balloons have been employed (e.g., Flener *et al.*, 2013; Figure 9.1). Flying at heights of less than 200 metres, a ground resolution of 1–5 cm can be obtained using standard cameras. Such approaches have been used to map channel bathymetry of the Durance River in France (Feurer *et al.*, 2008). The current ongoing advancement in 'drone' technology is likely to see further uptake and progress in this emerging field of remote sensing and river science. Terrestrially mounted sensors have also been employed successfully; for example, Resop and Hession (2010) used repeated terrestrial LiDAR surveys to measure bank retreat rates of 0.15 metres per annum on an 11 metre shoreline length of a creek in Virginia, USA. Similarly, terrestrial LiDAR has been used to map instream habitat complexity (Resop *et al.*, 2012). The use of field spectrometers especially to measure phenology across seasons and instrumented buoys to assist with calibration and interpretation of remotely sensed imagery is also likely to lead to greater capability with regard to information retrieval relevant to river science. Web cams with near real-time images of river water levels are also being used to observe floods, passage of woody debris, and map changes in channel bar morphology (MacVicar and Piégay, 2012).

Real-time imagery accessible through the Web is also attractive in terms of engaging public participation in river science.

Within river science it is now realised that when dealing with large or multiple river systems multi-scalar remote sensing (Whited *et al.*, 2011) is likely to be required with differing platforms and sensors all contributing invaluable information via upscaling or downscaling. In the future, riparian 'tree-mounted' web cams through to sophisticated hyper-spectral spaceborne sensors will all have a role to play depending upon application. Data fusion and the appropriate integration of differing types of data at multiple temporal and spatial scales will also be one of the keys to driving forward advancement in river science.

Advancing the subject of river science and remote sensing

Essentially, the modern era of riverine remote sensing that is now recognised as a sub-discipline of river science did not really emerge until the 1970s, and until the 1990s its scientific significance was modest. Post 1990 to the present the growth and development of the subject has been rapid, impressive and important. Reviews in the early 1990s highlighting the potential of river remote sensing included that of Muller (1992), Muller *et al.*, (1993), and Milton *et al.* (1995). Using the Web of Knowledge search tool as an index of research activity, the number of publications pre 1990 was less than 1 per year in the subject of 'remote sensing and rivers', in the 1990s this rose to 33 and post 2000 attained a value of 222. Post-millennium reviews detailing the contribution of remote sensing to river science include that of Marcus and Fonstad (2010), while the growth of the subject has also seen a research monograph on the subject edited and published (Carbonneau and Piégay, 2012). In terms of post-1990 advances, the list is broad ranging and spectacular. In terms of existing capability, for example, there have been improvements in the range of water quality parameters detectable (although often inappropriate to detection on small rivers), floodplain inundation mapping and accuracy of channel planform change estimation. Significant new capability has also emerged in river bathymetric mapping, measurement of 3D channel morphology, substrate type recognition and bed material size mapping, aquatic vegetation mapping and floodplain vegetation structure and community composition. These advances have been brought about by a mix of improvements in sensors, range of platforms used and advances in image analysis and the expertise of a small dedicated scientific community. In terms of sensors, some of the advances have been made by examining wavelengths outside the traditionally used visible wavelengths. As such, information is now available on the temperature variability of not only surface waters but also land surfaces, shedding new light on thermal habitat heterogeneity for fish and other animals (Torgerson *et al.*, 2001; Tonella *et al.*, 2010). Using acoustic devices mounted on rafts the submerged soundscapes of rivers have also been identified (Tonella *et al.*, 2010).

The work undertaken is reflected in the huge growth of journal publications over the period 1990 to 2010. The period 1990 to 2010 can rightly be thought of as decades of the 'emergence and development of riverine remote sensing'. The next two decades are likely to be seen as the period of 'application of river remote sensing to science and management'. The application of remote sensing to river management has been sporadic and primarily focused on basic land cover mapping. Studies, however, have shown the potential for assessment of such phenomena as bed degradation below dams (Huang *et al.*, 2009), bridge detection (Luo *et al.*, 2007) and ecosystem service delivery (Large and Gilvear, 2014). Web cams are also providing 24/7 coverage of channel response to dam removal in the case of the landmark removal of the Elwha dam in the USA.

As the subject of river science unfolds over the forthcoming years and decades, not only will the quality of remotely sensed data be enhanced but the historical archive will lengthen and improve our understanding of the temporal dynamics over medium and long-term timescales. As such the contribution of remote sensing to the river science and management of the future will be immense.

State of the science

Image analysis opportunities and challenges

Image radiometric and geometric rectification

Manual mapping of river features from imagery has a long history in the geographical sciences but few studies mapped large tracts of river. More recently a wealth of studies reporting the use of automated classification of river features at the reach scale are apparent, with reasonable to good accuracies (Marcus *et al.*, 2003; Carbonneau *et al.*, 2004; Wright *et al.*, 2000). In the case of mapping stream networks at the catchment scale, image processing is likely to be more complicated. To obtain full coverage imagery it is likely that imagery will have to be captured over a number of hours or on different days, creating differences in illumination and position relative to solar azimuth. The correction of airborne and satellite data for atmospheric effects is critical to the success of image classification approaches. This is especially the case when the imagery is collected over a number of hours or days as would certainly be the case with airborne imagery and large river systems. Bi-directional reflectance and lighting variations across images poses a problem when mosaicking aerial images together for river networks. Standardising image contrast can be problematic and require histogram matching using edge differences or possibly using specifically placed ground targets. There are several approaches to the atmospheric correction of airborne and satellite imagery. The dark-pixel subtraction method has been widely used and, as it is an image-based procedure, it has the benefit of being easy to implement. However, as the dark-pixel correction is largely scene-dependent, it may not be appropriate for the correction of multi-temporal imagery covering differing geographical locations. The empirical line method is also widely used and, provided reflectance spectra of several ground-based reference targets are available, it can be used to correct airborne and satellite at-sensor-radiance to a standardised property such as reflectance. Obviously, such approaches are not required for satellite sensors, whose dedicated atmospheric correction models can provide atmospherically corrected remote sensing products. However, where such products are not available, the development of commercially available atmospheric correction models, such as ATCOR and MODTRAN, means that multi-temporal, multi-sensor, imagery products can be atmospherically corrected by the user (Yuen and Bishop, 2004). The latest generation of the atmospheric correction models, such as FLAASH, now require only minimal information on the nature of the atmosphere to enable an effective correction (Alder-Golden *et al.*, 2005; Schaepman *et al.*, 2005) and as such there is hope for the future in terms of river-network-wide application where the nature of the atmosphere at the time of acquisition is unknown.

Sun angle and viewing geometry effects can also cause problems for accurate image classification, particularly where canopy or terrain shadows occur. These effects can be minimised by acquiring data under comparable conditions, but there are problems unique to rivers. Water is sensitive to sun–target–sensor geometry and in the case of sinuous channels this poses a problem that will need special attention. Indeed, glare from specular reflectance can obscure all information. In some instances, topographical correction models can be used to correct for the effects of terrain shadows (Ekstrand, 1996; Soenen, *et al.*, 2005; Wu *et al.*, 2008) and such approaches might be applicable to removing cast shadow from riparian canopies. Conyers and Fonstad (2005) have recently developed such a technique for shadow removal on river water. The information collected by on-board GPS-based navigation systems means many airborne and satellite products can often be automatically geo-rectified and geo-registered without the need for manual registration to a geographic coordinate system. In many cases, however, imagery may still need rectification and ground control points may be necessary. In areas of human occupation common control points visible on the ground can often be used (Konrad *et al.*, 2008), but in natural settings ground-based targets may need to be placed (Lejot *et al.*, 2007). The experience with unmanned airborne vehicles is that geo-rectification can be complex even with targets (e.g., Lejot *et al.*, 2007).

Image classification and feature recognition

Fluvial systems provide a challenging environment in which to classify features of interest from imagery, in that they can be submerged or are only evident to an 'expert' observer by taking into account subtle morphological, sedimentological or vegetative clues. For example, robust repeatable methods for identification of bankfull channel capacity in the field via a break of slope or vegetation limit have proved difficult. However, in some cases a 'birds-eye' view, by showing large-scale patterns, may facilitate identification of features whether by manual or automated methods. Manual methods of identification can be effective if a protocol for recognition is developed and universally applied. However, coverage of a large river system at the necessary spatial resolution may make it unrealistic for very large river networks. Approaches to the automated classification of airborne and satellite imagery have advanced tremendously, over recent years from basic hard-boundary classification algorithms, such as minimum distance and maximum likelihood classification (Thomson *et al.*, 1998; Bryant and Gilvear, 1999), to more advanced soft-boundary approaches, such as linear spectral mixture modelling and support vector machine classifiers (Brown *et al.*, 1999; Brown *et al.*, 2000; Luo *et al.*, 2007). Thus, Carbonneau *et al.* (2004) used textural analysis to map D50 grain size in gravel bed rivers. These new approaches offer increased classification power and accuracy in terms of mapping rivers, and can also be used to provide a sub-pixel level of classification where required (e.g., when analysing data of a coarse spatial resolution), although in a fluvial environment this is unlikely to be that useful except in the case that large tracts of exposed sediment or open water exist. In most instances, the collection of adequate ground-reference data for the training of algorithms is vital to the success of any classification approach, particularly where artificial-learning-based algorithms are being employed (e.g., neural networks). Sufficient ground-reference data must also be available for the validation of the classified image. In a limited number of cases, techniques have been developed that are not reliant upon field-based measurements. Fonstad and Marcus (2005) have developed a method for automated mapping of water depth and applied the technique to the Nuaces River in Texas using 33 digital orthophoto quadrangles. Similarly, Carbonneau *et al.* (2004), using a set of 3 cm resolution airborne digital imagery covering the full 80 km of the Sainte-Marguerite River in Quebec, Canada, automatically derived high-resolution measurements of flow depth, substrate size and flow velocity, and the image analysis techniques once calibrated are fully automated and can be run without user intervention for the entire image set. In the case of large river systems there is therefore usually the need for parallel field campaigns, and given the fact that field-based reference and validation data is best collected at the time of image acquisition, more than one field team may be necessary with availability at short notice as a weather window emerges. A variation on the use of ground reference data for algorithm training purposes is the collation of a spectral library (typically using in situ reflectance spectra of the classes of interest), which can be subsequently used to perform a spectral-matching-type classification (e.g., Kutser *et al.*, 2006; van der Meer, 2006). Such feature specific recognition algorithms may be required to identify attributes such as woody debris. Such approaches, however, would still require adequate ground reference data for validation purposes.

River science and remote sensing sub-disciplines

This section presents brief resumes and examples of the capability, nature, potential and challenges of remote sensing of rivers across the key sub-disciplines of river science (Figure 9.3). It gives a glimpse into the potential of remote sensing to contribute to the development of river science.

Hydrology

Remote sensing of river hydrology has primarily focused on two areas: floodplain inundation and roughness mapping, and channel discharge estimation (Table 9.1). Remotely sensing flood inundation has a long history. Here we explore some recent and notable studies and advances over the last two decades. Alsdorf *et al.* (2000) mapped water level changes within flooded Amazonian riparian forest by using the double bounce returns of water and vegetation surfaces using spaceborne interferometric synthetic aperture radar (SAR). The processed data showed that within 20 km of the main channel water level recession occurred at rates of 7–11 cm per day, while at distances of 80 kilometres the rate was only 2–5 centimetres per day. Such knowledge is useful for many areas of river science, perhaps most importantly in relation to providing empirical data related to the flood pulse concept. A rapidly emerging theme is quantification of floodplain roughness (e.g., Straatsma and Baptiste, 2008; Wilson and Atkinson, 2007). Antonarakis *et al.* (2008a) used object-based classification of LiDAR data. Forzieri *et al.* (2011) took this further and fused Quickbird multi-spectral imagery with airborne laser scanning data to

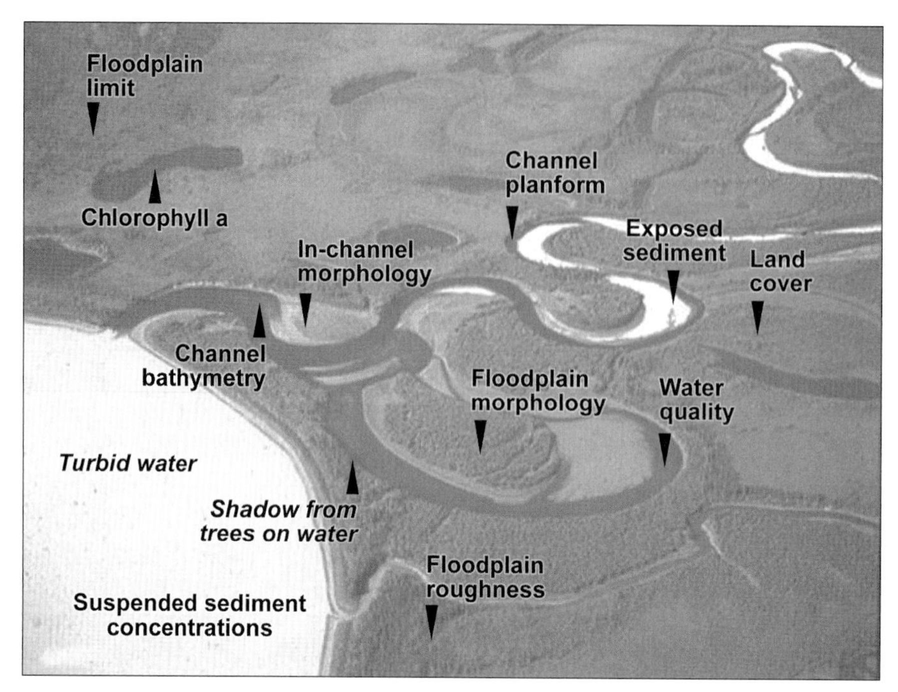

Figure 9.3 An oblique aerial view of a tributary entering the River Yukon with illustrations of the capability and challenges (in italics) of remote sensing in terms of capturing information relevant to river science.

examine roughness with riparian systems. The methodology is repeatable and could be used to examine change with time of the effects of, for instance, floodplain forest regeneration on flood flow attenuation. There is considerable scope for significant advancement of remote sensing in shedding light on the flow attenuation and water residence effects of alluvial forests and floodplain wetlands (Figure 9.4). Such knowledge has obvious importance in illustrating the ecosystem service value of river corridors in a natural or semi-natural state in terms of reducing downstream flooding.

Water quality

Remote sensing of water quality has a long and established record in marine, estuarine and lake environments. The spatial resolution of sensors in the past has thwarted the adoption of remote-sensing-based water quality algorithms for the river environment. There are still issues on small and medium width rivers and where shallow water results in reflectance from the substrate and hence an additional issue compared to deep water environments.

Disentangling the effects of multiple water quality parameters can also be difficult, but spectral unmixing has been used with success in rivers to determine parameters such as chlorophyll a, suspended sediment, water colour and dissolved organic carbon (Table 9.1). Chlorophyll a has been the most widely derived water quality parameter. Examples of good correlations of field measurement of chlorophyll a with remotely sensed data have been achieved on large rivers in the USA (Shafique *et al.*, 2001; Olmanson *et al.*, 2013). Turbidity is another parameter that appears to be easily detectable, for example using the AVRI

Table 9.1 Recent examples (post 2000) of the application of remote sensing to and river science (hydrology, water quality, geomorphology, and ecology).

River location	Scientific purpose	Imagery platform, type, and scale	Reference
(a) Hydrology			
Amazon floodplain	Mapping of the nature of inundation	JERS I	Alsdorf *et al.* (2007)
River Meuse in The Netherlands	Mapping of the extent of floodplain inundation	SAR	Bates and DeRoo (2000)
Sieve River, Italy	Floodplain forest hydraulic roughness	Quickbird and airborne laser scanning	Forzieri *et al.* (2011)
	Floodplain extent	Synthetic Aperture Radar	Horritt *et al.* (2001)
(b) Water quality			
Yangtze River, China	Suspended sediment	Landsat-7 ETM+	Wang *et al.*, (2009)
	Water temperature		Torgerson *et al.*, (2001)
Great Miami River, Ohio	Chlorophyl a	Hyperlion and CASI (1 metre resolution)	Shafique *et al.* (2001)
Rivers draining to Lake Taihu, China	Water clarity	Landsat-7 ETM+	Zhao *et al*, (2011)
River Kolyma, Russia	Dissolved organic carbon	Landsat-7 ETM+	Griffen *et al.* (2011)
(c) Geomorphology			
Waimakariri River, New Zealand	Channel morphology and DEM production in large braided systems	Aerial Photography; 0.5 metre resolution	Lane, (2000)
Soda Butte Creek, Montana and Cache Creek Wyoming, USA	In-channel hydro-geomorphic units including coarse woody debris	Airborne multi-spectral (blue, green, red and infra-red band widths); 1 metre resolution	Wright *et al.* (2000)
North Pacific Rim rivers; North America and Russia	Juvenile salmonid habitat	Landsat TM; Quickbird and airborne multi-spectral imagery	Whited *et al.*, (2013)
Platte River, Nebraska, USA	Channel bed topography	LiDAR	Kinzel *et al.* (2007)
Rhone River, France	Channel morphology and substrate	Drone and digital camera (5–14 cm resolution)	Lejot *et al.*, (2007)
River Ribble, North West England	Percentage clay, silt and sand in inter-tidal sediments	Airborne thematic mapper data; 2 metre pixel size	Rainey *et al.* (2000)

(*continued overleaf*)

Table 9.1 (*Continued*)

River location	Scientific purpose	Imagery platform, type, and scale	Reference
Saint Marguerite River, Canada	Substrate particle size	Airborne hyper-spectral data	Carbonneau *et al.* (2004)
Madre de Dios, Peru	Floodplain geomorphology and ecosystem structure	Landsat ETM+; JERS1; Radar C band.	Hamilton *et al.* (2006)
Yuba River, Califormia	Floodplain morphology and overbank sedimentation	Photogrammetry and aerial photography	Ghoshal *et al.* (2010)
Mekong River, Thailand – Lao PDR	Bank erosion	Aerial photography and Spot 5	Kummu *et al.* (2008)
Stroubles Creek, Virginia	Erosion on a 11 metre bank face	Terrestrial LiDAR	Resop and Hession (2010)
(d) Ecology			
Amazon, Brazil	Crocodile and caiman nesting habitat	Landsat	Villamarin *et al.* (2011)
Amazon, Brazil	Aquatic macrophyte cover in floodplain wetlands	Radarset-1. MODIS	Silva *et al.* (2010)
Roseg and Tagliamento floodplains, European Alps	Floodplain thermal heterogeneity	Airborne thermal infra-red	Tonella *et al.* (2011)
San Marguerita River, Canada	Juvenile salmon densities	Helicopter mounted digital photography	Hedger *et al.* (2006)
Sacramento-San Joaquin rRver delta, California	Submerged and emergent aquatic macrophytes	Airborne hyperspectral imagery	Santos *et al.* (2009)
Millingerwaard floodplain, River Rhine, Netherlands	Net primary productivity	Hymap	Kooistra *et al.* (2008)
River Waal, Netherlands	Floodplain vegetation land use	CASI and LiDAR	Geerling *et al.* (2007)

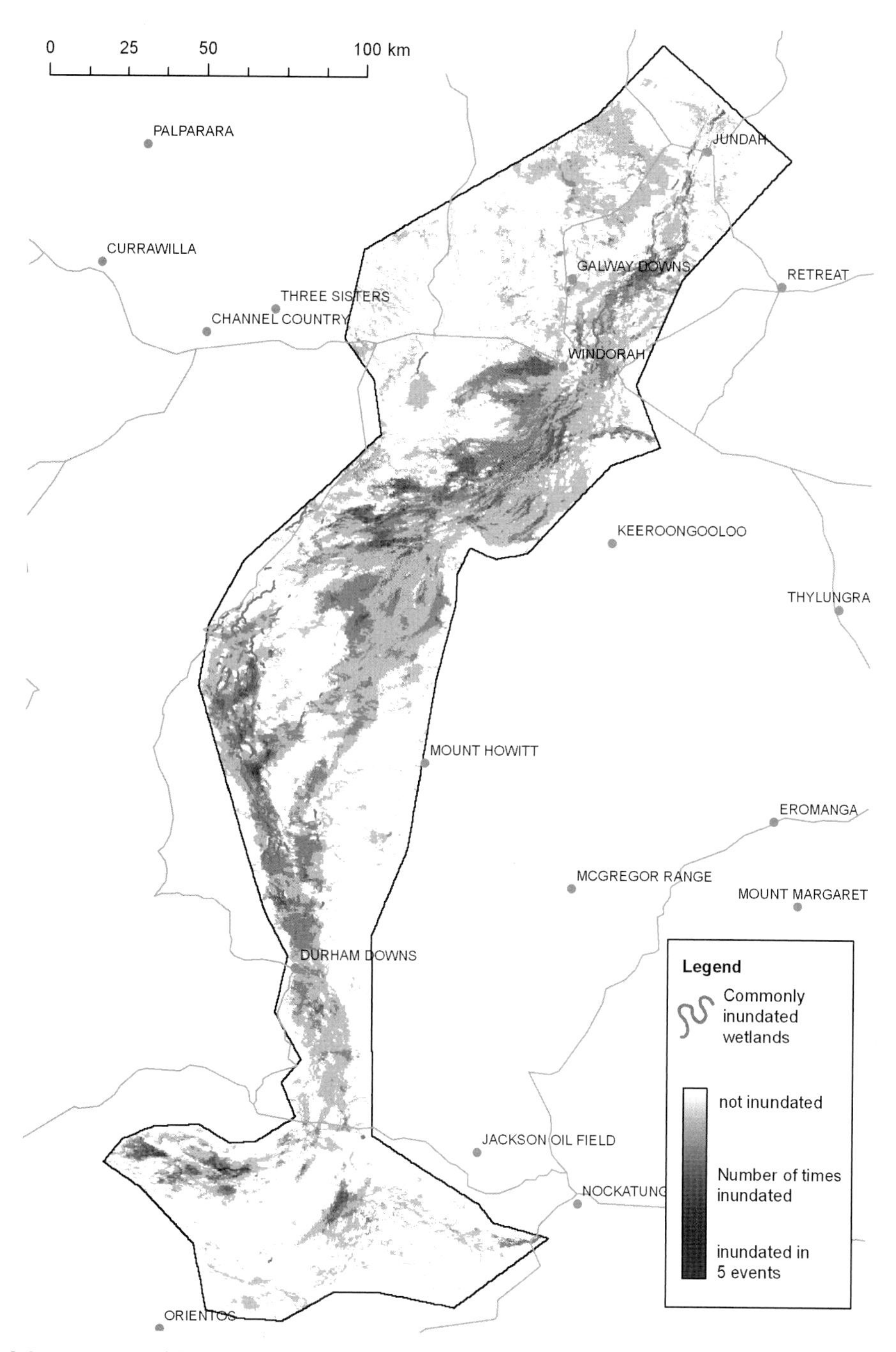

Figure 9.4 Frequency of floodplain inundation in relation to known wetlands on the Cooper Creek reach of the Murray–Darling as deduced from MODIS data and Normalised Difference Vegetation Cover. Image analysis undertaken by Dr Steve Wealands, University of Melbourne. (*See colour plate section for colour figure*).

scanner (Huguenin *et al.*, 2004; Karaska *et al.*, 2004). Chen has used regression-based approaches to estimate turbidity from MODIS reflectance (Chen *et al.*, 2007), while Wang and Lu (2010) directly estimated suspended sediment concentrations on the Yangtze River using the same sensor. Recent advances in lake remote sensing in terms of identifying individual phytoplankton species also have the potential for use in river waters on medium- and large-sized rivers (Hunter *et al.*, 2008, 2010).

Geomorphology and physical habitat

The ability to capture information on in-channel and floodplain morphology is now highly advanced due to an improvement in sensor spatial and spectral resolution, better knowledge of the spectral characteristics of river channels and improvements in image analysis methods (Marcus and Fonstad, 2008). The widespread availability of hyperspectral data and LiDAR has been particularly important in this regard in relation to instream habitat (Table 9.1). Marcus (2002) was one of the first scientists to realise the potential of hyper-spectral imagery and thus mapped instream habitats in Yellowstone National Park. Using maximum likelihood supervised classification on a 1-metre resolution 128-band hyper-spectral data producer, accuracies of 85–91% were achieved. Gilvear *et al.* (2004a) also showed the potential of CASI data to map riverine habitats and extended its application in to estuarine environments. Progress with hyper-spectral imagery has been particularly significant in terms of mapping river bathymetry (Gilvear *et al.*, 2007; Marcus and Legleiter, 2008). Legleiter *et al.* (2004), using the Hydrolight software radiative transfer model (Mobley and Sundman, 2003), laid the foundation for the advancement by predicting the effects of channel morphology and sensor spatial resolution on imagery derived water depths. Their research suggests a ratio of red and green wavelengths is a stable correlate to bed depths when using multi-spectral and hyper-spectral data. Optimal band ratios only provide relative depths, however, and need to be converted to absolute depths by field survey validation or other methods (Fonstad and Marcus, 2005). It should be noted that such techniques only work on relatively shallow and clear waters. On large and turbid rivers bathymetric sounding from boats would be preferable.

Over the last decade, airborne and terrestrially mounted Light Imaging, Detection and Ranging (LiDAR) has also added to the tools available to river scientists to map above the water line river and floodplain topography. Thus, Thomas *et al.* (2005) advocate airborne laser scanning for riverbank erosion assessment, while Kinzel *et al.* (2007) mapped the bed topography of a shallow sand bed stream using an airborne derived LiDAR dataset. Charlton *et al.* (2003) mapped channel morphology on a gravel bed river in northeast England. Repeat LiDAR surveys of a river reach can also be used to accurately determine channel change and inferences made about sediment budgets (e.g., Fuller *et al.*, 2003). LiDAR is now being used by the UK Environmental Agency to map floodplain topography for flood hazard mapping. Recent advances in the integration of scanning LiDAR technology with CCD digital imaging technology has produced airborne technology with access to real-time orthoimaging systems. The NASA ATM is a conically scanning airborne laser altimeter system capable of acquiring a swath width of 250 m with a spot spacing of 1–3 m and vertical precision of 10–15 cm. The potential of this in geomorphological and floodplain research has been demonstrated by Garvin and Williams (1993) and Marks and Bates (2000), Antonarakis *et al.* (2008b) and. Legleiter (2012). A key issue with LiDAR data, even more so than in the field, is definition of the channel boundary and bankfull conditions. Nevertheless, both airborne and terrestrial LiDAR has great potential to examine bed morphology and in-channel heterogeneity (BrasingtonBrasington *et al.,* 2012). Water penetrating LiDAR is another potential means of capturing river bathymetry, but has not been extensively investigated within the river environment.

Ecology

At the species level, obviously the spatial resolution of remote sensing means that capturing the location of organisms is often impossible. However, mapping of vegetation communities is possible and location of species inferred from species–habitat relationships (Table 9.1). The previous two sections have demonstrated that there is the ability to map channel hydraulic features, such as pools and riffles, exposed sand and gravel bars and standing water on the floodplain. Mapping of river macrophyte stands (Silva *et al.,* 2010) is also achievable, and similarly riparian and floodplain vegetation communities (Bryant and Gilvear, 1999; Geerling *et al.,* 2007). For example, in relation to floodplain habitats Villamarin *et al.* (2011) were able to map the spatial distribution of crocodile nesting sites using ground-derived data on nesting–habitat relationships and remotely sensed data. Using a similar approach, Hedger *et al.* (2006) collected field data relating juvenile salmon densities to substrate D50 grain size and subsequently mapped this on the San Marguerite River in Canada by image processing of helicopter-derived high spatial resolution aerial photography.

Recent success has been achieved in building upon simple land-cover classification of riparian zones and floodplains (Gilvear *et al.,* 2004a; Gilvear *et al.,* 2004b; Goetz, 2006) to mapping species. Goodwin *et al.* (2005), working on the Murray–Darling using 80 cm spatial resolution CASI-2 data, found that spectral reflectance curves of individual species and supervised maximum likelihood classification indicated that turpentine (*Syncarpia glomulifera*), mesic vegetation (primarily rainforest species) and an amalgamated group of eucalypts could be readily distinguished. The discrimination of *S. glomulifera* was particularly robust, with consistently high classification accuracies. In a very different environment and in a different way Jones *et al.* (2011) have achieved success in mapping Japanese knotweed, an invasive riparian species in the UK, using object-based classification to infra-red and colour digital photography. Advances have also been made in mapping important ecological variables such as productivity (Kooistra *et al.,* 2008).

In-channel, Santos *et al.* (2009) used airborne hyper-spectral data to map in-channel macrophytes and assessed the effect of herbicide treatments used to manage these species from 2003 to 2007. Each year, submersed aquatic plant species occupied about 12% of the surface area of the Delta in early summer and floating invasive plant species occupied 2–3%. Our understanding of both riparian zones and instream ecology is also being made possible by mapping important variables such as temperature (Tonella *et al.,* 2011). Remote sensing of river ecology is an area where field scientists and remote sensors working together can make impressive steps forward, especially in terms of elucidating abiotic–biotic linkages.

Relevance of remote sensing to river science

All key components of the river ecosystem

Remote sensing is likely to play a significant and major contribution to river science in the future. The key contribution is that it offers the possibility of determining spatial patterns of channel morphology, instream hydraulic habitat, channel bed configuration and substratum composition, riparian habitat patch composition, structure and floodplain morphology and vegetation mosaics (Table 9.2). Measurement across a range of wavelengths offers the potential to examine water depths, water chemical composition, thermal properties, soil moisture variability, plant vigour, 3-dimensional form of river and floodplain surfaces, and river dynamics. In essence the capability is available to quantify all components of river ecosystems while remembering that the appropriate image analysis techniques and approaches for capturing this data are not always simple and disentangling the relative effect of differing components on a signal can be complex. This data is also usually fully geo-referenced with high levels of accuracy, and is non-invasive and thus repeatable.

There are few areas on earth now where imagery is not available and a simple scan of GoogleEarth and other virtual globes illustrates the diversity of river types globally. Large and Gilvear (2014), for example, have shown the possibility of assessing potential ecosystem service delivery of a variety of rivers from Russia to England from GoogleEarth data. Remote sensing provides the ability for researchers to examine the physical habitat of river networks for rivers that are highly inaccessible or at scales where a full survey on the ground would be impossible. Of course, remote sensing should be seen as a tool alongside conventional, more locally based field studies and thus not a total panacea. The real advance is that is it allows upscaling of results of intensive field studies. There is also huge scope to explore scaling by the coupling of remotely sensed data with differing levels of accuracy. Thus sub-centimetre terrestrial LiDAR data at the reach scale can be matched to sub-metre airborne LiDAR data of the river coupled to spaceborne radar data at less than 2 metre accuracy. Such a spectrum of scales potentially provides understanding of fluvial systems (Williams *et al.*, 2013).

Near real-time observations

Near real-time observations provide the opportunity to study rivers in action. For example, Hirpa *et al.* (2013) demonstrated the utility of remote sensing for real-time river discharge observation and 1–15 day forecasting on the rivers Ganges and Brahmaputra. Many web cams have been installed on rivers allowing real-time downloads of image showing current water level and channel morphology. Such capability provides opportunities to compare scenes before–during–after flood events and to mount intensive field campaigns immediately following observations of notable events and change. Thus it could lead to exciting new developments in terms of designing a temporal framework for sampling campaigns.

River dynamics

The temporal dimension of rivers has been highly documented, but datasets that provide evidence of the level and extent of dynamism are sparse and have poor temporal coverage. A number of researchers have stressed that survey information is a 'snapshot' in time and thus only net change

Table 9.2 A list of potential applications of remote sensing to river science with possible imagery requirements.

River science discipline	Typical platform(s) used	Typical sensor types
(a) Hydrology		
River network	Satellite	Radar
Flood inundation mapping	Satellite/airborne	Radar/aerial photography
Discharge estimation	Satellite	Radar
Soil moisture mapping	Satellite/airborne	Thermal mapper
(b) Water quality		
Suspended solids and turbidity	Satellite/airborne	Hyper-spectral/multi-spectral
Chlorophyl a	Satellite/airborne	Hyper-spectral
Dissolved organic carbon	Satellite/airborne	Hyper-spectral
(c) Geomorphology		
Mapping channel morphology	Satellite/ airborne/UAV	Aerial photography/multi-spectral/hyper-spectral
Mapping floodplain morphology	Satellite/airborne	LiDAR/aerial photography/multi-spectral/hyper-spectral
River morphology and bathmetry	Airborne/UAV	LiDAR/ multi-spectral/hyper-spectral
Channel planform change	Satellite/ airborne/UAV/terrestrial	Aerial photography/multi-spectral
Bank erosion	Satellite/airborne/UAV/terrestrial	Photography/LiDAR
Substrate type	Airborne/UAV/terrestrial	Multi-spectral/hyper-spectral
Bed material size	Airborne/UAV/terrestrial	Multi-spectral/hyper-spectral
Mapping instream habitats	Airborne/UAV/terrestrial	LiDAR/Multi-spectral/hyper-spectral
(d) Ecology		
Floodplain vegetation structure	Satellite/airborne	Radar/LiDAR
Vegetation productivity	Satellite/airborne	Hyper-spectral
Floodplain temperatures	Airborne	Thematic mapper
Species specific habitat types	Satellite/airborne/UAVs	Hyper-spectral
Invasive species	Satellite/airborne	Hyper-spectral
Submerged macrophytes	Satellite/airborne/UAVs	Hyper-spectral
Emergent macrophytes	Satellite/airborne/UAVs	Hyper-spectral

between two dates can be inferred. This, as discussed above, is certainly the case with the use of old aerial photographs in river science. To date, the length of the record has often been inadequate to fully characterise the dynamism of the riverscape in low energy, relatively stable river environments. One criticism of river studies to date has been the focus on unstable river reaches where rates of change can be detected, without due attention to reaches where rates of change are slower. With the possibility of high frequency remote sensing, better understanding of the real dynamism of rivers over the duration of flood events, seasonal, annual and decadal timescales is likely to become a reality. For example, in tidal rivers suspended sediment dynamics over the tidal cycle has been elucidated with the use of repeat remote sensing campaigns on both the flood and ebb tide (Wakefield *et al.*, 2011).

Spatial analysis and modelling

Remote sensing of river ecosystems is primarily about the mapping of physical habitat, water chemistry and associated processes. Currently the key issue within river science is coupling physical and biological data. Spatial analysis, of which the collection of remotely sensed data is a key component, provides a framework within which biological data, for example of fish productivity data gained from electrofishing, can be coupled to remotely sensed output of physical habitat. This remotely sensed physical habitat can also be multi-scalar, ranging from boulder complexity at the sub-reach scale through to mapping of geomorphic river types across the river network. This also allows nested multi-scalar modelling (e.g., Brasington, 2010) from the micro- to network scale. Moreover, remotely sensed datasets can encompass attributes such as channel slope, channel width, bed material grain size, riparian zone characteristics and floodplain land use and as such can be the catalyst for robust modelling of such processes and attributes as flood routing, ecosystem service delivery, and animal and plant species distributions. It also allows exploration of linkages between abiotic and biotic components at a scale appropriate for river management; which is not always the case with field-based experiments and observations.

Logistics and practicalities

The reality of remote sensing, when directed beyond the reach-scale and in non-ideal conditions, throws up a range of challenges. At the reach scale, tree canopies and the shadows cast obscuring features of interest, and in northern latitudes ice and snow cover can be an issue. Commissioning airborne imagery to cover all of a river network is costly and image processing can be challenging given differing sun angles. For large river reaches the volume of data can also present challenges in terms of data processing. In some climates finding the right climatological conditions for flying a campaign can be difficult, especially if there is the need to synchronise the survey with specific flow or tidal conditions. Users of remote sensing should be aware of these issues at the planning stage of any mapping or surveillance project.

Case study – application

Remote sensing of river channel physical form for the Sustainable Rivers Audit in the Murray–Darling Basin, Australia

The Murray–Darling Basin is both economically and environmentally important in

Australia, supporting 70% of the countries irrigated agriculture and more than 40% of the gross value of agricultural production (Davies *et al.*, 2010), along with 16 Ramsar protected wetlands (DSEWPaC, 2011). It is also one of the world's large river basins with a catchment area of 1,040,000 km^2, including regions of both arid and temperate climates. In 2000, with increasing investment in river and catchment restoration, the then Murray–Darling Basin Commission began work on the Sustainable Rivers Audit (SRA) to report on status and trend in river condition (Davies *et al.*, 2010). The SRA is concerned with surveillance monitoring rather than measuring compliance with standards or targets and it is focused on detecting and reporting the signs of change rather than the causes. The initial SRA reporting was based largely on field surveys of fish and aquatic macroinvertebrate communities over the period 2004–07 and some preliminary analysis of hydrological change in the basin's major rivers (Davies *et al.*, 2008). Following the initial reporting, the scope of the SRA was expanded to include, among other aspects, a condition assessment of river channel physical form based on a combination of remote sensing and modelling. A second report, including this component, highlighted river condition over the period 2008–10 (Davies *et al.*, 2012). This case study is reported here to highlight how remote sensing can be a useful tool in the context of river science and management and in terms of the requirements of a successful river remote sensing project in terms of data acquisition, data processing and interpretation.

The remote sensing was undertaken in 2009–10, using airborne LiDAR to survey river channels across the Murray–Darling Basin. A total of 1610 sites were selected for survey using a stratified random sampling procedure with 70 sites in each of 23 sub-catchments. Each site had a rectangular footprint centred on the sample point extending 2 km in the direction of the river alignment, and a width of 0.7 km to include the full width of the meander belt in many cases. For reliable positioning, flight lines are straight and there was hence the need to define a rectangular sample unit despite the sinuous nature of both the river channel and the floodplain. Each site was covered by two 577-m wide LiDAR swathes with 35% overlap. Trimble Harrier 56 and Harrier 68 LiDAR systems were used to collect full waveform LiDAR at a density of at least four outgoing pulses per square metre and with a vertical accuracy of the LiDAR for bare ground surfaces of 0.2 m after primary processing (Terranean, 2010). To provide this data resolution, flying height was 500 m, flying speed was 205 km/hour, scan angle was 60 degrees and scan and pulse rates were 76 Hz and 200 kHz respectively. For each sortie, LiDAR was recorded over at least four horizontal control points on vertical structures, such as buildings, and six vertical control points on hard bare flat surfaces, such as roads for testing positional accuracy and precision (Terranean, 2010). The aerial survey also included the capture of multi-spectral imagery that was orthorectified against LiDAR terrain surfaces. The LiDAR does not penetrate the water column, so flights were carried out during low-flow or cease-to-flow conditions if possible. Surveys were carried out over a seven-month period, including a four-month delay as a result of major floods across the basin.

There were two stages of data processing (Terranean, 2010). In the primary stage, raw instrument output was processed to generate survey points with spatial coordinates in three dimensions, the strength of the returned signal and a classification of the type of surface for each point (e.g., ground, vegetation, building or water). This stage also included an assessment of positional accuracy using the control points, which were all independently surveyed. The LiDAR points classified as ground were used to generate a 1-m Digital Elevation Model (DEM) for each site.

The SRA assessment included measures of the bankfull channel geometry that were estimated from these LiDAR surveys. In the secondary processing stage, up to eight operators concurrently worked to measure these channel cross-sectional attributes for each site using a purpose-built tool developed in the TNTmips object-oriented Spatial Modelling Language. The first step was to map the water surface using false colour imagery. Where the water surface is obscured in the imagery, the LiDAR ground points were used, with points mostly absent where there is surface water. The flow direction and approximate path of the river channel was manually drawn using a relief shaded DEM. Using an automated procedure, the channel centreline was generated along the lowest part of the channel or centre of surface water. Nineteen transects were drawn at right angles to this centreline along the site.

The identification of bankfull levels by operators unfamiliar and inexperienced in river morphology was the most challenging aspect in the secondary processing stage. It required a combination of automated and manual interpretation with subsequent quality checking by a fluvial geomorphologist for selected sites. The TNTmips tool analysed each transect profile to calculate the rate of change in lateral bank gradient with elevation. Maxima in this profile were mapped as candidate levels for bankfull. The operator used these maxima from multiple transects and the multi-spectral imagery and DEM viewed in 3D using anaglyph glasses to select the bankfull level (Terranean, 2010). Where the operator had difficulty identifying bankfull level, survey and imagery data for the site were passed to a fluvial geomorphologist who provided expert judgement.

For the SRA, river condition is assessed based on departure of the observed (i.e., current) conditions from a reference state defined as the condition expected in the absence of anthropogenic disturbances post-European settlement (Davies *et al.*, 2010). The metrics extracted from LiDAR surveys indicate the current state but not the reference state. Ideally, one might use reference sites, with little anthropogenic disturbance, to develop a model or reference condition. However, it is difficult to find undisturbed sites as channel changes have occurred throughout the basin in response to historic catchment and river disturbances. Instead, models were developed to estimate the SRA LiDAR sites in the absence of anthropogenic disturbances. A Boosted Regression Tree model was used because it can fit complex no-linear relationships, is not sensitive to outliers, and is considered to have superior predictive performance to traditional modelling approaches (Elith *et al.*, 2008). These models were fitted using the observed channel metrics derived from the LiDAR surveys and use GIS variables as predictors. These predictors included measures of anthropogenic disturbance including a range of catchment, hydrological and other human disturbances known to produce channel changes. The fitted model was subsequently used to predict the reference state of the river morphology by setting these measures of anthropogenic disturbance to zero. A cross-prediction approach was used to ensure data for sites at which the reference state was being modelled were not used to also fit the model. In addition, model uncertainties were evaluated for each sub-catchment in the MDB in a cross-validation procedure. Using these model uncertainties, a plausible range was established for the reference state and a score of 1 was assigned if the observed state was within this reference range. The score declined from 1 if the observed value fell below this range or increased above 1 if it fell above this range.

For all variables, many sites had observed conditions outside of the reference range. Approximately 40% of sites had an observed bankfull width greater than reference and a similar proportion had observed bankfull depth greater than reference. These results indicate widespread channel enlargement as a result of anthropogenic disturbances. Similarly, between 10 and 40% of sites had reduced channel variability compared to reference conditions, as indicated by bankfull width or depth variability, bank angle and concavity variability and bank complexity. These results indicate widespread channel simplification as a result of anthropogenic disturbance. An expert system was used to integrate these scores with the other scores used in the geomorphic assessment and aggregate results for valley and basin scale reporting (Davies *et al.*, 2012).

Despite the novelty of both the interpretation of remote sensing data and modelling components for the physical form assessment, the approach was successful. The scale of effort on LiDAR data capture and processing was considerable and it is unlikely that this effort can repeated at a frequency of much less than 10 years. Given the relatively slow nature of channel adjustments, it may not be sensible to resurvey channel conditions at the basin-scale for a decade or more in any case. However, the method produced well-documented and systematic protocols which will be repeatable when subsequent surveys are undertaken to establish trends in response to ongoing anthropogenic disturbances and river restoration.

Key messages

Remote sensing is a key component of twenty-first-century river science and when integrated with field data and within appropriate modelling frameworks can lead to integration and advancement of the science both across disciplines and across scales. It can be central to such fields of enquiry as ecohydraulics, ecogeomorphology and hydroecology. Remote sensing will provide optimal value to river science when conducted alongside field-based ground truth data and process-based measurements (Konrad *et al.*, 2008). In particular, remote sensing provides a key role in upscaling field-based reach-scale derived data to the river network scale and lends itself also to multi-scalar perspectives (Bertoldi *et al.*, 2011).

Twenty-first-century remote sensing provides a suite of powerful tools for examining river environments, but for their potential to be unlocked and maximised, appropriate platform and sensor choice is the key. Given the array of platforms operating across the electromagnetic spectrum, most achieving high spatial resolution, the potential for mapping of most terrestrial and aquatic river features of interest is a reality. Perhaps only the hyporheos currently precludes substantial information retrieval from remote sensing, and here ground penetrating radar may have potential. In many cases multi-sensor and multi-scalar approaches are likely to maximise information retrieval. In this latter mode it has the ability to provide global coverage at coarse spatial resolutions (<5 metre) and river-sector and reach-scale coverage at fine spatial scales (e.g., sub-metre and centimetre levels). A key role of river science is to provide the framework within which to integrate information at multiple scales obtained by remote sensing and its linkage to early surveys and historic and modern field data. Remote sensing is now established as one of the core sub-disciplines of river science and is central to advancing our understanding of river ecosystems.

In summary, remote sensing has a vital contribution to play in the field of river science, both in terms of better understanding of the functioning of river ecosystems, and in terms of application in surveillance linked to monitoring ecosystem health and impacts of human activities.

References

Adler-Golden, S.M., Acharya, P.K., Berk, A., *et al.* 2005. Remote bathymetry of the littoral zone from AVIRIS, LASH, and QuickBird imagery. *IEEE Transactions on Geoscience And Remote Sensing* 43, 337–347.

Alsdorf D., Bates P.D., Melack J., *et al.* 2007. Spatial and temporal complexity of the Amazon flood measured from space. *Geophysical Research Letters* 34: Article number L08402.

Alsdorf, D.E., Melack J.M., Dunne T., *et al.* 2000. Interferometric radar measurments of water level changes on the Amazon floodplain. *Nature* 404, 174–177.

Antonarakis A.S., Richards K.S. and Brasington J. 2008a. Object-based land cover classification using airborne LiDAR. *Remote Sensing of Environment* 112, 2988–2998.

Antonarakis A.S., Richards K.S. and Brasington J. 2008b. Retrieval of vegetative fluid resistance using airborne lidar. *Journal Geophysical Research - Biogeosciences* 113, G02S07.

Bates P.D. and De Roo A. P. J. 2000. A simple raster-based model for flood inundation simulation, *Journal of Hydrology* 236, 54–77.

Bertoldi W., Gurnell A.M. and Drake N.A. 2011. The topographic signature of vegetation development along a braided river: Results of a combined analysis of airborne lidar, color air photographs, and ground measurements. *Water Resources Research* 47, 6.

Brasington J., Vericat D. and Rychkov I. 2012. Modelling River Bed Morphology, Roughness and Surface Sedimentology using High Resolution Terrestrial Laser Scanning. *Water Resources Research* 48(11), W11519.

Brasington J. 2010. From grain to floodplain: hyperscale models of braided rivers. *Journal of Hydraulic Research* 48 (4), 52–53 Suppl. 4 2010.

Brown M., Gunn S.R. and Lewis H.G. 1999. Support vector machines for optimal classification and spectral unmixing. *Ecological Modelling* 120,167–179.

Brown M., Lewis H.G. and Gunn S.R. 2000. Linear spectral mixture models and support vector machines for remote sensing. *IEEE Transactions on Geoscience and Remote Sensing* 38, 2346–2360.

Bryant R.G. and Gilvear D.J. 1999. Quantifying geomorphic and riparian land cover changes either side of a large flood event using airborne remote sensing: River Tay, Scotland. *Geomorphology* 29, 307–321.

Carbonneau P.E. and Piégay H. 2012. *Fluvial Remote Sensing for River Science and Management*. Wiley.

Carbonneau P.E., Lane, S.N. and Bergeron, N.E. 2004. Catchment-scale mapping of surface grain size in gravel bed rivers using airborne digital imagery. *Water Resources Research* 40(7), DOI 10.1029/2003WR002759.

Carbonneau P.E., Lane, S.N. and Bergeron, N.E. 2006. Feature based image processing methods applied to bathymetric measurements from airborne remote sensing in fluvial environments. *Earth Surface Processes and Landforms* 31, 1413–1423.

Charlton M.E., Large A.R.G. and Fuller I.C. 2003. Application of airborne lidar in river environments: The River Coquet, Northumberland, UK. *Earth Surface Processes and Landforms* 28(3), 299–306.

Chen Z., Muller-Karger F.E. and Hu. C. 2007. Remote sensing of water clarity in Tampa Bay. *Remote Sensing of Environment* 109 (2), 207–220.

Conyers M.M. and Fonstad M.A. 2005. The unusual channel resistance of the Texas Hill Country and its effect on flood flow predictions. *Physical Geography* 26, 379–395.

Davies P.E., Harris J.H., Hillman T.J. and Walker, K.F. 2008. *Sustainable Rivers Audit Report 1: A report on the ecological health of rivers in the Murray-Darling Basin, 2004–2007*. Murray-Darling Basin Commission, Canberra, Australia.

Davies P.E., Harris J.H., Hillman T.J. and Walker J.F. 2010. The Sustainable Rivers Audit: assessing river ecosystem health in the Murray–Darling Basin, Australia. *Marine and Freshwater Research* 61, 764–777.

Davies, P.E., Stewardson M.J., Hillman T.J., *et al.* 2012. Sustainable Rivers Audit Report 2: The ecological health of rivers in the Murray–Darling Basin at the end of the Millennium Drought (2008–2010). Murray-Darling Basin Authority, Canberra, Australia. http://www.mdba.gov.au/what-we-do/mon-eval-reporting/sustainable-rivers-audit.

DSEWPaC. 2011. Directory of Important Wetlands in Australia. Australian Government Department of Sustainability, Environment, Water, Population and Communities, Canberra. <http://www.environment.gov.au/water/topics/ wetlands/database/diwa.html> (accessed 21.04.11).

Ekstrand, S. 1996. Landsat TM-based forest damage assessment: correction for topographic effects. *Photogrammetric Engineering and Remote Sensing* 62, 151–161.

Elith J., Leathwick J.R. and Hastie T. 2008. A working guide to boosted regression trees. *Journal of Animal Ecology* 77, 802–813.

Feurer D., Bailly J.S. and Peuch C., *et al.* 2008. Very high resolution mapping of river immersed topography by remote sensing. *Progress in Physical Geography* 32, 1–17.

Flener C., Vaaja M. and Jaakkola A., *et al.* 2013. Seamless mapping of river channels at high resolution using mobile LiDAR and UAV-Photography. *Remote Sensing* 5, 6382–6407.

Fonstad M.A. and Marcus W.A. 2005. Remote sensing of stream depths with hydraulically assisted bathymetry (HAB) models. *Geomorphology* 72, 320–339.

Forzieri G., Guanieri L. and Vivoni R., *et al.* 2011. Spectral ALS data fusion for different roughness parametrizations of forest floodplains. *River Research and Applications* 27, 826–840.

France M.J., Collins W.G. and Chindeley, T.R. 1986. Extraction of hydrological parameters from Landsat Thematic Mapper Imagery. *Proceedings of the 20th International Symposium on remote Sensing of Environment (ERIM)*, Nairobi, 1165–1173.

Fuller I.C., Large A.R.G. and Charlton M.E. *et al.* 2003. Reach-scale sediment transfers: An evaluation of two morphological budgeting approaches. *Earth Surface Processes and Landforms* 28, 889–903.

Garvin J.B. and Williams R. 1993. Geodetic airborne laser altimetry of Breidamerkurjokull and Skeidararjokull, Iceland and Jakobshavn, Greenland. *Annals of Glaciology* 17, 377–386.

Geerling G.W., Labrador-Garcia M., Clevers J.G.P.W., *et al.* 2007. Classification of floodplain vegetation by data fusion of spectral (CASI) and LiDAR data. *International Journal of Remote Sensing* 28, 4263–4284.

Ghoshal S., James L.A., Singer M.B. and Aalto R. 2010. Channel and floodplain change analysis over a 100 year period. Lower Yuba river, California. *Remote Sensing* 27, 1797–1825.

Gilvear D.J., Waters T. and Milner, A. 1998. Image analysis of aerial photography to quantify the effect of gold placer mining on channel morphology, Interior Alaska. In: *Landform Monitoring, Modelling and Analysis*, S. Lane, K. Richards and J. Chandler (Eds), Wiley Chichester, pp. 195–216.

Gilvear D.J, Tyler A. and Davids C. 2004a. Detection of estuarine and tidal river hydromorphology using hyper-spectral and LiDAR data: Forth estuary, Scotland. *Estuarine, Coastal and Shelf Science* 61, 379–392.

Gilvear D.J., Davids C. and Tyler A.N. 2004b. The use of remotely sensed data to detect channel hydromorphology; River Tummel, Scotland. *River Research and Applications* 20(7), 795–811.

Gilvear D.J., Hunter P. and Higgins T. 2007. An experimental approach to the measurement of the effects of water depth and substrate on optical and near infrared reflectance: a field-based assessment of the feasibility of mapping submerged instream habitat. *International Journal of Remote Sensing* 28, 2241–2256.

Goetz, S.J. 2006. Remote sensing of riparian buffers: Past progress and future prospects. *Journal of the American Water Resources Association* 42(1), 133–143.

Goodwin N., Turner R. and Meron R. 2005. Classifying Eucalyptus forests with high spatial and spectral resolution imagery: an investigation of individual species and vegetation communities. *Australian Journal of Botany* 53, 337–345.

Griffin C.G., Frey K.E., Rogan J. and Holmes R.M. 2011. Spatial and interannual variability of dissolved organic matter in the Kolyma River, East Siberia, observed using satellite imagery. *Journal of Geophysical Research – Biogeosciences* 116, 17–27.

Gurnell A.M., Downward S.R. and Jones R. 1994. Channel planform change on the River Dee meanders, 1876-1992. *Regulated Rivers* 9, 187–204.

Hamilton S.K., Kellndorfer J., Lehner B. and Tobler, M. 2006. Remote sensing of floodplain geomorphology as a surrogate for biodiversity in a tropical river system (Madre de Dios, Peru). *Geomorphology* 89, 23–38.

Hedger R.D., Dodson J.J., Bourque J.F., *et al.* 2006. Improving models of juvenile Atlantic salmon habitat use through high resolution remote sensing. *Ecological Modelling* 197, 505–511.

Hirpa F.A., Hopson T.M., and Groeve T.D., *et al.* 2013. Upstream satellite remote sensing for river discharge forecasting: Application to major rivers in South Asia. *Remote Sensing of Environment* 131, 140–151.

Horritt M.S., Mason D.C. and Luckman, A.J. 2001. Flood boundary delineation from Synthetic Aperture Radar imagery using a statistical active contour model. *International Journal of Remote Sensing* 22(13), 2489–2507.

Huang M., Gong J.H., Shi Z. and Zhang. L.H. 2009. River bed identification of check-dam engineering using SPOT-5 image in the HongShiMao watershed of the Loess Plateau, China. *International journal of Remote Sensing* 30, 1853–865.

Huguenin R.L., Wang M.H., Biehl R., *et al.* 2004. Automated subpixel photobathymetry and water quality mapping. *Photogrammetric Engineering & Remote Sensing* 70, 111–123.

Hunter P., Tyler A., Carvalho L., *et al.* 2010. Hyperspectral remote sensing of cyanobacterial pigments as indicators for cell populations and toxins in eutrophic lakes. *Remote Sensing of the Environment* 114, 1556–1577.

Hunter P., Tyler A., Presing M., *et al.* 2008. Spectral discrimination of phytoplankton colour groups: The effect of suspended particulate matter and sensor spectral resolution. *Remote Sensing of Environment* 112(4), 1527–1544.

Jones D., Pike S., Thomas M. and Murphy D. 2011. Object-based image analysis for detection of Japanese Knotweed (polygonacae) in wales (UK). *Remote Sensing* 3(2), 319–342.

Karaska M.A., Huguenin R.L., Beacham J.L., *et al.* 2004. AVIRIS measurements of chlorophyll, suspended minerals, dissolved organic carbon, and turbidity in the Neuse River, North Carolina. *Photogrammetric Engineering & Remote Sensing,* 70(1), 125–133.

Kinzel P.J., Wright C.W., Nelson J.M. and Burman, A.R. 2007. Evaluation of an experimental LiDAR for surveying a shallow, braided, sand-bedded river. *Journal of Hydraulic Engineering,* 133(7), 838–842.

Kleynhans C.J. 1996. A qualitative procedure for the assessment of the habitat integrity status of the Luvuvhu River (Limpopo system, South Africa). *Journal of Aquatic Ecosystem Health* 5, 41–54.

Kooistra L., Wamelink W., Schaepman-Strub G., *et al.* 2008. Assessing and predicting biodiversity in a floodplain ecosystem: Assimilation of net primary production derived from imaging spectrometer data into a dynamics vegetation model. *Remote Sensing of Environment* 112, 2118–2130.

Konrad C.P., Black R.W., Voss F., and Neale M.U. 2008. Integrating remotely acquired and field data to assess effects of setback levees on riparian and aquatic habitats in glacial melt water rivers. *River Research and Applications* 24, 355–372.

Kummu M., Lu X.X., Rasphone A., *et al.* 2008. Riverbank changes along the Mekong river: remote sensing detection in the Vientiane-Nong Khai area. *Quaternary International* 186, 100–112.

Kutser T., Miller I. and Jupp, D.L.B. 2006. Mapping coral reef benthic substrates using hyperspectral space-borne images and spectral libraries. *Estuarine Coastal and Shelf Science* 70, 449–460.

Large A.R.G. and Gilvear D.J. 2014. Using Google Earth, a virtual-globe imaging platform, for ecosystem services based river assessment. *River Research and Applications* 31, 406–421.

Lejot J., Delacourt C., Piégay H., *et al.* 2007. Very high spatial resolution imagery for channel bathymetry and topography from an unmanned mapping controlled platform. *Earth Surface Processes and Landforms* 32, 1705–1925.

Legleiter C.J., Roberts D.A., Marcus W.A. and Fonstad M.A. 2004. Passive remote sensing of river channel morphology and instream habitat: physical basis and feasibility. *Remote Sensing of Environment* 93, 493–510.

Legleiter C.J. 2012. Remote measurement of river morphology via fusion of LiDAR topography and spectrally based bathymetry. *Earth Surface Processes and Landforms* 37, 499–518.

Luo J., Ming D., Liu W., *et al.* 2007. Extraction of bridges over water from IKONOS panchromatic data. *International Journal of Remote Sensing* 28, 3633–3648.

Marcus W.A. 2002. Mapping of stream microhabitats with high spatial resolution hyperspectral imagery. *Journal of Geographical Systems* 4, 113–126.

Marcus W.A., Legleiter C.J., Aspinall R.J., *et al.* 2003. High spatial resolution hyperspectral mapping of in-stream habitats, depths, and woody debris in mountain streams. *Geomorphology* 55, 363–380.

Marcus W.A. and Fonstad M.A. 2008. Optical remote mapping of rivers at sub meter resolutions and watershed extents. *Earth Surface Processes and Landforms* 33, 4–24.

Marcus W.A. and Fonstad M.A. 2010. Remote sensing of rivers: the emergence of a sub-discipline in the river sciences. *Earth Surface Processes and Landforms* 35, 1867–1872.

Marks K. and Bates P. 2000. Integration of high-resolution topographic data with floodplain flow models. *Hydrological Processes* 14(11–12), 2109–2122.

MacVicar B. and Piégay H. 2012. Implementation and validation of video monitoring for wood budgeting in a wandering piedmont river. *Earth Surface Processes and Landforms* 31, 17–27.

Milton E.J., Gilvear D.J. and Hooper I.D. 1995. Investigating river channel changes using remotely sensed data. In: *Changing River Channels*. Gurnell A. and Petts G.E. (Eds). *Changing River Channels*. Wiley, Chichester, pp. 277–301.

Mobley, C.D. and Sundman L.K, 2003. Effects of optically shallow bottoms on upwelling radiances: Effects of inhomogeneous and sloping bottoms. *Limnology Oceanography* 48(1), 329–336.

Muller E. 1992. Evaluation de la bande TM5 pour la cartographie morpho-hydrogeologique de la moyenne vallee de la Garonne. Project SPOT4/MIR. CNES, Paris.

Muller E., Decamps H., Dobson M.K. 1993. Contribution of space remote sensing to river studies. *Freshwater Biology* 29, 301–312.

Olmanson L.G., Brezonik P.L. and Bauer M.E. 2013. Airborne hyperspectral remote sensing to assess spatial distribution of water quality characteristics in large rivers: the Mississippi River and its tributaries in Minnesota. *Remote Sensing of Environment* 130, 254–265.

Rainey, M.P., Tyler, A.N., Bryant, R.G., *et al.* 2000. The influence of surface and interstitial moisture on the spectral characteristics of inter-tidal sediments; implications for airborne image acquisition and processing, *International Journal of Remote Sensing* 21, 3025–3038.

Resop J.P. and Hession W.C. 2010. Terrestrial laser scanning for monitoring stream bank retreat: comparison with traditional surveying techniques. *Journal of Hydraulic Engineering* 136, 794–798.

Resop J.P., Kozarek J.L. and Hession W.C. 2012. Terrestrial laser scanning for delineating in-stream boulders and quantifying habitat complexity measures. *Photogrammetric Engineering and Remote Sensing* 78, 363–371.

Santos M.J., Khanna S., Hestir E.L. *et al.* 2009. Use of Hyperspectral remote sensing to evaluate efficacy of aquatic plant management. *Invasive Plant Science and Management* 2, 216–229.

Schaepman M.E., Koetz B., Schaepman-Strub G. and Itten K.I. 2005. Spectrodirectional remote sensing for the improved estimation of biophysical and -chemical variables: two case studies. *International Journal of Applied Earth Observation and Geoinformation* 6, 271–282.

Shafique N.A., Autrey B.C., Fulk F. and Cormier S.M. 2001. Hyperspectral narrow wavebands selection for optimizing water quality monitoring on the Great Miami River, Ohio. *Journal of Spatial Hydrology* 1(1), 1–22.

Silva T.S.F., Costa M.P.F. and Melack J.M. 2010. Spatial and temporal variability of macrophyte cover and productivity in the eastern Amazon floodplain: a remote sensing approach. *Remote Sensing of the Environment* 114, 1998–2010.

Soenen S.A., Peddle D.R. and Coburn, C.A. 2005. SCS+C: A modified sun-canopy-sensor topographic correction in forested terrain. *IEEE Transactions on Geoscience and Remote Sensing* 43, 2148–2159.

Straatsma, M.W. and Baptist M.J. 2008. Floodplain roughness parameterization using airborne laser scanning and spectral remote sensing. *Remote Sensing of Environment* 112(3), 1062–1080.

Terranean. 2010. LiDAR and multispectral remote sensing for the Murray-Darling Basin Sustainable Rivers Audit. Report by Terranean Mapping Technologies to Murray Darling Basin Authority, Canberra, Australia. http://www.mdba.gov.au/kid/files/2492D13_17559_LiDAR_multispectral_remote_sensing.pdf

Thomas D.P., Gupta, S.C., Bauer, M.E. and Klirchoff, C.E. 2005. Airborne laser scanning for riverbank erosion assessment. *Remote Sensing of the Environment* 95, 493–501.

Thomson A.G., Fuller R.M. and Eastwood, J.A. 1998. Supervised versus unsupervised methods for classification of coasts and river corridors from airborne remote sensing. *International Journal of Remote Sensing* 19, 3423–3431.

Tonella D., Acuña V., Lorang M.S., *et al.* 2010. A field-based investigation to examine underwater soundscapes of five common river habitats. *Hydrological Processes* 24, 3146–3156.

Tonella D., Acuña V., Uehlinger U., *et al.* 2011. Thermal heterogeneity in river floodplains. *Ecosystems* 13(5), 727–740.

Torgersen C.E., Faux R.N. and McIntosh B.A., *et al.* 2001, Airborne thermal remote sensing for water temperature assessment in rivers and streams. *Remote Sensing of Environment* 76, 386–398.

van der Meer, F. 2006. The effectiveness of spectral similarity measures for the analysis of hyperspectral imagery. *International Journal of Applied Earth Observation and Geoinformation* 8, 3–17.

Villamarin F., Marioni B., Thorbjarnarson J.B., *et al.* 2011. Conservation and management implications of nest-site selection of the sympatric crocodilians Melanosuchus niger and Caiman crocodilus in Central Amazonia, Brazil. *Biological Conservation* 144, 913–919.

Wang J.J. and Lu X.X. 2010. Estimation of suspended sediment concentrations using Terra Modis: an example from the Lower Yangtze River, China. *Science of the Total Environment* 408, 1131–1138.

Wang J.J., Lu X.X., Liew, S.C. and Zhou, Y. 2009. Retrieval of suspended sediment concentrations in large turbid rivers using Landsat ETM+: an example from the Yangtze River, China. *Earth Surface Processes and Landforms* 34, 1082–1092.

Wakefield R., Tyler A., McDonald P., *et al.* 2011. Estimating sediment and caesium-137 fluxes in the Ribble Estuary through time-series airborne remote sensing, *Journal of Environmental Radioactivity* 102(3), 252–261.

Westaway R.M., Lane S.N. and Hicks D.M. 2003. Remote survey of large scale braided, gravel bed rivers using digital photogrammetry and image analysis. *International Journal of Remote Sensing* 24, 795–815.

Whited D., Kimball J. and Stanford J. 2012. Estimation of juvenile salmon habitat in Pacific Rim rivers using scalable remote sensing and geospatial analysis. *River Research and Applications* 29, 135–148.

Wilson M.D. and Atkinson P. 2007. The use of remotely sensed land cover to derive floodplain friction coefficients for flood inundation modelling. *Hydrological Processes* 26, 3576–3586.

Winterbottom S.J. and Gilvear D.J. 1997. Quantification of channel bed morphology in gravel-bed rivers using airborne multispectral imagery and aerial photography. *Regulated Rivers: Research and Management* 13, 489–499.

Williams R.D., Brasington J., Vericat D. and Hicks D M. 2013. Hyperscale terrain modelling of braided rivers: fusing mobile terrestrial laser scanning and optical bathymetric mapping. *Earth Surface Processes and Landforms*. doi: 10.1002/esp.3437

Wright A., Marcus W.M. and Aspinall, R. 2000. Evaluation of multi-spectral imagery as a tool for mapping stream morphology. *Geomorphology* 33: 107–120.

Wu, J., Bauer, M.E., Wang, D. and Manson, S.M. 2008. A comparison of illumination geometry-based methods for topographic correction of QuickBird images of an undulant area. *ISPRS Journal of Photogrammetry and Remote Sensing* 63, 223–236.

Yuen, P.W.T. and Bishop, G. 2004. *Enhancements of target detection using atmospheric correction preprocessing techniques in hyperspectral remote sensing. Conference on Military Remote Sensing*, London, England.

Zhao D.H., Cay Y., Jiang H., *et al.* 2011. Estimation of water clarity in Taihu lake and surrounding rivers using Landsat imagery. *Advances in Water Resources* 34, 165–173.

CHAPTER 10

Monitoring the resilience of rivers as social–ecological systems: a paradigm shift for river assessment in the twenty-first century

Melissa Parsons[1], Martin C. Thoms[1], Joseph Flotemersch[2] and Michael Reid[1]
[1] *Riverine Landscapes Research Laboratory, Geography and Planning, University of New England, Armidale, Australia*
[2] *National Exposure Research Laboratory, U.S. Environmental Protection Agency, Ecological Exposure Research Division, Cincinnati, Ohio, USA*

Introduction

Sustainability has become a cornerstone of natural resource management throughout the world, including policy and legislation governing the management of water resources in European countries (e.g., EU Water Framework Directive 2000), the United States (e.g., Clean Water Act 1972), South Africa (e.g., National Water Act 1998) and Australia (e.g., Water Act 2007). Despite debate about its definition and application (Norton, 2005), sustainability has proven endearing as a philosophy for guiding the way that humans make use of resources supplied by the natural environment. In a nutshell, sustainability refers to human use of the environment and its resources to meet present needs, without compromising the ability of future generations to use the same environment to meet their needs (Chapin *et al.*, 2009). Sustainability implies that the environment provides resources for humans, while recognising that humans use the environment to maintain their well-being and, in doing so, may change the natural dynamics of ecosystems (Chapin *et al.*, 2009). The intention of sustainable natural resource management is to balance the current use of natural resources against the ecosystem's ability to continue to maintain or supply resources into the future. Implicit in sustainability is the notion that degradation reduces the ability of an ecosystem to supply resources. Given the rapidly diminishing ability of the world's ecosystems to support human wellbeing (Millennium Ecosystem Assessment, 2005) it is not surprising that many governments view sustainable use of the natural environment as an issue of significant national interest.

The all-pervading link between humans and their environment is the genesis of the term social–ecological systems (Walker and Salt, 2012). Two concepts have captured

River Science: Research and Management for the 21st Century, First Edition.
Edited by David J. Gilvear, Malcolm T. Greenwood, Martin C. Thoms and Paul J. Wood.

the attention of natural resource managers enthusiastic to embrace linked human–environmental dimensions of social–ecological systems: resilience thinking and ecosystem services. Resilience thinking evolved from concepts of geography (Chorley and Kennedy, 1971) and theories of complex adaptive systems that describe how the interactions of components within a system cause new conditions to which a system must adapt (Gunderson and Holling, 2002). Resilience is the amount of change a system can undergo (its capacity to absorb disturbance) and remain within the same regime with the same structure, function and feedbacks (Walker and Salt, 2006). Resilience-based natural resource management advocates ecosystem stewardship, where emphasis is placed on building the adaptive capacity of social and ecological systems to prevent transformations to undesirable states (Chapin *et al.*, 2009).

Ecosystem services are also a key part of social–ecological systems. Ecosystem services are the quantifiable or qualitative benefits of ecosystem functioning to the overall environment, including the products, services and other benefits humans receive from natural, regulated or otherwise perturbed ecosystems (Millennium Ecosystem Assessment, 2005). The services supplied by ecosystems fall into four categories: supporting services (biodiversity and nutrient cycling), regulating services (functions such as flood control and primary production), provisioning services (products such as fisheries and irrigated agriculture) and cultural services (non-material benefits such as ceremony and education). Ecosystem services are used in natural resource management to convey the fundamental link between ecosystem condition and human wellbeing (Millennium Ecosystem Assessment, 2005). Monetary values can be assigned to some ecosystem services based on their value to individuals, societies and nations (Heal, 2000). Ecosystem service valuation also helps to assess the costs and benefits of decisions about the use of natural resources (Johnston *et al.*, 2011; Kareiva *et al.*, 2011; Willemen *et al.*, 2012). Thus, resilience and ecosystem services both explicitly acknowledge that social and biophysical factors interact to cause substantial and undesirable changes in ecosystems, but at the same time, humans have the capacity to act to prevent or reverse undesirable changes in ecosystems.

Rivers are social–ecological systems. For centuries, humans have utilised rivers for freshwater, food, transport, power generation, building materials, religious ceremony and recreation (Figure 10.1). Costanza *et al.* (1997) valued the services provided by the world's ecosystems at US$33 trillion per year, of which 15% is contributed by freshwater wetlands, floodplains and lakes. It has been estimated that the ecosystem services supplied by floodplains are valued at US$3920 × 10^9 ha yr^{-1} compared to US$969 ha yr^{-1} for forests and US$92 ha yr^{-1} for cropland (Tockner and Stanford, 2002). In Australia, the ecosystem goods and services of the floodplain–river ecosystems of the Murray–Darling Basin have been valued at US$179,752 million per year (Thoms and Sheldon, 2000). These ecosystems also support a well-established and economically important agricultural industry that was valued at AU$15.9 billion per year in 2005–06, of which AU$5.5 billion was produced with the assistance of irrigation (ABARE, 2009). Despite these values, or perhaps because of them, humans have modified river ecosystems to enhance the provision of particular ecosystem services. Enhancing the provision of one service may,

Figure 10.1 Rivers are social–ecological systems, utilised by humans for services including (a) transport, (b) power generation, (c) food and (d) recreation. Photos: (a) canal bridge over the River Elbe, Germany (M. Parsons); (b) Mississippi River, USA (M. Parsons); (c) Zambezi River, Zimbabwe (M. Parsons); (d) Namoi River, Australia (M. Southwell/A. Matheson). (*See colour plate section for colour figure*).

however, lead to a reduction or elimination of another service (Rodrìguez *et al.*, 2005). For example, dams have been built to increase the availability of water for irrigated agriculture, to ensure water supply during droughts, to produce power and to mitigate the impacts of floods on human settlements. The introduction of dams can change the downstream quality and flow regime of rivers, with subsequent impacts on other ecosystem services such as biodiversity, fisheries, recreation and sediment transport (Dynesius and Nilsson, 1994). Modifying river ecosystems to enhance the provision of particular ecosystem services inevitably requires tradeoffs between different uses of the resource (Rodrìguez *et al.*, 2006). Such tradeoffs arise from management choices made by humans who 'intentionally or otherwise, change the type, magnitude and relative mix of services provided by ecosystems' (Rodrìquez *et al.*, 2005: p. 433). Part of the challenge in making these tradeoffs is to consider the sustainability of the river ecosystem – that is, how the utilisation of an ecosystem service now may influence the provision of that service into the future. Assessing tradeoffs, and deciding whether utilisation of ecosystem services is in the national interest, therefore requires consideration of both the ecological and social components of river ecosystems.

River assessment is the evaluation of river condition using surveys and other direct measures to determine the effects that human activities have on the structure and function of river ecosystems. River assessment commonly includes some type of monitoring mandated as part of government programs or legislation (Lindenmayer and Likens, 2010). Mandated monitoring tracks biological, chemical, hydrological and/or physical elements of river ecosystems through time to determine trends in river condition and detect environmental harm. The elements selected in any monitoring programme are generally chosen because they change in some way in response to human impacts and therefore can be used to infer deterioration or improvement in the condition of river ecosystems (Downes *et al.*, 2002). River assessment and monitoring has developed into a scientific discipline in its own right, endeavouring to empirically identify river ecosystem deterioration or improvement through the use of increasingly sophisticated sampling methods, statistical analyses and reporting tools (Rosenberg and Resh, 1993; Barbour *et al.*, 1999; Wright *et al.*, 2000; Bailey *et al.*, 2004; Hughes *et al.*, 2010). Indeed, advances made in river assessment and monitoring over the past 50 years have enhanced the protection of river ecosystems worldwide. However, river assessment and monitoring has focused on the biophysical elements of river ecosystems (cf. Norris and Thoms, 1999). Despite rivers being social–ecological systems, the social elements that influence river condition have largely been ignored in river assessment and monitoring programmes. We argue in this chapter that advancing national interests in river ecosystem sustainability in the twenty-first century will require river assessment programmes to pay greater attention to the linkages between social factors and the condition of biophysical elements of river ecosystems. First, we briefly describe the development of the major, biophysically-focused contemporary (post-1980s) river assessment and monitoring approaches. We then assess the utility of biophysical parameters for assessing rivers as social–ecological systems. We then develop a framework describing how the social and ecological components of river ecosystems can be included in river assessment programmes, based on principles of resilience thinking and strategic adaptive management.

A brief overview of contemporary river assessment and monitoring

The development of river assessment and monitoring has been described in a number of publications (e.g., Cairns and Pratt, 1993; Bonada *et al.*, 2006; Friberg *et al.*, 2011). Population growth and its increased concentration within urban areas during the industrial revolution resulted in increasing amounts of effluents discharged into local waterways (Bonada *et al.*, 2006). Health risks resulting from these exposures led to the development of bacteriological methods to monitor the concentrations and impacts of effluents (Bonada *et al.*, 2006). The turn of the twentieth century saw the emergence of the use of biological organisms such as plants, macroinvertebrates and fish in monitoring programmes (Kolkwitz and Marsson, 1909). Programmes have continued to evolve in content and approach (Buss *et al.*, 2015), and range in complexity from the least sophisticated programmes that may focus exclusively on a single element (e.g., water quality/chemistry) to integrated assessment programmes that monitor a suite of elements (e.g., water chemistry, physical habitat and biological assemblages).

Parameters used in river assessment and monitoring programmes

Chemical parameters provide direct measures of water quality and are often associated with legislated water quality standards. Water quality/chemical parameters can generally be split into two categories: field measures and laboratory measures. Field measures include dissolved oxygen, conductivity, turbidity, and pH. Temperature, while not a chemical measure, is also often collected in the field. Some instruments have probes for measuring other parameters (e.g., chlorophyll, nitrate), but the use of these is still under refinement. Laboratory measures are analysed from water samples collected in the field and transported to the laboratory. They can include common measures such as nutrients (e.g., total phosphorus and nitrogen) and simple cations and anions (e.g., sulfate and chloride). These analytes have established impacts and links to stressors, and their low analytical costs permit their analysis as part of routine monitoring. Less common laboratory measures include heavy metals, pesticides, aromatic and aliphatic hydrocarbons and emerging contaminants such as pharmaceuticals and personal care products. The cost of analysing these measures is usually high. Although technological improvements will likely reduce analytical costs, their present cost make them a lower priority in routine assessment and monitoring programmes without a clear objective for their use. Beyond water quality standards, measured water quality/chemical parameters can also be critical for helping characterise stressors and for interpreting biological assessment results.

Biological assemblages are the central focus of many assessment and monitoring programmes. Biological assemblages provide a direct measure of biological condition relative to biological integrity – a stated objective of, for example, the Clean Water Act of 1972 (USGPO, 1989) and the Water Framework Directive of the European Union (2000/60/EG, Abl. L 327 of 22.12.2000). In addition, biological assessments contribute to narrative water quality standards that are an important part of US state water-laws, and similarly, are essential for enforcement of the US Endangered Species Act (16 U.S.C. 1531–1544), Canada's Species at Risk Act (SARA; http://laws.justice.gc.ca/en/s‐15.3/text.html) and the European Union Habitats Directive (92/43/EEC, Abl. L 43 of 21.05.1992). Biota integrate the effects of multiple stressors in space and time (Rosenberg and Resh, 1993). These environmental sentinels provide a way of detecting stressors that may be so variable in time (e.g., pulses of metal effluent associated with storms) or space (e.g., bank erosion) that they are neither logistically nor economically feasible to monitor directly. For example, episodic pollutants cause mortality that is reflected in changes in community structure long after the event. Similarly, sediment inputs associated with spatially variable erosion will have impacts far from the source, helping to integrate this variability into a distinct biological response. A variety of organisms have been used for biological monitoring (e.g., Bonada *et al.*, 2006; Flotemersch *et al.*, 2006; Friberg *et al.*, 2011). The three most common are algae, macroinvertebrates and fish. Use of aquatic macrophytes has increased in recent years but is yet to be widely incorporated into biological monitoring programmes (Angradi, 2006).

Algae offer the advantage of being primary producers with rapid reproductive rates and short lifespans, which means they are indicators of short-term impact (Stevenson and Smol, 2003). They are sensitive to a variety of physical and chemical factors. As primary producers, many taxa are especially sensitive to nutrient pollution and will respond directly (Stevenson and Smol, 2003) which has led to their use in the development of nutrient criteria. Similarly, algae will likely respond more directly than other organisms to certain contaminants (e.g., herbicides). Sampling is relatively easy for many of the common algal taxa. In wadeable streams, sampling has primarily focused on periphyton or attached algae, especially diatoms (Stevenson and

Smol, 2003). In non-wadeable systems, the phytoplankton, or unattached free-floating taxa, may also provide an appropriate algal assemblage for use in assessment. Algae can be characterised in terms of both individual taxonomic change or in terms of whole assemblage biomass (or chlorophyll) response (Stevenson and Smol, 2003).

Benthic macroinvertebrates are the primary consumers in most systems and are an important link between primary resources and higher trophic levels, including many important recreational and commercial fish. Most macroinvertebrates are relatively sessile, which means they are excellent for evaluating site-specific impacts. They have a variety of lifecycles, with short-lived and long-lived taxa, and thus provide a way of integrating impacts over a variety of timescales (Rosenberg and Resh, 1993). Macroinvertebrates are relatively easy to identify to the family level and many are easy to identify to genus. Macroinvertebrate taxa vary in their tolerance to different stressors, providing information for interpreting cumulative stressor impacts through community assemblage structure (Rosenberg and Resh, 1993). Collection methods are relatively easy, straightforward and inexpensive. Wadeable stream sampling methods have focused primarily on the benthic invertebrates. However, large rivers may have a substantial zooplankton assemblage which is a useful indicator of water quality and physical stressors (e.g., Dettmers *et al.*, 2001; Steinberg and Condon, 2009).

Fish are included in assessment and monitoring programmes as they are a functionally diverse group that represent a variety of habitat uses. Their use in assessing the sustainability and biological integrity of water resources is discussed in detail by Simon (1999). Among their useful traits, fish are relatively longer lived and include many mobile species, so they can potentially integrate the effects of stressors over longer spatial and temporal scales. The environmental requirements and life histories of many fish species are well understood, meaning that the presence or absence of taxa can often be easily interpreted. Many fish species are consumed by humans and, therefore, they provide an assessment that is directly related to human health. In addition, many aquatic life uses are linked to fisheries, providing a direct measure of those uses. Fish are generally easy to collect and to identify to species. Most can be identified in the field and released, unharmed.

Physical habitat assessment examines the structural features of riverine environments that influence the structure and function of biological communities. Habitat and biological diversity are linked and the loss or damage of habitat is one of the principal stressors to biota (Raven *et al.*, 1998). When habitat assessment is combined with land use/land cover data for adjacent and catchment areas it is possible to draw an accurate picture of physical factors acting on a reach, to subsequently assist with initial stressor identification for impaired river sites. There are many habitat assessment approaches, ranging from methods designed to describe the geomorphic condition of streams per se to those designed to assess biotic habitat condition including the adjacent riparian habitat (see review by Parsons *et al.*, 2004). Recent approaches to the physical assessment of rivers have adopted a hydromorphology perspective, which emphasises that the interaction between the flow of water and channel form is key to river condition (Newson and Large, 2006; Vaughan *et al.*, 2009; Elosegi and Sabater, 2013).

Evolution of approaches to river assessment and monitoring

In addition to the different components used in river assessment and monitoring, philosophies and approaches to measuring and assessing impairment have changed with the evolution of the discipline (Figure 10.2). Initially, the concept of biological indicators of environmental condition developed out of the idea of saprobity – that is, the degree of organic pollution and resulting decrease in dissolved oxygen (Cairns and Pratt, 1993). Observations of the relationships between environmental pollution and taxon occurrence subsequently facilitated the derivation of biological indicators of different types of pollution (Cairns and Pratt, 1993). Knowledge of the pollution tolerances of individual taxa then led scientists to think about how communities might integrate the effects of pollution over time and how pollution was reflected by the abundance and composition of taxa within a community. Diversity indices from the field of community ecology have been used to monitor stream health (Rosenberg and Resh, 1993). However, diversity indices are often too general to decipher the many ways in which pollutants and other stressors influence community composition and abundance. This realisation stimulated the development of multi-metric approaches to monitoring throughout the 1980s and 1990s (Figure 10.2). The multi-metric approach combines pollution tolerance information with the functional, life-history and habitat context of taxa (Karr, 1981; Kerans and Karr, 1994). Multi-metric approaches continue to be used widely in many monitoring programmes, particularly in North America (e.g., Hughes *et al.*, 2010).

At around the same time as multi-metric approaches to monitoring emerged in North America, predictive modelling approaches were being developed in the United Kingdom (Figure 10.2: Wright *et al.*, 2000). Such models are able to predict the fauna that should be present at a test site based on its physical features, by matching the test site to a set of reference sites with similar physical features. The deviation of the observed test-site community from the expected reference community (O/E ratio) is a measure of the ecological status of a site (Bailey *et al.*, 2004). The RIVPACS models have subsequently been modified for use in national monitoring programmes in Australia and Canada (Wright *et al.*, 2000;

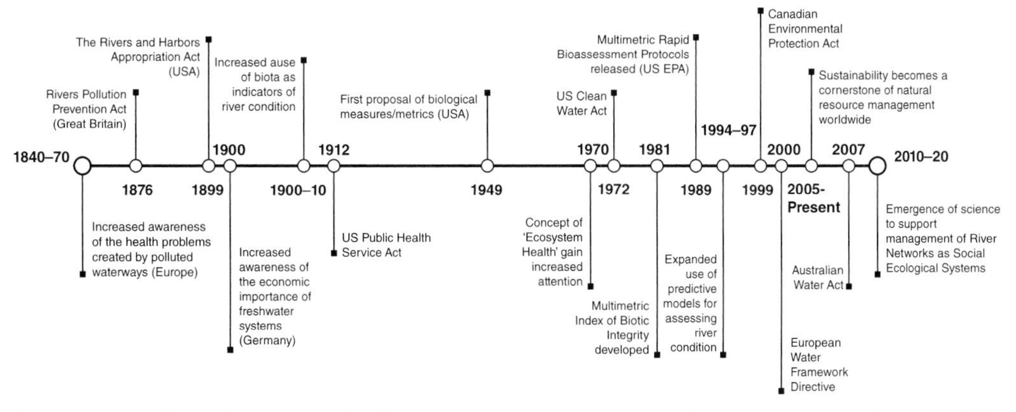

Figure 10.2 Timeline of important developments in river assessment and monitoring since the turn of the twentieth century.

Bailey *et al.*, 2004). Bayesian approaches are also emerging to detect degradation in river ecosystems (e.g., Webb and King, 2009).

Index-based, multi-metric and predictive modelling approaches use a singular biological element to assess and monitor river condition. There are few large-scale integrated assessment and monitoring programmes, despite compelling evidence for the impacts of catchment degradation and direct channel modifications on river ecosystems (e.g., Meybeck, 2003). Integrated assessment programmes combine a suite of elements (e.g., water chemistry, physical habitat, geomorphology, hydrology, multiple biological assemblages) to give a broader indication of the state of river ecosystems. Prominent examples of larger-scale integrated programmes include the US Environmental Protection Agency's National Rivers and Streams Assessment (NRSA: USEPA, 2011), the European Union's Water Framework Directive (Heiskanen *et al.*, 2004) and the Sustainable Rivers Audit (SRA) approach used in the Murray–Darling Basin in Australia (Davies *et al.*, 2012).

The NRSA is a study of the condition of flowing waters in the USA. The NRSA is a component of USEPA's larger National Aquatic Resource Surveys programme that samples all aquatic resources of the nation on a rotational basis (i.e., rivers and streams, lakes, wetlands or coastal waters). The NRSA sampling design includes sites from small streams to large rivers and is designed to answer three key questions: (i) what percentage of the nation's rivers and streams are in good, fair and poor condition for key indicators of ecological and human health; (ii) what is the relative importance of key stressors such as nutrients and habitat condition and (iii) what are the trends in stream condition when compared to previous studies (USEPA, 2011)? In the 2008–09 NRSA, 2400 sites were sampled to represent the condition of rivers and streams across the country: 1200 in each of the two categories of waters (wadeable and non-wadeable). The NRSA measures a wide variety of variables intended to characterise the chemical, physical and biological condition. These include water chemistry, nutrients, chlorophyll-a, sediment enzymes, enterococci, fish tissue, physical habitat characteristics and biological assessments including sampling of periphyton, benthic macroinvertebrates and fish community. The study also includes sampling for pharmaceuticals and personal care products in selected urban waters.

Australia's Sustainable Rivers Audit (SRA) is another example of a large-scale integrated assessment and monitoring programme. The SRA is designed to represent functional and structural links between ecosystem components, biophysical condition and humans within the rivers of the Murray–Darling Basin, Australia (Davies *et al.*, 2012). Environmental metrics derived from field-collected data or modelling are combined as indicators of condition for five themes – hydrology, fish, macroinvertebrates, vegetation and physical habitat. Condition indicator ratings are combined using expert systems rules to indicate ecosystem health. The SRA is underpinned by a series of conceptual models for each theme and overall ecosystem structure and function of rivers within the Murray–Darling Basin. It also utilises a reference condition – an estimate of condition in the absence of no significant human intervention in the landscape – and this provides a benchmark for comparisons.

Despite the rich contribution of assessment and monitoring programmes to protecting and conserving rivers, progress in the discipline has essentially been associated with the development of increasingly sophisticated methods, indicators of condition and communication interfaces, or the integration of technological advancements such as modelling and remote sensing into assessment and monitoring programmes. There is also emerging concern that monitoring occurs for its own sake, to serve rigid and sometimes outdated legislative requirements, rather than being used in a proactive way to guide the management and sustainability of ecosystems (e.g., Lindenmayer and Likens, 2010). However, the discipline of natural resource management has made significant philosophical and conceptual advancements over the previous decade in response to rapid and widespread changes in ecosystems (Chapin *et al.*, 2009). This transition has moved from ideas of ecosystems as stable, unchanging and able to be managed for single resources or species, towards ideas where ecosystems are seen as coupled social–ecological systems (Chapin *et al.*, 2009). Under this social–ecological view, changes in ecosystems manifest from interconnections among the physical, ecological and social components of ecosystems at multiple scales, rather than a component in isolation. Thus, there is a need to scrutinise how current methods, tools and approaches in river assessment and monitoring can serve recent social–ecological approaches to natural resource management which require assessment of the ways in which human capital can be utilised to manage the sustainable provision of river ecosystem services.

Monitoring and assessing rivers as social-ecological systems

The conceptual hiatus in river assessment and monitoring comes at an opportune time because many natural resource management agencies are revising their policies and programmes to align with concepts of rivers as social–ecological systems, often using a resilience framework (Benson and Garmestani, 2011; NRC, 2012). We envisage that 10 years from now, in 2024, ideas of rivers as social-ecological systems will be well-established in government policy, legislation and natural resource management programmes. Thus, monitoring will not only be able to detect human impacts on biophysical aspects of river ecosystems, but will be able to assess the social–ecological resilience of river ecosystems, and the ability of society to transform and adapt in the face of change. But what are the characteristics of resilience in social–ecological systems and how might these characteristics be monitored and assessed? Can the biophysical parameters and approaches currently used in assessment and monitoring programmes provide information about the resilience of river ecosystems, or would new approaches be needed?

From a natural resource management perspective, resilient systems are those which can absorb external shocks and disturbances while maintaining the ability to supply ecosystem services to society without further degradation to the resource (Chapin *et al.*, 2009). Associated with resilience thinking is a set of concepts describing the mechanisms by which a system can change to a different regime, or the attributes that help a system absorb disturbance and prevent regime shifts (Table 10.1). Explaining in detail each of the resilience terms outlined in Table 10.1 is beyond the scope of this chapter. Instead, we refer readers to the foundation literature of resilience thinking (Gunderson and Holling, 2002; Walker and Salt, 2006; Chapin *et al.*, 2009; Gunderson *et al.*, 2010; Walker and Salt, 2012). What the resilience concepts in

Table 10.1 Key terms and concepts of resilience thinking.

Term	Definition	Function	Reference
Diversity	Diversity is the different kinds of components that make up a system. Two types of diversity are important for resilience: functional diversity (diversity in the range of functional groups that a system depends on) and response diversity (diversity of the range of different response types existing within a functional group).	Diversity is a major source of future options and the source of the capacity for a system to absorb and respond to change and disturbance.	Walker and Salt, 2006; Walker and Salt, 2012
Ecological variability	Variability of ecological processes in space and time.	Ecosystems vary in time and space across multiple scales. Variability drives many ecosystem processes and structures.	Walker and Salt, 2006
Modularity	A modular system consists of loosely interacting groups of tightly interacting individuals.	Over-connected systems are susceptible to shocks and transmit shocks rapidly.	Walker and Salt, 2012
Tight feedbacks	The secondary effects of a direct effect of one variable on another that cause a change in the magnitude of that (first) effect. A positive feedback enhances the effect; a negative effect dampens it.	Feedbacks can signal when threshold changes are being approached.	Scheffer, 2009; Walker and Salt, 2012
Social capital	The capacity of a society to act in a cohesive way to guide its future and deal with crises.	Social capacity attributes such as trust, networks and norms provide people with the capacity to absorb or respond to change or external shocks.	Berkes and Folke, 1998; Walker and Salt, 2006; Walker and Salt, 2012
Innovation	The capacity of a system to learn, adapt, experiment, develop rules locally and embrace change.	Innovative systems are able to embrace change and adapt easily to external shocks.	Walker and Salt, 2006
Overlap in governance	Institutions that include redundancy in their governance structures and a mix of common and private property rights.	Redundancy increases the flexibility of a system to absorb and respond to change or external shocks.	Walker and Salt, 2006

Regime	A regime is a set of states that a system can exist in and still maintain the same basic structure and function.	The social–ecological system can exist in multiple states within a regime because of both natural variation and anthropogenic influences. Regime shifts occur when the social–ecological system crosses a threshold into an alternative regime.	Walker and Salt, 2012
Thresholds or tipping points	Critical level of one or more controlling variables that, when crossed, triggers an abrupt change in the system.	Change across thresholds places the system into an alternative regime with different controlling variables. Change can be into desirable or undesirable regimes.	Chapin *et al.*, 2009; Scheffer, 2009
Slow variables	Variables that strongly influence social–ecological systems but remain relatively constant over years to decades.	Trajectories of change in social–ecological systems are often controlled by a small number of slow variables. These variables are often overlooked or taken for granted in resource management.	Chapin *et al.*, 2009
Adaptive cycle	Cycle of social–ecological systems through various phases of disruption and renewal.	The adaptive cycle has four phases: rapid growth, conservation, release and reorganisation. The vulnerability and resilience of social–ecological systems changes through the phases of the adaptive cycle.	Chapin *et al.*, 2009; Walker and Salt, 2012; Cork, 2010

(continued overleaf)

Table 10.1 *(Continued)*

Term	Definition	Function	Reference
Adaptability	Capacity of human actors (individuals and groups) to respond to, create and shape variability and change in the state of the system.	Adaptability of humans provides capacity to manage resilience and avoid crossing thresholds into undesirable regimes.	Chapin *et al.*, 2009; Walker and Salt, 2012
Transformability	The capacity to create a new system with new controlling variables when economic, social or ecological conditions make the existing system untenable.	Transformations are often triggered by crises, but during such times, adaptive societies can consider novel solutions and create new systems.	Walker and Salt, 2012
Specified resilience	The resilience of some part of the system to particular kinds of disturbance.	Identifies known and possible thresholds between alternate system states.	Walker and Salt, 2012
General resilience	The capacities or attributes of a system that allows it to absorb disturbances of all kinds.	Such attributes help to stay away from thresholds and maintain a large and safe operating space for the ecosystem.	Walker and Salt, 2012
Social–ecological system	System with interacting and interdependent physical, biological and social components.	A social–ecological perspective acknowledges that humans are part of natural ecosystems.	Chapin *et al.*, 2009
Ecosystem services	Ecosystem services are the benefits people derive from ecosystems.	Provisioning services are the products obtained from ecosystems. Regulating services are the benefits obtained from regulation of ecosystem processes. Cultural services are the non-material benefits obtained from ecosystems. Supporting services are the services necessary for the production of the provisioning, regulating and cultural services.	Millennium Ecosystem Assessment, 2005

Table 10.1 do highlight is that a resilience paradigm requires assessment and monitoring of the social and ecological attributes of the system that confer resilience and the mechanisms by which a system undergoes a regime shift to a new state – parameters such as thresholds, diversity, variability, social capital, modularity, slow variables, adaptability and transformability.

To evaluate whether biophysical parameters can provide information about river ecosystem resilience, we reviewed a small set of nine commonly measured chemical (pH, turbidity), biological (EPT index, family richness, observed/expected ratio), habitat (native vegetation extent, instream habitat quality index) and hydrological (high flow index, discharge volume) parameters against the attributes that confer resilience and the mechanisms by which a system undergoes a regime shift to a new state (Table 10.2). Two biological parameters (number of families of fish or macroinvertebrates, O/E ratio fish/macroinvertebrates) can provide information about current biological diversity (Table 10.2). Several other parameters have potential to provide information about system resilience, but would need to be re-analysed to better inform the resilience attribute. For example, turbidity is a slow variable linked to regime shifts from a clear-water macrophyte-dominated regime towards a turbid-water phytoplankton-dominated regime in lakes and wetlands (e.g., Scheffer, 2009). Long-term turbidity monitoring data could be used to inform changes at or near thresholds of system change, although detecting thresholds can require mathematical analysis of long-term data (Biggs *et al.*, 2009). In general, however, the majority of biological, physical and chemical parameters did not provide any information about the attributes needed to determine river ecosystem resilience (Table 10.2). Biophysical parameters do not measure social diversity, economic diversity, feedbacks, social capital, innovation, governance, adaptive cycles, adaptability or transformability (Table 10.2). Thus, the resilience of river ecosystems cannot fully be determined from the biophysical parameters commonly measured in assessment and monitoring programmes.

Methods developed over 50 years of river assessment and monitoring focus on detecting the impacts of human activities on river ecosystems. Mandated assessment and monitoring programmes include biophysical parameters to measure adherence to standards (e.g., water quality standards) or to determine the general state of the ecosystem in relation to human pressures (e.g., biological assessment). Such mandated monitoring is required to protect human life and aquatic ecosystems. However, assessment and monitoring under a resilience approach requires a slightly different emphasis – that of being able to detect, with some confidence, whether rivers are maintaining their resilience, and their capacity to supply ecosystem services to society. Under a resilience framework the social and ecological components of the system must both be considered. Assessment and monitoring therefore shifts from detecting human impacts and recommending management actions, towards determining how human management of the system can adapt and maintain a viable, functioning ecosystem that brings multiple benefits to society.

Faced with this paradigm shift, what parameters should be measured to assess the resilience of rivers as social-ecological systems? Our review (Table 10.2) highlights that social, economic and biophysical parameters are needed to assess the resilience of rivers. Social and economic

Table 10.2 Utility of commonly used biophysical parameters for monitoring attributes of river ecosystem resilience. Y = Yes, N = No, P = Potentially (i.e., monitoring data may be used to calculate or advise a resilience attribute).

Attributes under a resilience approach	Utility of biophysical parameters commonly measured in assessment and monitoring programmes for a resilience approach								
	pH	Turbidity	EPT	No. families fish/ macroinvertebrates	O/E ratio fish/ macroinvertebrates	% Native vegetation	Habitat quality index	High flow index	Discharge volume
Biological diversity	N	N	P	Y	Y	P	P	N	N
Landscape diversity	N	N	N	N	N	P	P	N	N
Social diversity	N	N	N	N	N	N	N	N	N
Economic diversity	N	N	N	N	N	N	N	N	N
Ecological variability	N	N	P	P	P	P	P	P	P
Modularity	N	N	N	P	N	N	N	N	N
Tight feedbacks	N	N	N	N	N	N	N	N	N
Social capital	N	N	N	N	N	N	N	N	N
Innovation	N	N	N	N	N	N	N	N	N
Overlap in governance	N	N	N	N	N	N	N	N	N
Regime shift	P	P	P	N	P	P	N	N	P
Thresholds or tipping points	P	P	P	P	P	P	P	P	P
Slow variables	P	P	N	P	N	P	N	P	P
Adaptive cycle	N	N	N	N	N	N	N	N	N
Adaptability	N	N	N	N	N	N	N	N	N
Transformability	N	N	N	N	N	N	N	N	N
Ecosystem services	P	P	N	P	N	P	N	P	P

parameters are available and have been used to assess the adaptive capacity of communities within the Murray–Darling Basin in Australia (ABARE-BRS, 2010). However, it is still early days in the development of the indicators needed to fully assess the social-ecological resilience of river ecosystems. It is likely to be an ongoing endeavour to determine how best to assess and monitor resilience as robustly and empirically as we now assess and monitor the biophysical aspects of rivers. Lessons learned in the discipline of biophysical assessment and monitoring about metrics, sampling methods, modelling, reference conditions and detection of impact will, no doubt, provide important lessons for developing new ways to assess and monitor rivers as social–ecological systems.

A framework for monitoring and assessing rivers as social–ecological systems

Frameworks are used widely in many disciplines as a means to organise ideas, understand systems, identify direct cause and effect, and to link and guide decisions about system management (Dollar *et al.*, 2007). Frameworks can fail in these functions, especially in complex systems, because of the lack of recognition of the hierarchical organisation of ideas, contexts and methods. The 'why', 'what' and 'how' components of any framework are often mixed together, resulting in disordered levels of organisation and logic. In this section, we present a framework for monitoring and assessing rivers as social–ecological systems that organises components into logical levels.

In the previous section we argued that the biophysical parameters measured in mandated monitoring programmes are not always a good fit to assess rivers as social–ecological systems. Biophysical parameters provide information about the ecological state of a river ecosystem, but not about river resilience in the face of change, nor about the social state of a river ecosystem or catchment. So how can assessment and monitoring programmes become contextually relevant under the newer resilience-based approaches to river management? We propose a framework that places resilience at the core of assessment and monitoring (Figure 10.3). Under a resilience paradigm, assessment and monitoring is conducted to detect the capacity of the system to absorb disturbance without changing state, and would include both social and ecological indicators of resilience. Operationalising the assessment and monitoring of resilience is achieved through strategic adaptive management. We believe that the framework encapsulates new ideas of social–ecological systems that are at the forefront of natural resource management and applies them to the assessment and monitoring of rivers. The framework helps to shift assessment and monitoring towards questions that are scientifically related to ecosystem resilience and which can test policy and resource management options about the effects of human activities on sustainability (Lindenmayer and Likens, 2010).

The first tier of the framework places river ecosystem resilience as the central aim of assessment and monitoring programmes (Figure 10.3). In other words, resilience forms the principle that guides the other tiers of the framework. Without this first tier anchor, the methods and tools of assessment and monitoring have no context. It is important to note that resilience is the principle we have used in this chapter to be the central aim of assessment and monitoring, although other principles may be used in

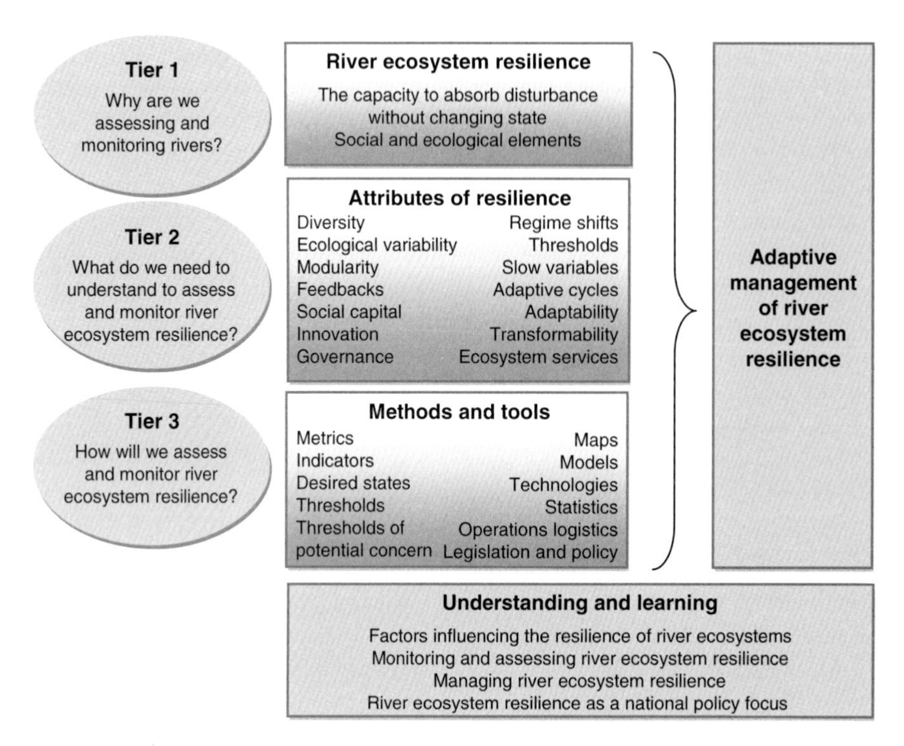

Figure 10.3 A framework for monitoring and assessing rivers as social–ecological systems.

the top tier, such as sustainability, productivity or ecosystem services. The top tier of the framework should ideally be linked to government policy because the outcomes of assessment and monitoring inform policy decisions. The key is that the first tier of the framework sets the context for assessment and monitoring – what are we assessing and monitoring, and subsequently managing towards? New assessment and monitoring programmes should be underpinned by an explicit consideration of what the programme aims to report on, and what information the reporting provides for policy and legislative outcomes.

The second tier of the framework defines the attributes of resilience to include when assessing and monitoring river ecosystem resilience (Figure 10.3). The first tier sets the context – in this case resilience – but the second tier expands on the specific attributes that need to be understood to assess and monitor river ecosystem resilience. Assessing and monitoring the resilience of river ecosystems requires attention to the social, economic and ecological attributes that confer resilience in river ecosystems. As discussed in the previous section, many of these attributes are very different from the biophysical parameters traditionally used in monitoring programmes.

The third tier of the framework describes how the specific attributes of resilience will be measured (Figure 10.3). This is the technical domain in which models, metrics, tools and procedures are developed and tested. The position of the technical domain in the framework differs from what often occurs in the development of river assessment and monitoring programmes, where the technical components become the programmatic focus rather than being a means to inform a higher level context such as resilience. It is not until the context and attributes have been set that suitable technical elements can be determined. New technical elements may need to be derived because many existing river assessment and monitoring elements and tools are not suited to a resilience context.

Applying the framework for monitoring and assessing rivers as social–ecological systems

The assessment and monitoring of river ecosystem resilience has few precedents. Thus, there is no standard formula for designing a resilience context into new or existing river assessment and monitoring programmes. There are, however, excellent guidelines for assessing resilience in spheres of natural resource management such as rangeland management or catchment management. Biggs *et al.* (2015) set out seven principles for building resilience and sustaining ecosystem services in social–ecological systems, including maintaining diversity and redundancy, managing connectivity, encouraging learning and experimentation and promoting polycentric governance systems. Likewise, in their book *Resilience Practice*, Walker and Salt (2012) describe three broad steps to the practice of resilience-based approaches to natural resource management. The first step is to describe the system, including the scales bounding the system, the people and governance structures associated with the system, the important values of the system, and the drivers of system change. The second step is to assess general and specified resilience, to determine how the system is behaving and to identify the system properties of greatest concern. The third step is to manage resilience using adaptive management and adaptive governance approaches. Adaptive management is based on principles of 'learning by doing' (Kofinas, 2009) involving two-way feedback between management policy and the state of the resource (Berkes and Folke, 1998). Adaptive management treats resource management policies as experiments from which managers can learn (Holling, 1978; Allan and Stankey, 2009). Thus, it is expected that some type of monitoring is required to track the state of the resource in relation to the management policies being applied. In the case of resilience, management policies must be able to track trends in the state of resilience attributes through time.

Strategic Adaptive Management (SAM) provides an ideal foundation for monitoring and assessing river ecosystem resilience because it can integrate the attributes of resilience within an adaptive management philosophy. Based on principles of active adaptive management, SAM was pioneered on river systems in South Africa's Kruger National Park (Biggs and Rogers, 2003; Pollard *et al.*, 2011) and is now used more widely across South Africa's national parks and catchment management authorities (Roux and Foxcroft, 2011). Strategic adaptive management offers a framework for natural resource management and decision making in situations characterised by variability, uncertainty, incomplete knowledge and multiple stakeholders, thus making it ideal to assess and monitor rivers as social–ecological systems. Three key tenets form the basis for the management and decision-making process in strategic adaptive management. First, adaptive planning is used to 'build a sense of common purpose among all relevant stakeholders and to develop a collective roadmap for getting from a current (usually undesirable) reality to a more desirable social-ecological system' (Roux and Foxcroft, 2011). A collective vision statement informs the delineation of an objectives hierarchy. Measurable targets are set for each objective, and monitored using Thresholds of Probable Concern representing the boundaries of acceptable change in the system (Rogers and Biggs, 1999). Second, adaptive implementation involves incorporating the outcomes of the adaptive planning process into organisations, such as through developing monitoring programmes that link to measurable targets (Roux and Foxcroft, 2011). Third, adaptive evaluation ensures that learning occurs throughout all steps of SAM.

The integration of strategic adaptive management and resilience thinking is necessary because, even though each is strong on its own, when applied to the problem of assessment and monitoring of rivers as social–ecological systems, there are gaps in each that can be strengthened by principles from the other approaches. Resilience thinking presents a useful social–ecological approach for understanding resilience in ecosystems, but it does not have a strong operational and implementation procedure. Strategic adaptive management provides an excellent operational procedure for managing resilience in ecosystems, but so far it has been applied to managing ecosystems where biodiversity conservation is the main goal. Thus, integration of the principles from each approach will provide a powerful and cutting-edge basis for the components of a framework that can be applied for assessing and monitoring resilience in river ecosystems.

Conclusion

Rivers have an important role in the development and continued prosperity of many countries through, inter alia, the ecosystem services they provide. Despite the use of the best available biophysical information and the investment of much time and effort, many assessment and monitoring initiatives have been ineffective in guiding governments to conserve the long-term sustainability of river ecosystems. The consequences of unrestrained development of catchment resources provide challenges for natural resource managers and river scientists on issues of equity of use between various consumers of valuable ecosystem services. Thus, in the face of likely increased conflicts associated with continued use of the ecosystem services provided by rivers, decision-making processes will need to engage social communities in learning about values and tradeoffs (Bikangaga *et al.*, 2007; Howard, 2008; Jackson *et al.*, 2008). Insufficient consideration of the importance of social–ecological systems is compounded by inappropriate approaches to the assessment and monitoring of river ecosystems. Notwithstanding mandated monitoring as a stipulated requirement of government legislation or a political directive, many programmes monitor for monitoring sake and have become self-serving within government departments and academic monitoring communities (Lindenmayer and Likens, 2010). While assessment reports can be useful for producing coarse-level summaries of changes in resource condition, many assessment and monitoring programmes cannot identify mechanisms influencing change or contribute to decision-making processes about resource use tradeoffs within complex social–ecological systems.

Adaptive approaches, based on concepts of resilience and thus incorporating social and ecological components of river ecosystems, represent a significant and beneficial paradigm shift for the discipline of river assessment and monitoring. River monitoring and assessment has a long tradition that has served river conservation well. Yet, in the last 10 years or so, many conceptual advances have been made about our understanding of the structure and function of river ecosystems (cf. Thorp *et al.*, 2008) but few about monitoring and assessing rivers as integrated systems of biophysical and social components. Biophysical parameters that are commonly measured only target the ecological part of social–ecological systems. River assessment and monitoring traditionally takes the view that humans have some impact on the system, and therefore the biophysical parameters used in assessment and monitoring programmes are designed to detect impact. These biophysical parameters do not detect resilience. Recent paradigms of social–ecological systems move away from the view of humans as stressors and view humans as users of ecosystem services. Resilient systems are able to continue to supply the ecosystem services necessary for human wellbeing. Assessment and monitoring therefore requires a focus on assessing the resilience of ecosystems and the supply of ecosystem goods and services.

A basis for the components of a best practice framework for assessing and monitoring the resilience of river ecosystems has been outlined. It is important to reiterate that the framework represents the underlying philosophy of an approach for monitoring and assessing resilience in river ecosystems. Implementing the framework will require the development of management, policy and governance structures and attitudes required to manage for resilience in river ecosystems. This step is not simple or easy. As Rogers (2006) points out, the real river management challenge lies in developing a collective understanding, and integration, within and between scientists, citizens and management agencies. However, the experiences of biodiversity management in the rivers of Kruger National Park suggest that this can be achieved though participatory and cooperative approaches and a commitment to long-term co-learning.

Future research needs associated with the adoption and implementation of a framework for monitoring and assessing resilience in river ecosystems would seek to fill gaps in knowledge in the second and third levels of the framework (Figure 10.3). Research gaps are not only associated with the understanding of social–ecological resilience in river ecosystems, but also with the institutional aspects of adopting a resilience approach to managing river ecosystems. Indeed, it is important to note that strategic adaptive management provides an approach that can deal with uncertainties and incompleteness of knowledge. We do not have to know everything to start the process. Commitment to an adaptive, participatory, learning-by-doing approach ensures that a common goal unites all stakeholders and builds trust and cooperation. Conceptual modelling and the use of thresholds of potential concern give options for moving forward in the face of limited understanding.

However, it is also important that research takes place within the context of the highest-level component – managing for river ecosystem resilience in social–ecological systems, or in other words, maintaining the capacity of river ecosystems to absorb human and natural disturbance without changing to a different state. This is so that research carried out by different groups or disciplines is conducted with the same guiding principles in mind: those of resilience thinking, social–ecological systems and strategic adaptive management. We suggest that future research needs would focus on the following areas:

- Understanding how biophysical aspects of river ecosystems (heterogeneity and variability, scale, flux and cycling, slow variables and multiple stable states) are linked to social aspects of river ecosystems. For example, how do stakeholders view variability and heterogeneity in river ecosystems? How is variability and heterogeneity perceived in river policy settings, by citizens and by governments?
- Defining the role of science in the formation and adaptation of water policy and legislation. For example, how do stakeholders view scientists and vice versa? How can science be better translated and communicated into the social, economic and policy arenas? How can scientists communicate uncertainty to the public and policy-makers? What modes of research practice drive scientists and where do these fit into water management activities?
- Operationalising thresholds (actual or potential) that could switch a river ecosystem into an alternate state. This would require understanding of the type and scale of operation of the threshold. The concept of thresholds is most powerful when coupled with a monitoring procedure that can relate the state of the system to upper and lower limits of an acceptable state. A trial application of a thresholds approach to monitoring aquatic systems, similar to that used in strategic adaptive management, would indicate the practical utility of the threshold concept on the ground in river ecosystems.
- Gauging the mood of communities of practice (such as legislators, scientists, managers and politicians) and citizens to change in the way that river ecosystems are managed: what are the attitudes to change of different stakeholders? Would a change to the management of resilience in river ecosystems be viewed favourably or unfavourably by stakeholders? A multi-stakeholder forum, similar in form to the Australian 2020 summit, could provide an avenue for the adoption of consensus-type decision-making processes that have proven successful in the implementation of strategic adaptive management in South Africa.
- Evaluating the preparedness of governance and institutional arrangements – to implement a river ecosystem management approach based on maintaining the social and biophysical resilience of river ecosystems. For example, how would current laws and policies support or impede the adoption of resilience principles? Is a learning-by-doing approach possible in different institutional structures? What capacity exists within organisations to adapt to new paradigms of water resource management?

Resilience is gaining prominence in personal, public and institutional discourse across many topics, including natural resource management, natural disaster management, the economy and human health. Yet far from being a buzzword, resilience has the philosophical applicability to capture the responses of humans to their increasingly unpredictable and changing world. Applying concepts of resilience in the field of river management will require some means to assess and monitor resilience, and to be able to track not just the state of a system, but its remaining ability to absorb disturbance and provide ecosystem services. In essence, successful implementation of the resilience paradigm into the river management arena will require major shifts in thinking among all stakeholders: scientists, citizens and management agencies. It will be a major undertaking, but current

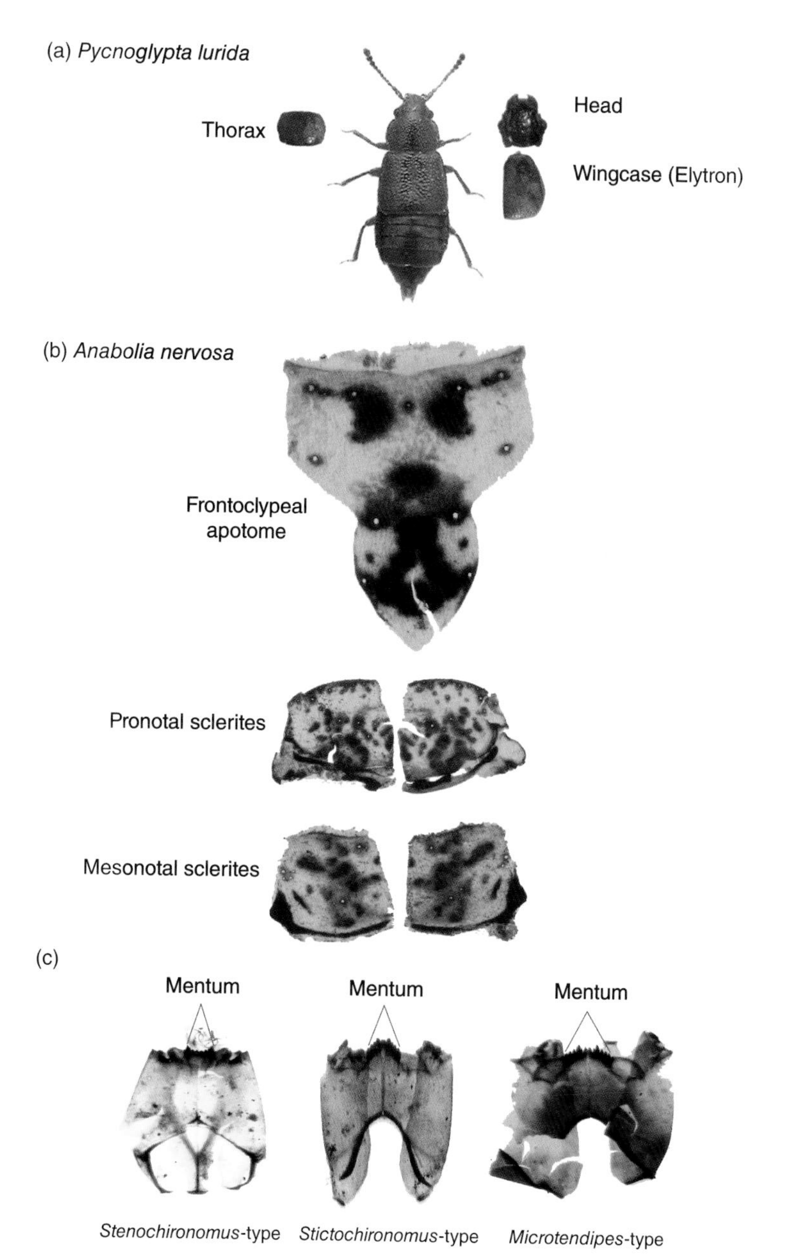

Figure 4.3 Sub-fossil fragments of beetle, caddisfly and non-biting midges collected from palaeochannel deposits from the River Trent floodplain. (a) *Pycnoglypta lurida* (Coleoptera: Staphylinidae), a rove beetle found in Late-Glacial deposits but no longer present in UK. (b) *Anabolia nervosa* (Trichoptera: Limnephilidae), a cased-caddisfly of rivers, lakes and ponds but not of temporary waterbodies. (c) Head capsules of non-biting midges *Stenochironomus*-type (a wood miner of lentic habitats), *Stictochironomus*-type (associated with macrophytes in littoral zone) and *Microctendipes*-type (common in lentic water-bodies) (Diptera: Chironomidae). Note the differing pattern of the toothed mentum.

River Science: Research and Management for the 21st Century, First Edition.
Edited by David J. Gilvear, Malcolm T. Greenwood, Martin C. Thoms and Paul J. Wood.

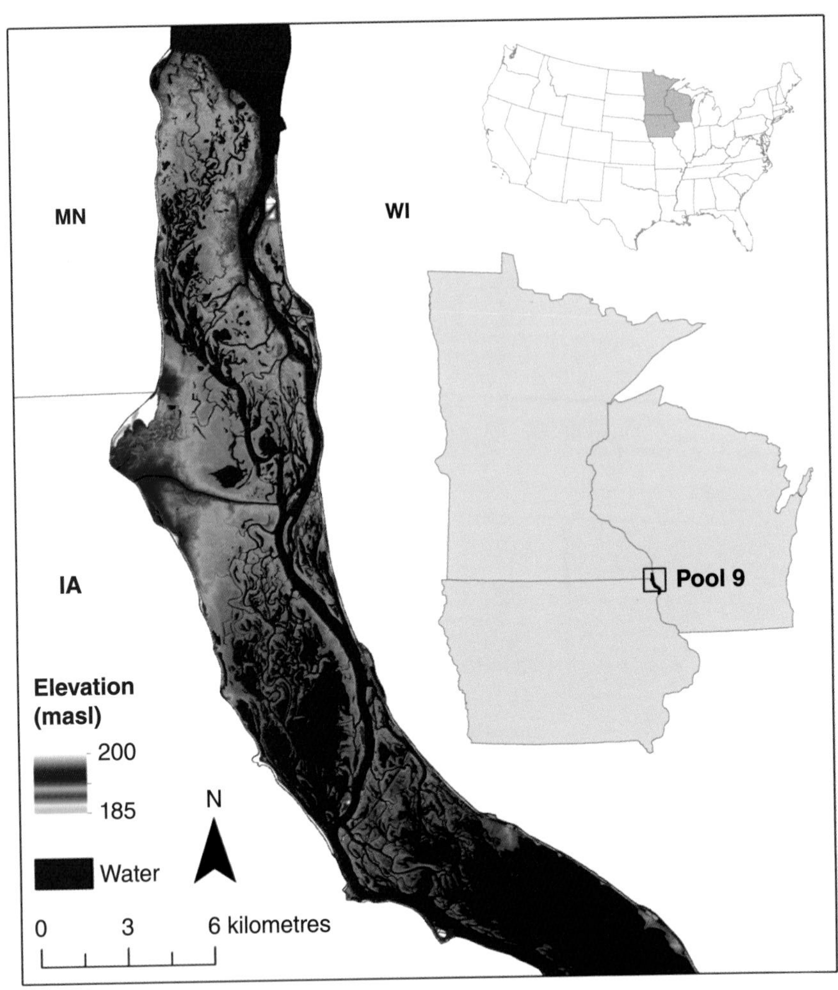

Figure 6.9 Regional location of Pool 9 of the Upper Mississippi River and detail of floodplain topography in its upper reaches. Scown *et al.*, 2015. Reproduced with permission from Elsevier.

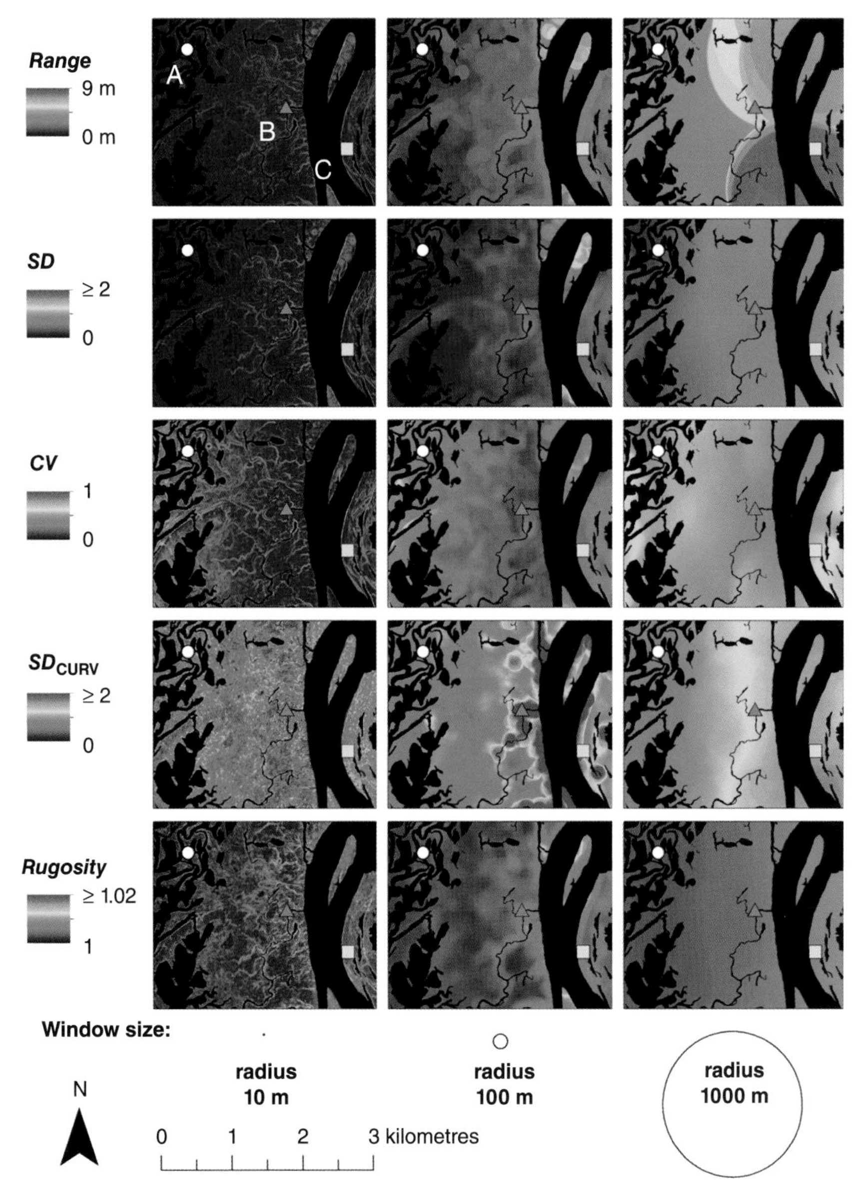

Figure 6.11 (a) Local surface metric results for three window sizes in the 3 × 2.5 km sample section of floodplain and locations of the three example cells.

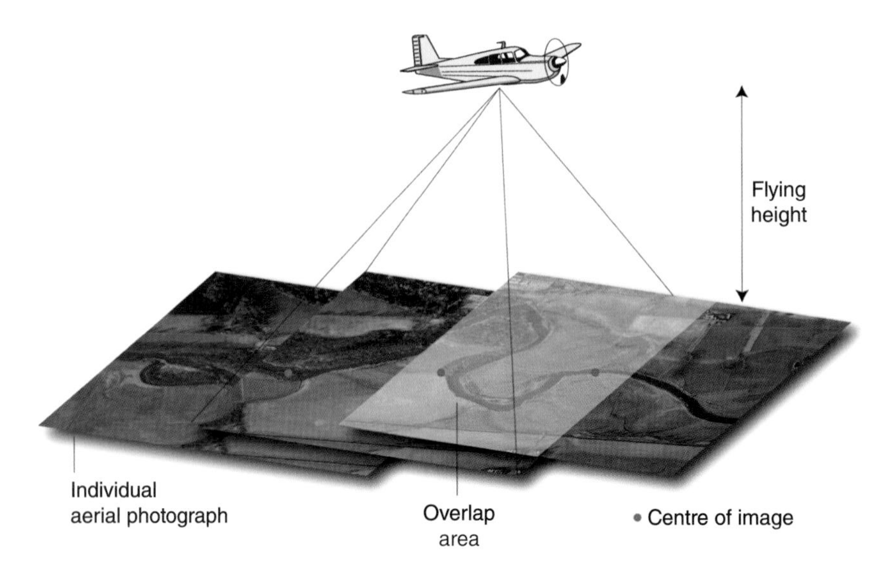

Figure 9.2 A mosaic of aerial photographs taken from a plane showing the river corridor of the River Tay, Scotland.

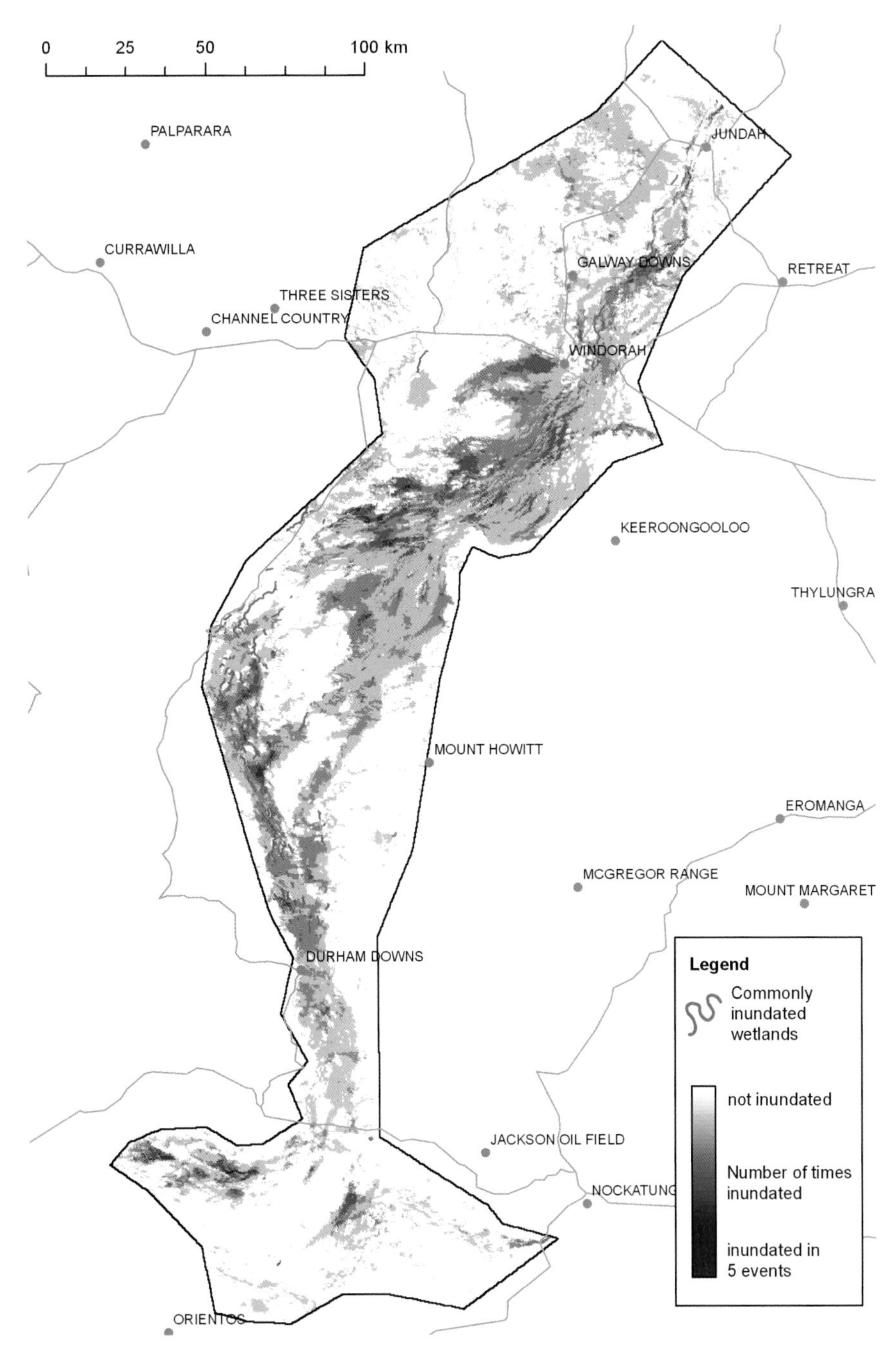

Figure 9.4 Frequency of floodplain inundation in relation to known wetlands on the Cooper Creek reach of the Murray–Darling as deduced from MODIS data and Normalised Difference Vegetation Cover.

Figure 10.1 Rivers are social–ecological systems, utilised by humans for services including (a) transport, (b) power generation, (c) food and (d) recreation. Photos: (a) canal bridge over the River Elbe, Germany (M. Parsons); (b) Mississippi River, USA (M. Parsons); (c) Zambezi River, Zimbabwe (M. Parsons); (d) Namoi River, Australia (M. Southwell/A. Matheson).

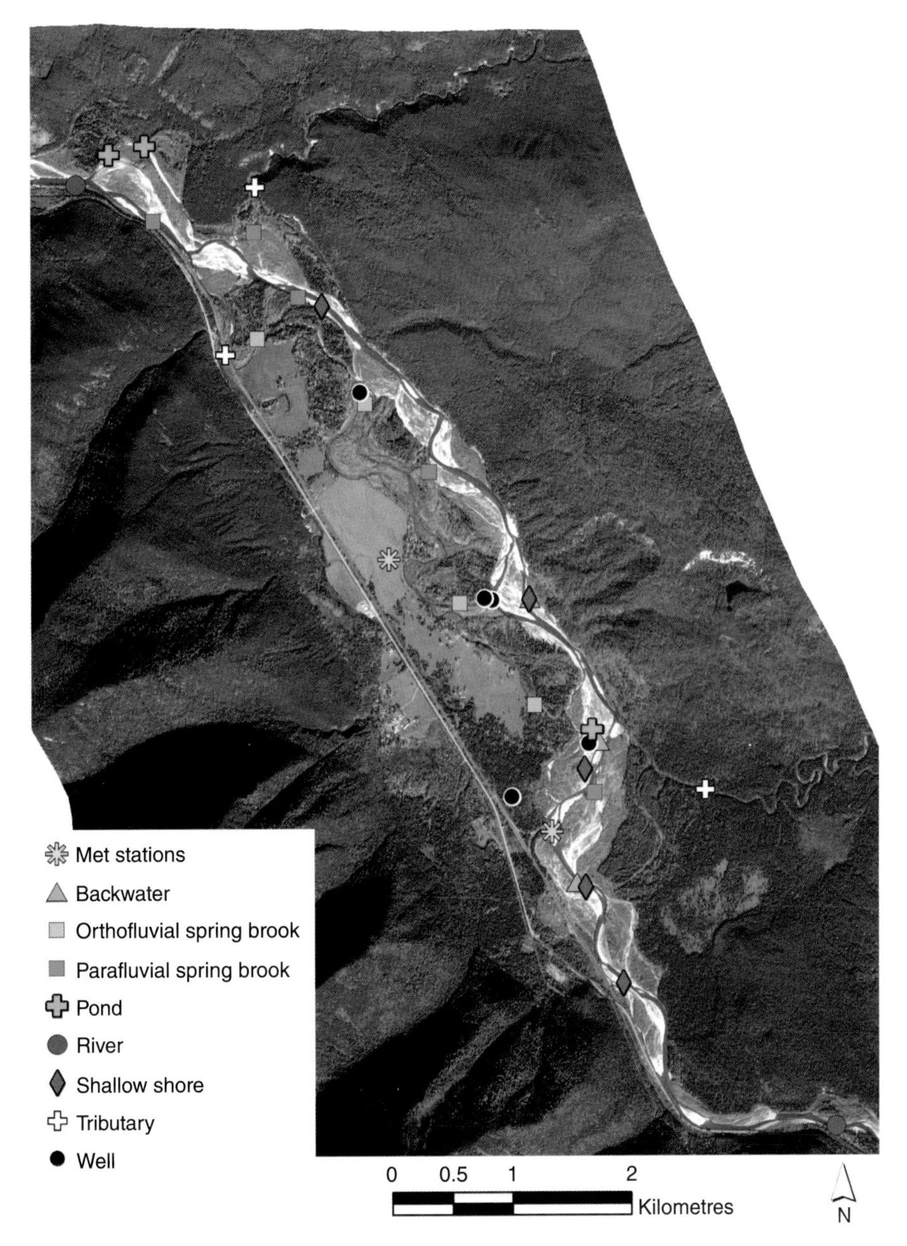

Figure 13.1 Locations of sampling sites (habitat types as keyed in the inset) of the Nyack floodplain of the Middle Fork Flathead River, Montana. The base layer is a multispectral satellite image obtained October, 2004.

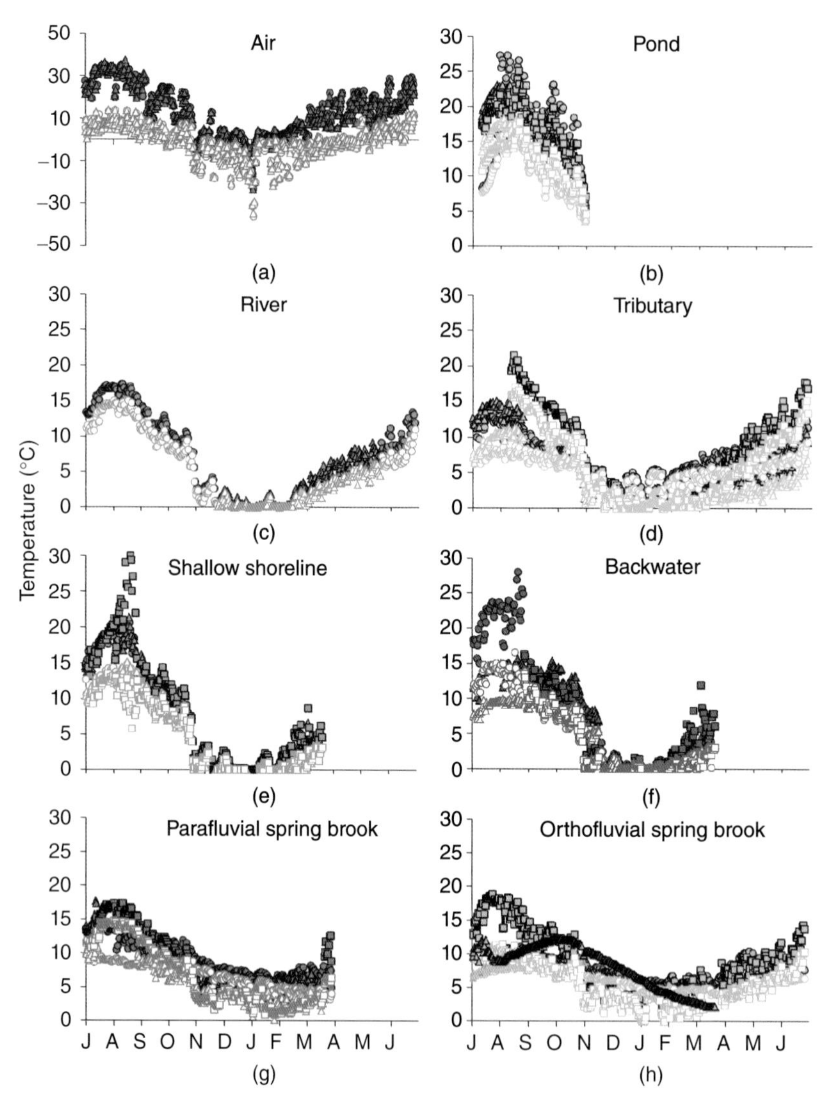

Figure 13.3 Annual temperature patterns (2003–04) for maximum (white symbols) and minimum (grey symbols) daily temperatures for Nyack floodplain sites. Sites are from different habitats as shown in Figure 1: (a) air ($n = 2$), (b) pond ($n = 3$), (c) river ($n = 2$), (d) tributary ($n = 3$), (e) shallow shoreline ($n = 3$), (f) backwater ($n = 3$), (g) parafluvial spring brook ($n = 3$), and (h) orthofluvial spring brook ($n = 3$). Separate sites within a habitat are represented by different symbols, the circle, triangle or square.

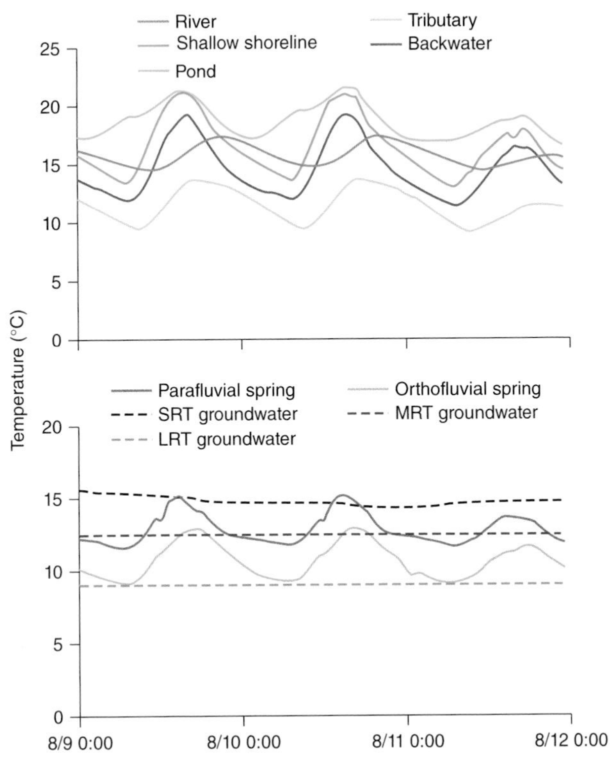

Figure 13.4 Average hourly temperatures in habitats of the Nyack floodplain during the hottest period of the year (10–12 August 2003).

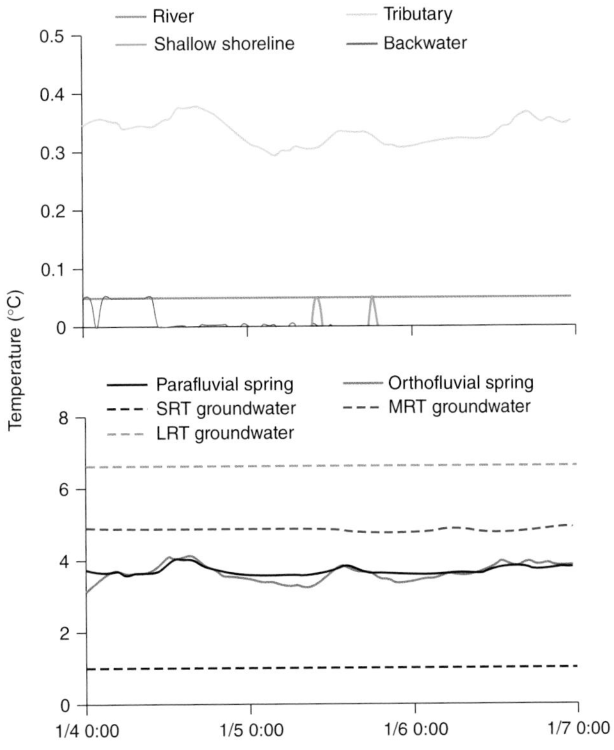

Figure 13.5 Average hourly temperatures in habitats of the Nyack floodplain during the coldest period of the year (4–7 January 2004).

Figure 14.8 (a) Visual (top) and corresponding infrared image (bottom) taken on 21 May 2010 15:02 just downstream of site 6a. (b) Visual (top) and corresponding infrared image (bottom) taken on 16 June 2010 15:09 downstream of site 6c. (c) Visual (top) and corresponding infrared image (bottom) taken on 16 June 2010 15:18 at cross-section 4a/b. (Note: White housings on visual images mark contain water temperature loggers, and show monitoring p0osition within channel. Channel width is about 4 m and flow is from right to left. Vantage point and scale of visual and infrared pictures are not exactly the same.

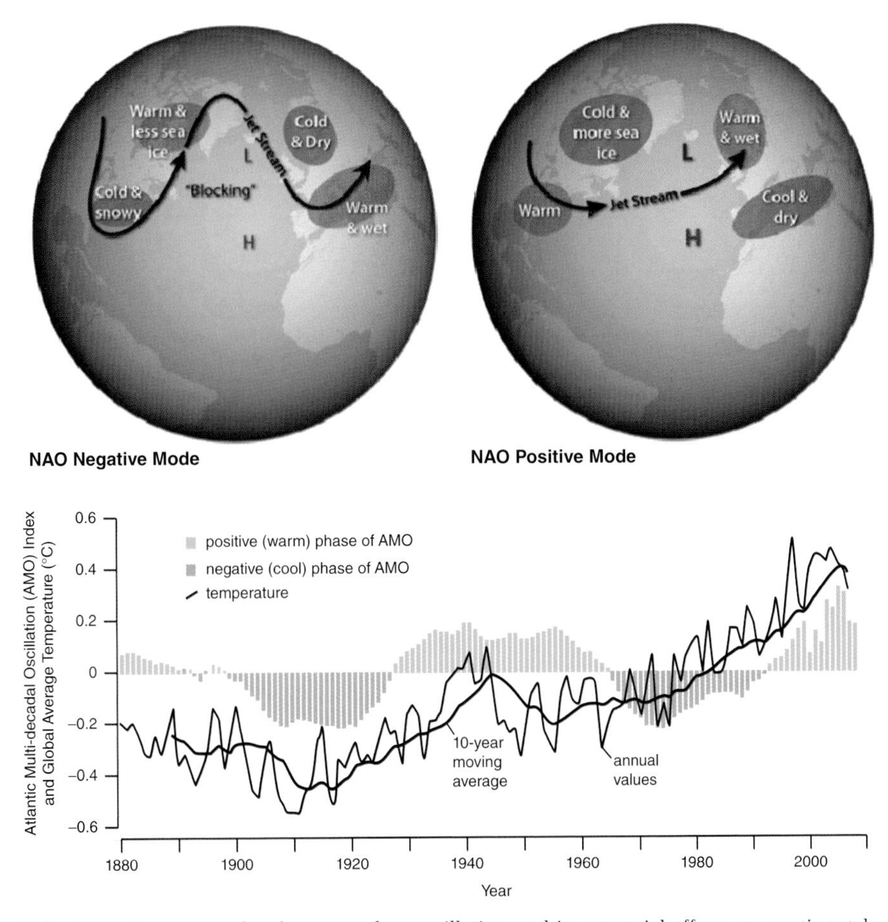

Figure 15.1 Generalised example of a sea surface oscillation and its potential effects on continental weather patterns (top). AMO index showing the warm (above line bars) and cool periods (below line bars) with global average temperature superimposed (bottom). Top graphic adapted from AIRMAP by Ned Gardiner and David Herring, NOAA. Bottom graphic created by Michon Scott, National Snow and Ice Data Center.

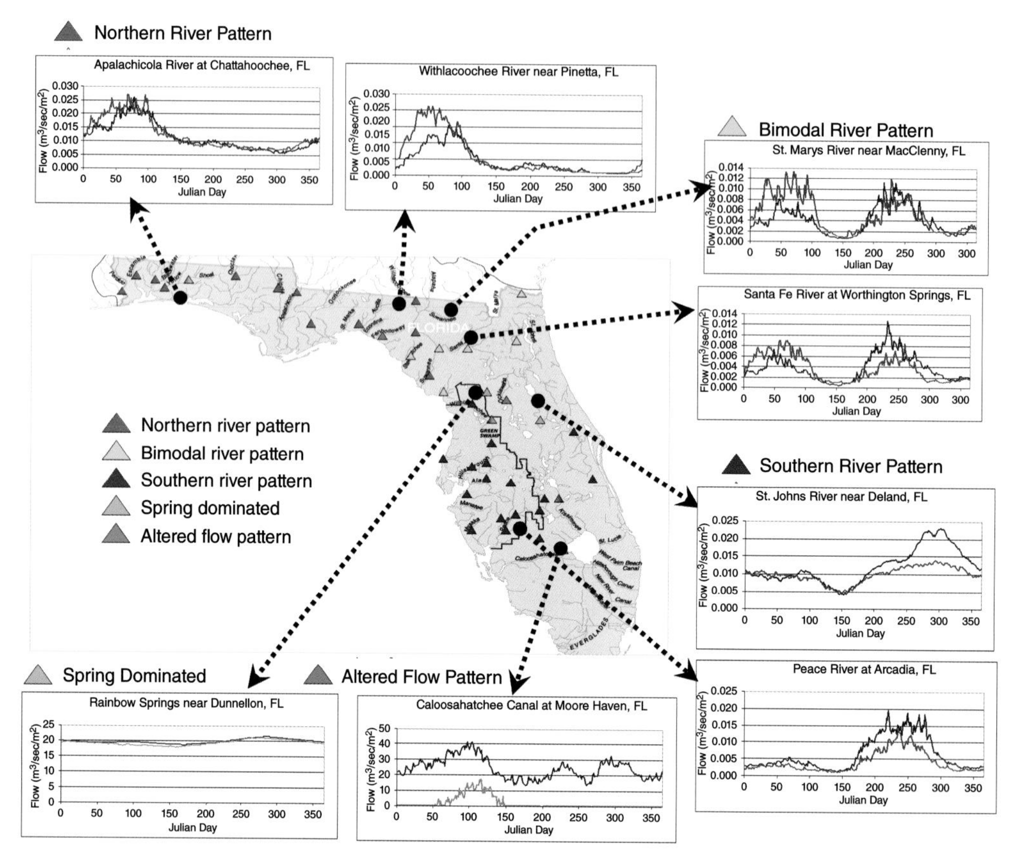

Figure 15.2 Examples of geographic differences in how the AMO can influence both the volume and, independently, the seasonality of seasonal baseline flow rates in the rivers of Florida. (Graphics adapted from M. Kelly and J.A. Gore, 2008.) Blue lines show the warm AMO period (1940–69) while green lines indicate the cooler AMO period (1970–99). Note that the effect of the AMO is different, and even reversed, depending on the geographic location of the river basin. Data from the USGS National Water Information service.

Figure 17.2 Condit Dam on the White Salmon River, Washington State, USA. This dam measured 38 m high and 144 m wide, and provided 80,000 MWh of power generation per year. The dam also resulted in the loss of 53 km of salmonid habitat. Breach of the dam occurred on October 26, 2011, and it was subsequently removed. (Photo by A. Yeakley.)

Figure 17.3 Capture and relocation of Chinook salmon on the lower White Salmon River. Prior to removal of the Condit Dam, personnel from the US Fish and Wildlife Service captured some 679 fall Chinook salmon adults, and relocated them upstream of the dam. (Photo by A. Yeakley.)

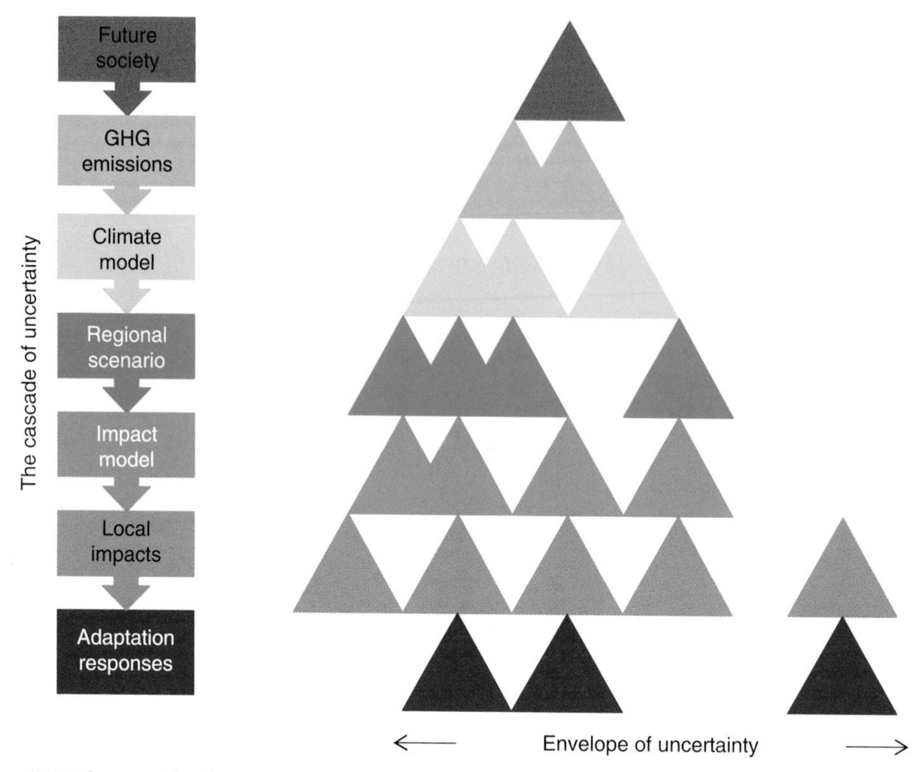

Figure 18.1 The cascade of uncertainty proceeds from different socio-economic and demographic pathways, their translation into concentrations of atmospheric greenhouse gas (GHG) concentrations, expressed climate outcomes in global and regional models, translation into local impacts on human and natural systems, and implied adaptation responses. The number of triangles at each level symbolises the growing number of permutations and hence the expanding envelope of uncertainty. For example, even relatively reliable hydrological models yield very different results depending on the methods (and data) used for calibration. Missing triangles represent incomplete knowledge or sampling of uncertainty. Adapted from Wilby and Dessai (2010).

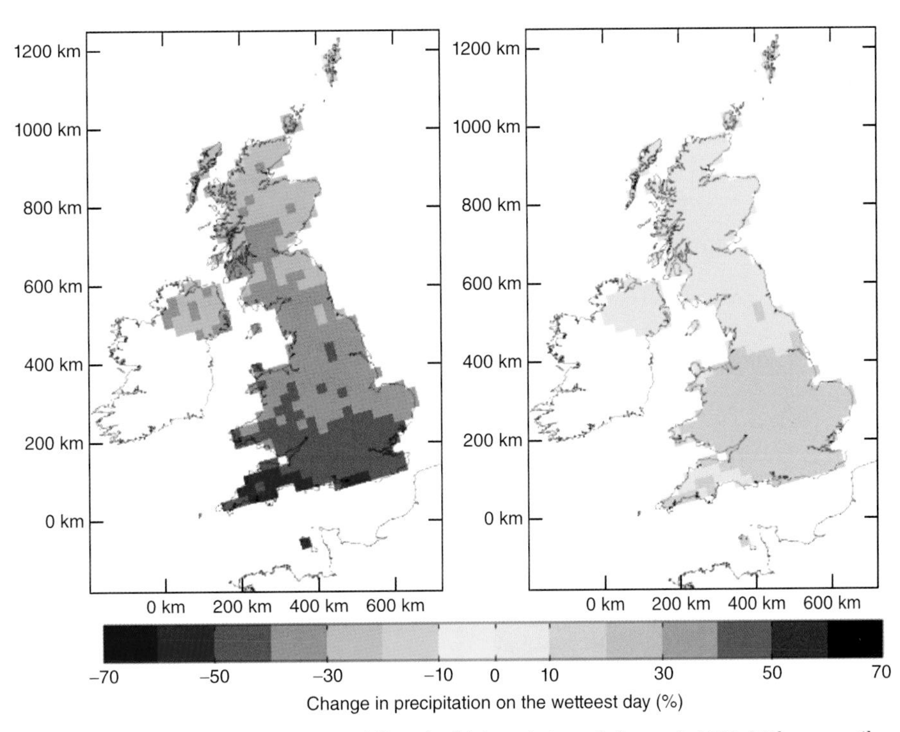

Figure 18.3 UKCP09 changes in summer rainfall under high emissions (left panel: A1FI, 10th percentile ensemble member) and low emissions (right panel: B1, 90th percentile ensemble member) by the 2050s.

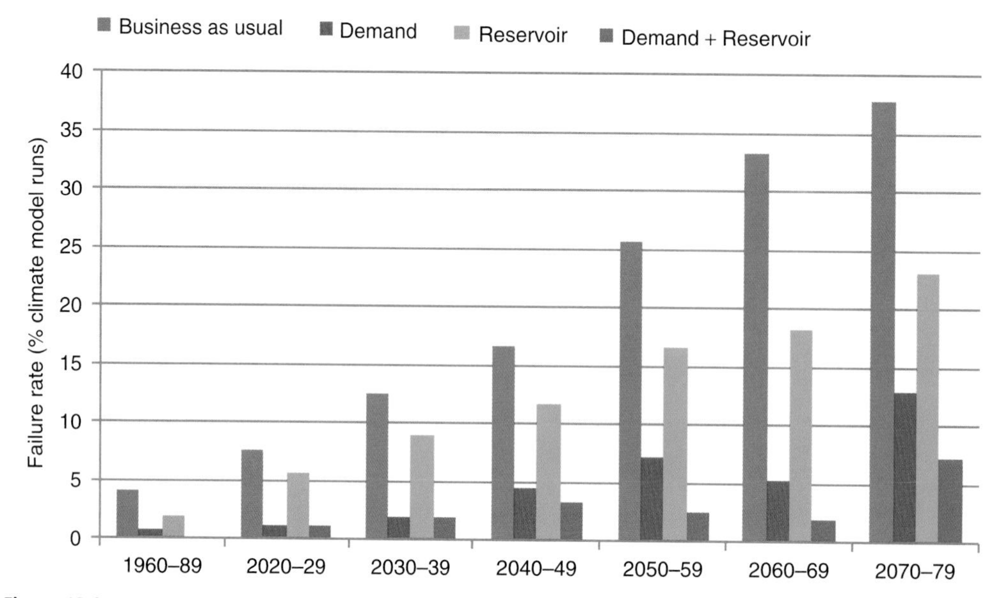

Figure 18.6 Percentage of model runs with single year water supply failure under various strategies for a water supply zone in e Devon under SRES A1B emissions. Adapted from Lopez *et al.* (2009).

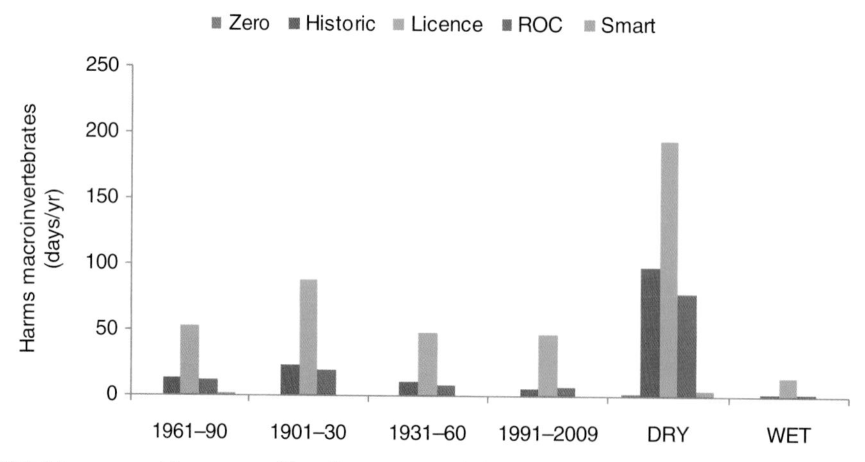

Figure 18.8 Mean annual frequency of low flows (<224 ml/d) that are harmful to macroinvertebrate communities in the River Itchen under various abstraction license conditions (Zero, Historic, Licence, Existing Review of Consents (ROC), Smart), precipitation variability (1961–90, 1901–30, 1931–60, 1991–2009) and climate change projections for the 2020s (DRY and HOT). The Smart license takes water from the environment when it is least harmful, and imposes hands-off flow conditions under very dry conditions. Source: Wilby *et al.* (2011).

political, biophysical, social and economic conditions suggest that there is a window of opportunity for a mindset shift to start to occur – essentially a rare release to a reorganisation phase in social, economic, biophysical and political adaptive cycles.

Acknowledgements

The authors thank Professor Michael Delong (Winona State University) for hosting us during the preparation of the manuscript. MT was supported by a Fulbright Senior Scholarship. The views expressed in this article are those of the author(s) and do not necessarily reflect the views or policies of the US EPA.

References

Allan, C. and Stankey, G.H. 2009. *Adaptive Environmental Management: A practitioner's guide*. CSIRO Publishing: Melbourne, Australia.

Angradi, T.R. (ed.). 2006. Environmental Assessment and monitoring Program: Great River Ecosystems, Field Operations Manual. EPA/620/R-06/002. US Environmental Protection Agency, Washington, DC., USA. Available from: http://www.epa.gov/emap/greatriver/fom.html.

Australian Bureau of Agricultural and Resource Economics (ABARE). 2009. Socio-economic context for the Murray-Darling Basin. Report to the Murray Darling Basin Authority, MDBA Technical Report Series, Basin Plan: BP02. Murray Darling Basin Authority: Canberra, Australia.

Australian Bureau of Agricultural and Resource Economics – Bureau of Rural Sciences (ABARE-BRS). 2010. Indicators of community vulnerability and adaptive capacity across the Murray-Darling Basin – a focus on irrigation in agriculture. Australian Government: Canberra, Australia.

Bailey, R.C., Norris, R.H. and Reynoldson, T.B. 2004. *Bioassessment of Freshwater Ecosystems: Using the Reference Condition Approach*. Kluwer: Massachusetts, USA.

Barbour, M.T., Gerritsen, J., Snyder, B.D. and Stribling J. B. 1999. Rapid bioassessment protocols for use in streams and wadeable rivers: periphyton, benthic macroinvertebrates and fish. Second edition. EPA 841-B-99-002. US Environmental Protection Agency, Office of Water, Washington, DC, USA.

Benson, M.H. and Garmestani, A.S. 2011. Embracing panarchy, building resilience and integrating adaptive management through a rebirth of the National Environmental Policy Act. *Journal of Environmental Management* 92: 1420–1427.

Berkes, F. and Folke, C. (eds). 1998. *Linking Social and Ecological Systems: Management practices and social mechanisms for building resilience*. Cambridge University Press: Cambridge, UK.

Biggs H.C. and Rogers K.H. 2003. An adaptive system to link science, monitoring and management in practice. In: du Toit J.T., Rogers K.H. and Biggs H.C. (eds). *The Kruger Experience: Ecology and Management of Savanna Heterogeneity*. Island Press: Washington, DC, USA.

Biggs, R., Carpenter, S.R. and Brock, W.A. 2009. Turning back from the brink: detecting an impending regime shift in time to avert it. *Proceedings of the National Academy of Sciences* 106: 826–831.

Biggs, R., Schlüter, M. and Schoon, M.L. (eds). 2015. *Principles for Building Resilience: Sustaining Ecosystem Services in Social-ecological Systems*. Cambridge University Press: Cambridge, UK.

Bikangaga, S., Pia Picchi, M., Focardi, S. and Rossi, C. 2007. Perceived benefits of littoral wetlands in Uganda: focus on the Nabugabo wetlands. *Wetlands Ecology and Management* 15: 529–535.

Bonada, N., Prat, N., Resh, V.H. and Statzner, B. 2006. Developments in aquatic insect biomonitoring: A comparative analysis of recent approaches. *Annual Review of Entomology* 51: 495–523.

Buss, D.F., Carlisle, D.M., Chon, T.S., *et al.* 2015. Stream biomonitoring using macroinvertebrates around the globe: a comparison of large-scale programs. *Environmental Monitoring and Assessment* 187: 4132. DOI: 10.1007/s10661-014-4132-8.

Cairns, J. and Pratt, J.R. 1993. A history of biological monitoring using benthic macroinvertebrates. In: *Rosenberg, D.M.* and Resh, V.H. (eds) *Freshwater Biomonitoring and Benthic Macroinvertebrates*. Chapman and Hall: New York, USA, pp. 10–27.

Chapin, F.S., Kofinas, G.P. and Folke, C. (eds). 2009. *Principles of Ecosystem Stewardship: Resilience Based Natural-resource Management in a Changing World*. Springer: New York, USA.

Chorley, R.J. and Kennedy, B.A. 1971. *Physical Geography: A Systems Approach.* Prentice Hall: London, UK.

Cork, S. 2010. Resilience of social-ecological systems. In: Cork, S (ed.). *Resilience and Transformation: Preparing Australia for Uncertain Futures.* CSIRO Publishing: Melbourne, Australia.

Costanza, R., d'Arge, R., de Groot, R., *et al.* 1997. The value of the world's ecosystem services and natural capital. *Nature* 387: 253–260.

Davies, P.E., Stewardson, M.J., Hillman, T.J., *et al.* 2012. *Sustainable Rivers Audit 2: The ecological health of rivers in the Murray-Darling Basin at the end of the Millennium Drought (2008–2010). Volume 1.* Murray Darling Basin Authority: Canberra, Australia.

Dettmers, J.M., Wahl, D.H., Daniel, A.S. and Gutreuter, S. 2001. Life in the fast lane: fish and foodweb structure in the main channel of larger rivers. *Journal of the North American Benthological Society* 23(2): 255–265.

Dynesius, M. and Nilsson, C. 1994. Fragmentation and flow regulation of river systems in the northern third of the world. *Science* 266: 753–762.

Dollar, E.S.J., James, C.S., Rogers, K.H. and Thoms, M.C. 2007. A framework for interdisciplinary understanding of rivers as ecosystems. *Geomorphology* 89: 147–162.

Downes, B.J., Barmuta, L.A., Fairweather, P.G., *et al.* 2002. *Monitoring Ecological Impacts: Concepts and Practice in Flowing Waters.* Cambridge University Press: Cambridge, UK.

Elosegi, A. and Sabater, S. 2013. Effects of hydromorphological impacts on river ecosystem functioning: a review and suggestions for assessing ecological impacts. *Hydrobiologia* 712: 129–143.

Flotemersch, J.E., Stribling, J.B. and Paul, M.J. 2006. *Concepts and Approaches for the Bioassessment of Non-Wadeable Streams and Rivers.* EPA/600/R-06/127. U.S. Environmental Protection Agency: Cincinnati, OH.

Friberg, N., Bonada, N., Bradley, D.C., *et al.* 2011. Biomonitoring of human impacts in freshwater ecosystems: the good, the bad and the ugly. *Advances in Ecological Research*, 44: 1–68.

Gunderson, L.H. and Holling, C.S. (eds). 2002. *Panarchy: Understanding Transformations in Human and Natural Systems.* Island Press: Washington, DC, USA.

Gunderson, L.H., Allen, C.R. and Holling, C.S. (eds). 2010. *Foundations of Ecological Resilience.* Island Press: Washington DC, USA.

Heal, G. 2000. Valuing ecosystem services. *Ecosystems* 3: 24–30.

Heiskanen, A.-S., van de Bund, W., Cardoso, A.C. and Nõges, P. 2004. Towards good ecological status of surface waters in Europe – interpretation and harmonisation of the concept. *Water Science and Technology* 49: 169–177.

Holling, C.S. 1978. *Adaptive Environmental Assessment and Management.* Wiley: Chichester, UK.

Howard, J.L. 2008. The future of the Murray River: Amenity re-considered? *Geographical Research* 46: 291–302.

Hughes, D.L., Brossett, M.P., Gore, J.A. and Olson, J.R. (eds). 2010. *Rapid Bioassessment of Stream Health.* CRC Press: Boca Raton, FL, USA.

Jackson, S., Stoeckl, N., Straton, A. and Stanley, O. 2008. The changing value of Australian tropical rivers. *Geographical Research* 46: 275–290.

Johnston, R.J., Segerson, K., Schults, E.T., Besedin, E.Y. and Ramachandran, M. 2011. Indices of biotic integrity in stated preference valuation of aquatic ecosystem services. *Ecological Economics* 70: 1946–1956.

Kareiva, P., Tallis, H., Ricketts, T.H., Daily, G.C. and Polasky, S. (eds). 2011. *Natural Capital: Theory and Practice of Mapping Ecosystem Services.* Oxford University Press: Oxford, UK.

Karr, J.R. 1981. Assessment of biotic integrity using fish communities. *Fisheries* 6: 21–27.

Kerans, B.L. and Karr. J.R. 1994. A Benthic Index of Biotic Integrity (B-IBI) for rivers of the Tennessee Valley. *Ecological Applications* 4: 768–785.

Kofinas, G.P. 2009. Adaptive co-management in social-ecological systems. In: Chapin, F.S., Kofinas, G.P. and Folke, C. (eds). *Principles of Ecosystem Stewardship: Resilience Based Natural-resource Management in a Changing World.* Springer: New York, USA, pp. 77–102.

Kolkwitz, R. and M. Marsson. 1909. Ökologie der tierischen Saprobien. Beiträgu zur Lehre von der biologischen Gewäasserbeurteilung. *Internationale Revue der Gesamten Hydrobiologie und Hydrogeograpie* 2: 1–52. [Translated 1967]. Ecology of animal saprobia. In: Keup, L.E., Ingram, W.M. and Machenthum, K.M. (eds). *Biology of Water Pollution.* Federal Water Pollution Control Administration. Washington, DC, USA, pp. 485–495.

Lindenmayer, D.B. and Likens, G.E. 2010. *Effective Ecological Monitoring.* CSIRO Publishing: Melbourne, Australia.

Meybeck, M. 2003. Global analysis of river systems: From Earth system controls to Anthropocene syndromes. *Philosophical Transactions of the Royal Society, B: Biological Sciences* 358: 1935–1955.

Millennium Ecosystem Assessment. 2005. *Ecosystems and Human Well-being: Synthesis*. Island Press: Washington DC, USA.

Natural Resources Commission (NRC). 2012. Framework for assessing and recommending upgraded catchment action plans. Natural Resources Commission Document No. D11/0769. New South Wales Government, Sydney, Australia.

Newson, M.D. and Large, A.R.G. 2006. 'Natural' rivers, 'hydromorphological quality' and river restoration: a challenging new agenda for applied fluvial geomorphology. *Earth Surface Processes and Landforms* 31: 1606–1624.

Norris, R.H. and Thoms, M.C. 1999. What is river health? *Freshwater Biology* 41: 197–209.

Norton, B.G. 2005. *Sustainability: A Philosophy of Adaptive Ecosystem Management*. The University of Chicago Press: Chicago, USA.

Parsons, M., Thoms, M.C. and Norris, R.H. 2004. Development of a standardized approach to river habitat assessment in Australia. *Environmental Monitoring and Assessment* 98: 109–130.

Pollard, S., du Toit, D. and Biggs, H.C. 2011. River management under transformation: the emergence of strategic adaptive management of river systems in the Kruger National Park. *Koedoe* [Online]. doi:10.4102/koedoe.v53i2.1011

Raven, P.J., Holmes, N.T.H., Dawson, F.H. and Everard, M. 1998. Quality assessment using river habitat survey data. *Aquatic Conservation: Marine and Freshwater Ecosystems* 8: 477–499.

Rodrìguez, J.P., Beard, T.D., Agard, J., *et al.* 2005. Interactions among ecosystem services. In: Carpenter, S.R., Pingali, P.L., Bennett, E.M. and Zurek, M.B. (eds). *Ecosystems and Human Well Being: Scenarios, Volume 2. Findings of the Scenarios Working Group of the Millennium Ecosystem Assessment*. Island Press: Washington DC, USA.

Rodrìguez, J.P., Beard, T.D., Bennett, E.M., *et al.* 2006. Trade-offs across space, time, and ecosystem services. *Ecology and Society* 11(1): 28 [online]. URL: http://www.ecologyandsociety.org/vol11/iss1/art28/.

Rogers, K.H. 2006. The real river management challenge: integrating scientists, stakeholders and service agencies. *River Research and Applications* 22: 269–280.

Rogers, K.H. and Biggs, H. 1999. Integrating indicators, endpoints and value systems in strategic management of the rivers of the Kruger National Park. *Freshwater Biology* 41: 439–451.

Rosenberg, D.M. and Resh, V.H. (eds). 1993. *Freshwater Biomonitoring and Benthic Macroinvertebrates*. Chapman and Hall: New York, USA.

Roux, D.J. and Foxcroft, L.C. 2011. The development and application of strategic adaptive management within South African National Parks. *Koedoe* [Online] doi:10.4102/koedoe.v53i2.1049

Scheffer, M. 2009. *Critical Transitions in Nature and Society*. Princeton University Press: Princeton, NJ, USA.

Simon, T.P. (ed.). 1999. *Assessing the Sustainability and Biological Integrity of Water Resources using Fish Communities*. CRC Press: Boca Raton, FL, USA.

Steinberg, D.K. and Condon, R.H. 2009. Zooplankton of the York River. *Journal of Coastal Research* 57: 66–79.

Stevenson, R.J. and Smol, J.P. 2003. Use of algae in environmental assessments. In: Wehr, J.D. and Sheath, R.G. (eds). *Freshwater Algae in North America: Classification and Ecology*. Academic Press: San Diego, California, USA, pp. 775–804.

Thoms, M.C. and Sheldon, F. 2000. Australian lowland river systems: An introduction. *Regulated Rivers: Research and Management* 16: 375–383.

Thorp J.H., Thoms M.C. and Delong M. 2008. *The Riverine Ecosystem Synthesis: Towards Conceptual Cohesiveness in River Science*. Elsevier: London, UK.

Tockner, K. and Stanford, J.A. 2002. Riverine flood plains: present state and future trends. *Environmental Conservation* 29: 308–330.

USEPA (United States Environmental Protection Agency). 2011. National Rivers and Streams Assessement Fact Sheet. http://water.epa.gov/type/watersheds/monitoring/upload/2011may_nrsa_fact_sheet.pdf). Accessed 31 October 2012.

USGPO. 1989. Federal Water Pollution Control Act (33 U.S.C. 1251 et seq.) as amended by a P.L. 92-500. In: Compilation of Selected Water Resources and Water Pollution Control Laws, Printed for use of the Committee on Public Works and Transportation U.S. Government Printing Office: Washington, DC.

Vaughan, I.P., Diamond, M., Gurnell, A.M., *et al.* 2009. Integrating ecology with hydromorphology: a priority for river science and management. *Aquatic Conservation: Marine and Freshwater Ecosystems* 19, 113–125.

Walker, B. and Salt, D. 2006. *Resilience Thinking*. Island Press: Washington DC, USA.

Walker, B. and Salt, D. 2012. *Resilience Practice*. Island Press: Washington DC, USA.

Webb, J.A. and King, E.L. 2009. A Bayesian hierarchical trend analysis finds strong evidence for large-scale temporal declines in stream ecological condition around Melbourne, Australia. *Ecography* 32: 215–225.

Willemen, L., Veldkamp, A., Verburg, P.H., *et al.* 2012. A multi-scale modelling approach for analysing landscape service dynamics. *Journal of Environmental Management* 100: 86–95.

Wright, J.F., Sutcliffe, D.W. and Furse, M.T. (eds). 2000. *Assessing the Biological Quality of Freshwaters: RIVPACS and Other Techniques*. Freshwater Biological Association: Cumbria, UK.

PART 2
Contemporary river science

CHAPTER 11

Faunal response to fine sediment deposition in urban rivers

Paul J. Wood[1], Patrick D. Armitage[2], Matthew J. Hill[1], Kate L. Mathers[1] and Jonathan Millett[1]

[1] *Centre for Hydrological and Ecosystem Sciences, Department of Geography, Loughborough University, Loughborough, UK*

[2] *Freshwater Biological Association, Wareham, Dorset, UK*

Introduction

'River science' is a holistic discipline bringing together stakeholders and others with an interest in the management, conservation and scientific study of lotic waters (Gilvear *et al.*, Chapter 1). Many of the natural and social scientists and engineers actively engaged in river science endeavour to characterise the natural variability of riverine ecosystems and gain an understanding as to how anthropogenic activities, modifications and management practices have affected instream hydrology, geomorphology, ecology and ecosystem service provision. More importantly, much of contemporary applied research seeks to quantify, and in many instances redress, many of the perceived (or real) negative effects of anthropogenic activities (Moggridge *et al.*, 2014; Francis, 2012).

Some of the starkest examples of anthropogenic modifications to riverine systems can be found in both historic and contemporary urban expansion (Hoggart and Francis, 2014). Although urbanised river reaches represent a relatively small proportion of total stream length, their physical location within areas of high population density means that they are some of the most frequently experienced by the wider population. It is anticipated that around 70% of the global population will reside in urban centres by 2050 (Garcia-Armisen *et al.*, 2014) and that in many regions this will put increasing pressures on existing watercourses. Urbanisation has had dramatic effects on rivers primarily as a result of the channel modifications required to facilitate urban development (straightening or adjustment of the channel to enable residential or industrial developments), land drainage and flood management. These changes, however, often lead to the degradation of instream communities when compared to natural or semi-natural systems (Moggridge *et al.*, 2014; Francis 2014). The effects of urbanisation on lotic ecosystems are of global importance, with large numbers of river reaches affected and many more rivers at risk through the increasing demands that growing populations place on riverine resources.

Many urban rivers with a historical legacy of anthropogenic management and

River Science: Research and Management for the 21st Century, First Edition.
Edited by David J. Gilvear, Malcolm T. Greenwood, Martin C. Thoms and Paul J. Wood.

modification are fundamentally different to their former state. In Europe, many of these rivers are classified as Heavily Modified Water Bodies (HMWB), or in some instances as Artificial Water Bodies (AWB) where new channels or configurations have been created (EU, 2000; Liefferink *et al.*, 2011). It is widely recognised that many HMWB in urban areas provide important economic services in terms of access to commercial ports and for flood mitigation and alleviation (Francis, 2012). Within Europe, the designation of some urban rivers as HMWBs reflects the anthropogenic requirements placed upon contemporary watercourses and the social and economic benefits that the services they provide bring. It is also a recognition that it may not be practically or economically possible to modify the existing channel or remove artificial structures (EU, 2000). As a result, it may be unrealistic of river scientists to expect HMWB urban rivers to function in a similar manner to 'natural' rivers even if it were possible.

Many of the changes to urban rivers have resulted in alterations to the delivery of water and sediment to the channel, and in many instances this has resulted in the mobilisation of large volumes of fine sediment (typically referred to as particles < 2 mm in size). These sediments are transported and deposited within river channels, often resulting in the modification of benthic substratum composition and instream habitat characteristics (Collins *et al.*, 2011). The impact of increased fine sediment loading as a result of agricultural practices, urban development and channel management activities for flood defence purposes, have been widely acknowledged but poorly quantified (Burdon *et al.*, 2013; Jones *et al.*, 2015).

Fine sediment deposition and infiltration into the bed of lotic ecosystems (sedimentation, siltation and colmation) has been widely recognised as one of the most important causes of degradation within lotic ecosystems (Wood and Armitage, 1997; Jones *et al.*, 2012). The diffuse nature of fine sediment inputs, however, means that it is not always possible to identify the sources or routes by which sediment is delivered to the channel (Collins *et al.*, 2011; Koiter *et al.*, 2013). Instream sedimentation can have direct and indirect impacts on all trophic levels from algae (Yamada and Nakamura, 2002) and macrophytes (O'Hare *et al.*, 2011), through to benthic invertebrates (Rabeni *et al.*, 2005) and fish (Walters *et al.*, 2003). It is also widely anticipated that fine sediment impacts will be exacerbated by climatic driven changes to rainfall and runoff regimes (Scheurer *et al.*, 2009).

The modification of benthic substratum composition through sedimentation often leads to the homogenisation of benthic habitats (Longing *et al.*, 2010; Descloux *et al.*, 2013). This loss of habitat heterogeneity may be exacerbated in urban locations by the construction of artificial channels with largely impervious beds and banks, and where substrates may be significantly modified and degraded compared to natural streams (Paul and Meyer, 2001; King *et al.*, 2011). In highly modified urban channels there may also be a strong interaction between fine sediment and other contaminants associated with runoff from adjacent land (Larned *et al.*, 2006; Suren and McMurtie, 2005; Von Bertrab *et al.*, 2013).

A range of field and channel experiments have replicated natural and modified riverine systems to determine the influence of increased fine sediment on invertebrate drift (Larsen and Ormerod, 2010), the survivorship of individual taxa to sediment deposition (Molinos and Donohue, 2009) and burial by selected particle sizes (Wood *et al.*, 2005).

In this chapter, we quantify the influence of increasing sediment input – that is sediment loading – on the benthic invertebrate community inhabiting an artificial channel with an impervious concrete bed. This approach provided highly controlled conditions but also reflected channel and habitat characteristics typical of many highly modified and managed urban streams. We hypothesised that increasing quantities of un-cohesive fine sediment (<125 μm and <1 mm) input would result in a reduction in invertebrate abundance, number of taxa and diversity (Shannon–Wiener Diversity index) and an increase in dominance (Berger–Parker Dominance index). Two sets of experiments were undertaken coinciding with summer baseflow conditions (June–July) and late autumn (November) when discharge was naturally higher.

Study site

The study was conducted on Wood Brook, a small stream on the western edge of the town of Loughborough (Leicestershire, UK). This peri-urban river drains agricultural land, rising at an altitude of 200 m and flows into the River Soar, a tributary of the River Trent. The study site was located on a trapezoidal artificial bypass channel of a small reservoir (Nanpantan reservoir) approximately 100 m in length, with a surface area of 4 ha and capacity of 136,000 m^3 (Greenwood *et al.*, 2001). The bed of the channel was concrete-lined, which was partially covered with gravel and sand from a semi-natural reach upstream (Figure 11.1). We selected an artificial channel with a concrete bed since it provided a uniform channel width, slope and homogeneous substratum, typical of many modified urban streams. This allowed the direct influence of fine sediment input on the invertebrate community to be observed in isolation from confounding factors associated with instream refugia, water quality or sediment quality issues. In addition, the narrow channel width facilitated the collection of samples without physical disturbance of the channel bed. Under summer baseflow conditions the channel had a wet width of 0.5 m (mean depth 0.06 m and mean flow velocity 0.43 m s^{-1} over the summer study period), which increased to 0.75 m during typical autumn flow conditions (mean depth 0.11 m and mean flow velocity 0.62 m s^{-1} over the autumn study period).

Method

Three fine sediment applications to the channel bed (treatments) were examined during the experiment: (i) 1 kg m^{-2}; (ii) 3 kg m^{-2} and (iii) 5 kg m^{-2}, plus control patches where no sediment was applied. The fine sediments used in the experiments were graded and comprised pre-sieved un-cohesive sand with a particle size range between <125 μm and <1 mm. The sediments were applied uniformly over eight 3-m long sections (patches) of the channel bed using a 1-mm mesh sieve (with at least 3 m between patches). Sediment treatments were replicated (8 patches – 2 replicates of 4 treatments) and their longitudinal position on the channel randomised so that treatment could be examined independently of location within the channel. Following the application of the fine sediments to the experimental patches, the sand was clearly visible on the surface of the substratum. Preliminary investigations

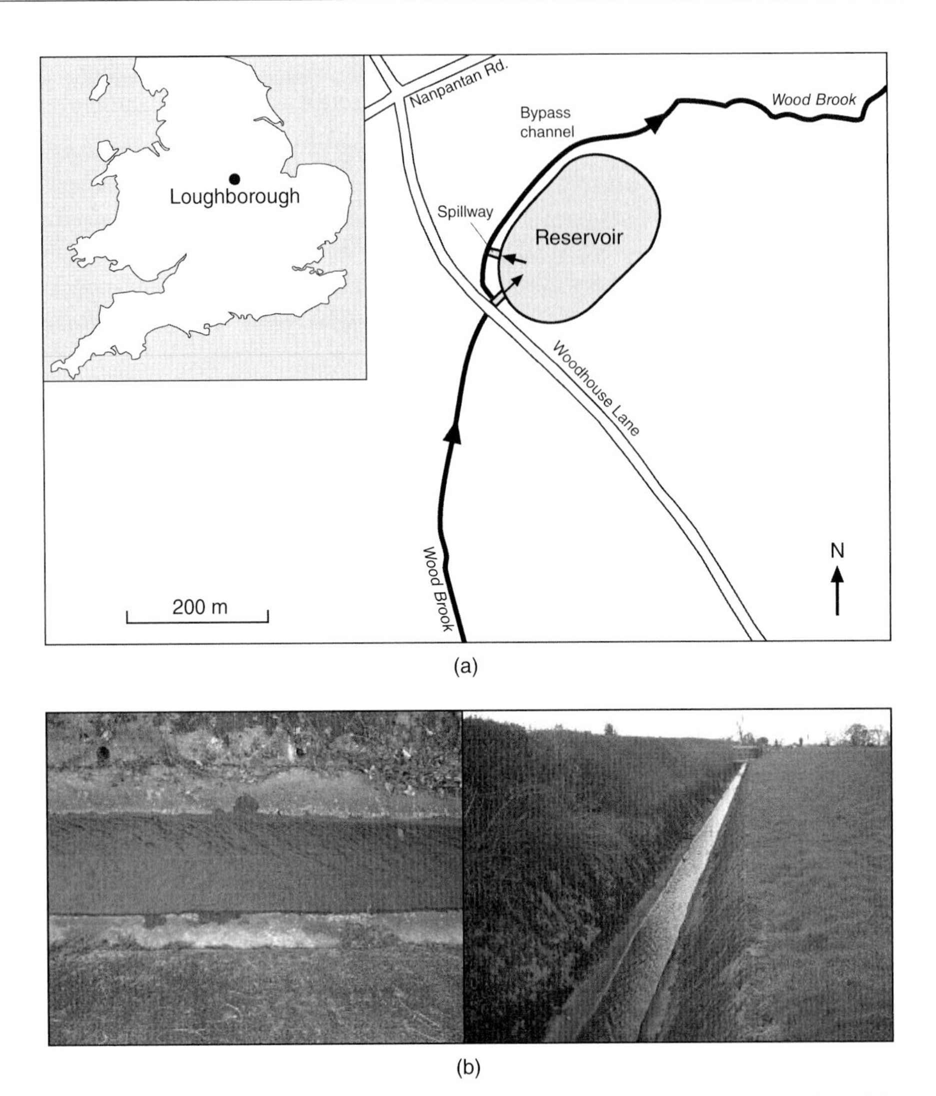

Figure 11.1 Location of the study site and photographic details of the artificial bypass-channel used during the investigation: (a) The Wood Brook study site (Loughborough, UK); and (b) photographs of the concrete-lined channel used during the sedimentation experiments (wetted channel width is approximately 0.75 m).

indicated that sediment input remained visible on the experimental patches until the next significant rainfall event caused an increase in discharge, which mobilised and redistributed the sediments on the concrete bed. The experiments, therefore, commenced following a 7-day period of stable flow (river flow following 7 days without significant precipitation or runoff) and were terminated at the onset of the next significant rainfall event. This minimised the potentially confounding influence of flow variability.

Benthic invertebrates were collected using a modified Surber sampler during each survey (150 mm × 200 mm frame fitted with a 90 μm mesh net). Samples were collected over a 30-second time period using a flat scraper, which facilitated the collection of the benthos and fine sediments simultaneously within the sample frame. For the summer baseflow experimental period, benthic samples were collected from all sediment treatment and control patches (8 patches, 2 invertebrate samples per patch and 5 sampling occasions; $n = 80$) immediately before the application of fine sediment (day 0), 24 hours following the application of sediment (day 1), and subsequently on day 5, day 10 and day 15. The size of the patches used (3 m long) allowed the collection of benthic samples whilst avoiding areas sampled on previous occasions. The experiment ceased when a summer storm occurred leading to an increase in stream discharge and the erosion of sediment on day 17. For the period of the autumn experiment, replicate benthic samples were collected from all sediment treatment and control patches (8 patches, 2 invertebrate samples per patch and 4 sampling occasions; $n = 64$) immediately prior to the application of fine sediment (day 0), 24 hours following the application of sediment (day 1) and subsequently on day 3 and day 5. The experiment ended following a significant rainfall/runoff event 7 days after the application of the fine sediments.

All samples were preserved in the field with 10% formaldehyde solution and returned to the laboratory for processing and identification. Invertebrate abundance was quantified and all individuals were identified to the lowest taxonomic level possible, usually species or genus, except Oligochaeta (order), cased caddisfly larvae from the family Hydroptilidae and some Diptera larvae, which occurred at low abundances (<5 individuals per experiment) and were identified to family level. All fine sediment within the samples were retained and dried in an oven at 60 °C until a constant weight was recorded. The sediments were passed through a 1 mm mesh sieve and retained on a 125 μm sieve to determine the mass of fines present (within the size range applied during the experiments) prior to sediment treatment (pre-treatment control samples and control patches) and during each subsequent survey following its application (sediment treatments and control patches).

Data analysis

The Shannon–Wiener Diversity index and the Berger–Parker Dominance index were derived using the programme Species Diversity and Richness IV (Seaby and Henderson, 2006). Macroinvertebrate community abundance, the abundance of individual taxa and number of taxa were derived from the raw data. Invertebrate community and individual taxa abundance data were $\log_{10}$ transformed and tested for heteroscedacity (or homogeneity of variance; Levene's test) and normality. To determine if the response of the community varied between seasons we combined the datasets for community abundance, number of taxa, Shannon–Wierner Diversity index, and the Berger–Parker Dominance index. To do this we recorded the sampling time relative to the length of study, rather than absolute times. This required removing data for day 10 from the summer dataset. For summer day 0=1, day 2=1, day 5=3 and day 15=4; for autumn day 0=1, day 1=1, day 3=3 and day 5=4, and are subsequently referred to as sampling time 1–4. Data were analysed in IBM SPSS Statistics v.20.0 (IBM Corp.) using a Linear Mixed Model (LMM) to test for main effects and factorial combinations of the effect of time from sediment application, sediment treatment and season. We included sampling time as a repeated measure using a compound symmetry covariance structure;

each patch was treated as an independent replicate. Different taxa were present in the summer and autumn experiments and consequently individual taxa abundances were therefore analysed separately for the summer and autumn experiments using a Repeated Measures General Linear Model in IBM SPSS Statistics v.20.0 (IBM Corp.). In this analysis we tested for differences due to the main effects of sediment treatments, differences at different sampling dates and the interaction between these two main effects. Where the assumption of spherecity was not fulfilled, a Greenhouse–Geisser correction was applied. Differences between treatments within the main effects of sediment treatment and sampling date were tested using a Protected Fisher's LSD test.

Results

A total of 39 taxa were recorded during the summer low flow experimental period with Chironomidae representing the greatest richness (17 taxa) and abundance (>75% of individuals) recorded throughout the study period (Table 11.1). A total of 28 taxa were recorded during the autumn higher flow experimental period (Table 11.1). Chironomidae represented the greatest taxa richness (12 taxa) and abundance of individuals throughout the study period (>60% of individuals).

The amount of sediment retained on the bed during the summer (average of 131 g m^{-2} $\pm$ 1.17) was higher than that recovered during the autumn (average of 84.35 g m^{-2} $\pm$ 1.17) with a marked increase in fine sediment recorded for the 5 kg m^{-2} treatment during the summer (Figure 11.2). When both seasons were considered, there was no effect of sedimentation (treatment) on community abundance (Table 11.2), although abundance was slightly higher in summer (Figure 11.3a and Figure 11.4a). There was a significant effect of sediment treatment on the number of taxa (Table 11.2) but there was no clear or consistent pattern over time or for treatments (Figure 11.3b and Figure 11.4b). The number of taxa recorded was typically higher during the summer (mean = 8.2 $\pm$ 0.18) than the autumn (mean = 5.9 $\pm$ 0.18). Sedimentation resulted in a significant effect on the Shannon–Wiener Diversity index for treatments and time (Figure 11.3c and 11.4c, Table 11.2). On average the Shannon–Wiener Diversity index was lower in summer than autumn. The Berger–Parker Dominance index was higher for increased sediment loads and the pattern differed between summer and autumn periods (Figure 11.3d and Figure 11.4d).

When individual taxa were considered for the summer baseflow sampling period, the responses were variable and taxa specific (Table 11.3), with some displaying no clear response (e.g., *Potomopyrgus antipodarum, Baetis rhodani, Tinodes waeneri, Cricotopus trifasciatus* and *Polypedilum* sp.), some a significant reduction in abundance within patches subject to increased sedimentation, such as the trichopteran *Hydroptila* sp. ($F_{3,12} = 56.10, p < 0.001$ – Figure 11.5a), and others an increase in abundance for higher sediment applications, such as the tanypod chironomid larva *Macropelopia nebulosa* ($F_{3,12} = 34.31, p < 0.001$; Figure 11.5b).

When individual taxa were considered for the autumn sampling period, responses were variable and taxa specific (Table 11.3), with some displaying little or no response (e.g., *Tinodes waeneri, Rhyacophila dorsalis, Rheotanytarsus* sp. and *Tanytarsini* spp.) and others displaying either a moderate increase or reduction in abundance on patches subject to increased sedimentation,

Table 11.1 Taxa recorded and their relative frequency of occurrence during the sedimentation experiments for the summer baseflow and autumn experimental periods. R – Rare (<5 individuals in all samples); P – present (mean abundance <10 individuals per-sample and absent from >50% of samples); C – Common (>10 individuals per-sample and present in the majority of samples); A – Abundant (>25 individuals per-sample and present in the majority of samples).

Taxon	Summer experiment	Autumn experiment
Potamopyrgus antipodarum	A	A
Sphaeriidae	P	
Oligochaeta	A	A
Asellus aquaticus	C	R
Gammarus pulex	C	A
Nemurella picteti	C	R
Isoperla grammatica	R	R
Baetis rhodani	A	A
Ephemerella ignita	C	R
Heptagenia semicolorata	R	R
Hydroptila sp.	A	R
Athripsodes bilineatus	P	
Rhyacophilia dorsalis	C	C
Tinodes waeneri	C	C
Hydropsyche siltalai	P	C
Elmis aenea	C	C
Hydranea sp.	P	
Brillia bifidus	A	A
Cricotopus trifasciatus	A	A
Eukiefferiella brevicalcar (*cf*)	R	
Eukiefferiella claripennis	A	P
Eukiefferiella ilkleyensis	R	
Macropelopia nebulosa	A	C
Metriocnemus sp.	A	A
Micropsectra sp.	A	A
Orthocladius thienemanni (*cf*)	R	
Orthocladius / Cricotopus spp.	A	A
Parametriocnemus stylatus	C	R
Parasmittia sp.	R	
Polypedilum sp.	A	A
Prodiamesa olivacea	C	C
Rheotanytarsus sp.	A	A
Tanytarsini spp.	A	A
Tvetenia calvescens (*cf*)	C	
Ceratopogonidae	P	
Stratyomyidae	R	
Psychodidae	P	
Simuliidae	C	C
Tipulidae	C	C

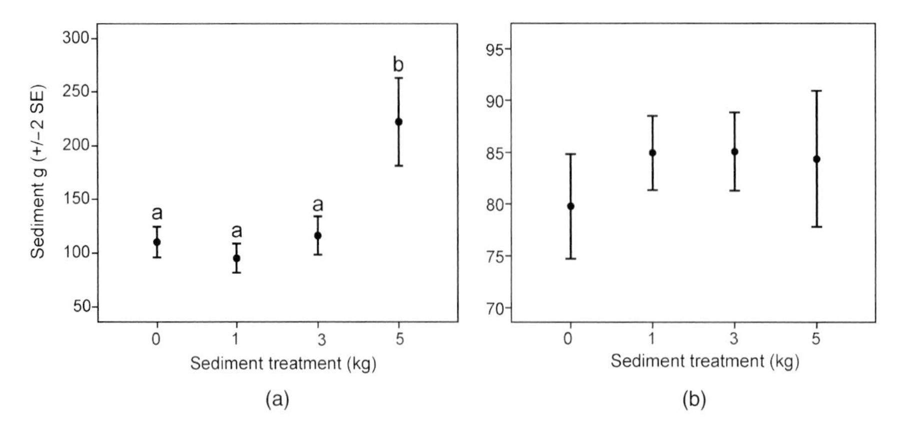

Figure 11.2 Mean (+/– 2SE) mass of sediment (g m^{-2}) for different sedimentation treatments during the summer baseflow period (a) and during the autumn period (b). The results of post-hoc tests (Protected Fisher's LSD test) are indicated on the graphs. Treatments with a different letter are significantly different ($p < 0.05$).

Table 11.2 Result of univariate linear mixed model analysis of the effects of fine sediment treatment, time after application of sediment treatment (sampling time) and their interaction with macroinvertebrate community metrics for the combined summer and autumn datasets.

Effect	d.f.	Community abundance (log)		Shannon–Wiener diversity		Number of taxa		Berger–Parker dominance	
		F	*P*	F	*P*	F	*P*	F	*P*
Sediment treatment	3,24	562.905	<0.001	198.691	<0.001	79.736	<0.001	121.889	<0.001
Time	3,24	8.944	<0.001	4.020	.011	4.605	.005	3.292	.025
Sediment treatment * Time	9,72	10.887	<0.001	2.798	.047	0.536	.659	4.890	.004

such as the ephemeropteran *Baetis rhodani* ($F_{3,12} = 3.90$; $p < 0.05$ – Figure 11.6a) and the amphipod shrimp *Gammarus pulex* ($F_{3,12} = 3.85$; $p < 0.05$ – Figure 11.6b).

Discussion

It is increasingly recognised that relatively few 'natural' rivers remain unaffected or unaltered by anthropogenic activities (Francis, 2014). In many locations, especially urban areas, anthropogenic activities have become the dominant force driving river channel and hydrological change (Pastore *et al.*, 2010). Although the date of the onset of the Anthropocene (period/epoch of anthropogenic forcing) is still debated (Lewis and Maslin, 2015), it is clear that the future management of riverine systems needs to take account of the changes that this brought with it. As a result, there is a need to develop greater understanding of the spatial and temporal dynamics of rivers under contemporary management regimes, especially those that have been heavily modified, regulated and/or urbanised and where no natural analogues exist. There is a clear need to manage urban, heavily modified or artificial watercourses so that they support biodiversity, are resilient to perturbations and fulfil legislative requirements, such as the Good Ecological Potential for the EU Water Framework Directive. This may

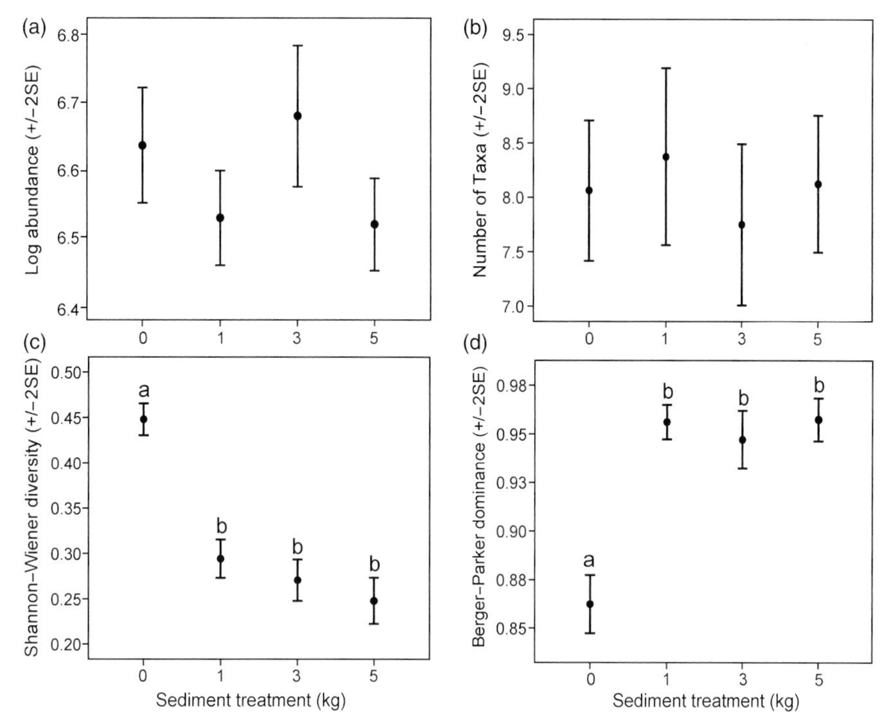

Figure 11.3 Mean (+/− 2SE) Log-community abundance (a), number of taxa (b), Shannon–Wiener Diversity index (c), Berger–Parker dominance index for fine sediment treatments during the summer baseflow period (d). The results of post-hoc tests (Protected Fisher's LSD test) are indicated on the graphs. Points with a different letter are significantly different ($p < 0.05$).

require a new approach to river management and one that recognises and reconciles the value (physically, biologically, socially and economically) of anthropogenically dominated and created systems.

Instream sedimentation has been identified as a major cause of underlying changes to benthic invertebrate communities across the globe (Donohue *et al.*, 2003; Kemp *et al.*, 2011; Wagenhoff *et al.*, 2012) especially in urban settings (Taylor and Owens, 2009; Walters *et al.*, 2003). However, the temporal resolution over which the community changes and the mechanisms through which individual taxa respond remains poorly studied (Wood *et al.*, 2005; Molinos and Donohue, 2009). Our results indicate community response occurs within the short timeframe of the experimental period following sediment treatment compared to control patches. A number of experimental studies have documented rapid responses of benthic fauna and communities to increased suspended sediment concentrations (Crosa *et al.*, 2010) and sediment deposition (Culp *et al.*, 1986; Ramezani *et al.*, 2014), although the majority of studies have examined changes over longer (week–month) time periods (e.g., Larsen *et al.*, 2011), or compared sites along a fine sediment gradient (Kreutzweiser *et al.*, 2005; Wagenhoff *et al.*, 2011; Descloux *et al.*, 2014). The rapid rate of change may reflect the highly modified and artificial concrete-lined channel studied that is common in many urban rivers.

We hypothesised that the invertebrate community response would reflect the

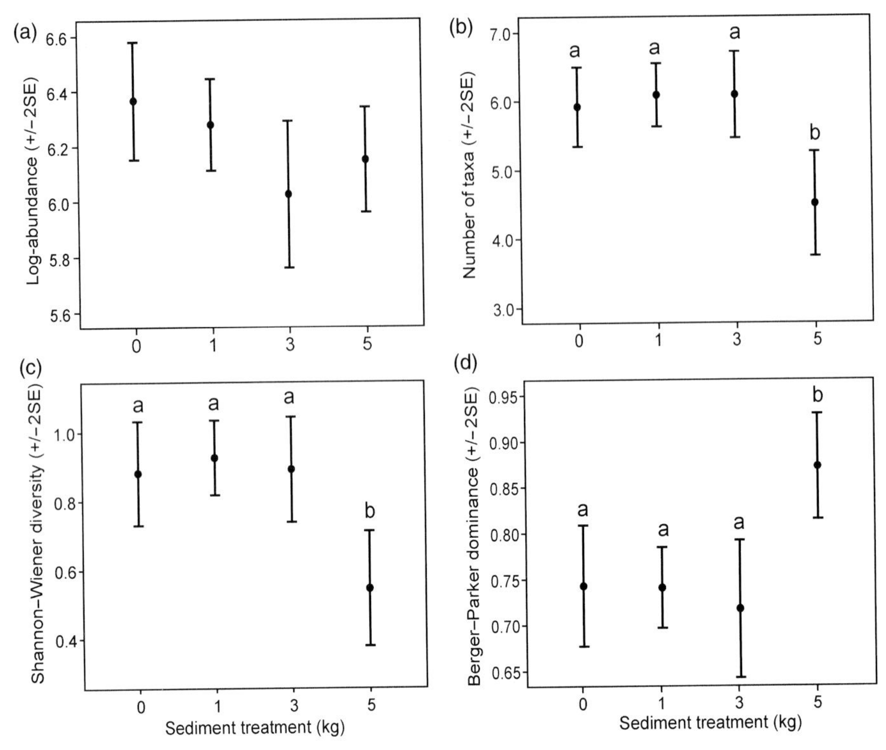

Figure 11.4 Mean (+/- 2SE) Log-community abundance (a), number of taxa (b), Shannon–Wiener Diversity index (c), Berger–Parker dominance index for fine sediment treatments during the autumn period (d). The results of post-hoc tests (Protected Fisher's LSD test) are indicated on the graphs. Points with a different letter are significantly different ($p < 0.05$).

increasing load of fine sediment on experimental patches. During periods of stable summer baseflow, sediment treatment resulted in changes to the Shannon–Wiener Diversity and Berger–Parker Dominance indices compared to control patches. However, during the autumn, only the greatest sediment treatment (5 kg m^{2-1}) resulted in any effect. The results demonstrate that there is not a simple linear effect of increasing sediment load that leads to a reduction of invertebrate abundance, number of taxa and diversity (Shannon–Wiener Diversity index) and an increase in dominance (Berger–Parker Dominance index) and our hypothesis can therefore be rejected. However, the results of the experiment do indicate that the invertebrate community and individual taxa respond in a variety of ways – reflecting a gradient, stepped or seasonally varied response.

The results of the experiments also indicate that similar levels of fine sediment input at different times of year resulted in different responses. The differences recorded between the two sampling periods partly reflects increased efficiency of sediment transport and erosion processes associated with the increased discharge during the autumn (e.g., hydraulic forces such as shear stress) (Culp *et al.*, 1986; Johnson *et al.*, 2009). During the summer baseflow experimental period, discharge was lower and limited sediment erosion following

Table 11.3 Taxa recorded during the summer baseflow and autumn sedimentation experimental periods displaying a significant response to sediment input, no significant response and the nature of any trend in the pattern recorded. Only those taxa that were Common (>10 individuals per-sample and present in the majority of samples) or abundant (>25 individuals per-sample and present in the majority of samples) were included in the analysis; ns – no significant response; ** significant response ($p < 0.05$ – non-parametric ANOVA/Kruskal Wallis); ↑ – trend of increasing mean abundance on patches subject to sedimentation; ↓ – trend of decreasing mean abundance on patches subject to sedimentation; = no trend in mean abundance or abundance approximately equal; and – abundance of taxa too low to determine any pattern or trend.

Taxon	Summer experiment (significance/trend)	Autumn experiment (significance/trend)
Potamopyrgus antipodarum	ns =	ns =
Oligochaeta	** ↑	** ↑
Asellus aquaticus	ns =	
Gammarus pulex	ns ↓	** ↓
Nemurella picteti	ns ↓	
Baetis rhodani	ns =	** ↓
Ephemerella ignita	ns	
Hydroptila sp.	** ↓	
Rhyacophilia dorsalis	ns =	ns =
Tinodes waeneri	ns =	ns =
Hydropsyche siltalai		ns ↓
Elmis aenea	ns ↓	ns =
Brillia bifidus	ns ↑	ns ↑
Cricotopus trifasciatus	ns =	ns ↓
Eukiefferiella claripennis	ns =	
Macropelopia nebulosa	** ↑	ns =
Metriocnemus sp.	ns ↓	ns =
Micropsectra sp.	ns =	ns ↑
Orthocladius / Cricotopus spp.	ns =	ns =
Parametriocnemus stylatus	ns =	
Polypedilum sp.	ns =	ns ↑
Prodiamesa olivacea	** ↓	** ↓
Rheotanytarsus sp.	ns =	ns =
Tanytarsini spp.	ns ↑	ns =
Tvetenia calvescens (*cf*)	ns =	
Simuliidae	** ↓	ns =
Tipulidae	ns =	ns =

its input. As a result, fine sediments were widely distributed throughout the system and reflect conditions of high sediment availability recorded within many low gradient/lowland and urban river systems during periods of baseflow (Wood and Armitage, 1999; Miserendino *et al.*, 2008).

The response of the invertebrate community to fine sediment addition during the autumn period was different to that recorded during summer baseflow conditions. In particular, the effect on the invertebrate community did not persist, even though the increased mass of fine

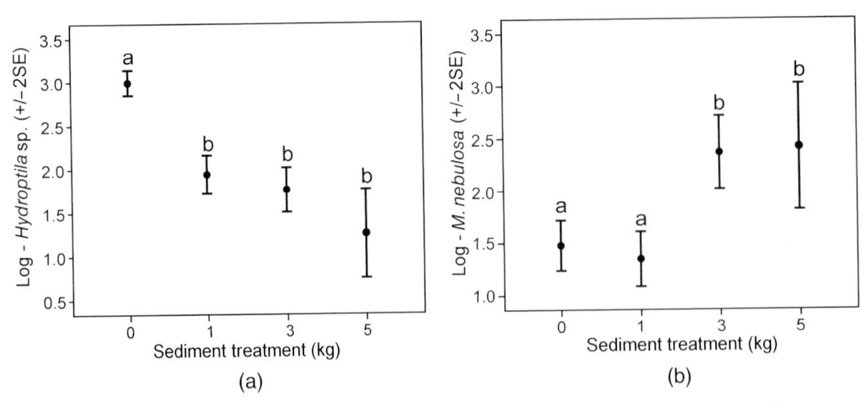

Figure 11.5 Mean (+/− 2SE) Log-*Hydroptila* sp. – Trichoptera: Hydroptilidae abundance (a) and Log-*Macropelopia nebulosa* – Diptera: Chironomidae: Tanypodinae abundance (b) for different fine sediment treatments during the summer baseflow period. The results of post-hoc tests (Protected Fisher's LSD test) are indicated on the graphs. Points with a different letter are significantly different ($p < 0.05$).

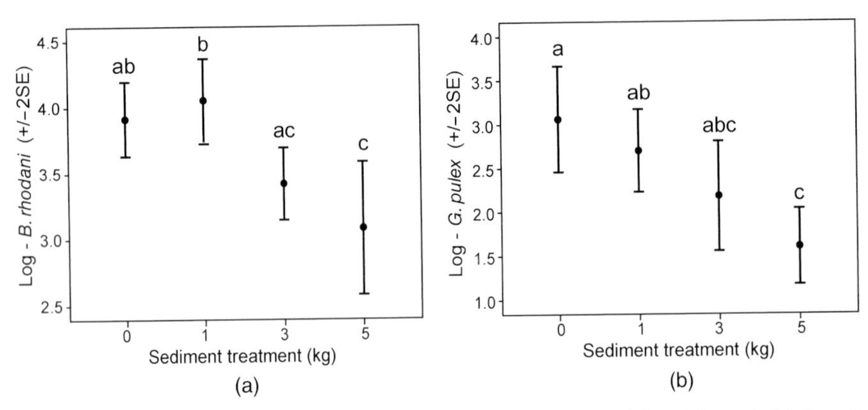

Figure 11.6 Mean (+/− 2SE), Log-*Baetis rhodani* – Ephemeroptera: Baetidae (a), and (b) Log-*Gammarus pulex* – Crustacea: Amphipoda: Gammaridae abundance for fine sediment treatments during the autumn experimental period. The results of post-hoc tests (Protected Fisher's LSD test) are indicated on the graphs. Points with a different letter are significantly different ($p < 0.05$).

sediment on the bed could still be visibly observed up to 5 days after the sediment was applied. The increased discharge and more efficient sediment transport processes during the autumn period resulted in increased erosion of fine sediments.

The experiments undertaken provide clear evidence that the timing of sediment input to a river channel can potentially have a strong influence on the response of the benthic community and individual taxa (Meyer *et al.*, 2008) and that the effects may be accentuated in anthropogenically modified channels (Dunbar *et al.*, 2010). Channel management activities coinciding with summer baseflow conditions in urbanised channels, such as dredging and cutting of marginal vegetation and instream macrophytes, may mobilise large volumes of fine sediment that may be re-deposited on the channel bed a short distance downstream (Doeg and Koehn, 1994; Sarriquet *et al.*, 2007). The effect of these activities could therefore persist until sediment remobilisation takes place associated with increased discharge or floods later in the year. In rivers subject to significant anthropogenic management activities and where instream habitats have been significantly modified due to urbanisation, there is evidence that significant volumes of fine sediment are stored within the main channel (Dunbar *et al.*, 2010). Event-based channel management activities, timed to occur prior to subsequent increases in discharge (natural floods or freshets from reservoir releases), could potentially mitigate some of the most deleterious impacts associated with fine sediments. This may help facilitate the downstream transport and flushing of fines, and may prevent the sediments simply being re-deposited a short distance downstream (Meyer *et al.*, 2008; Sarriquet *et al.*, 2007).

Conclusion

Invertebrate communities of anthropogenically-modified channels are often impoverished compared to those of natural systems (Beavan *et al.*, 2001), and thus taxa inhabiting urbanised channels may be particularly vulnerable to fine sediment loading. A greater knowledge regarding aquatic invertebrate community composition and taxon-specific responses with regards to urban land use is required to enable effective monitoring and management of lotic systems (Collins *et al.*, 2011; Kemp *et al.*, 2011). This may require acceptance that urban and highly modified rivers, and the novel communities they support, are fundamentally different and that these should be accepted and even celebrated as part of the wider services the systems provide (Moyle, 2014). The results of the current study of a concretised channel demonstrated a rapid and variable response of the instream invertebrate community and taxa to increased sedimentation. The results also indicate that there may not be simple linear ecological responses to increasing fine sediment loads within channels and that the invertebrate community and taxon-specific responses probably vary as a function of river discharge. Significant advances have been made recently in the development of biomonitoring tools which quantify fine sediment impacts on instream communities and which facilitate identification of vulnerable locations within river channels (e.g., Relyea *et al.*, 2012; Turley *et al.*, 2014, 2015). In addition to the application of these tools, future experimental investigations are required to explore natural and anthropogenically-modified systems under a range of flow conditions. This will help with the quantification of temporal and spatial effects of sedimentation on instream communities and enable the development of management strategies, which mitigate fine sediment effects and where appropriate provide the tools to monitor the effectiveness of river restoration measures.

Acknowledgements

The authors gratefully acknowledge the support of Severn Trent Water for granting permission to undertake research at the site and their cooperation at all stages of the work. We thank Tom Haselhurst and Richard Hutton for field assistance and Mark Szegner for help with the preparation of the figures. We thank Prof. Stephen Rice and Dr Chris Extence for valuable discussions relating to the results of this experiment.

References

Beavan, L., J. Sadler & C. Pinder. 2001. The invertebrate fauna of a physically modified urban river. *Hydrobiologia* 445: 97–108.

Burdon F. J., A. R. McIntosh & J. S. Harding. 2013. Habitat loss drives threshold response of benthic invertebrate communities to deposited sediment in agricultural streams. *Ecological Applications* 23: 1036–1047.

Collins, A. L., P. S. Naden, D. A. Sear, *et al.* 2011. Sediment targets for informing river catchment management: international experience and prospects. *Hydrological Processes* 25: 2112–2129

Crosa, G., E. Castelli, G. Gentili & P. Espa. 2010. Effects of suspended sediments from reservoir flushing on fish and macroinvertebrates in an alpine stream. *Aquatic Sciences* 72: 85–95.

Culp, J. M., F. J. Wrona & R. W. Davies. 1986. Response of stream benthos and drift to fine sediment deposition versus transport. *Canadian Journal of Zoology* 64: 1345–1351.

Descloux, S., T. Datry & P. Marmonier. 2013. Benthic and hyporheic invertebrate assemblages along a gradient of increasing streambed colmation by fine sediment. *Aquatic Sciences*: DOI: 10.1007/s00027-013-0295-6.

Descloux, S., T. Datry & P. Usseglio-Olatera. 2014. Trait-based structure of invertebrates along a gradient of sediment colmation: Benthos versus hyporheos responses. *Science of the Total Environment* 466: 255–276.

Doeg T. J. & J. D. Koehn. 1994. Effects of draining and desilting a small weir on downstream fish and macroinvertebrates. *Regulated Rivers: Research and Management* 9: 263–278.

Donohue I., E. Verheyen & K. Irvine. 2003. In situ experiments on the effects of increased sediment loads in littoral rocky shore communities in Lake Tanganyika, East Africa. *Freshwater Biology* 48: 1603–1616.

Dunbar M. J., M. L. Pedersen, D. Cadman, *et al.* 2010. River discharge and local-scale physical habitat influence macroinvertebrate LIFE scores. *Freshwater Biology* 55: 226–242.

EU. 2000. Directive 2000/60/EC of the European Parliament and of the Council of 23 October 2000 establishing a framework for Community action in the field of water policy, 22/12/2000. *Official Journal L* 327: 1–73.

Francis R. A. 2012. Positioning urban rivers within urban ecology. *Urban Ecosystems* 15: 285–291.

Francis R. A. 2014. Urban rivers: novel ecosystems, new challenges. *Wiley Interdisciplinary Reviews: Water* 1: 19–29.

García-Armisen T., Ö. İnceeoğlu, N. K. Ouattara, *et al.* 2014. Seasonal variations and resilience of bacterial communities in a sewage polluted urban river. *PLOS ONE* 9: e92579.

Greenwood, M. T., M. A. Bickerton & G. E. Petts. 2001. Assessing adult Trichoptera communities of small streams: a case study from Charnwood Forest, Leicestershire, UK. *Aquatic Conservation: Marine and Freshwater Ecosystems* 11: 93–107.

Hoggart S. P. G. & R. A. Francis. 2014. Use of coir rolls for habitat enhancement of urban river walls. *Fundamental and Applied Limnology* 185: 19–30.

IBM Corp. Released 2011. *IBM SPSS Statistics for Windows, Version 20.0.* Armonk, NY: IBM Corp.

Johnson, M. F., I. Reid, S. P. Rice & P. J. Wood. 2009. The influence of net-spinning caddisfly larvae on the incipient motion of fine gravels: an experimental field and laboratory flume investigation. *Earth Surface Processes and Landforms* 34: 413–423.

Jones, J. I., J. F. Murphy, A.L. Collins, *et al.* 2012. The impact of fine sediment on macro-invertebrates. *River Research and Applications*: DOI: 10.1002/rra.1516

Jones, I., I. Growns, A. Arnold *et al.* 2015. The effects of increased flow and fine sediment on hyporheic invertebrates and nutrients in stream mesocosms. *Freshwater Biology*: DOI: 10.1111/fwb.12536.

Kemp, P., D. Sear, A. Collins, et al. 2011. The impacts of fine sediment on riverine fish. *Hydrological Processes* 25: 1800–1821.

King, R. S., M. E. Baker, P. F. Kazyak & D. E. Weller. 2011. How novel is too novel? Stream community thresholds at exceptionally low levels of catchment urbanization. *Ecological Applications* 21: 1695–1678.

Kreutzweiser D. P., S. S. Capell & K. P. Good. 2005. Effects of fine sediment inputs from a logging road on stream insect communities: a large-scale experimental approach in a Canadian headwater stream. *Aquatic Ecology* 39: 55–66.

Koiter, A. J., P. N. Owens, E. L. Petticrew & D. A. Lobb. 2013. The behavioural characteristics of sediment properties and their implications for sediment fingerprinting as an approach for identifying sediment sources in river basins, *Earth Science Reviews* 125: 24–42.

Larned, S. T., A. M. Suren, M. Flanagan, *et al.* 2006. Macrophytes in urban stream rehabilitation: Establishment, ecological effects, and public perception. *Restoration Ecology* 14: 429–440.

Larsen S. & S. J. Ormerod. 2010. Low-level effects of inert sediments on temperate stream invertebrates. *Freshwater Biology* 55: 476–486.

Larsen S., G. Pace and S. J. Ormerod. 2011. Experimental effects of sediment deposition on the structure and function of macroinvertebrate assemblages in temperate streams. *River Research and Applications* 27: 257–267.

Lewis S. L. & M. A. Maslin. 2015. Defining the Anthropocene. *Nature* 519: 171–180.

Liefferink, D., M. Wiering & Y. Uitenboogaart. 2011. The EU Water Framework Directive: A multi-dimensional analysis of implementation and domestic impact. *Land Use Policy* 28: 712–722.

Longing, S. D., J. R. Voahell, C. A. Dolloff & C. N. Roghair. 2010. Relationships of sediment and benthic macroinvertebrate assemblages in headwater streams using systematic longitudinal sampling at the reach scale. *Environmental Monitoring and Assessment* 161: 517–530.

Meyer E. I., O. Niepagenkemper, F. Molls & B. Spanhoff. 2008. An experimental assessment of the effectiveness of gravel cleaning operations in improving hyporheic water quality in potential salmonid spawning areas. *River Research and Applications* 24: 119–131.

Miserendino, M. L., C. Brand & C. Y. Di Prinzio. 2008. Assessing urban impacts on water quality, benthic communities and fish in stream on the Andes Mountains, Patagonia (Argentina). *Water Air and Soil Pollution* 194: 91–110.

Moggridge H. L., M. J. Hill & P. J. Wood. 2014. Urban aquatic ecosystems: the good the bad and the ugly. *Fundamental and Applied Limnology* 185: 1–6.

Molinos, J. G. & I. Donohue. 2009. Differential contribution of concentration and exposure time to sediment dose effects on stream biota. *Journal of the North American Benthological Society* 28: 110–121.

Moyle, P. B. 2014. Novel aquatic ecosystems: the new reality for streams in California and other Mediterranean climate regions. *River Research and Applications* 30: 1335–1344

O'Hare, J. M., A. M. Gurnell, M. J. Dunbar, *et al.* 2011. Physical constrains on the distribution of macrophytes linked with flow and sediment dynamics in British rivers. *River Research and Applications* 27: 671–683.

Pastore, S. K., M. B. Green, A. Munoz-Hernandez, *et al.* 2010. Tapping environmental history to recreate America's colonial hydrology. *Environmental Science and Technology* 44: 8798–8803.

Paul, M. J. and J. L. Meyer. 2001. Streams in the urban landscape. *Annual Review of Ecology and Systematics* 32: 333–365.

Rabeni, C. F., K. E. Doisy & L. D. Zweig. 2005. Stream invertebrate community functional responses to deposited sediment. *Aquatic Sciences* 67: 395–402.

Ramezani, J., L. Rennebeck, G. P. Closs & C. D. Matthaei. 2014. Effects of fine sediment addition and removal on stream invertebrates and fish: a reach-scale experiment. *Freshwater Biology* 59: 2584–2604.

Relyea, C. D., W. Minshall & R. J. Danehy, 2012. Development and validation of an aquatic fine sediment biotic index. *Environmental Management* 49: 242–252.

Sarriquet, P. E., P. Bordenave & P. Marmonier, 2007. Effects of bottom sedimentation restoration on interstitial habitat characteristics and benthic macroinvertebrate assemblages in a headwater stream. *River Research and Applications* 23: 815–828.

Scheurer, K, C. Alewell, D. Banniger & P. Burkhardt-Holm. 2009. Climate and land-use changes affecting river sediment and brown trout in alpine countries – a review. *Environmental Science and Pollution Research* 16: 232–242.

Seaby, R. M. & P. A. Henderson, 2006. *Species Diversity and Richness Version 4.* Pisces Conservation Ltd., Lymington, England.

Suren, A. M. & S. McMurtrie. 2005. Assessing the effectiveness of enhancement activities in urban streams: II. Responses of invertebrate communities. *River Research and Applications* 21: 439–453.

Taylor K. G. & P. N. Owens. 2009. Sediments in urban river basins: a review of sediment-contaminant dynamics in an environmental system conditioned by human activities. *Journal of Soil and Sediments* 9: 281–303

Turley, M. D., G. S. Bilotta, C. A. Extence & R. E. Brazier. 2014. Evaluation of a fine sediment biomonitoring tool across a wide range of temperate rivers and streams, *Freshwater Biology* 59: 2268–2277

Turley, M.D., G. S. Bilotta, T. Krueger, *et al.* 2015. Developing an improved biomonitoring tool for fine sediment: Combining expert knowledge and empirical data. *Ecological Indicators* 54: 82–86.

Von Bertrab, M.G., A. Krein, S. Stendera, *et al.* 2013. Is fine sediment deposition a main driver for the composition of benthic macro inverterbrate assemblages? *Ecological Indicators* 24: 589–598.

Wagenhoff A., C. R. Townsend & C. D. Matthaei, 2012. Macroinvertebrate responses along broad stressor gradients of deposited fine sediment and dissolved nutrients: a stream mesocosm experiment. *Journal of Applied Ecology* 49: 892–902.

Wagenhoff A., C. R. Townsend, N. Phillips & C. D. Matthaei. 2011. Subsidy-stress and multiple-stressor effects along gradients of deposited fine sediment and dissolved nutrients in a regional set of streams and rivers. *Freshwater Biology* 56: 1916–1936.

Walters D. M., D. S. Leigh & A. B. Bearden, 2003. Urbanization, sedimentation, and the homogenization of fish assemblages in the Etowah River Basin, USA. *Hydrobiologia* 494: 5–10.

Wood P. J. & P. D. Armitage, 1997. Biological effects of fine sediment in the lotic environment. *Environmental Management* 21: 203–217.

Wood, P. J. & P. D. Armitage, 1999. Sediment deposition in a small lowland stream - management implications. *Regulated Rivers: Research and Management* 15: 199–210.

Wood, P. J., J. Toone, M. T. Greenwood & P. D. Armitage, 2005. The response of four lotic macroinvertebrate taxa to burial by sediments. *Archiv für Hydrobiologie* 163: 145–162.

Yamada. H. & F. Nakamura, 2002. Effects of fine sediment deposition and channel works on periphyton biomass in the Makomanai River, northern Japan. *River Research and Applications* 18: 481–493.

CHAPTER 12

Characterising riverine landscapes; history, application and future challenges

Victoria S. Milner[1], David J. Gilvear[2] and Martin C. Thoms[3]
[1] *Institute of Science and the Environment, University of Worcester, Worcester, UK*
[2] *School of Geography, Earth and Environmental Sciences, University of Plymouth, Plymouth, UK*
[3] *Riverine Landscapes Research Laboratory, Geography and Planning, University of New England, Australia*

Introduction

Advancing our understanding of the structure and functioning of riverine landscapes is a cornerstone of river science (Thorp *et al.*, 2008). This requires empirical data and models of how environmental and ecological patterns relate to biological, chemical and physical processes, and their interactions across spatial and temporal scales that occur at different organisational complexities. Patterns are an indication of the natural spatial and temporal heterogeneity within ecosystems (Levin, 1992). Understanding how riverine landscapes function as ecosystems relies on our ability to capture this heterogeneity across meaningful and interpretable scales (Underwood *et al.*, 2000; Thorp *et al.*, 2008). Imposing order on natural systems, including riverine landscapes, is inherently complex due to their dynamism and high spatiotemporal variability across longitudinal, lateral, vertical and temporal dimensions (Ward, 1989). Classification has a long history in science and has been widely used in different aspects of river science, such as conservation (i.e., Margules and Pressey, 2000; Nel *et al.*, 2009), river management (Rosgen, 1994; Brierley and Fryirs, 2005), and in identifying natural and anthropogenic patterns of biological and physical concerns. Characterisation, by comparison, is a process of describing the distinctive features of a landscape or a river system, whereas classification is a process of ordering objects or environmental variables into groups based on shared characteristics. Classification involves three discrete components: taxonomy, typology and allocation. Taxonomy is an objective procedure consisting of ordering objects into classes based on their measured characteristics, whereas a typology is a subjective, judgemental process of identifying different classes (Newson *et al.*, 1998). Taxonomists have referred to these two processes as *natural* and *special* classifications (Sneath and Snokal, 1973). The classification of animals, as undertaken by Linneus, into species is regarded as a natural classification. However, in river science, landscape characterisations or classifications founded on typologies are more common, such as the geographic cycle of Davis (1899), the River Continuum Concept (RCC) by

River Science: Research and Management for the 21st Century, First Edition.
Edited by David J. Gilvear, Malcolm T. Greenwood, Martin C. Thoms and Paul J. Wood.

Vannote *et al.* (1980) and the Montgomery and Buffington (1997) typology developed for mountain drainage basins in the Pacific Northwest, USA.

Regardless of the terms used, characterisation and classification processes aim to organise, simplify and understand the natural forms and processes within environmental systems (Juracek and Fitzpatrick, 2003). If groups of river systems or river reaches with similar character reflect fluvial forms and processes that resulted in their groupings, we can start to improve and understand the fundamental laws controlling the behaviour of the objects classified (Portt *et al.*, 1989). In other words, river scientists can move towards predicting a river system's morphology and behaviour at a range of scales, from knowledge of its flow and sediment regime and boundary conditions, to support channel management, conservation and restoration projects (Rosgen, 1994). Characterisation and classification of river systems also improve the process of communication by providing a consistent framework for monitoring of instream and riparian conditions, and reporting the status to regulatory authorities (Melles *et al.*, 2012).

Traditionally, river characterisations/classifications have adopted a bottom-up, constructivist approach whereby scientists have training and expertise in 'reading' the landscape (e.g., Davis, 1899; Schumm, 1963, 1977). Common to this approach is identifying landforms, understanding their morphodynamics, interpreting the evolution and history of the riverine landscape, and considering the interaction of these features at different scales (Brierley *et al.*, 2013). Successful bottom-up approaches to the characterisation/classification of riverine landscapes requires suitable training and a conceptual and theoretical background in the chosen subject, whether it be geomorphology or stream ecology. Since the beginning of the twenty-first century, a notable transition has been an increase in top-down approaches to river characterisation/classification. This change has occurred because of an increasing demand for information on channel types and classes by river managers and local practitioners, and the availability of high-quality remote sensing data/technologies and the use of multivariate statistics. Thus, there has been a rapid development of desktop river characterisation/classification approaches. The ease and accessibility of top-down approaches can potentially lead to a reduction in the theoretical and conceptual understanding of the character of riverine landscapes and the features being classified. Users of characterisation/classification schemes (using either a bottom-up or top-down approach) need to be aware that while characterisation/classification is useful if applied to a suitable problem, it is on its own a limited tool (Kondolf *et al.*, 2003).

In this chapter, we review the history, application and future challenges of river classification. We advocate river characterisation/classification should not simply improve our understanding of patterns and processes, but also extend our knowledge of river science both conceptually and theoretically and be applicable within an interdisciplinary domain. Specifically, we identify a chronology of geomorphic-based river system characterisation into four distinct periods: the pioneer, the consolidation, application and the river science phase. The chronology of geomorphic-based river system characterisation highlights a trend from bottom-up, constructivist approaches to top-down, reductionist approaches within river characterisation/classification. Examples of river characterisation/classification approaches used for science and management applications are described that typify the latter phase. Finally, we identify the future challenges facing river characterisations/classifications, and conclude by emphasising the importance of spatiotemporal scales, the value of using remote sensing technologies, and discuss future priorities.

A chronology of geomorphic based river system characterisation

A large number of river classifications, typologies and characterisation systems have been developed in fluvial geomorphology since the late nineteenth century (Milner *et al.*, 2013). Approaches to characterisation/classification include observational, empirical, conceptual and statistical. The various approaches reflect the many disciplines involved, the large number and range of variables used, the different purposes of classification, and the challenge of simplifying highly variable and diverse datasets from natural systems (Kondolf, 1995; Juracek and Fitzpatrick, 2003; Kondolf *et al.*, 2003). This section provides an outline to the progression from river characterisation to classification from the late nineteenth century to the twenty-first century through identification of key influential developments within fluvial geomorphology. We recognise four distinct phases of characterisation/classification: the pioneer, consolidation, application, and river science phase, but acknowledge the fuzzy boundaries and overlap inherent between phases.

The pioneer phase (ca. 1900s–1970s)

The pioneer phase was primarily characterised by field sketching and an observational approach to river systems. An early physical landscape characterisation was the geographic cycle of Davis (1899), which divided river systems in an evolutionary cycle of youthful, mature and old – the spatial equivalent of upland, lowland and coastal (Newson *et al.*, 1998). Inherent within this phase, geomorphologists and biologists recognised the importance of river systems as a physical and biological continuum from river source to mouth. Biological classifications were developed based on different fish species, such as salmon, trout and the grayling/barbel zone (Huet, 1954; Pennack, 1971, Hawkes, 1975). Within fluvial geomorphology, Horton (1945) developed the use of stream ordering (as modified by Strahler, 1952, 1957), emphasising the importance of the whole river network, and thus incorporating a crude measure of scale. Leopold and Wolman (1957) distinguished straight, meandering and braided channel patterns based on relationships between slope and discharge. This pioneering work was important in linking fluvial form and processes, and provided a platform for subsequent work. For example, the pattern-based approach of Leopold and Wolman (1957) was later expanded to include anastomosing channels (Smith and Smith, 1980; Knighton and Nanson, 1993; Makaske, 2001), and anabranching channels (Nanson and Knighton, 1996). Another pivotal process-based classification within this phase is that of Schumm's (1963, 1977), which uses channel stability (stable, eroding or depositing beds) and the dominance of sediment transport (mixed load, suspended load or bedload) to divide rivers into source, transfer and depositional zones. Within this pioneering phase, characterisation was typified by observation, field sketching and the recognition of channel features by trained geomorphologists, who aimed to increase understanding of river systems.

The consolidation phase (1970s–ca. 1990s)

During the consolidation phase, there was a sharp increase in form- and process-based approaches to river characterisation and classification, particularly in North America and Europe. A transition occurred from studying rivers at a large to a smaller scale, through the inclusion of morphological complexity at the reach scale. In Canada, Kellerhals *et al.* (1972, 1976), Galay *et al.* (1973) and Mollard (1973) proposed descriptive classification systems for a wide range of stream morphologies, using a combination of channel patterns, channel islands, bar forms and degree of lateral activity to define a variety of channel types. These Canadian classification systems highlight a shift from characterising rivers by large-scale planform (as in the previous phase) to using channel bedforms, and in part, a recognition of the continuum of channel types as compared to a few distinct types.

The importance of physical habitat to aquatic biota was also increasingly recognised. Vannote *et al.* (1980) developed the River Continuum Concept (RCC), showing longitudinal changes in the physical structure of a river with associated changes in invertebrate, macrophyte and fish communities. Within this period, including physical habitat within river characterisation/classification schemes continued at multiple hierarchical scales. An early hierarchical classification was proposed by Warren (1979) comprising 11 levels ranging from a regional scale ($>10\,km^2$) to a microhabitat scale ($<1\,m^2$) based on climate, substrate, water chemistry, biota and culture. Frissell *et al.* (1986) extended Warren's classification by including spatially nested levels of resolution, such as the river network, valley segment, reach, morphological unit and microhabitat. This schema has been subsequently updated to include river zones (cf. Figure 2.2 in Chapter 2). Frissell's *et al.* (1986) stream classification was specifically developed for use on second- and third-order channels in forested mountain environments (Van Niekerk *et al.*, 1995), but represented an important advancement by incorporating both source and processes of development, form and pattern within each hierarchical level (Naiman, 1998).

The application phase (1990s–ca. 2005)

A main feature of this phase was the application of hierarchical river classifications and typologies for management purposes. In the USA and New Zealand, the Rosgen Classification System (Rosgen; 1985, 1994, 1996) has been widely advocated as a tool for river restoration. The classification identifies 94 stream types based on combinations of channel slope, entrenchment, sinuosity, width–depth ratios and substrate, which are grouped into seven major categories, from steep cascading channels, gully systems to pool–riffle morphologies (Rosgen, 1994). Despite wide-scale usage within the USA, the RCS has received heavy criticism as a predictor of fluvial process and channel form (Miller and Ritter, 1996; Doyle and Harbor, 2003). The first hierarchical, process-based typology to gain widespread acceptance was produced by Montgomery and Buffington (1997) in the Pacific Northwest, USA. The typology addresses morphological attributes to the relative ratio of sediment supply to transport capacity, identifying three dominant channel substrates: bedrock, alluvium

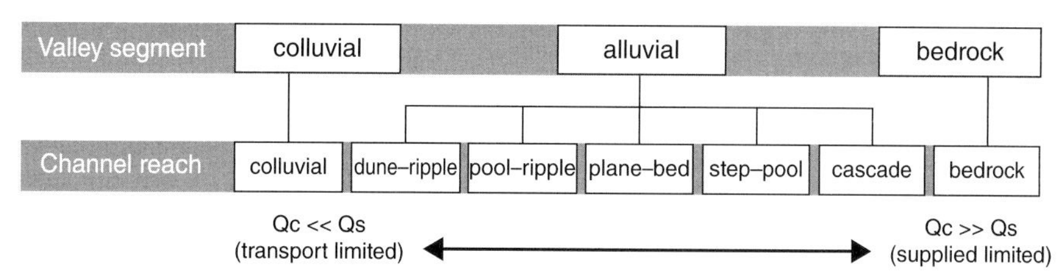

Figure 12.1 Channel types of Montgomery and Buffington shown as a function of transport capacity to relative sediment supply. Montgomery and Buffington, 1997. Reproduced with permission from The Geological Society of America.

and colluvium, and seven channel types (Figure 12.1). In Scotland, an environmental regulatory authority (the Scottish Environment Protection Agency, SEPA) modified the Montgomery and Buffington (1997) typology as a basis for river typing and regulation of river engineering.

This phase was also characterised by the recognition of monitoring improvements or degradations in the physical habitat of a river, particularly for environmental planning, appraisal and impact assessment (Raven *et al.*, 2002). In the UK, the River Habitat Survey (RHS) was developed by the Environment Agency of England and Wales to assess the character and habitat quality of watercourses based on their physical structure (Raven *et al.*, 1997). Data is collected at 10 equidistant 'spot-checks' along a 500 m reach, independent of river size (Wilkinson *et al.*, 1998). The RHS generates two indices/scores that can be used in subsequent assessments of both environmental and biological quality: the Habitat Quality Assessment (HQA) and Habitat Modification Score (HMS). The HQA is an indicator of the physical habitat heterogeneity, whereas the HMS details the level of anthropogenic-induced modification on the 500 m reach. In France, the Système d'Evaluation de la Qualité du Milieu Physique (SEQ-MP), developed by the Agence de l'Eau Rhine-Meuse, allocated stream reaches into one of five categories (from excellent to very poor) based on their physical habitat characteristics using a map- and field-based approach (Agence de l'Eau Rhin-Meuse, 1996). A similar field-based approach by the Länderarbeitsgemeinschaft Wasser (LAWA-vor-Ort) in Germany was also developed to document changes in the physical habitat of rivers. The method identifies the structural quality of small- and medium-size rivers through linking channel features to processes (LAWA, 2000). The approach uses 25 attributes based on stream course development, longitudinal profile, substate, cross-section profile, bank and riparian structure. Stream reaches are classified into a seven tier classification ranging from 1 (unchanged) to 7 (completely changed; Raven *et al.*, 2002). In Australia, physical habitat assessments have also been widely employed. The River Styles framework (of Brierley and Fryirs, 2000, 2005) has been applied in many coastal catchments of New South Wales as a tool for guiding river restoration. The approach classifies channel types, evaluates the physical condition of rivers, and prioritises restoration activities (Chessman *et al.*, 2006).

The river science phase (contemporary application)

Inherent within the river science phase is a trend of integrating different disciplines within river appraisal, typing and characterisation schemes. In the Bega River basin in New South Wales, Australia, Chessman *et al.*, (2006) used the River Styles framework to explain the spatial distribution of macrophyte and macroinvertebrate assemblages (at family level) due to differences in geomorphic character, behaviour and condition. Thomson *et al.* (2004), also working in New South Wales, contrasted macroinvertebrate community composition (at family level) and physical habitat characteristics of pools and runs for three River Styles: a gorge, a bedrock-controlled channel with a discontinuous floodplain, and a meandering gravel-bed channel. Differences in invertebrate fauna were found for pools between the three River Styles, but no distinctions were evident for runs. Both studies indicate the usefulness of geomorphic typologies in explaining some spatial variation in biological assemblages at the reach scale, although differences at a catchment scale are probably due to other factors of altitude, water temperature, and to biological processes such as dispersal mechanisms, predation and colonisation (Milner *et al.*, 2015).

The use of multivariate statistics, remote sensing and image-based methods for reconnaissance, characterisation and linking physical habitat to aquatic biodiversity have grown rapidly in the past decade (Carbonneau and Piégay, 2012). Schmitt *et al.* (2007), for example, developed a quantitative morphodynamic typology of rivers in the French Upper Rhine basin using ordination methods and multivariate statistics. The typology was derived using agglomerative hierarchical clustering of 31 quantitative and qualitative variables from 187 field sites. Principal component analysis (PCA) and multiple correspondence analysis separated the field sites into seven groups, but overlap and variability was present within and between groups. In a similar suite of studies, Thoms *et al.* (2004, 2007) and Harris *et al.* (2009) used geomorphic derived variables from a GIS approach and a hierarchical analysis to characterise floodplain sedimentation patterns and the ecohydrology of stream networks, within the Murray–Darling Basin, Australia. The approach used is presented as a case study later in the chapter.

Other remote sensing tools employed for physical characterisation of river systems include using colour aerial photography and Airborne Thematic Mapper (ATM) multispectral imagery (bands 1 to 7; 420 to 900 nm) to map available gravel-bed habitat for improving lamprey populations (*Lampetra fluviatilis*; Gilvear *et al.*, 2008), and using digital elevation models (DEM) and multispectral imagery to derive landscape metrics (i.e., catchment area, channel elevation, density of hydrojunctions or nodes) that can be used to rank rivers in relation to salmon productivity (Luck *et al.*, 2010). Technological advances in remote sensing tools can now characterise river catchments via automated grain size mapping. Dugdale *et al.*, (2010) used very high-resolution aerial imagery to generate grain size maps for entire catchments. The study developed an 'aerial photosieving' method, which utilises very high-resolution aerial imagery proposed for grain size map production to visually measure particle sizes (Dugdale *et al.*, 2010). The method is intended to reduce field-based data collection, which is often costly and problematic in remote areas. Low-altitude remote sensing using kites, blimps, unmanned aerial vehicles (UAV) and systems (UAS) is also being increasingly used to characterise smaller scale features within river systems (Carbonneau *et al.*, 2012). An UAS, such as the Draganflyer X6, a small and lightweight (1 kg) rotary-winged system, flown at low altitudes (<60 m) can collect hyperspatial resolution imagery to quantify fluvial topography, woody debris and hydraulic habitat (Woodget *et al.*, 2014). The advances in remote sensing techniques can decrease field collection, and allow river researchers and managers to capture changes in fluvial forms and processes across large scales, on geomorphically dynamic river systems.

Geomorphic-based river characterisation case studies

No single characterisation/classification scheme can satisfy all possible purposes, nor can it encompass the multitude of river landscapes. Geomorphic-based characterisations are undertaken at a range of spatial scales and, regardless of the approach and methods used, they must be based on conceptual frameworks, underpinned by defensible scientific principles. Scientific principles act as a framework to guide the process of identifying common channel types and their distinguishing features, as well as allocating channel types to an existing characterisation. Two important principles for the characterisation of river systems are: first, characterisation must be undertaken at scales appropriate for the context in which they are to be used or for the questions being asked. Riverine landscapes are the result of processes operating at multiple scales (Parsons and Thoms, 2007), and teasing apart regional and local effects requires appropriate stratification of sites, along with the selection of variables at the correct scale for the study. Second, characterisation should ideally be based on a holistic range of variables, which are relevant to the physical character of the river system. Consequently, knowledge of the concepts of hierarchy theory is important (Dollar *et al.*, 2007). Groups of interest must be identified based on the self-emergence of groups of similar character, rather than groups being imposed or inherited from other studies or locations. Each characterisation scheme has its own inherent focus or context to study rivers and their character, which is not universal for all studies. Characterisation approaches must, therefore, evolve to become more objective. The following case studies; at scales from networks to reaches, highlight the use of these principles.

Network-scale characterisation in the Murray–Darling Basin, Australia

GIS-based approaches have been used to characterise the stream network of catchments within the Murray–Darling Basin (e.g., Thoms *et al.*, 2004, 2007; Harris *et al.*, 2009). Thoms *et al.* (2004, 2007) used a top-down approach to characterise floodplain sedimentation patterns at three different spatial scales: the channel (10^3 km), floodplain process zone (10 km) and geomorphic unit (10^2 km). The study used a 1:100,000 scale digital streamline layer of the stream network and sites were selected at 5 km intervals along the network. At each site, 15 geomorphic variables were extracted from various digital data of the basin using automated GIS modules to define the physical character of the riverine landscape. A hierarchical analysis (e.g., rivers × process zones × geomorphic units × variable numbers) identified groups of sites with similar physical character. Data were classified using a flexible unweighted pair-group method with arithmetic averages (UPGMA) fusion strategy (as recommended by Belbin and McDonald, 1993). An association matrix was developed using the Gower measure, which is range standardised and suitable for data measured on different units (Belbin, 1993). Groups of sites with similar physical character were chosen via several optimisation routines that select the minimum number of site groups with the greatest relative similarity. These statistical groups were imposed onto the river network and termed river types or functional process zones. Additionally, a SIMPER (SIMilarity PERcentages) analysis was used to determine which geomorphic variables contributed to the within-group similarity of the different river types. Identification of these variables can be used to construct a nomenclature for river types or functional process zones within the stream network.

Harris *et al.* (2009) used this top-down approach to characterise 1152 km of stream network within the Ovens River catchment, in the Murray–Darling Basin. Six river types emerged in the Ovens River network with 82% similarity (Figure 12.2). The spatial distribution and a SIMPER analysis generated a description of each river type. River Types 1 and 2, located in the upland regions of the stream network, were characterised by highly constrained valley settings with relatively steep down valley and valley side slopes (Figure 12.2). However, River Type 1 occupied a lower down valley slope range compared to River Type 2, which was associated with upland, constrained valleys. River Types 3 and 4 were characterised by relatively open valleys and well-developed

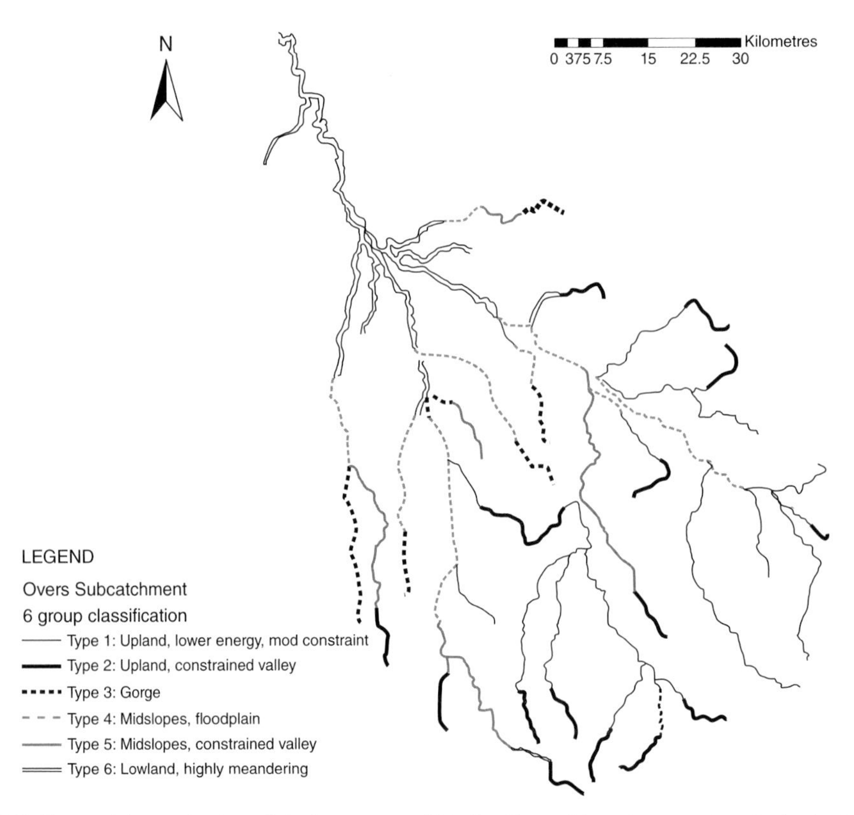

Figure 12.2 The spatial organisation of six river types within the Ovens River stream network, Australia. Harris *et al.*, 2009. Reproduced with permission from Taylor & Francis.

floodplains. River Type 4 had lower down valley slopes, and thus lower stream energies than River Type 3, and was positioned in the mid to lower regions of the stream network catchment. River Type 3 was found in mid-regions only (Figure 12.2). River Type 5 was located in the mid to upper regions of the catchment and was characterised by relatively constrained valley widths, steep valley side slopes, and moderate down valley slopes and energy. In comparison, River Type 6 was dominated by wider valley widths, extensive floodplain surfaces, meandering channels and was found in the lower regions of the stream network (Figure 12.2).

The characterisation exercise also used several ecological community metrics (i.e., richness, composition and diversity) to measure the physical structure and morphological diversity of river types in the Ovens River. River networks were viewed as a community of river types, with a 'river type' being analogous to a 'species' in ecology. Thus, the overall diversity of a community of river types within a stream network can be determined and individual components of diversity, such as abundance, evenness and richness (Harris *et al.*, 2009). In this study, diversity was measured at the network scale, where richness was calculated as the number of river types present, abundance as the total channel length of each river type, and evenness was measured using the Simpsons Evenness Index. The index provides a value between 0 and 1, representing the overall distribution of channel lengths between different river types. When an evenness value approaches 1, channel lengths are more evenly distributed between river types. A combined diversity measure for river types within the Ovens River network was measured using the Shannon–Wiener diversity index.

The diversity of river types within the Ovens River varied markedly. River Types 1, 2 and 5 possessed diversity values greater than 2.0, with River Types 3 and 4 having diversity values of 1.92 and 1.73 respectively. River Type 6 had the lowest diversity value. Overall, the composition of the different river types varied in abundance, richness and evenness (Table 12.1). The most abundant river type was River Type 1, with a

Table 12.1 Composition of the six river types in the Ovens stream network, Australia. The indices in this table are commonly used for determining the diversity of ecological communities, but have been used to characterise river morphology. Abundance represents the proportion and total length of each river type within the stream network, richness is the number of segments per river type, and evenness indicates the distribution of channel segment lengths in each river type.

River type	Description	Abundance (%)	Total channel length (km)	Richness (no. of individual segments)	Evenness (Simpson's evenness index)	Diversity (Shannon–Wiener value)
1	Upland, lower energy, moderate valley constrained	25.8	298	14	0.83	2.24
2	Upland constrained valley	13.9	160	16	0.94	2.77
3	Gorge	7.5	86.7	8	0.83	1.92
4	Midslopes floodplain	17.5	202	9	0.78	1.73
5	Midslopes constrained	15.9	183	9	0.87	2.11
6	Lowland highly meandering	19.3	222	3	0.16	0.35

total channel length of 298 km, followed by River Type 6, 4, 5, 2 and 3. Regarding the number of individual segments comprising each river type (richness), River Type 2 had 16 individual segments with an average segment length of 21 km. Next was River Type 1, while River Type 4 and 5 had similar richness. River Type 3 contained the second lowest richness with eight segments and River Type 6 recorded the lowest richness with only three individual river segments and an average length of 74 km. Evenness values between river types ranged between 0.16–0.94 and five of the six types had an evenness value above 0.78 (Table 12.1), hence segments were similar in length. River Type 6 had the lowest evenness value, corresponding to the small number of very long segments associated with this type. River Type 2 was the most even, suggesting a high number of individual segments with similar channel lengths. River Types 1, 3, 4 and 5 had similar evenness values. The abundance, evenness and richness values suggest several clusters of river types in the Ovens River (Table 12.1). The first cluster includes River Type 1, 4 and 5, which are characterised by high abundance and evenness. The second cluster comprises River Types 2 and 3 of lower abundance values. Lastly, River Type 6 was an outlier due to low richness and evenness values.

Reach-scale typing on the River Dee, Scotland

Identifying how physical habitat character and the behaviour of a river effects aquatic biota in natural settings are important if the influence of human disturbance is to be understood (Chessman *et al.*, 2006). In hierarchical geomorphic typologies, fluvial features within channel types or classes are positioned within a larger-scale framework whereby variables such as valley setting and discharge constrain their functioning and behaviour (Frissell *et al.*, 1986; Hawkins *et al.*, 1993). Differences in topography (slope and lateral confinement) and hydrology (stream power) influence flow characteristics, dictate transport, re-working and storage of bed sediments and create varying combinations of geomorphic units within channel types. These geomorphic units often differ hydraulically and sedimentologically, and can be viewed as a mosaic of internally similar patches (or microhabitats) nested within each channel type (Li *et al.*, 2012). Stream biota, especially macroinvertebrates, respond to hydraulic and physical habitat conditions, which characterise geomorphic units (Braaten and Berry, 1997). As channel types comprise varying assemblages of geomorphic units, it is reasonable to expect that aquatic biota will also differ between channel types within the same climatic and biogeographical limits (Thomson *et al.*, 2004).

In the River Dee and adjacent Allt a'Ghlinne Bhig catchment in the Cairngorm Mountains, Scotland, Milner (2010) measured physical habitat differences of 41 reaches. Stream reaches were classified into common channel types in the Montgomery and Buffington (1997) typology (i.e., bedrock, step–pool, plane–bed, and pool–riffle) and intermediate types (i.e., plane–riffle and wandering) due to their dominance in the study area. Field surveys assessed stream morphology via measurements of channel cross-sectional geometry, channel bed slope, water depth, grain size and mean column velocity. One hundred measurements of water depth, grain size and mean column velocity (at 0.6 depth) were sampled at equidistant locations in a zigzag pattern per reach (see Milner, 2010, for fieldwork procedure). Macroinvertebrates were collected in autumn and spring (i.e., September 2007 and April 2008) using sweep and kick techniques. Samples were collected in all geomorphic units (e.g., pools, riffles, glides) present at a reach to include physical habitat heterogeneity, but the duration of kick sampling in each unit was relative to the spatial coverage within the reach. Each 3-minute kick sample was spatially representative of the geomorphic units comprising the reach.

Box plots visually revealed variations in channel slope and velocity between common and transitional channel types (Figure 12.3). Channel slope demonstrated a linear trend across channel types from step-pool through to bedrock reaches, to other alluvial channels of wandering and pool–riffle reaches. Step–pool reaches were located on steeper slopes with a large range of velocities indicating a heterogeneous hydraulic environment. Differentiation in slope and velocity between the alluvial channel types of plane–bed, plane–riffle, pool–riffle and wandering reaches was less marked (Figure 12.3). The transitional alluvial reaches occupied a smaller range in slope values, but possessed varied velocity conditions (Figure 12.3). Differences in physical habitat characteristics (i.e., channel slope, velocity, grain size and water depth) between channel types were tested for significance using a one-way analysis of similarity (ANOSIM, Clarke, 2003). ANOSIM produces a Global-R test statistic that contrasts the similarities among samples within groups (i.e., within a channel type) with the similarities between groups (i.e., between channel types) (Clarke and Warwick, 2001). A total of 9999 multi-response permutation permutations were used. Physical habitat variables differed significantly among channel types ($R = 0.48$, $p = 0.001$), and between most pairwise comparisons of channel types. The largest difference was between step–pool and pool–riffle ($R = 0.55$, $p = < 0.001$), bedrock ($R = 0.64$, $p = < 0.01$), plane–bed ($R = 0.56$, $p = < 0.01$) and plane–riffle samples ($R = 0.39$, $p = < 0.05$). Physical habitat traits in bedrock reaches also differed from all alluvial channel types (all p values < 0.05). Differentiation in physical habitat was less evident between the alluvial channel types with only plane–bed and pool–riffle samples been statistically different ($R = 0.2$, $p = <0.05$). The ANOSIM

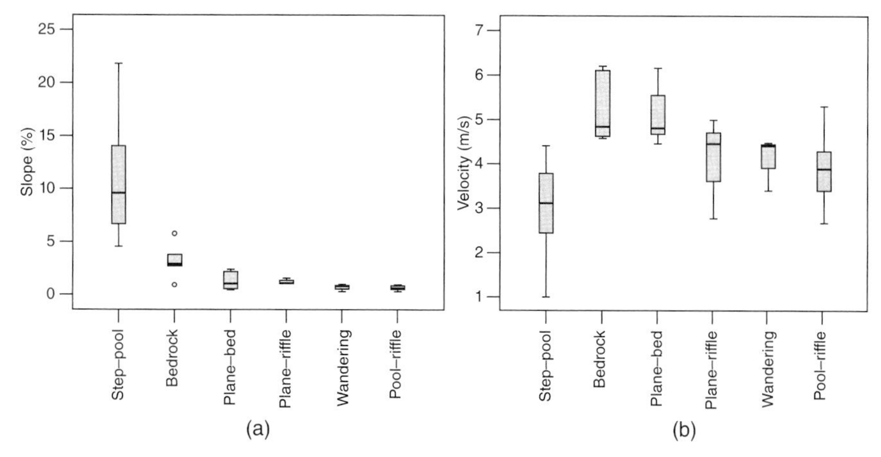

Figure 12.3 Boxplots for (a) channel slope and (b) velocity data according to channel type. Data was derived from reaches in the upper River Dee and Allt a'Ghlinne Bhig catchments, Scotland. Milner, 2010.

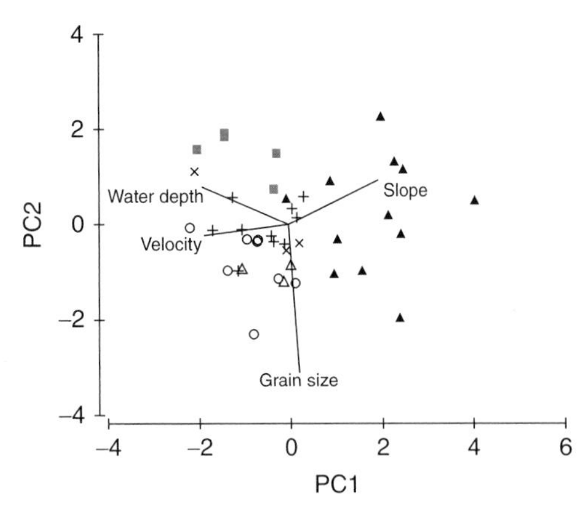

Figure 12.4 PCA ordination of physical habitat variables from reaches in the upper River Dee and Allt a'Ghlinne Bhig catchments, Scotland. Channel types are ▲ step–pool, ■ bedrock, o plane–bed, △ plane–riffle, × wandering and + pool–riffle reaches. Milner, 2010.

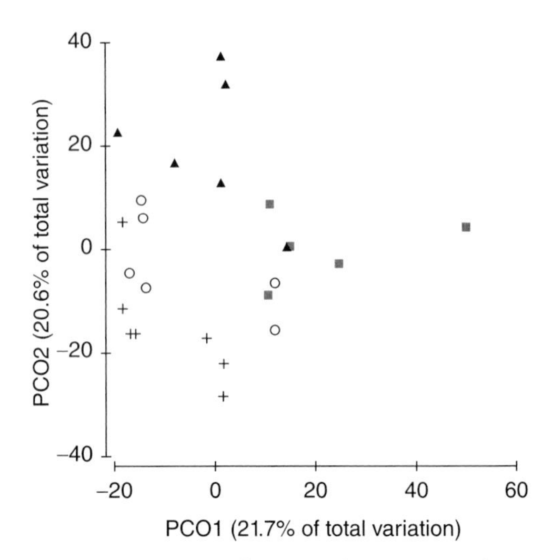

Figure 12.5 PCoA ordination of macroinvertebrate abundance data from reaches in the upper River Dee and Allt a'Ghlinne Bhig catchments, Scotland. Channel types are ▲ step–pool, ■ bedrock, ○ plane–bed, and + pool–riffle reaches. Milner *et al.*, 2015. Reproduced with permission from CSIRO.

analyses on the physical habitat traits are visually supported by a PCA ordination (Figure 12.4).

Milner *et al.* (2015) also investigated whether individual channel types (i.e., step–pool, bedrock, plane–bed, and pool–riffle) support a distinct invertebrate fauna. Principal coordinates analysis (PCoA; Gower, 1966) was carried out to visualise spatial patterns in macroinvertebrate abundance data (Figure 12.5). One-way PERMANOVA (a test for locational differences between pre-defined groups) revealed macroinvertebrate assemblages varied significantly across the geomorphic typology (F-ratio = 3.6, p = 0.001), and between all pairwise combinations of channel types (all combinations $p < 0.05$). The Indicator Value method (IndVal; the indicator value of a species; Dufrêne and Legendre, 1997) revealed differing abundances of common taxa discriminated channel types, with step-pool and pool-riffle reaches containing the most indicator taxa. The baetids *Alainites muticus* and *Baetis rhodani*, *Leuctra inermis* (Leuctridae) and *Brachyptera risi* (Taeniopterygidae) characterised step–pool reaches, whilst riffle beetles *Limnius* and *Oulimnius*, *Caenis rivulorum* (Caenidae) and *Leuctra hippopus* (Leuctridae) were indicative of pool–riffle reaches (Milner *et al.*, 2015). Higher abundances of two less common taxa, *Siphonoperla torrentium* (Chloroperlidae) and *Baetis fuscatus* (Baetidae) typified plane–bed channels. Bedrock reaches had lower taxonomic richness and abundance, especially of common mayflies and stoneflies, and contained one weak positive indicator, *Protonemura meyeri* (Nemouridae). The findings of the study indicate that in catchments with similar water quality, physical habitat distinctions in common channel types at the reach scale reflect biological diversity. Fluvial geomorphology via geometry and sediment transport/supply conditions influence physical habitat characteristics and affects macroinvertebrate distributions at the reach scale. Geomorphic typologies can, therefore, provide a valuable tool for researchers and managers, particularly if variables can be obtained remotely, but sampling strategies need to capture current sediment conditions and habitat heterogeneity (Milner *et al.*, 2015).

River classifications: future challenges

River characterisation and scale

Rivers are natural hierarchical systems that can be resolved into different levels of organisation (Dollar *et al.*, 2007). According to hierarchy theory, each level within a hierarchical organisation (e.g., network, reach or microhabitat) is a discrete unit of the levels above and an agglomeration of lower levels. Separate levels can be distinguished by different rates of processing and morphological character. Therefore, higher levels in the hierarchy provide a constraint on lower levels in the organisation, especially that level immediately above the level under investigation, while lower levels in the organisation have an upward influence through emergent properties. This recognition of top-down constraints and bottom-up influences has implications for undertaking river characterisation. The spatial scale of the river characterisation, classification or typology dictates the resolution and the selection of variables. Thus, the suite of variables used will depend upon the scale of the characterisation, for example the network, reach or the microhabitat scale. The scale of observation inflicts differing limitations on system structure, form and function. Moreover, relationships between spatiotemporal scales and their influence on geomorphic and ecological functioning also depend on the purpose of the enquiry (Brierley and Fryirs, 2005). For instance, dispersal mechanisms, predator–prey, and species interactions occur at a different spatial scale to geomorphic processes that govern channel morphology (Brierley and Fryirs, 2005). A future challenge for river classifications is to use the spatial scale of the observations and experiments to the phenomena under investigation. A difficulty is measuring linkages and capturing variability present at a multitude of spatial scales that describes both broad-scale trends and local or short-term variations.

Besides the spatial scale used in a river classification, typology or characterisation scheme, the length of survey strongly influences observed links between stream morphology and channel processes, response potential and habitat characteristics (Montgomery and Buffington, 1997). Common scales of analysis include fixed channel lengths (e.g., as used in the RHS), multiple channel widths and downstream hydraulic geometry (DHG) relationships. DHG is the most commonly used conceptual framework in fluvial geomorphology for explaining variations in river form and process across catchment scales (Fonstad and Marcus, 2010). DHG suggests as discharge increases with distance downstream, channel morphology responds exponentially with cross-sectional area and velocity increasing, coupled with grain size decreasing and a rise in sediment storage (Robert, 2003). DHG has been widely used by river restoration and flood management planners to identify dimensions of a 'natural' river in relation to channel morphology and drainage area (Fonstad and Marcus, 2010). For example, Pasternack *et al.* (2004) used hydraulic geometry measurements within a 2D hydrodynamic model to predict reach-scale spawning gravel replenishment on the lower Mokelumne River, California. Despite the wide usage of DHG in restoration modelling, the approach has received criticism for excluding habitat heterogeneity (Fonstad and Marcus, 2010). However, application of DHG within river characterisations and classifications may offer a useful and strong theoretical basis to describe river form and processes at relevant multiple scales.

Techniques and technological advances

In the last two decades, the increasing use of high-resolution, catchment extent river data has become a valuable tool for examining channel processes and forms across a multitude of spatial scales (Fonstad and Marcus, 2010). The main approaches to multi-resolution data gathering includes intensive field surveys, extensive surveys from boats fitted with precision instruments, and remote sensing techniques (Fonstad and Marcus, 2010). Intensive field surveys produce detailed, high-resolution data, but are restricted in geographical extent, and may be only suitable for smaller spatial scales. In contrast, extensive surveys from boats using sonar or acoustic Doppler velocimeters (e.g., Parsons *et al.*, 2005) and remote sensing technologies provide high-resolution surveys across a large number of sites, continuously along river lengths and over broad spatial scales at the landscape level (Fonstad and Marcus, 2003). A challenge for users of river classifications is the conflict between obtaining a high-resolution dataset covering a limited area or using a remote sensing approach, which generates a high-resolution dataset spanning basin extents, but with a loss of precision.

The availability of increasingly high resolution topographic datasets has encouraged use of inductive data mining approaches in fluvial geomorphic applications from regulators, practitioners and researchers. Techniques and technologies used to map river forms and processes at multiple scales include multivariate statistics (Thoms *et al.*, 2007), fuzzy classification (Legleither and Goodchild, 2005), feature recognition approaches (Molloy and Stepinski, 2007) and various sensors mounted on blimps, drones, helicopters, aeroplanes and satellites. At local and intermediate scales, hand-held thermal imaging cameras have been used to understand the effects of water column and streambed thermal refugia on endangered mussels in the Delaware River, USA (Briggs *et al.*, 2013), and to characterise the thermal regime and mixing zones of groundwater inflows to brook trout streams as part of a restoration project (USGS, 2014). At a landscape scale, Handcock *et al.* (2012) used thermal infrared remote sensing to detect spatial patterns of radiant temperature at the water's surface Applications include describing the thermal heterogeneity in river floodplains to measure habitat diversity and determining whether water temperature regimes meet management guidelines for coldwater fishes such as salmon and trout. Other remote sensing technologies, including high-resolution aerial photos and LiDAR data, have allowed subaqueous sediment size mapping at a network scale (i.e., tens of kilometres; Carbonneau *et al.*, 2004; Verdu *et al.*, 2005). These technological advances allow river ecosystems to be mapped at multiple spatial scales, and permit detailed surveys of abiotic and biotic variables. Depending on the purpose of the scheme, not all river characterisations or classifications may require high-resolution data on river systems, but these techniques and the availability of hyper-scale data presentation will help to connect channel forms and processes with the ecological dynamics of aquatic biota, which represents a significant advancement in the field.

There are a wide range of physical classification schemes available to characterise riverine landscapes. The scale at which these exercises occur is changing and ranges from entire river networks through to micro-habitats, and large datasets are being generated as a consequence. Parallel with these advances is the need to develop quantitative techniques to measure spatial heterogeneity of riverine landscapes. Analysis techniques used by community ecologists can potentially provide measures of the physical diversity within riverine landscapes. At the network scale for example, the river network can be viewed as a community of river types; with a 'river type' being analogous to a biotic 'species'. Richness is the number of disjunct river stretches within a river type, abundance is the total length of stream channel within a river type, and evenness is the distribution of channel lengths between river segments. Measurement of these variables enables the calculation of a combined diversity value for each river type and for the whole catchment. The manipulation of ecological diversity metrics for application in the physical realm has not been used previously to define physical diversity in rivers (apart from Harris *et al.*, 2009). These techniques can be transferred to any river network, and the outputs can be compared both within and between stream communities. The physical diversity of a stream network can be associated with stream habitats, which can enable inferences about biological diversity. Although the relationship between physical heterogeneity in a stream network and biological diversity is recognised (Thorp *et al.*, 2008), further research is needed to determine how measures of river network diversity described here relate directly to biological communities.

Linking channel form and processes to ecological dynamics

Integrating fluvial forms and processes with ecological functioning is a key priority for river science and management (Vaughan *et al.*, 2009). Where channel types are delineated on processes and form (e.g., on the balance between transport capacity to sediment supply), difference in geomorphic dynamics should engender biotic differences (Milner *et al.*, 2015). Here, geomorphic classifications should provide a valuable tool to link form and processes to biological structure at different spatial scales. Montgomery *et al.* (1999) showed channel type affected Pacific salmonid spawning distributions in rivers in the Pacific Northwest. Spawning Pacific salmonids chose 'response' reach types of pool–riffle and plane–bed characterised by a high sediment supply to transport capacity ratio. The timing and depth of channel bed mobility within the two reach types influenced survival of buried eggs (Montgomery *et al.*, 1999). In the Allt a'Ghlinne Bhig and Girnock Burn, Scotland, Moir *et al.*, (2004) similarly found Atlantic salmon (*Salmo salar*) preferentially selected pool–riffle and transitional pool–riffle/plane–bed, and avoided 'transport' reach types of plane-bed and step–pool. The studies indicate using channel types within a process-based typology is potentially useful for predicting the spatial distribution of spawning activity within a catchment. A future challenge for geomorphic typologies is to merge the range of techniques (as described above) and technological advances across different spatiotemporal scales whilst incorporating ecological functioning.

Conclusion

This chapter has reviewed the theory, history, application and future challenges of river classification, and has demonstrated the significant advances that have occurred in the field and in real-world applications. Advances have mainly happened due to improved knowledge of river ecosystem functioning and the emergence of technologies, especially remote sensing and data processing capabilities. However, if river classification/geomorphic typologies are to be integrated with ecological functioning, small-scale and physical/biological processes need to be included with channel type/class designation or within sampling designs for ecological applications. Therefore, we recommend future river classification/geomorphic typologies need to choose the most appropriate spatial scale for the observations of the phenomena under investigation. We advocate a paradigm shift within river science and its application, which explicitly tackles the issues of multiple spatiotemporal scales, survey length and heterogeneity. Spatiotemporal scales need to be viewed in a 3D perspective and not solely a 2D patch-matrix framework. The use and availability of remote sensing technologies and the increasing use of multivariate statistics and fuzzy clustering of channel typing should underpin this paradigm shift. Such a transition will improve conceptual understanding of river ecosystems from a hydromorphic and eco-geomorphic perspective, and improve river management, restoration and conservation activity.

Acknowledgements

We wish to thank an anonymous reviewer for their useful comments and suggestions, which greatly helped to improve the chapter.

References

Agence de l'Eau Rhin-Meuse. 1996. *Outil d'évaluation de la qualité du milieu physique – synthèse*. Metz, France.

Belbin L. 1993. Environmental representativeness: regional partitioning and reserve selection. *Biological Conservation* **66**: 223–230.

Belbin L, McDonald C, 1993. Comparing three classification strategies for use in ecology. *Journal of Vegetation Science* **4**: 341–348.

Braaten PJ, Berry CR. 1997. Fish associations with four habitat types in a South Dakota prairie stream. *Journal of Freshwater Ecology* **12**: 1522–1529.

Brierley GJ, Fryirs K. 2000. River Styles, a Geomorphic Approach to Catchment Characterization: Implications for River Rehabilitation in Bega Catchment, New South Wales, Australia. *Environmental Management* **25** (6): 661–679.

Brierley GJ, Fryirs KA. 2005. *Geomorphology and River Management: Applications of the River Styles Framework*. Blackwell Publications: Oxford, UK.

Brierley GJ, Fryirs KA, Cullum C, *et al.* 2013. Reading the landscape: Integrating the theory and practice of geomorphology to develop place-based understandings of river systems. *Progress in Physical Geography* **37** (5): 601–621.

Briggs MA, Voytek EB, Day-Lewis FD, *et al.* 2013. The hydrodynamic controls on thermal refugia for endangered mussels in the Delaware River. *Environmental Sciences and Technology* **47** (20): 11423–11431.

Carbonneau PE, Lane SN, Bergeron NE. 2004. Catchment-scale mapping of surface grain size in gravel bed rivers using airborne digital imagery. *Water Resources Research* **40** (7): W07202.

Carbonneau PE, Piégay H. 2012. Introduction: the growing use of imagery in fundamental and applied river sciences. In: Carbonneau PE and Piégay H. (eds) *Fluvial Remote Sensing for Science and Management*. John Wiley & Sons Ltd: Chichester, pp. 1–18.

Carbonneau PE, Piégay H, Lejot J, *et al.* 2012. Hyperspatial imagery in riverine environments. In: Carbonneau PE and Piégay H. (eds) *Fluvial Remote Sensing for Science and Management*. John Wiley & Sons Ltd: Chichester, pp. 163–191.

Chessman BC, Fryirs KA, Brierley GJ. 2006. Linking geomorphic character, behaviour and condition to fluvial biodiversity: implications for river management. *Aquatic Conservation: Marine and Freshwater Ecosystems* **16**: 267–288.

Clarke KR. 1993. Non-parametric multivariate analyses of changes in community structure. *Australian Journal of Ecology* **18**: 117–143.

Davis WM. 1899. The geographical cycle. *Geographical Journal* **14**: 481–504.

Dollar ESJ, James CS, Rogers KH, Thoms MC, 2007. A framework for interdisciplinary understanding of rivers and ecosystems. *Geomorphology* **89**: 147–162.

Doyle MW, Harbor JM. 2003. A scaling approximation of equilibrium time - scales for sand bed and gravel bed rivers responding to base level lowering. *Geomorphology* **54**: 217–223.

Dufrêne M, Legendre P. 1997. Species assemblages and indicator species: the need for a flexible asymmetrical approach. *Ecological Monographs* **67**: 345–366.

Dugdale SJ, Carbonneau PE, Campbell D. 2010. Aerial photosieving of exposed gravel bars for the rapid calibration of airborne grain size maps. *Earth Surface Processes and Landforms* **35**(6): 627–639.

Fonstad MA, Marcus WA. 2010. High-resolution, basin-extent observations of fluvial forms and implications for process understanding. *Earth Surface Processes and Landforms* **35**: 680–698.

Frissell CA, Liss WJ, Warren CE, Hurley MD. 1986. A hierarchical framework for stream habitat classification: Viewing streams in a watershed context. *Environmental Management* **10**: 199–214.

Galay VJ, Kellerhals R, Bray DI. 1973. Diversity of river types in Canada. *Fluvial Processes and Sedimentation. Proceedings of the Hydrology Symposium*. National Research Council of Canada: 217–250.

Gilvear DJ, Sutherland P, Higgins T. 2008. An assessment of the use of remote sensing to map habitat features important to sustaining lamprey populations. *Aquatic Conservation: Marine and Freshwater Ecosystems* **18**: 807–818.

Gower JC. 1966. Some distance properties of latent root and vector methods used in multivariate analysis. *Biometrika* **53**: 325–338.

Handcock RN, Torgersen CE, Cherkauer KA, *et al.* 2012. Thermal infrared remote sensing of water temperature in riverine landscapes. In: Carbonneau PE and Piégay H (eds) *Fluvial Remote Sensing for Science and Management*. John Wiley & Sons Ltd: Chichester, pp. 85–113.

Harris CA, Thoms MC, Scown MA. 2009. The ecohydrology of stream networks. *International Association of Hydrological Sciences* **328**: 127–138.

Hawkes HA. 1975. River zonation and classification. In: Whitton BA (ed.) *River Ecology*. Blackwell: London, pp. 312–374.

Hawkins CP, Kershner JL, Bisson PA, *et al.* 1993. A hierarchical approach to classifying stream habitat features. *Fisheries* **18**: 3–12.

Horton RE. 1945. Erosional development of streams and their drainage basins: hydro-physical approach to quantitative morphology. *Geological Society of America Bulletin* **56**(3): 275–370.

Huet M. 1954. Biologie, profiles en long et en travers des eaux courants. *Bulletin Français de Pisciculture* **175**: 41–53.

Juracek KE, Fitzpatrick FA. 2003. Limitations and implications of stream classification. *Journal of the American Water Resources Association* **39**(3): 659–670.

Kellerhals R, Neill CR, Bray DI. 1972. Hydraulic and geomorphic characteristics of rivers in Alberta. *Research Council of Alberta, River Engineering and Surface Hydrology Report* **72**: 1–52.

Kellerhals R, Church M, Bray DI. 1976. Classification and analysis of river processes. *Journal of the Hydraulics Division ASCE* **102**: 813–829.

Knighton AD, Nanson GC. 1993. Anastomosis and the continuum of channel pattern. *Earth Surface Processes and Landforms* **18**: 613–625.

Kondolf GM. 1995. Geomorphological stream channel classification in aquatic habitat restoration: uses and limitations. *Aquatic Conservation: Marine and Freshwater Ecosystems* **5**: 127–141.

Kondolf GM, Montgomery DR, Piégay H, Schmitt L. 2003. Geomorphic classification of rivers and streams. In Kondolf, GM and Piégay, H (eds) *Tools in Fluvial Geomorphology*. John Wiley & Sons Ltd: Chichester.

LAWA. 2000. Gewässerstrukturgütebewertung in der Bundesrepublik Deutschland. *Verfahren für kleine und mittelgroβe Flieβgewässer*. Bund/Länder-Arbeitsgemeinschaft Wasser: Berlin.

Legleiter CJ, Goodchild MF. 2005. Alternative representations of instream habitat: Classification using remote sensing, hydraulic modeling, and fuzzy logic. *International Journal of Geographical Information Science* **19**(1): 29–50.

Leopold LB, Wolman MG. 1957. River channel patterns: braided, meandering and straight. *U.S. Geological Society Professional Paper* **282-B**: 39–85.

Levin SA. 1992. The problem of pattern and scale in ecology. *Ecology* **73**(6): 1943–1967.

Li F, Chung N, Mi-Jung B, *et al.* 2012. Relationships between stream macroinvertebrates and environmental variables at multiple spatial scales. *Freshwater Biology* **57**: 2107–2124.

Luck MN, Maumenee D, Whited J, *et al.* 2010. Remote sensing analysis of physical complexity of north Pacific rim rivers to assist wild salmon conservation. *Earth Surface Processes and Landforms* **35** (11): 1330–1343.

Makaske B. 2001. Anastomosing rivers: a review of their classification, origin and sedimentary products. *Earth Surface Review* **53**: 149–196.

Margules CR, Pressey RL. 2000. Systematic conservation planning. *Nature* **405**: 243–253.

Melles SJ, Jones NE, Schmidt B. 2012. Review of theoretical developments in stream ecology and their influence on stream classification and conservation planning. *Freshwater Biology* **57**: 415–434.

Miller JR, Ritter JB. 1996. Discussion. An examination of the Rosgen classification of natural rivers. *Catena* **27**: 295–299.

Milner VS. 2010. *Assessing the performance of morphologically based river typing in Scotland using a geomorphological and ecological approach.* Unpublished PhD thesis, University of Stirling, Stirling, UK.

Milner VS, Gilvear DJ, Willby NJ. 2013. An assessment of variants in the professional judgement of geomorphologically based channel types. *River Research and Applications* **29**: 236–249.

Milner VS, Willby NJ, Gilvear DJ, Perfect C. 2015. Linkages between reach scale physical habitat and invertebrate assemblages in upland streams. *Marine and Freshwater Research* **66**: 438–448.

Moir HJ, Gibbins CN, Soulsby C, Webb JH. 2004. Linking channel geomorphic characteristics to spatial patterns of spawning activity and discharge use by Atlantic salmon (*Salmo salar L.*). *Geomorphology* **60**: 21–35.

Mollard JD. 1973. Air photo interpretation of fluvial features. *Fluvial Processes and Sedimentation.* Research Council of Canada: 341–380.

Molloy I, Stepinski TF. 2007. Automatic mapping of valley networks on Mars. *Computers & Geosciences* **33**(6): 728–738.

Montgomery DR, Beamer EM, Pess GR, Quinn TP. 1999. Channel types and salmonid spawning distribution and abundance. *Canadian Journal of Fisheries and Aquatic Sciences* **56**: 377–877.

Montgomery DR, Buffington JM. 1997. Channel-reach morphology in mountain drainage basins. *Geological Society of America Bulletin* **109**: 596–611.

Naiman RJ. 1998. Biotic stream classification. In: Naiman, RJ, Bilby RE (eds) *River Ecology and Management Lessons from the Pacific Coastal Ecoregion.* Springer: New York, pp. 97–119.

Nanson GC, Knighton AD. 1996. Anabranching rivers: their causes, character and classification. *Earth Surface Processes and Landforms* **21**(3): 217–239.

Nel JL, Roux DJ, Abell R, *et al.* 2009. Progress and challenges in freshwater conservation planning. *Aquatic Conservation: Marine and Freshwater Ecosystems* **19**: 474–485.

Newson MD, Clark MJ, Sear DA, Brookes A. 1998. The geomorphological basis for classifying rivers. *Aquatic Conservation: Marine and Freshwater Ecosystems* **8**: 415–430.

Pasternack GB, Wang CL, Merz J. 2004. Application of a 2D hydrodynamic model to reach-scale spawning gravel replenishment on the lower Mokelumne river, California. *River Research and Applications* **20**: 205–225.

Parsons DR, Best JL, Orfeo O, *et al.* 2005. Morphology and flow fields of three-dimensional dunes, Rio Parana, Argentina: Results from simultaneous multibeam echo sounding and acoustic Doppler current profi ling. *Journal of Geophysical Research – Earth Surface* **110**(F4): F04S03.

Parsons ME, Thoms MC. 2007. Hierarchical patterns of large woody debris distribution and macroinvertebrate-environment associations in river ecosystems. *Geomorphology* **89**: 127–146.

Pennack RW. 1971. Towards a classification of lotic habitats. *Hydrobiologia* **38**: 321–334.

Portt C, King SW, Hynes HBN. 1989. *A Review and Evaluation of Stream Habitat Classification Systems and Recommendations for the Development of a System for Use in Southern Ontario*. Ontario Ministry of Natural Resources: Downsview, Ontario.

Raven PJ, Fox P, Everard M, *et al.* 1997. River Habitat Survey: a new system for classifying rivers according to their habitat quality. In: Boon, PJ and Howell, DL (eds) *Freshwater Quality: Defining the Indefinable?* The Stationery Office: Edinburgh, pp. 215–234.

Raven PJ, Holmes NTH, Charrier P, *et al.* 2002. Towards a harmonized approach for hydromorphological assessment of rivers in Europe: a qualitative comparison of three survey methods. *Aquatic Conservation: Marine and Freshwater Ecosystems* **12**: 405–424.

Robert A. 2003. *River Processes: An Introduction to Fluvial Dynamics*. Arnold: London, UK.

Rosgen DL. 1985. A stream classification system. In: Johnsson RR, Zeibell CD, Patton DR, *et al.* (eds) *Riparian Ecosystems and Their Management: Reconciling Conflicting Uses United States Forest Service*. General Technical Report M-120, Rocky Mountain Forest and Range Experimental Station: Fort Collins, Colorado, pp. 91–95.

Rosgen DL. 1994. A classification of natural streams. *Catena* **22**: 169–199.

Rosgen DL. 1996. *Applied River Morphology*. Wildland Hydrology: Pagosa Springs, Colorado, USA.

Schmitt L, Maire G, Nobelis P, Humbert J. 2007. Quantitative morphodynamic typology of rivers: a methodological study based on the French Upper Rhine basin. *Earth Surface Processes and Landforms* **32**: 1726–1746.

Schumm SA. 1963. *A Tentative Classification of Alluvial River Channels*. US Geological Survey Circular 477, Washington, DC.

Schumm SA. 1977. *The Fluvial System*, John Wiley and Sons Ltd: New York, USA.

Smith DG, Smith ND. 1980. Sedimentation in anastomosed river systems: examples from alluvial valleys neat Banff, Alberta. *Journal of Sedimentary Petrology* **50**(1): 157–158.

Sneath PHA, Snokal RR. 1973. *Numerical Taxonomy: The Principles and Practice of Numerical Classification*. W.H. Freeman: San Francisco.

Strahler AN. 1952. Hypsometric (area-altitude) analysis of erosional topology. *Geological Society of America Bulletin* **63**(11): 1117–1142.

Strahler AN. 1957. Quantitative analysis of watershed geomorphology. *Transactions of the American Geophysical Union* **8**(6): 913–920.

Thoms MC, Hill SM, Spry MJ, *et al.* 2004. The geomorphology of the Darling River. In: Breckwodt R, Boden R, Andrews J (eds) *The Darling*. The Murray Darling Basin Commission: Australia, pp. 68–105.

Thoms MC, Parsons ME, Foster JM. 2007. The use of multivariate statistics to elucidate patterns of floodplain sedimentation at different spatial scales. *Earth Surface Processes and Landforms* **32**(5): 672–686.

Thomson JR, Taylor MP, Brierley GJ. 2004. Are River Styles ecologically meaningful? A test of the ecological significance of a geomorphic river characterization scheme. *Aquatic Conservation: Marine and Freshwater Ecosystems* **14**, 25–48.

Thorp JH, Thoms MC, Delong M. 2008. *The Riverine Ecosystems Synthesis: Towards Conceptual Cohesiveness in River Science*. Academic Press: New York, USA.

Underwood AJ, Chapman MG, Connell SD. 2000. Observations in ecology: you can't make progress on processes without understanding the patterns. *Journal of Experimental Marine Biology and Ecology* **250**: 97–115.

USGS. 2014. Thermal Imaging Camera Use: Brook Trout Restoration Activities by the USGS Virginia Water Science Center. Available online: http://water.usgs.gov/ogw/bgas/thermal-cam/vawsc. html [Accessed 15/10/2014].

Van Niekerk AW, Heritage GL, Moon BP. 1995. River classification for management: the geomorphology of the Sabie River in the eastern Transvaal, *South African Geographical Journal* **77** (2): 68–76.

Vannote RL, Minshall GW, Cummins KW, *et al.* 1980. The river continuum concept. *Canadian Journal of Fisheries and Aquatic Sciences* **37**: 130–137.

Vaughan IP, Diamond M, Gurnell AM, *et al.* 2009. Intergrating ecology with hydromorphology: a priority for river science and management. *Aquatic Conservation: Marine and Freshwater Ecosystems* **19**: 113–125.

Verdu JM, Batalla RJ, Martinez-Casasnovas JA. 2005. High-resolution grain-size characterization of gravel bars using imagery analysis and geo-statistics. *Geomorphology* **72** (1–4): 73–93.

Ward JV. 1989. The four-dimensional nature of lotic ecosystems. *Journal of the North American Benthological Society* **8**: 2–8.

Warren CE. 1979. *Toward classification and rationale for watershed management and stream protection.* Report No. EPA-600/3-79-059. United States

Environmental Protection Agency, Corvallis, Oregon.

Wilkinson J, Martin J, Boon PJ, Holmes NTH. 1998. Convergence of field survey protocols for SERCON (System for Evaluation Rivers for Conservation), and RHS (River Habitat Survey). *Aquatic Conservation: Marine and Freshwater Ecosystems* **8**: 579–596.

Woodget AS, Carbonneau PE, Visser F, Maddock IP. (2014) Quantifying submerged fluvial topography using hyperspatial resolution UAS imagery and structure from motion photogrammetry. *Earth Surface Processes and Landforms*. DOI: 10.1002/esp.3613.

CHAPTER 13

Thermal diversity and the phenology of floodplain aquatic biota

Jack A. Stanford, Michelle L. Anderson, Brian L. Reid, Samantha D. Chilcote and Thomas S. Bansak
Flathead Lake Biological Station, University of Montana, Polson, Montana, USA

Introduction

Heat is a primary determinant of the distribution, growth and reproduction of ectothermic biota in stream ecosystems because ectothermic metabolism is constrained by environmental temperature dynamics (Cummins, 1974; Vannote and Sweeney, 1980; Ward and Stanford, 1982; Poole and Berman, 2001; and many others). Indeed the majority of stream biodiversity is composed of ectothermic vertebrates and invertebrates and they display a wide range of adaptations to maximise fitness (i.e., attain a positive life history energy balance, sensu Hall *et al.*, 1992). Many stream invertebrates undergo diapause during periods when temperatures are too cold or too hot for effective metabolism and many, especially stream insects, have very specific temperature cues to break diapause and for metamorphosis to adults (terrestrial flying–mating stage) (Ward and Stanford, 1982). Because growth and maturation is strongly determined by temperature in relation to foraging, predator avoidance and other habitat considerations, the physiological time (active period for growth and maturation) required to hatch and grow to maturity can therefore usually be described in terms of the heat budget or accumulation of thermal units (e.g., degree-days) of the habitat during the growth period. If an array of habitats that have different temperature patterns are available, mobile ectotherms can move from one habitat to another to maximise growth. For example, juvenile coho (*Oncorhynchus kisutch*) salmon will feed on sockeye (*O. nerka*) salmon eggs in cold spring brooks preferred for sockeye spawning within the channel network of floodplains and move to warmer habitats for resting and digestion, which maximises metabolism (Armstrong and Schindler, 2013).

Heterogeneous thermal patterns are expected to characterise floodplain habitats and greatly influence the distribution of biota (Tockner *et al.*, 2000; Ward *et al.*, 2002) for the reasons described above. Interactions of flooding, sediment transport, surface–groundwater exchange, deposition of drift wood, and vascular plant succession create a dynamic habitat template or mosaic on river floodplains. Thus, through time, floodplains are shaped by nonlinear

River Science: Research and Management for the 21st Century, First Edition.
Edited by David J. Gilvear, Malcolm T. Greenwood, Martin C. Thoms and Paul J. Wood.

processes that create a shifting habitat mosaic (SHM hereafter) (Stanford *et al.*, 2005; Latterell *et al.*, 2006) that usually encompass a wide range of temperature patterns.

The SHM includes the components of the main river channel, often a network of channels, and a suite of lateral or off-channel habitats in parafluvial (annually flooded and broadly scoured with localised depositional areas) and orthofluvial (flooded only by infrequent over bank flow and broadly depositional with localised scoured areas) zones (Stanford *et al.*, 2005). River water flowing in permanent surface channels interacts with lateral water bodies during over bank flooding. Floodplain topography and stratigraphy of bed sediments influence the duration of aquatic habitat inundation and connectivity above and below ground (Poole *et al.*, 2002). Thus, lateral floodplain habitats are expected to display great thermal heterogeneity due to temporal and spatial variability in surface and groundwater connectivity and flow rates, as well as from shading by riparian vegetation (Arscott *et al.*, 2001; Malard *et al.*, 2001; Arrigoni *et al.*, 2008).

In large alluvial floodplains, groundwater is derived from the large aquifers associated with the extensive deposits of fluvial bed sediments. Water from surface channels penetrates into porous bed sediments at the upstream end of the floodplain, forming an expansive hyporheic zone that may include the entire alluvial aquifer if recharge is predominately from the river (Stanford *et al.*, 2005; Boulton *et al.*, 2010). The degree of surface–groundwater exchange is influenced by aquifer stratigraphy, channel geomorphology and river discharge (Mertes, 1997; Poole *et al.*, 2006; Helton *et al.*, 2014). During periods of high river discharge, aquifer storage increases, the water table rises, and effluent groundwater inundates low-lying depressions that may be either be connected or disconnected from the surface channel network (Mertes, 1997; Poole *et al.*, 2002). The opposite occurs as river discharge decreases; surface water drains laterally and vertically from the floodplain. Daily, seasonal and annual temperature patterns in the aquifer are usually substantially less variable in comparison to surface waters owing to the buffering effect of river water flux through the bed sediments along flow paths of varying length and depth (Hoehn and Cirpka, 2006; Arrigoni *et al.*, 2008).

Thus, on expansive floodplains, water is constantly moving through a wide range of materials, all of which have different thermal conductivities and different exposure to solar radiation. This is especially true for floodplains in mountain valleys with topographic complexity and variation in aspect. Therefore, a huge range in temperature dynamics may occur within the SHM of expansive floodplains, owing to high variation in surface and groundwater heat exchange (Anderson 2005).

An expansive alluvial floodplain on the Middle Fork Flathead River in Montana (USA), known as the Nyack Flood Plain Research Natural Area (Nyack, hereafter), has been the focus of intensive study into drivers and dynamics of its complex SHM (e.g., Stanford *et al.*, 2005; Whited *et al.*, 2007 – for a complete list of Nyack papers please see the website of the Flathead Lake Biological Station – http://flbs.umt.edu). Seasonal flow patterns in the river and associated aquifer dynamics are strongly linked to habitat distribution and structure and associated biodiversity (e.g., Mouw *et al.*, 2009). Within this chapter, we describe annual, seasonal and daily thermal patterns (magnitude, frequency, duration, timing and rate of change in temperatures – *sensu* Olden and Naiman, 2010) across the Nyack SHM in relation to well-documented temperature criteria for a suite of aquatic biota commonly occurring among Nyack habitats. We specifically sought to examine the following thermal characteristics: (i) the nature and diversity of temperature dynamics within aquatic habitats across the Nyack floodplain and (ii) habitat-specific temperature patterns recorded in the Nyack system in relation to published thermal tolerances for a suite of commonly occurring floodplain organisms. The purpose was to demonstrate that a wide variety of thermal niches occur in the Nyack SHM, which promotes biodiversity and may be anticipated to occur on all expansive river floodplains.

Methods

Study area

The Nyack (Figure 13.1) is located in the 2300 km^2 catchment of the Middle Fork Flathead River, a fifth-order river in northwest Montana. The flood plain is 10-km long by 2-km wide, and situated between bedrock-constrained canyons. These knick points and the lateral mountain slopes precisely delimit the floodplain boundaries. Annual peak discharge occurs during snowmelt in May or June and varies from 105 m^3 s^{-1} (1-year return interval) to the largest flood on record, which exceeded 2600 m^3 s^{-1} in 1964 (Whited *et al.*, 2007). Over bank discharge of the main channel is 465 m^3 s^{-1} (1.5-year return interval) and baseflows of 11–17 m^3 s^{-1} occur with the onset of very cold winter temperatures that leads to ice cover in October–February. Surficial gravel and cobble bed sediments are deep (100+m) at the upper end and shallow (<25 m) at the lower end of the system owing to glacial history. These bed sediments are entirely saturated during over bank flow; the water table elevation contracts as discharge in the river declines. The aquifer is fed almost entirely by the river; hillslope phreatic flow is minimal (Stanford *et al.*, 2005). Hence, the entire aquifer is essentially a hyporheic zone (*sensu* Boulton *et al.*, 2010) with highly variable flow paths related to the heterogeneity of the bed sediments; water residence time from the point of penetration (downwelling) to effluent sites ranges from < 1 hour in short (<10 m) flow paths (short residence time, SRT) through main channel gravel bars, to 1.5 years for flux through the entire aquifer (long residence time, LRT) (Helton *et al.*, 2014). In addition to the flowing channels, surface water habitats occur at points of topographic intersection with the water table (top of the aquifer). For example, spring brooks and ponds commonly occur in flood channels and may be ephemeral if they are located above the baseflow elevation of the water table.

Habitat types and instrumentation

We classified Nyack habitats following the approach of Stanford *et al.* (2005) and selected a suite of sampling sites (Figure 13.1): two main (primary) river channel sites, three shallow shoreline sites along the edge of the main river channel, three backwaters (sometimes called alcoves) that typically occurred at the confluence of the main river channel and seasonally active flood channels, three parafluvial (annually flooded) spring brooks, three orthofluvial (rarely flooded) spring brooks, three parafluvial ponds, and three perennial tributaries that drain into the main channel. We also sampled the alluvial aquifer at three sites via monitoring wells drilled into the aquifer. Each of the wells was classified as

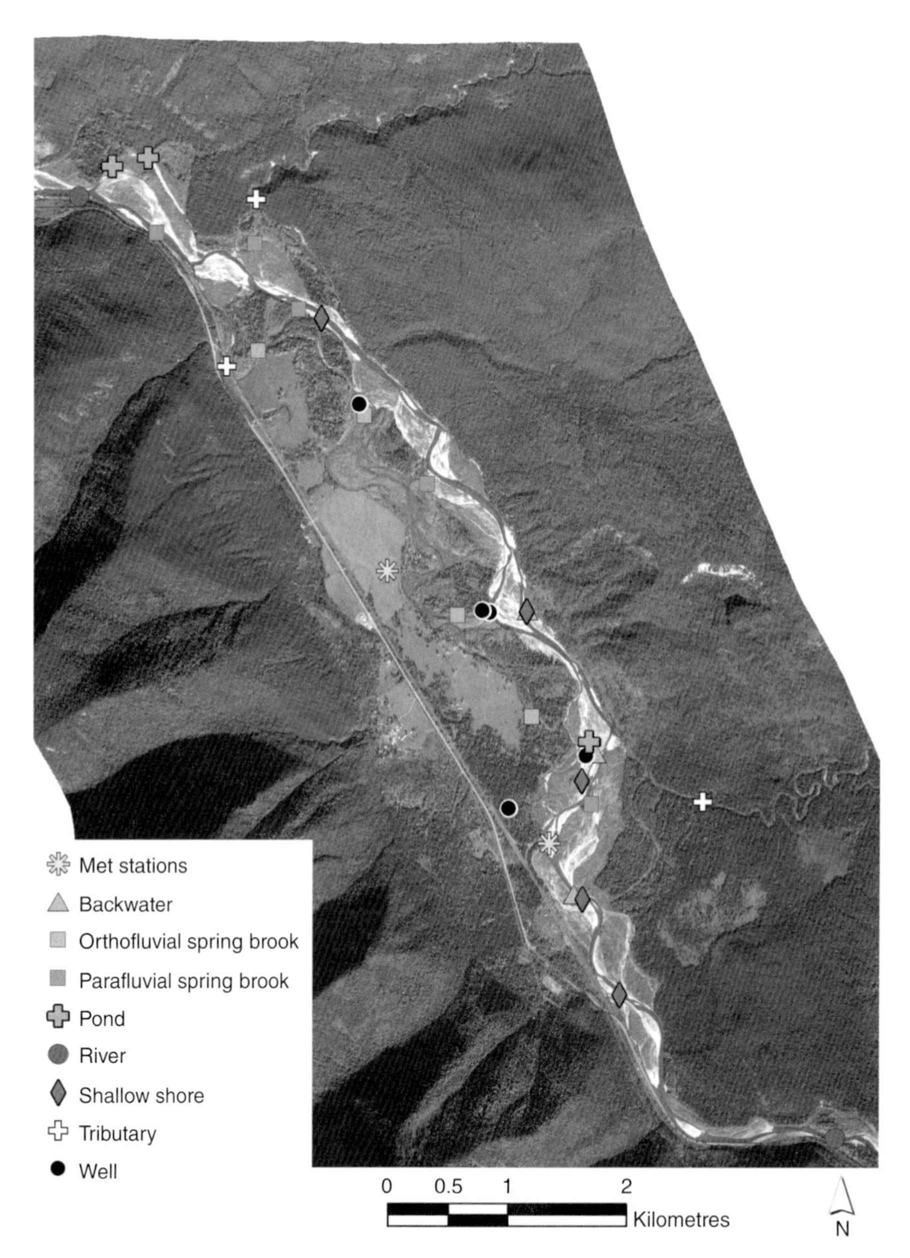

Figure 13.1 Locations of sampling sites (habitat types as keyed in the inset) of the Nyack floodplain of the Middle Fork Flathead River, Montana. The base layer is a multispectral satellite image obtained October, 2004. *(See colour plate section for colour figure.)*

having either short (SRT), medium (MRT) or long (LRT) residence time of groundwater, based upon time lags between peak annual temperatures in the wells and the main river channel (Helton *et al.*, 2014).

Two meteorological stations were installed on the floodplain, one on a parafluvial bench with a canopy of 20-year cottonwood trees, the other an open area on an orthofluvial bench (Figure 13.1). Sensors

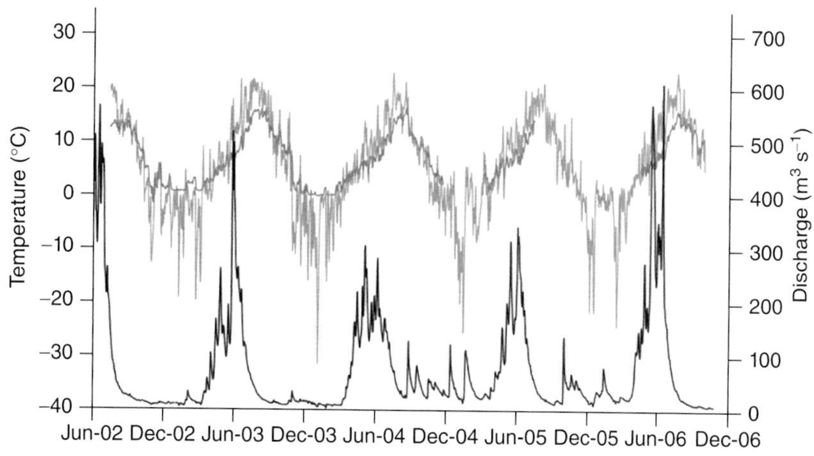

Figure 13.2 Annual patterns (2002–06) for daily average air temperature (upper gray line) and main river channel (bold line) at the Nyack floodplain in relation to discharge of the river (lower black line). Discharge data are from USGS gauge # 12358500, which strongly correlates with river stage measures at Nyack.

and data loggers collected hourly values of air temperature, humidity, precipitation, wind speed and direction, precipitation and soil temperature and moisture at 50 cm depth. Water temperatures were obtained in time series (usually at 1-hour intervals for long-term patterns; 10-minute intervals for diel patterns) with a variety of thermistor-loggers (Vemco Minilog TR, Onset Stowaway, and iButton). Calibration and field placement followed Malard *et al.* (2001) and Johnson *et al.* (2005); aquatic sensors were immersed 1–2 m below surface. Some records contained data gaps owing to exceeding logger capacities or loss of sensors. Short breaks in the data record were filled by interpolation from temperature values at the site before and after the data gap. In a few cases of longer (days to a week or more) lapses, we replaced missing values using linear regression modelling of data from another site within the same habitat type. Models were considered adequate to generate replacement data if $R^2 \geq 0.70$.

Temperature data analyses

From the hourly data collected in the field, we compiled annual, seasonal, monthly and daily (diel) temperature metrics within habitat types (Figure 13.1) in relation to river flow for the period July 2003 to July 2004 (a longer record existed at the weather stations, Figure 13.2), the most complete period of record at Nyack for all habitat types. Temperature metrics included daily maximum and minimum, average daily ($\overline{T}_\mathrm{d} = \frac{\sum_{i=1}^{n} X_i}{n}$), daily temperature pulse ($\mathrm{TP} = T_{\max} - T_{\min}$), rate of thermal change ($\mathrm{TC} = (T_{t2} - T_{t1})/\mathrm{hour}$), and degree-days ($\mathrm{DD} = \frac{\sum_{i=1}^{n} \overline{T_d}}{n}$).

Data were parsed into 6-week intervals in summer (15 July–31 August), autumn (1 October–15 November 15), winter (15 January–29 February) and spring (1 April–15 May).

Tolerance of selected biota to temperature exposure by habitat type

These metrics describing thermal exposure of biota were examined for departures from published thresholds delimiting optimal, stressful and lethal temperatures for various life stages of several commonly occurring

Table 13.1 Selected Nyack floodplain organisms and authorities for thermal thresholds. O = Optimum temperatures (°C), C_H = Critical high, C_L = Critical low, DD = degree-days

Organism	O	C_H	C_L	DD
Amphibian				
Anaxyrus boreas (Boreal toad)				
– tadpole (Beiswenger, 1978; Carey *et al.*, 2005)	26–30	38–40	4–10	1200–1800
– adult (Carey *et al.*, 2005; Lillywhite *et al.*, 1973)				
Fishes				
Salvelinus fontinalis (brook trout)				
– egg (Hokanson *et al.*, 1973; Humpesch, 1985; Baird *et al.*, 2002)	2–9	13–15	<1	220–672
– adult (Fry *et al.*, 1946; McCormick *et al.*, 1972; Lee and Rinne, 1980; Eaton and Scheller, 1996; Ott and Maret, 2003; McMahon *et al.*, 2007)	12–16	23–30	3	
Salvelinus confluentus (bull trout)				
– egg (Fraley and Shepard, 1989; McPhail and Murray, 1979; Gould, 1987)	2–6	10	1.2	350–1000
– juvenile to adult (McCormick *et al.*, 1972; Fraley and Shepard, 1989; Jakober *et al.*, 1998; Selong *et al.*, 2001; Gamett, 2002; Ott and Maret, 2003; McMahon *et al.*, 2007)	9–14	19–28	<1	
Oncorhynchus clarki lewisi (cutthroat trout)	13	13	4	
– egg (Hubert and Gern, 1995; Wagner *et al*,. 2006)	10–17	19–24	1–3	630–800
Insects				
Ameletus spp. Ephemeroptera (Pritchard and Zloty, 1994; Chilcote, 2004)	18–30*	nd	<0	250–900
Pteronarcella badia Plecoptera (Stanford, 1975)	7–13*	20?	<0	2300

*These species have specific temperature cues for egg hatching and emerge.

fish, amphibian and invertebrate species at Nyack (Table 13.1). Observing exposure to stressful or lethal high temperatures was straightforward, but the timing and variability of heat exposure within an annual cycle also strongly influences species-specific development (Sweeney and Vannote, 1978; Harper and Peckarsky, 2006; Ward and Stanford, 1982). Accordingly, we examined annual variation in degree-day (DD) accumulations for each habitat type in relation to the lifecycle of focal organisms for which DD criteria and hatching and emergence cues were explicit from rearing studies (references in Table 13.1). Winter thresholds were problematic because 0 °C is lethal to most riverine biota, but only if the water freezes solid and some insects can survive freezing for short periods. However, while river temperature may register at near zero for long periods of subzero air temperatures, the water column does not freeze owing to the kinetic energy of turbulent flow and warming from groundwater inflows. Thus, in well-oxygenated rivers with expansive alluvial aquifers, such as Nyack, biota will survive rigorous winter conditions but growth may be limited, especially for vertebrates. Moreover, most organisms are mobile and may simply move to warmer habitats, either in the river or in shallow groundwater, as water temperatures approach zero. Indeed, Chilcote (2004) observed overwinter survival of invertebrates and trout in ice-covered parafluvial ponds owing to groundwater flux.

Results

Temperature data analyses

Annual hydrologic and temperature variation

River flow and air temperatures followed seasonal patterns (Figure 13.2) typical of Rocky Mountain rivers with strong spring snowmelt hydrographs; maximum average daily air (22 ± 1 °C) and water (15 ± 1 °C) temperatures occurred in July and August, 5–6 weeks after the annual peak in river discharge. The river was at or near 0 °C during winter baseflows each year, and the water surface was partially or entirely frozen over during extended cold periods. Winter air temperatures across years in the dataset routinely fell below −10 °C for 4–8 weeks, typically falling below −20 °C at least once per year. Anchor ice formed during initial very cold periods in zones of unsaturated flow, owing to the water table elevations occurring below the river bottom and to super-cooled shoreline substratum, but no aquatic habitats froze into the bed sediments during the winter. Across years, river temperatures presented a stable annual wave form compared to the highly variable air temperature pattern (Figure 13.2). The 4-year river record indicated that a one year period (2003–04) was fairly typical of the long-term pattern for all habitats (Figure 13.2).

Seasonal temperature variation by habitat type

Air and soil temperatures in the riparian zone also followed seasonal patterns (Figures 13.3a–b, Table 13.2), but soil temperatures were moderated relative to air temperatures (i.e., summer soil temperatures at 50 cm depth were dampened by about 10 °C and did not fall much below zero in winter owing to snow cover). Soil and air maximum daily temperatures at both terrestrial sites typically were 25–30 °C from late June through August, with air temperatures above 35 °C rarely persisting for more than a few days. The daily temperature pulse or amplitude (TP) was most extreme in summer in terrestrial habitats, with TP values of 22.7 °C in air and 10.8 °C in soil (50 cm depth). In the colder months, minimum daily air temperatures frequently fell well below 0 °C in the autumn and winter, with temperatures below freezing for 63 days. During autumn, average minimum daily temperatures at soil sites remained above freezing (3.79 °C), but fell below freezing (−0.03 °C) in winter (Table 13.2). Similarly, temperature patterns in the aquatic habitats were strongly seasonal (Figure 3.3, Table 13.2). Ponds, tributaries, shorelines and backwaters were warmer in summer and fall than the main river channel owing to shallow water, longer water residence time and increased heating by solar radiation. Spring brooks were cooler in summer and warmer in winter relative to the river owing to buffering effect of the groundwater sources (aquifer discharges). Seasonal rates of thermal change (TC) for shallow shorelines, backwaters and tributaries varied from 1.12 to 1.64 °C/h, a much greater rate of change than observed for river and spring brook sites: 0.24 and 0.85 °C/h, respectively. Seasonal temperatures in the aquifer were considerably buffered in

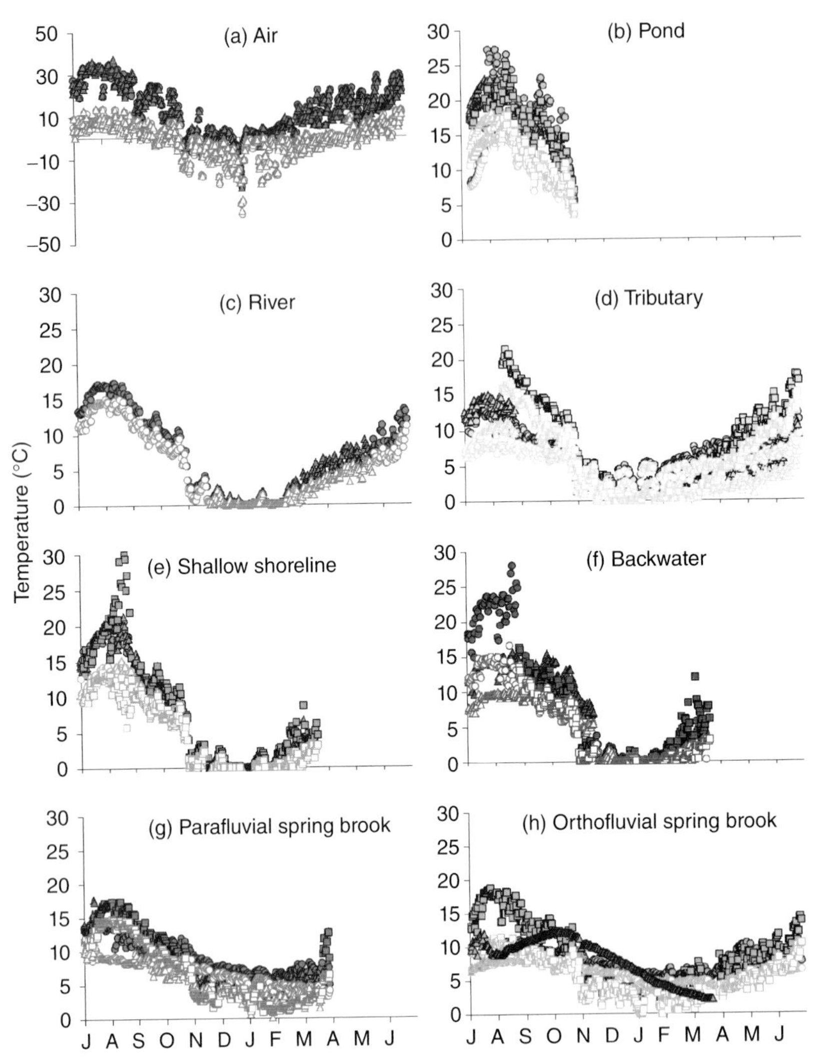

Figure 13.3 Annual temperature patterns (2003–04) for maximum (white symbols) and minimum (grey symbols) daily temperatures for Nyack floodplain sites. Sites are from different habitats as shown in Figure 1: (a) air ($n = 2$), (b) pond ($n = 3$), (c) river ($n = 2$), (d) tributary ($n = 3$), (e) shallow shoreline ($n = 3$), (f) backwater ($n = 3$), (g) parafluvial spring brook ($n = 3$), and (h) orthofluvial spring brook ($n = 3$). Separate sites within a habitat are represented by different symbols, the circle, triangle or square. *(See colour plate section for colour figure.)*

relation to surface waters (Table 13.2); the magnitude of groundwater thermal buffering was related to the residence time (time of contact with bed sediments) from the point of downwelling in the river to the position of the monitoring well (Helton *et al.*, 2013) . Temperature in short residence time (SRT) wells closely tracked the river with slight variation, whereas, temperatures in monitoring wells with medium and long residence times in the aquifer (MRT, LRT flow paths) were similar to the spring brooks that derive water from the longer flow paths (longer residence time) within the aquifer.

Table 13.2 Seasonal average daily temperatures ($\overline{T}_d$°C ± 1 SD) in floodplain aquatic habitats. RT indicates residence time, nd indicates no data available; P and O refer to parafluvial and orthofluvial areas of the flood plain.

Habitat	N	Summer		Autumn		Winter		Spring	
Air	2	15.55	(±4.40)	−1.28	(±7.15)	−2.46	(±6.76)	9.17	(±4.05)
Soil	2	17.82	(±4.54)	2.42	(±3.77)	0.04	(±0.41)	11.09	(±4.26)
River	1	13.35	(±1.98)	3.50	(±3.50)	1.48	(±1.43)	6.70	(±1.87)
Shallow shoreline	2–3	13.50	(±2.46)	2.89	(±3.52)	1.12	(±1.33)	nd	
Backwater	3	12.71	(±1.85)	3.42	(±3.59)	1.14	(±1.39)	nd	
Tributary	3	11.14	(±1.18)	4.20	(±2.87)	2.75	(±1.24)	7.00	(±1.77)
P. spring brook	3	11.58	(±1.35)	6.27	(±1.65)	4.24	(±1.00)	nd	
O. spring brook	2–3	10.24	(±0.45)	7.35	(±1.68)	4.07	(±0.51)	6.94	(±1.11)
Short RT well	2	13.26	(±2.02)	3.11	(±3.29)	1.50	(±1.3)	7.02	(±1.96)
Medium RT well	1	11.89	(±0.65)	6.86	(±1.87)	4.42	(±0.34)	6.52	(±1.87)
Long RT well	2	9.50	(±1.56)	9.67	(±1.53)	4.15	(±1.16)	3.81	(±0.98)
Ponds	3	15.72	(±2.38)	nd		nd		nd	

Diel temperature variation by habitat type

Temperatures of the surface waters exhibited greater diel variability during hot, mid-summer compared to cold, mid-winter periods; whereas, daily temperature pulse in all three groundwater sites was less than half a degree throughout the year (Figure 13.4). Summer river temperatures varied 3–4 °C daily, whereas the other habitats were more or less varied depending on water volume and groundwater influence (Figure 13.3). In winter, soils and all aquatic habitats had little or no diel temperature periodicity, remaining just above or at freezing (Figures 13.2 and 13.5).

Degree days by habitat type

Annual accumulation of degree-days (DD) varied between seasons and habitats (Figure 13.6). Annual accumulation of degree-days was substantially higher and accumulated more uniformly across seasons in spring brooks and the monitoring wells. Greater DD in groundwater-dominated habitats was related to long winter periods where temperatures remained well above freezing; whereas habitats with greater exposure to atmospheric heat exchange remained near freezing.

Species-specific thermal tolerances across habitat types

Exceedance of biotic thermal thresholds

Nyack aquatic habitats, even some ponds, are ideal habitat for trout because temperatures rarely exceeded 16 °C, and were generally optimal for egg incubation and hatching. Nonetheless, Table 13.3 demonstrates that suboptimal exposures did occur. This does not mean that trout were excluded from those habitats, but it is likely that thermal stress would stimulate migration to more favourable habitats, and consistently very cold winter habitats probably reduce growth of the fishes (Figure 13.7). Indeed, we have observed that invasive, nonnative brook trout tend to overwinter in groundwater influenced habitats, especially parafluvial ponds, because they are less tolerant of sustained cold periods compared to native cutthroat and bull trout. However,

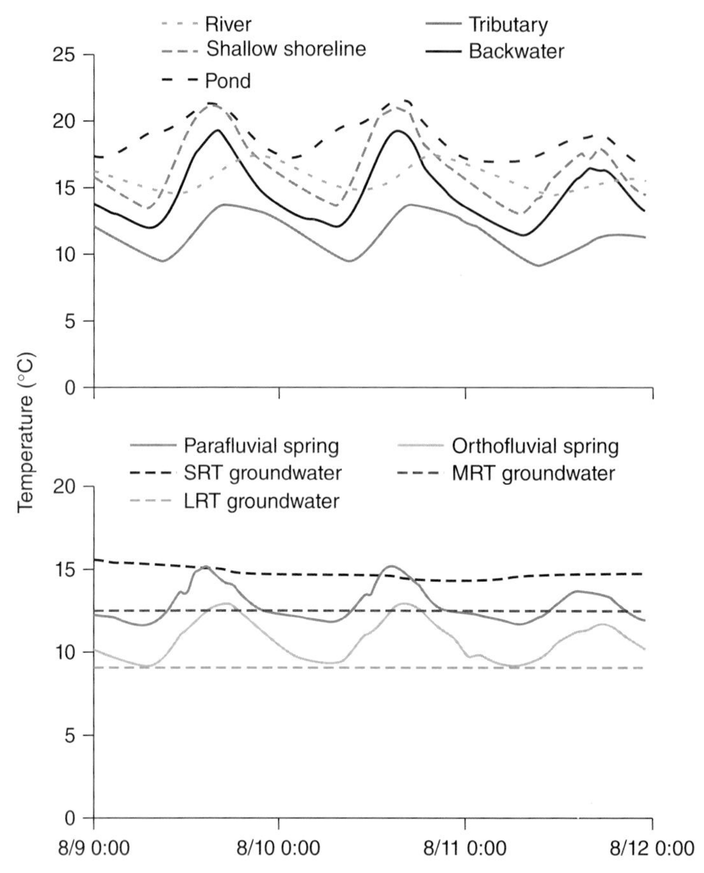

Figure 13.4 Average hourly temperatures in habitats of the Nyack floodplain during the hottest period of the year (10–12 August 2003). *(See colour plate section for colour figure.)*

on the hottest day of the year, river and shallow shoreline sites only exceeded the optimal trout growth threshold for 5–6 h; other aquatic habitats remained within in the optimal range for growth, although the spring brooks were at the lower end of the thermal preference (Figure 13.7). On the coldest winter day and for most of the mid-winter period (Figures 13.2 and 13.5), only the spring brooks consistently maintained temperatures throughout the day above values thought to allow trout growth (Figure 13.7). The other focal vertebrate, the boreal toad (*Anaxyrus boreas*, sometimes called western toad), was considerably habitat constrained. Boreal toads spawn and develop only in warm backwaters and ponds; all other habitats consistently exceeded critical low thresholds (Table 13.3). Chilcote (2004) demonstrated that boreal toads spawn only in the warmest parafluvial ponds with sand–silt substratum and minimal groundwater flux (warmest pattern for ponds in Figure 13.3).

The two aquatic insects for which we had published temperature criteria (Table 13.1) could survive in all habitats. However, the stonefly *Pternarcella badia* was not found in ponds, being more stenothermic than the mayfly in the genus *Ameletus* that were abundant in all habitats in the Nyack system, perhaps because more than one

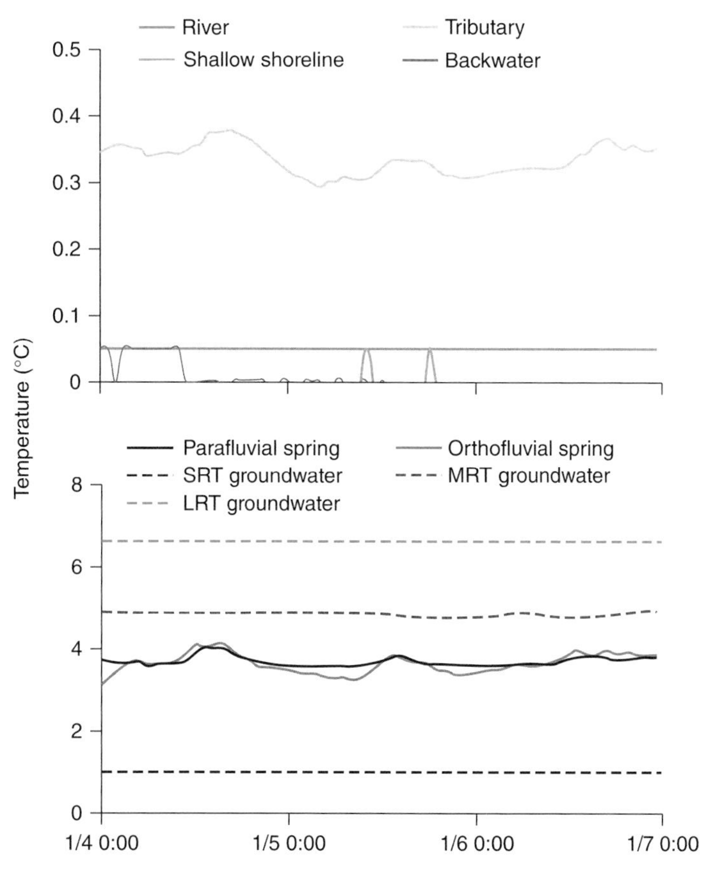

Figure 13.5 Average hourly temperatures in habitats of the Nyack floodplain during the coldest period of the year (4–7 January 2004). *(See colour plate section for colour figure.)*

Ameletus species were present. Moreover, invertebrate distributions were segregated by hatching and emergence cues linked to degree-days for growth as presented next.

Degree-day (DD) thresholds

We plotted DD accumulations for surface water habitats in relation to published DD thresholds required for focal species to hatch from eggs and grow to the adult stage (Figure 13.8). The cumulative curves vary in relation to the season that the calculation was initiated and temperature thresholds for focal biota. For example, cutthroat trout deposit eggs in the spring; whereas bull and brook trout spawn in the fall. However, in both instances, the top two panels in Figure 13.8 indicate that required DD were attained in all habitats except ponds, which are not shown because we did not have winter data. However, Figure 13.3 demonstrates that ponds warm quickly in summer and are inhabited by species such as boreal toads and a wide variety of insects, especially Ephemeroptera and Diptera that may produce a generation in a matter of days. Chilcote (2004) showed that some ponds remain ice free through the winter owing to continuous groundwater flux. Therefore, fall spawning trout, notably brook trout, survive and reproduce in ponds if they are colonised by adult trout during spates when

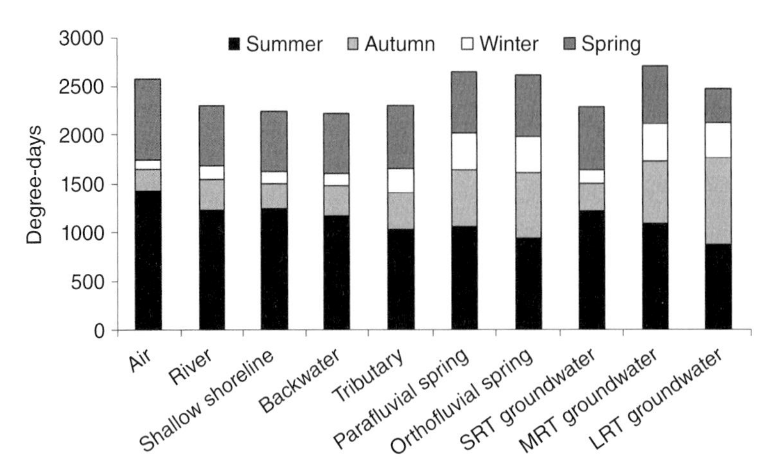

Figure 13.6 The seasonal accumulation of degree-days in aquatic habitat sites on the Nyack floodplain over a 1-year period (2003–04). Habitats as in Figure 13.1. Spring season degree-day data for shallow shorelines and backwaters were unavailable due to flooding by the main river channel, and thus were approximated using the river data.

ponds are connected to the channel network (parafluvial ponds typically become isolated except for groundwater flux at baseflows). The river contains a broad array of insect species, such as *Pteronarcella badia*, that are univoltine (one year lifecycle) – emerging as adults and ovipositing eggs in spring or early summer, with nymphs hatching immediately and growing throughout the fall and winter for 2200–2300 DD (Figure 13.8). This DD threshold was attained in all habitats, but Stanford (1975) showed that *P. badia*, like most aquatic insects (Ward and Stanford 1982), has a precise temperature cue of 12 °C for emergence. They will not emerge unless the cue occurs after the DD threshold has been attained. Thus their lifecycle is described by the summer DD pattern, and while they occasionally occur in other habitats, they are only abundant in the river channel.

Discussion

For the floodplain as a whole, temperature variation among aquatic habitats in summer was quite extreme with some ponds reaching 28 °C, while on the same day, points within the aquifer and some orthofluvial spring brooks temperatures were only 5–10 °C (Figures 13.3 and 13.4). Thus a floodplain temperature gradient of at least 20 °C exists on hot summer days at Nyack for aquatic habitats. Surface temperatures of bare gravel bars may reach 50 or 60 °C with direct sun exposure and at the highest air temperatures, whereas shaded soil temperatures were < 15 °C on the same days; so the terrestrial gradient is far greater than the aquatic gradient but both are highly deterministic for distribution of biota.

The profound floodplain temperature gradient is created by the spatial and temporal variation in processes that buffer the influence of solar insolation and air temperature on habitat-specific water temperatures. Groundwater flux and shading by riparian vegetation are key processes, in addition to ambient air temperatures, that regulate water temperature. Even though the spatial position of the various habitats types may

Table 13.3 Seasonal thermal constraints on selected aquatic vertebrates in Nyack floodplain habitats. O = optimal temperatures, S_L = low suboptimal temperatures, S_H = high suboptimal temperatures, C_L = temperatures at or below a low critical threshold, C_H = temperatures at or above a high critical threshold, determined by comparing $\overline{T}_d$ patterns in habitats to constraints given in references listed in Table 13.1. SU = summer, A = autumn, W = winter, SP = spring.

	Anaxyrus boreas	*Salvelinus fontinalis*		*Salvelinus confluentus*		*Oncorhynchus clarki lewisi*		*Cottus* spp.
Life stage **Growth period**	**Tadpole** **2–3 mo** **SP, SU, A**	**Egg** **2–5 mo** **A, W**	**Adult** **2–4 yr** **All year**	**Egg** **2–3 mo** **A, W**	**Juvenile–adult** **3–5 yr** **All year**	**Egg** **SP, SU**	**Adult** **SP, SU, A, W**	**Adult** **SP, SU, A, W** **<1 °C** **9–14 °C** **25–26 °C**
river	S_L (SU) C_L (A, SP)	O (A) S_L (W)	O (SU) S_L (A, SP) C_L (W)	O (A) S_L (W)	O (SU) S_L (A, W, SP)	S_H (SP) C_H (SU)	O (SU) S_L (A, SP) C_L (W)	O (SU) S_L (A, W, SP)
shallow shoreline	S_L (SU) C_L (A)	O (A) S_L (W)	O (SU) C_L (A, W)	O (A) C_L (W)	S_H (SU) S_L (A, W)	C_H (SU)	O (SU) C_L (A, W)	O (SU) S_L (A, W)
backwater	S_L (SU) C_L (A)	O (A) S_L (W)	O (SU) S_L (A) C_L (W)	O (A) C_L (W)	O (SU) S_L (A, W)	C_H (SU)	O (SU) C_L (A,W)	O (SU) S_L (A, W)
tributary	S_L (SU) C_L (A, SP)	O (A, W)	S_L (SU, A, SP) C_L (W)	O (A,W)	O (SU) S_L (A, W, SP)	S_H (SP, SU)	O (SU) S_L (A, SP) C_L (W)	O (SU) S_L (A, W, SP)
parafluvial spring brook	S_L (SU) C_L (A)	O (A, W)	O (SU) S_L (A, W)	O (A,W)	O (SU) S_L (A,W)	S_H (SU)	O (SU) S_L (A, W)	O (SU) S_L (A, W)
orthofluvial spring brook	C_L (SP, SU, A)	O (A, W)	S_L (SU, A, W, SP)	O (W) S_H (A)	O (SU) S_L (A, W, SP)	S_H (SP, SU)	O (SU) S_L (A, W, SP)	O (SU) S_L (A, W, SP)
ponds	O (SU) C_L (A, W, SP)	O (A)	O (A, W) S_H (SU)	nd	S_H (SU)	C_H (SU)	O (SU)	S_H (SU)

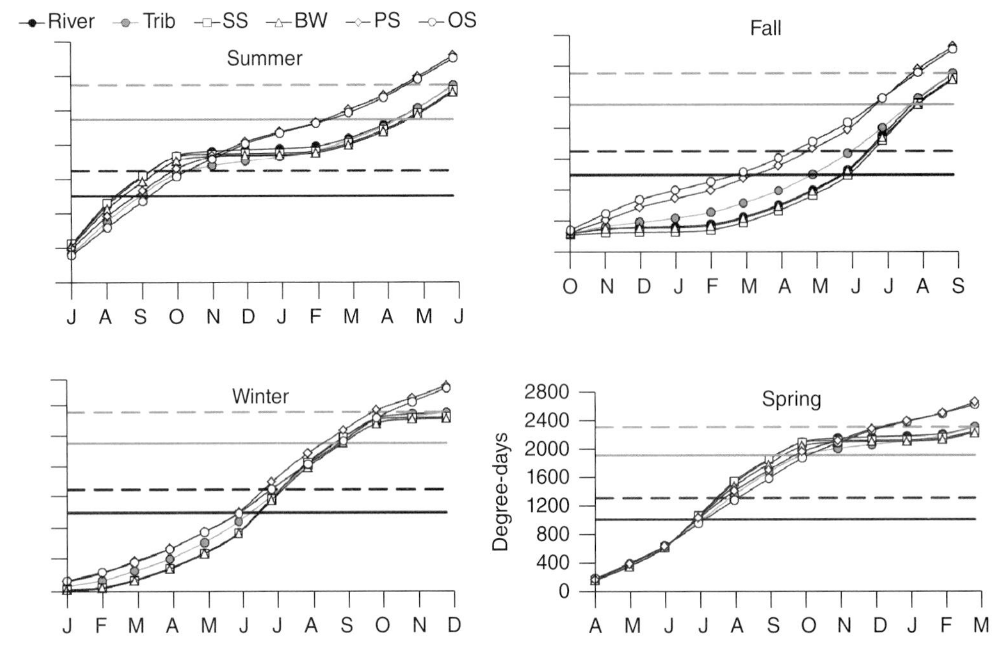

Figure 13.7 Cumulative degree-days for aquatic habitat sites at Nyack floodplain over a 1-year period (2003–04). Separate panels indicate different seasons as a starting point for degree-day calculation. Horizontal lines indicate significant thermal thresholds for development of Nyack floodplain aquatic organisms: eggs of all trout species hatch by 1000 dd (solid black), *Pisidium* clam embryos develop by 1300 dd (dashed black), *Anaxyrus boreas* tadpoles and aquatic insects with rapid development such as *Ametetus, Neophylax*, Nemouridae and some Baetidae, and Heptagenaiidae emerge by 1900 DD (solid grey), and aquatic insects that emerge by 2300 DD such as *Pteronarcella, Taenionema* and other Baetidae and Heptagenaiidae (dashed grey).

shift around over time in relation to cut and fill alluviation by flooding (Stanford *et al.*, 2005), habitats are sufficiently hydrologically interconnected above and below ground to allow species to find their preferred thermal environments, unless climate or anthropogenic alterations interfere with maintenance of the SHM. Channel straightening or inundation of floodplains by dams results in tremendous loss of habitat from a thermal perspective. Climate warming is also likely to be gradually altering mean winter and summer temperatures, which will have cascading effects such as changing oxygen saturation, which in turn potentially can induce much larger or prolonged areas of oxygen depletion in the aquifer and thereby limiting biota.

The key point is that thermal heterogeneity among habitats promotes adaptation to particular temperature patterns by particular species in order to maintain a positive life history energy balance of the population (Hall *et al.*, 1992). Indeed, for ectotherms the heat budget (DD) of the habitat, coupled with evolved hatching and emergence cues, synchronises growth and maturation of the population (e.g., Harper and Peckarsky, 2006) and allows many species to coexist within habitats, which maximises biodiversity at the floodplain scale. Moreover, many aquatic insects, such as winter stoneflies, grow readily at or near 0 °C (e.g., Cather and Gaufin, 1976), whereas fish and amphibians cannot (e.g., Figure 13.7). Thus, thermally driven adaptations to maximise fitness are

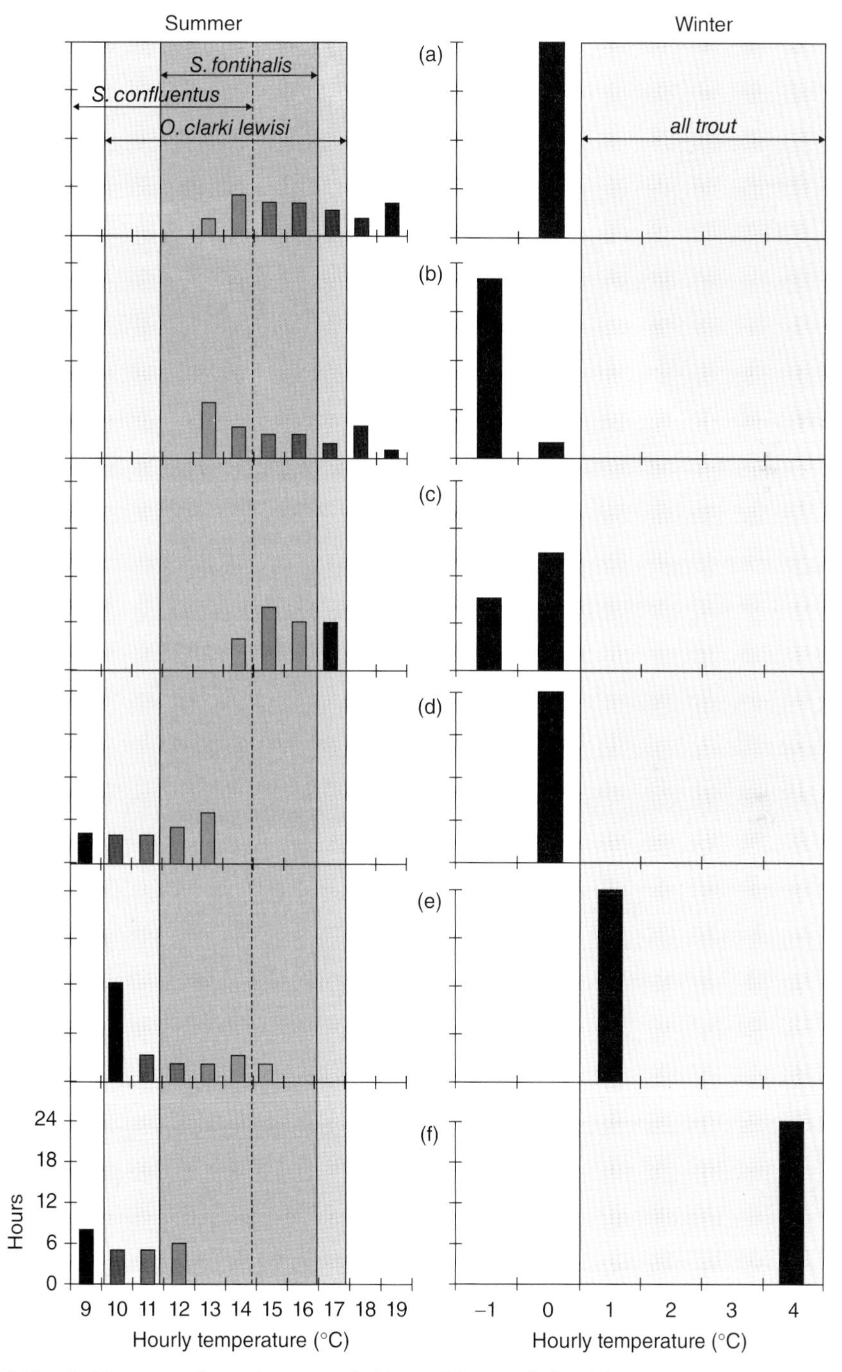

Figure 13.8 Hourly histograms for surface water habitats of the Nyack floodplain over a 24-hour period in summer and winter. Habitats are as follow: (a) river, (b) shallow shoreline, (c) backwater, (d) tributary, (e) parafluvial spring brook and (f) orthofluvial spring brook. Shaded areas with arrows indicate either optimal (summer) or critical low (winter) thermal ranges for adult trout activity and growth.

expressed by each species in the floodplain habitat catena. While in general all habitats including the aquifer have a heat budget that allows for most invertebrate lifecycles, hatching and emergence cues coupled with DD thresholds sort out thermal niches among the habitat types; whereas vertebrates can optimise metabolism by moving among habitats with different temperature patterns.

Aquatic insects are notably abundant (at least 350 species) at Nyack (our group, *unpubl. data*) owing in large part to adaptations to thermal heterogeneity. Lifecycle periodicity varies from a few days in some warm adapted species (e.g., *Ameletus*) to two years (e.g., the caddisfly *Archtopsyche grandis*; Hauer and Stanford, 1982) or more (e.g., the stonefly *Pteronarcys californica*; Stanford, 1975). Insect species sequentially emerge as flying or crepuscular adults from February to October annually, each occupying a specific thermal niche, which maintains a wide diversity of larval size classes between species in the river at any one point in time, thereby reducing competition for food and cover. In this regard, the cold and warm adapted species are segregated on the floodplain, somewhat similar to distributions along altitudinal gradients from cold, high mountains to warmer lowlands (e.g., Ward, 1986; Hauer *et al.*, 2000).

Maximum annual mean temperatures in the river were only slightly above the published optimal temperature ranges for the fishes that we examined. Even on hot summer days, all aquatic habitats of the floodplain provided some hours within the optimal thermal range for Nyack fishes. But fish are especially mobile and will find optimal temperatures if more suitable floodplain habitats are accessible (e.g., not blocked by beaver dams or human revetments). Indeed, as noted above, Armstrong and Schindler (2013) showed that juvenile coho salmon forage on eggs of sockeye salmon that spawn in cold groundwater discharge areas of spring brooks, and after foraging, the coho move to warmer habitats for digestion and growth. Likewise, after foraging widely on benthic algae and detritus in ponds, tadpoles of boreal toads congregate at the warmest margins of ponds presumably to maximise growth (Chilcote, 2004). One might therefore expect substantial inter- and intraspecific competition for optimal thermal habitats, suggesting a more robust understanding of specific adaptations to thermal diversity is needed, particularly among shallow waters where temperatures can change substantially over a diel cycle (e.g., Figure 13.4) and in relation to solar insolation, the colour of bed substratum, and water flux (Dale and Gillespie, 1977).

Winter conditions at Nyack present significant challenges for all floodplain biota owing to extremely cold air temperatures and icing for sustained periods (Figure 13.3). Terrestrial insects, boreal toads and other amphibians move or burrow deep into the soils where the aquifer maintains temperatures above freezing. Likewise, the hyporheic zone is a refugium for aquatic biota (Boulton *et al.*, 2010), although the aquifer community is distinctly different than in the river benthos (Stanford *et al.*, 1994). Average minimum soil temperatures in autumn and winter were potentially dangerous to adult toads attempting to overwinter in shallow burrows; but virtually nothing is known about how floodplain amphibians overwinter. Juvenile fishes are very abundant in winter-warm spring brooks at Nyack, suggesting immigration to avoid very cold temperatures in the river, as has been shown, for coho salmon, in Alaskan floodplain rivers (Malison *et al.*, 2014).

Given that temperature exerts a substantial selective force on ectotherms by placing constraints on metabolic efficiency, specific life history traits or thresholds evolve. The Thermal Equilibrium Hypothesis (Vannote and Sweeney, 1980) suggests that stability and life stage synchrony of aquatic insect populations that compose geographical distributions of each species is a dynamic balance of thermal, metabolic and phenologic (life history) constraints. Our data support that view. In addition within the thermal optimum range, a tradeoff exists between increased metabolic activity and increased mortality at higher temperatures (Lillywhite *et al.*, 1973; Dwyer and Kramer, 1975; Selong *et al.*, 2001; Cabanita and Atkinson, 2006). Growth and reproduction of populations may be desynchronised in aquatic ecosystems with minimal thermal variation, such as glacier-fed streams that remain continuously very cold (Knispel and Satori, 2006) and cave streams where temperature variation is buffered by phreatic insulation (López-Rodríguez and Tierno de Figueroa, 2013). Likewise at Nyack, temperatures are buffered (winter warm, summer cool – Figure 13.4) in the alluvial aquifer and, therefore, while mean annual temperature in the aquifer may be similar to many surficial habitats, aquifer DD accumulation may be 2–3 times greater and with little thermal amplitude (fewer cues). This causes desynchronisation and morphological variation in the lifecycles of hyporheic stoneflies (Stanford *et al.*, 1994).

The organisms we assessed were occasionally exposed to suboptimal thermal conditions in all of the habitats. But lethal conditions for most focal species occurred only in parafluvial ponds that became very warm in summer. Likewise, warm water species found in warm ponds were not present in other habitats (our group, *unpubl. data*). Biota with a wide distribution among habitats are expected to have broad temperature tolerances (e.g., *Ameletus*), or are able to utilise behavioural thermoregulation within habitats, or are very mobile and able to move from habitat to habitat as thermal stress is encountered. Less mobile organisms are expected to assemble in the most thermally favourable habitats, which may drive significant differences in community assemblages and food web structure among habitat types. Temperature can further modify food web structure by influencing the cycling of organic matter (Anderson and Sedell, 1979), feeding activity of mobile predators (Kishi *et al.*, 2005), and the prevalence of parasites in the ecosystem (Heinonen *et al.*, 1999). Species should persist from year to year in specific habitat types, unless a particular habitat is modified by flooding or humans in ways that causes temperature patterns to change.

We conclude that temperature, along with flooding and associated materials fluxes and plant succession, is a master variable that defines the SHM of the Nyack and other expansive alluvial floodplains. Extreme environmental temperature variation and responsive adaptations to optimise life history energy balance allows many ectothermic species to coexist, and explains why floodplains like Nyack tend to be 'hot spots' of biodiversity within regional landscapes.

Acknowledgements

This paper is dedicated to our long-time colleague Professor Dr. Geoff Petts who has conducted a notably distinguished career in river ecology contributing many studies similar to this one. All authors contributed to the study design. MLA, BLR, TSB and

SDC collected the data. JAS and MLA analysed the data and wrote the paper. We thank Dan Fagre of the United States Geological Survey for providing weather stations at Nyack. This work was supported by the USA National Science Foundation Grant No. EAR-0120523 ('Biocomplexity in the Environment – Dynamic Controls on Emergent Properties of River Flood Plains), but opinions, findings and conclusions or recommendations herein do not necessarily reflect the views of the National Science Foundation.

References

Anderson, M. P. 2005. Heat as a ground water tracer. *Ground Water* 43: 951–968.

Anderson, N. H. and J. R. Sedell. 1979. Detritus processing by macroinvertebrates in stream ecosystems. *Annual Review of Entomology* 24: 351–377.

Armstrong, J. B. and D. E. Schindler. 2013. Going with the flow: spatial distributions of juvenile coho salmon track an annually shifting mosaic of water temperature. *Ecosystems* 16: 1429–1441.

Arrigoni, A. S., G. C. Poole, L. A. K. Mertes, *et al.* 2008. Buffered, lagged or cooled? Disentangling hyporheic influences on temperature cycles in stream channels. *Water Resources Research* 44: 1–13.

Arscott, D. B., K. Tockner and J. V. Ward. 2001. Thermal heterogeneity along a braided floodplain river. *Canadian Journal of Fisheries and Aquatic Sciences* 58: 2359–2373.

Baird, H. B., C. C. Krueger and D. C. Josephson. 2002. Differences in incubation period and survival of embryos among brook trout strains. *North American Journal of Aquaculture* 64: 233–241.

Beiswenger, R. E. 1978. Responses of bufo tadpoles (Amphibia, Anura, Bufonidae) to laboratory gradients of temperature. *Journal of Herpetology* 12: 499–504.

Boulton, A. J., T. Daltry, T. Kasahara, *et al.* 2010. Ecology and management of the hyporheic zone: stream-groundwater interactions of running waters and their floodplains. *Journal of the North American Benthological Society* 29: 26–40.

Cabanita, R. and D. Atkinson. 2006. Seasonal time constraints do not explain exceptions to the temperature size rule in ectotherms. *Oikos* 114: 431–440.

Carey, C., P. S. Corn, M. S. Jones, *et al.* 2005. Factors limiting the recovery of boreal toads (*Bufo boreas*). In: M. Lanoo (ed.) *Amphibian Declines: The Conservation Status of United States Species*. University of California Press: Berkeley, California, pp. 222–236.

Cather, M. R. and A. R. Gaufin. 1976. Comparative ecology of three *Zapada* species of Mill Creek, Wasatch Mountains, Utah (Plecoptera: Nemouridae). *American Midland Naturalist* 95: 464–471.

Chilcote, S. D. 2004. *The ecology of parafluvial ponds on a Rock Mountain floodplain*. Ph.D. Dissertation. University of Montana. Missoula, Montana, USA.

Cummins, K. W. 1974. Structure and function of stream ecosystems. *BioScience* 24: 631–641.

Dale, H. M. and T. Gillespie. 1977. Diurnal fluctuations of temperature near the bottom of shallow water bodies as affected by solar radiation, bottom color and water circulation. *Hydrobiologia* 55: 87–92.

Dwyer, W. P. and R. H. Kramer. 1975. The influence of temperature on scope for activity in Cutthroat Trout, *Salmo clarki*. *Transactions of the American Fisheries Society* 104: 552–554.

Eaton, J. G. and R. M. Scheller. 1996. Effects of climate warming on fish thermal habitat in streams of the United States. *Limnology and Oceanography* 41:1109–1115.

Fraley, J. J. and B. B. Shepard. 1989. Life history, ecology and population status of migratory bull trout (*Salvelinus confluentus*) in the Flathead Lake and river system, Montana. *Northwest Science* 63: 133–143.

Fry, F. E. J., J. S. Hart and K. F. Walker. 1946. Lethal temperature relations for a sample young speckled trout, *Salvelinus fontinalis*. Pbl. Ont. Fish. Res. Lab. No. 66; University of Toronto Student Biology Serial. No. 54, University of Toronto Press.

Gamett, B. L. 2002. The relationship between water temperature and bull trout distribution and abundance. Masters Thesis. University of Utah, Logan, Utah.

Gould, W. R. 1987. Features in the early development of bull trout (*Salvelinus confluentus*). *Northwest Science* 61: 264–268.

Hall, C. A. S., J. A. Stanford and F. R. Hauer. 1992. The distributional abundance of organisms as a consequence of energy balances along multiple environmental gradients. *Oikos* 65: 377–390.

Harper, M. P. and B. L. Peckarsky. 2006. Emergence cues of a mayfly in a high-altitude stream ecosystem: potential response to climate change. *Ecological Applications* 16: 612–621.

Hauer, F. R. and J. A. Stanford. 1982. Ecological responses of hydropsychid caddisflies to stream regulation. *Canadian Journal of Fisheries and Aquatic Sciences* 39: 1235–1242.

Hauer, F. R., J. A. Stanford, J. J. Giersch and W. H. Lowe. 2000. Distribution and abundance patterns of macroinvertebrates in a mountain stream: an analysis along multiple environmental gradients. *Verhein Internationale Vereinigung Limnologie* 27: 1485–1488.

Heinonen, J., J. V. K. Kukkonen and I. J. Holopainen. 1999. The effects of parasites and temperatures on the accumulation of xenobiotics in a freshwater clam. *Ecological Applications* 9: 475–481.

Helton, A. M., G. C. Poole, R. A. Payn, *et al.* 2014. Relative influences of the river channel, floodplain surface, and alluvial aquifer on simulated hydrologic residence time in a montane river floodplain. *Geomorphology* 205: 17–26.

Helton, A. M., G. C. Poole, R. A. Payn, *et al.* 2013. Scaling flow path processes to fluvial landscapes: An integrated field and model assessment of temperature and dissolved oxygen dynamics in a river-floodplain-aquifer system. *Journal of Geophysical Research – Biogeosciences* 117: G00N114.

Hoehn, E. and O. A. Cirpka. 2006. Assessing residence times of hyporheic groundwater in two alluvial floodplains of the Southern Alps using water temperature and tracers. *Hydrology and Earth System Services* 10: 553–563.

Hokanson, K. E. F., J. H. McCormick, B. R. Jones and J. H. Tucker. 1973. Thermal requirements for maturation, spawning, and embryo survival of brook trout, *Salvelinus fontinalis*. *Journal of the Fisheries Research Board of Canada* 30: 975–984.

Hubert, W. A. and W. A. Gern. 1995. Influence of embryonic stage on survival of cutthroat trout exposed to temperature reduction. *Progressive Fish-Culturist* 57: 326–328.

Humpesch, U. H. 1985. Inter- and intra-specific variation in hatching success and embryonic development of five species of salmonids and *Thymallus thymallus*. *Archives für Hydrobiologie* 104:129–144.

Jakober, M. J., T. E. McMahon, R. F. Thurow and C. G. Clancy. 1998. Role of stream ice on fall and winter movements and habitat use by bull trout and cutthroat trout in Montana headwater streams. *Transactions of the American Fisheries Society* 127: 223–235.

Johnson, A. N., B. R. Boer, W. W. Woessner, *et al.* 2005. Evaluation of an inexpensive small-diameter temperature logger for documenting ground water-river water interactions. *Ground Water Monitoring and Remediation* 25: 68–74.

Kishi, D., M. Murakami, S. Nakano and K. Maekawa. 2005. Water temperature determines strength of top-down control in a stream food-web. *Freshwater Biology* 50: 1315–1322.

Knispel, S. and M. Sartori. 2006. Egg development in the mayflies of a Swiss glacial floodplain. *Journal of the North American Benthological Society* 25: 430–443.

Latterell, J. J., J. S. Bechtold, T. C. O'Keefe, *et al.* 2006. Dynamic patch mosaics and channel movement in an unconfined river valley of the Olympic Mountains. *Freshwater Biology* 51: 523–544.

Lee, R. M. and J. N. Rinne. 1980. Critical thermal maxima of five trout species in the southwestern United States. *Transactions of the American Fisheries Society* 109: 632–635.

Lillywhite, H. B., P. Licht and P. Chelgren. 1973. The role of behavioral thermoregulation in the growth energetics of the toad *Bufo boreas*. *Ecology* 54: 375–383.

López-Rodríguezl, M. J. and J. M. Tierno de Figueroa. 2013. Life in the dark: on the biology of the cavernicolous stonefly (*Protonemura gevi* Insecta, Plecoptera). *The American Naturalist* 180: 684–691.

Malard, F., A. Mangin, U. Uehlinger and J. V. Ward. 2001. Thermal heterogeneity in the hyporheic zone of a glacial floodplain. *Canadian Journal of Fisheries and Aquatic Sciences* 58: 1319–1335.

Malison, R. L., M. S. Lorang, D. C. Whited and J. A. Stanford. 2014. Beavers (*Castor canadensis*) influence habitat for juvenile salmon in a large Alaskan river floodplain. *Freshwater Biology* 59:1229–1246.

McCormick, J. H., K. E. Hokanson and B. R. Jones. 1972. Effects of temperature on growth and survival of young brook trout, *Salvelinus fontinalis*. *Journal of the Fisheries Research Board of Canada* 29: 1107–1112.

McMahon, T. E., A. V. Zale, F. T. Barrows, *et al.* 2007. Temperature and competition between bull trout and brook trout: a test of the elevation refuge hypothesis. *Transactions of the American Fisheries Society* 136: 1313–1326.

McPhail, J. D. and C. B. Murray. 1979. The early life history and ecology of Dolly Varden (*Salvelinus malma*) in the upper Arrow Lakes. Report to B. C. Hydro and Ministry of Environment, Fisheries Branch, Nelson, British Columbia.

Mertes, L. A. K. 1997. Documentation and significance of the perirheic zone on inundated floodplains. *Water Resources Research* 33: 1749–1762.

Mouw, J. E. B., J. A. Stanford and P. B. Alaback. 2009. Influences of flooding and hyporheic exchange on floodplain plant richness and productivity. *River Research and Applications* 25: 929–945.

Olden, J. D. and R. J. Naiman. 2010. Incorporating thermal regimes into environmental flows assessments: modifying dam operations to restore freshwater ecosystem integrity. *Freshwater Biology* 55: 86–107.

Ott, D. S. and T. R. Maret. 2003. Aquatic assemblages and their relation to temperature variables of least-disturbed streams in the Salmon River basin, Idaho, 2001. U. S. Geologic Survey, Water-Resources Investigative Report 03-4076.

Poole, G. C. and C. H. Berman. 2001. An ecological perspective on in-stream temperature: natural heat dynamics and mechanisms of human-caused thermal degradation. *Environmental Management* 27: 787–802.

Poole, G. C., J. A. Stanford, C. A. Frissell and S. W. Running. 2002. Three-dimensional mapping of geomorphic controls on flood-plain hydrology and connectivity from aerial photos. *Geomorphology* 48: 329–347.

Poole, G. C., J. A. Stanford, S. W. Running and C. A. Frissell. 2006. Multiscale geomorphic drivers of groundwater flow paths: subsurface hydrologic dynamics and hyporheic habitat diversity. *Journal of the North American Benthological Society* 25: 288–303.

Pritchard, G. and J. Zloty. 1994. Life histories of two *Ameletus* mayflies (Ephemeroptera) in two mountain streams: the influence of temperature, body size and parasitism. *Journal of the North American Benthological Society* 13: 557–568.

Selong, J. H., T. E. McMahon, A. V. Zale and F. T. Barrows. 2001. Effects of temperature on growth and survival of bull trout, with application of an improved method for determining thermal tolerance in fishes. *Transactions of the American Fisheries Society* 130: 1026–1037.

Stanford, J. A. 1975. *Ecological studies of Plecoptera in the upper Flathead and Tobacco Rivers, Montana*. Ph.D. Dissertation. University of Utah.

Stanford, J. A., M. S. Lorang and F. R. Hauer. 2005. The shifting habitat mosaic of river ecosystems. *Verhandlungen Internationale Vereinigung fur Theoretische und Angewandte Limnologie* 29: 123–136.

Stanford, J. A., J. V. Ward and B. K. Ellis. 1994. Ecology of the alluvial aquifers of the Flathead River, Montana. In: J. Gibert, D. L. Danielopol, J. A. Stanford (eds) *Groundwater Ecology*. Academic Press: San Diego, USA, pp. 367–390.

Sweeney, B. W. and R. L. Vannote. 1978. Size variation and the distribution of hemimetabolous aquatic insects: two thermal equilibrium hypotheses. *Science* 200: 444–446.

Tockner, K., F. Malard and J. V. Ward. 2000. An extension of the flood pulse concept. *Hydrological Processes* 14: 2861–2883.

Vannote, R. L. and B. W. Sweeney. 1980. Geographic analysis of thermal equilibria: a conceptual model for evaluating the effect of natural and modified thermal regimes on aquatic insect communities. *The American Naturalist* 115: 667–695.

Wagner, E. J., R. E. Arndt and R. Roubidoux. 2006. The effect of temperature changes and transport on cutthroat trout eggs after fertilization. *North American Journal of Aquaculture* 68: 235–239.

Ward, J. V. 1986. Altitudinal zonation in a Rocky Mountain stream. *Archiv für Hydrobiologie. Supplementband. Monographische Beiträge A* 74: 133–199.

Ward, J. V. and J. A. Stanford. 1982. Thermal response in the evolutionary ecology of aquatic insects. *Annual Review of Entomology* 27: 97–117.

Ward, J. V., K. Tockner, D. B. Arscott and C. Claret. 2002. Riverine landscape diversity. *Freshwater Biology* 47: 517–539.

Whited, D. C., M. S. Lorang, M. J. Harner, *et al.* 2007. Hydrologic disturbance and succession: Drivers of floodplain pattern. *Ecology* 88: 940–953.

CHAPTER 14

Microthermal variability in a Welsh upland stream

Laura Gangi[1,2,3], David M. Hannah[1] and Markus Weiler[2]

[1] *School of Geography, Earth and Environmental Sciences, University of Birmingham, Birmingham, UK*
[2] *Chair of Hydrology, Faculty of Environment and Natural Resources, University of Freiburg, Freiburg, Germany*
[3] *Forschungszentrum Juelich GmbH, Juelich, Germany*

Introduction

Stream temperature can be spatially heterogeneous at a range of scales from the (sub-) reach to the catchment (Carrivick *et al.*, 2012; Imholt *et al.*, 2013) and has significant ecological implications (Petts, 2000). Aquatic organisms depend on suitable habitat conditions, including a defined water temperature range (e.g., Hynes, 1970; Coutant, 1977). Various studies have highlighted the importance of water temperature distribution within channels for providing thermal refugia for freshwater fish (Peterson and Rabeni, 1996; Torgersen *et al.*, 1999; Ebersole *et al.*, 2001). The arrival of low-cost temperature sensors has facilitated the accurate and reliable monitoring of water temperature at multiple sites over long time periods (Webb *et al.*, 2008). This technology has permitted river temperature to be observed at different spatial scales: longitudinal and vertical profiles (Evans and Petts, 1997; Hannah *et al.*, 2009; Krause *et al.*, 2011) as well as wider reach-scale variations (Hawkins *et al.*, 1997; Ebersole *et al.*, 2003; Brown and Hannah, 2008). Furthermore, temperature studies at the river sector to reach-scale have been advanced over the last decade by the use of satellite-based remote thermal sensing (e.g., Torgersen *et al.*, 2001; Madej *et al.*, 2006; Cristea and Burges, 2009) and fibre-optic distributed temperature sensing (e.g., Krause *et al.*, 2012; Selker *et al.*, 2006). However, there remains a paucity of research on microthermal gradients, including lateral or vertical water temperature contrasts within channels over distances of centimetres to metres, and results are often inconsistent between different studies. Clark *et al.* (1999) examined the microthermal heterogeneity in small groundwater-dominated streams in Dorset, UK, and detected lateral temperature differences of up to 7 °C between the channel margin and the main body of flow, which were assumed to be of ecological significance. In contrast, Rutherford *et al.* (1993) demonstrated that temperature within the lower Waikato River, New Zealand, was uniform transversely (<0.03 °C); and Carling *et al.* (1994) reported that temperature was only up to 2 °C higher in dead zones of bends in parts of the River Severn near

River Science: Research and Management for the 21st Century, First Edition.
Edited by David J. Gilvear, Malcolm T. Greenwood, Martin C. Thoms and Paul J. Wood.

Shrewsbury, central England. Recently, ground-based infrared (IR) imagery was applied to investigate small-scale temperature variability in fluvial systems in a spatially continuous way (Cardenas *et al.*, 2008). In this study, it was reported that thermal heterogeneity during low stages was associated with the occurrence of biological and morphological in-stream structures and was captured well via a hand-held IR camera. Ground-based IR thermography was applied also to detect and quantify groundwater influence at the stream bed based on the temperature difference between stream water and groundwater (Schuetz and Weiler, 2011). We are unaware of other studies involving ground-based IR imagery of river channels; thus the potential of this method to detect local temperature patterns within smaller streams with broad coverage has still to be explored.

This study addresses these research gaps. We undertook a 10-week field campaign of temporally continuous point measurements of water temperature supplemented by spatially continuous temperature monitoring via hand-held thermal IR camera on two separate days for a Welsh upland stream. The aims were threefold: (i) to detect potential spatial heterogeneity of stream temperature at the micro-scale within the study reach; (ii) to explain spatiotemporal temperature dynamics related to hydrometeorological conditions and reach characteristics; and (iii) to evaluate the potential of ground-based thermal images to capture temperature variations at the micro-scale.

Methodology

Study area and site

The study catchment is located in Plynlimon, mid-Wales, UK, and forms part of the upper basin of the River Severn (Figure 14.1a, b). The field site is located at 298 m above sea level (asl) with catchment elevations varying from 620 m asl in the south-west to 290 m asl in the east. The catchment is underlain by Ordovician and Silurian mudstones, shales and greywackes (Neal *et al.*, 1997). The soil is made up largely of stagno-podzols; but peats, brown earths and gleys are also present. The predominant land use is pasture, moorland and forestry with coniferous plantation located mainly in the southern part of the catchment. Rainfall averages about 2500 mm yr^{-1} and mean annual air temperature is 7.3 °C in the catchment.

The Afon Llwyd is a small upland tributary of the Afon Clywedog, which is dammed by the Clywedog Reservoir. The study site on the Afon Llwyd is located approximately 1 km upstream of its entrance to the reservoir. However, impacts on the river flow regime related to the downstream impoundment are negligible due to the steep gradient of the channel (0.6%). At the study site, the average bankfull channel width is around 5 m, while the distance of the Afon Llwyd to its source is about 5 km with a drained catchment size of 7.5 km^2. The mean annual runoff is 0.42 m^3 s^{-1}. However, the flow regime is flashy with peak flows > 5 m^3 s^{-1}, which is common for headwater catchments in Plynlimon (Neal *et al.*, 1997).

The Afon Llwyd study reach has been used in two previous studies. Earlier work investigated the effects of -gravel–bed riffle–pool sequences on riparian hydrology (Emery, 2003) and the flow paths of saturated and unsaturated water in the adjacent floodplain (Bradley *et al.*, 2010). However, no stream temperature research had been conducted in this reach until now. From the late 1960s, much research was conducted in the adjacent Plynlimon catchments with a focus on water balance differences between

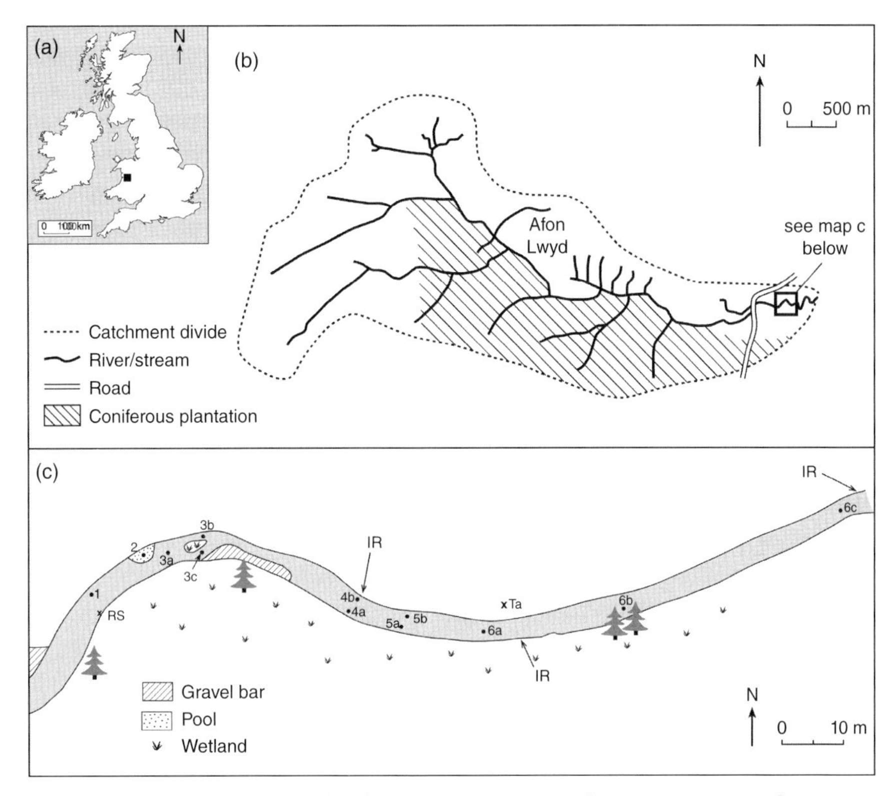

Figure 14.1 Study reach on Afon Llwyd with respective positions of water temperature loggers 1 to 6c, air temperature logger (Ta), water level sensor (RS) and infrared imaging (IR).

forested and deforested catchments (e.g., Kirby *et al.*, 1991).

Data collection: in situ temperature measurements

Water temperature, air temperature as well as water levels were measured over 10 weeks from 21 May 2010 to 5 August 2010. Table 14.1 details instruments and measurements. Water temperature was recorded in situ every 5 min by Tinytag TG-1400 temperature data-loggers at 12 distinct positions within the study reach (Figure 14.1c). The 12 positions are referred to subsequently as positions 1, 2, 3a, 3b, 3c, 4a, 4b, 5a, 5b, 6a, 6b, 6c. Temperature loggers were placed in white plastic radiation shields, which were open at two ends. The housings were fixed just above the stream bed with the openings oriented parallel to the stream flow direction, enabling unhindered water flow through the sensor housing. Position 1 was installed at the inlet of the study reach while 2 was positioned within a small pool (Figure 14.1c). Positions 3a, 3b and 3c comprised a section where the stream is dissected by a small mid-channel bar. Loggers at positions 4a and 4b were installed to capture potential shading effects from the north-facing and south-facing channel bank. Loggers at positions 5a and 5b lay within a stream section where the flow velocity is reduced generally compared to other stream sections. Loggers at position 6a, 6b and 6c were installed to monitor potential shading effects of coniferous trees standing at the south-side channel bank in line with logger position 6b. Air temperature

Table 14.1 Monitored variables and instrumentation.

Parameter	Instrument	Location	Accuracy
Air temperature	Tinytag TG-4100 temperature data logger	riverbank, 0.75 m above water surface	0.2 °C
Water temperature	Tinytag TG-4100 temperature data logger	0.05 m above streambed	0.2 °C
Water level	TruTrack WT-HR 1500 water height data logger	0.015 m above streambed	0.001 m

was recorded every 5 min with a Tinytag TG-1400 temperature data-logger at the northern riverbank in close vicinity to position 6a. All temperature loggers were cross-calibrated before field deployment (Hannah *et al.*, 2009).

Water levels were measured every 15 min by a TruTrack WT-HR 1500 water height data-logger close to position 1 at the study site inlet. Discharge was estimated by downscaling data recorded at the Centre for Ecology and Hydrology gauging station 54022 (Plynlimon flume) according to the catchment areas of the Afon Llwyd (7.5 km^2) and Plynlimon flume (8.7 km^2). Estimated runoff was highly correlated with observed water levels at Afon Llwyd over the study period ($r = 0.936$). Precipitation was measured by a tipping bucket at the Environment Agency-operated weather station, Dolydd, which is located about 250 m south-west of the study site.

The study reach was surveyed with a LEICA TC800 total station according to the manufacturer's manual. The survey covered the different positions of the temperature loggers and the water level sensor as well as the shape of the channel and in-stream structures such as gravel bars and pools.

Data analysis: in situ temperature measurements

Prior to analyses, data were checked for inconsistencies and gaps through visual inspection of time series plots or generation of cumulative and differences plots for comparable data. Occasional spurious values were removed and when possible the gaps were filled through linear interpolation or by linear regression derived from a corresponding time series where correlation analysis between time series exhibited $r > 0.9$.

Water temperature measured at the different positions within the channel was compared via visual inspection of time series, by comparison of statistical values and by generating box-and-whisker plots. For statistical analysis, summary statistics (such as daily mean water temperature, daily minimum/maximum values and daily ranges of water temperature) were computed. Box-and-whisker plots allowed inter-site comparison by summarising the median, minimum, maximum, upper and lower quartiles based on 5 min temperature data. Water temperature was analysed on a diurnal basis to examine potential spatial temperature patterns over the course of the day.

To determine the effect of stream thermal capacity on spatial temperature variation, summary statistics calculated individually for an extended low-flow period from 18 June 2010 to 28 June 2010 and a high-flow period from 15 July 2010 to 25 July 2010 were compared. The respective low-flow and high-flow periods were chosen based on the discharge hydrograph of the study period. To set the stream thermal dynamics in a hydrometeorological context, daily water temperature was correlated with daily air temperature and discharge. Pearson's product moment correlation coefficient (r) was used as a measure of correlation. Statistics are only presented if significant at $p < 0.05$. Unless stated otherwise, all correlations were significant at $p < 0.05$ level.

Thermal imaging approach

Thermal radiation emitted from surfaces can be detected remotely by IR sensors (Anderson and Wilson, 1984). For water surfaces, the IR imaging technique is sensitive to the upper 0.1 mm of the water column. To complement in situ measurements of stream temperature at discrete points, ground-based IR thermography using the portable thermographic system (INFRATEC VarioCAM hr) was conducted in the early afternoon on 21 April 2010 and again on 16 June 2010. The camera was mounted on a tripod located at the bank of the stream. Thermal images included 640 × 480 pixels and covered a spectral range of 7.5–14 μm. The detected radiant temperature had an absolute accuracy of 1.5 °C and the resolution of temperature was 0.08 °C. In addition to the IR pictures, corresponding visual images of the monitored sections were taken (1.3 MP). As the main focus of image interpretation was the distribution of water temperature, the emissivity in all the images was considered constant at 0.96, which is a value in the mid-range of published values for water surfaces (Anderson and Wilson, 1984).

The main cross-sections of in situ stream temperature measurements, as well as various structures within the stream such as such as vegetation, riffles and gravel bars, were examined via IR thermometry. Measurements were conducted between 13:30 and 15:30 on 21 April 2010 and from 14:30 until 16:30 on 16 June 2010. Meteorological conditions were similar on both recording days and were characterised by dry weather with transient cloud cover. Air temperature, relative humidity and wind speed were measured on site using a Kestrel 3000 pocket weather meter. Effects on temperature measurements related to air temperature and relative humidity were taken into account as air temperature and relative humidity data were input into the camera settings.

Radiant water surface temperature is only representative of the water column temperature when the water column is sufficiently mixed (Torgersen *et al.*, 2001). Measurements of vertical thermal profiles within the water column at different stream sections using a mercury-in-glass thermometer (±0.1 °C accuracy) revealed no thermal stratification. To estimate the accuracy of the measured radiant temperature, monitored stream temperature was compared against manual spot measurements of water temperature (kinetic water temperature) below the water surface. Differences between radiant and kinetic water temperature were < 0.2 °C. For image review, InfraTec IRBIS 3 software was used.

Results

Hydroclimatological context

The total amount of rainfall during the 10-week study period from 21 May 2010 to 5 August 2010 was 262 mm, which is approximately one-tenth of usual total annual precipitation. The highest daily totals in precipitation were observed on 20 July 2010, with in total 55.6 mm rainfall occurring (Figure 14.2). A dry period without any rainfall took place from 15 June 2010 to 23

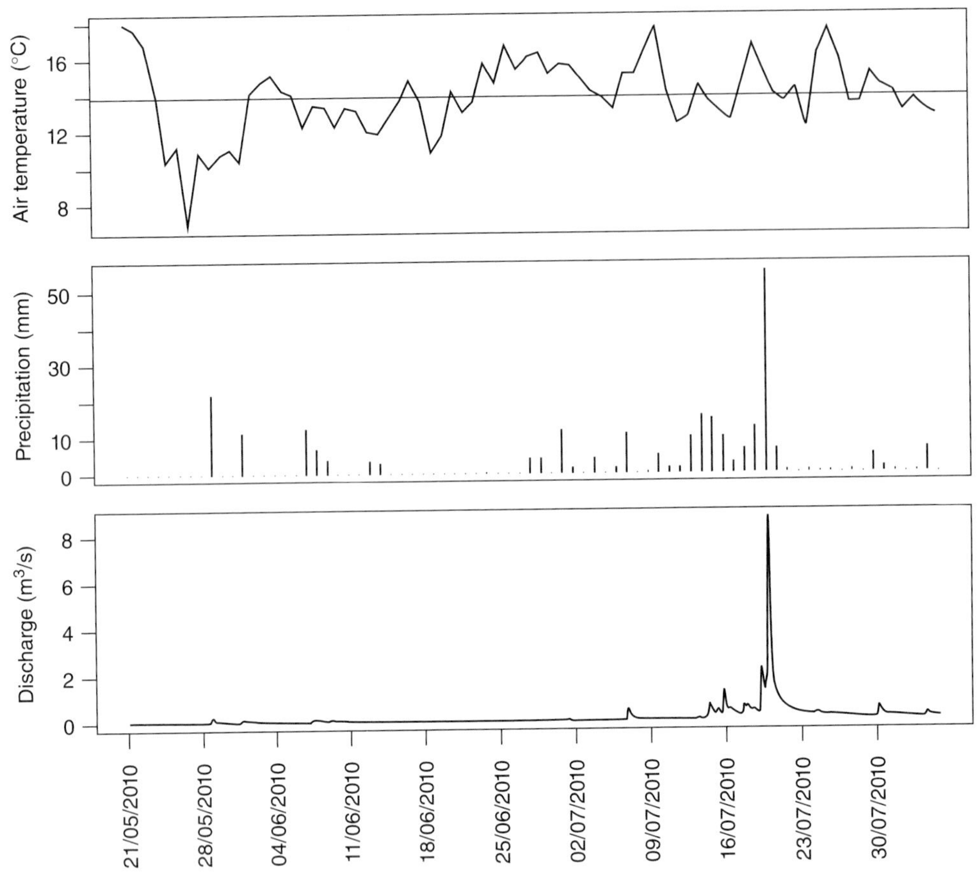

Figure 14.2 Mean daily air temperature, daily precipitation and discharge over the 10-week study period in summer 2010.

June 2010. The discharge hydrograph of the monitoring period reflects the precipitation pattern (Figure 14.2). Consequently, most of the study period was characterised by low flows followed by maximum flows with up to > 8 m^3 s^{-1} occurring from 15 July 2010 to 23 July 2010. Mean daily discharge over the monitoring period was 0.20 m^3 s^{-1}, which is about half the magnitude of the mean annual discharge of 0.43 m^3s^{-1}. Over the full study period, discharge was inversely correlated significantly with water temperature at all sites, yielding r-values > 0.242.

Air temperature showed clear diurnal fluctuations and averaged 13.88 °C over the study period with a standard deviation (Std) of 2.02 °C (Figure 14.2). Mean daily minimum (maximum) air temperature was 7.35 °C (19.42 °C) with a mean diurnal temperature range of 12.07 °C. Air temperature was significantly correlated with water temperature at all sites with $r > 0.713$. Figure 14.3 illustrates the relationship between daily water column temperature at position 1 and air temperature, revealing that water column temperature generally exceeded air temperature over the study period. The slope of the relationship is 0.59.

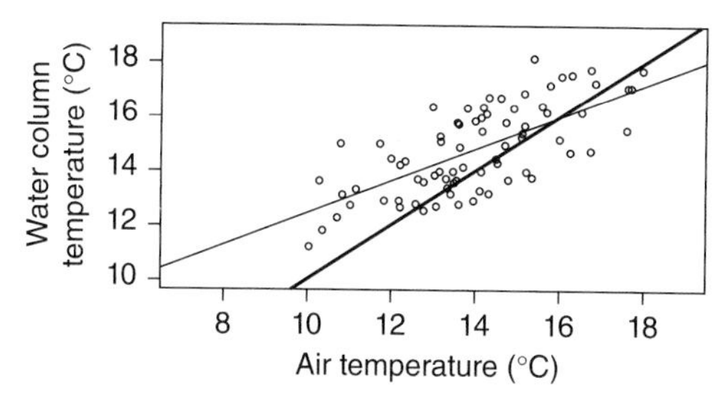

Figure 14.3 Relationship between air temperature and water temperature at position 1 based on mean daily values in summer 2010 including regression line (thin black line) and 1:1 line (bold black line).

In situ measurements of water temperature

Since water temperatures measured with sensor 4a diverged substantially from other sensors suddenly from 20 July 2010 onwards, this sensor was excluded from further analysis. Figure 14.4 shows daily maximum, mean and minimum water temperature at the different positions within the stream and discharge over the study period. Fluctuations of daily minimum, mean and maximum water temperature measured at the different channel positions mirrored each other over the study period. However, maximum and minimum temperature at position 5a were as much as 0.65 °C lower compared to temperature recordings at the other positions throughout the time period from 20 June 2010 to 1 July 2010. This coincided with an extended low-flow period (Figure 14.2). Averaged over the study period, measured water temperature was highly correlated between all sites ($r >$ 0.980) and mean values averaged over all loggers displayed standard variations that were below the accuracy of measurement (i.e., < 0.2 °C; Table 14.2). Mean daily water temperature, averaged over all temperature loggers, fluctuated around 14.81 °C ± 0.05 °C (Table 14.2). Daily minimum water temperature was on average 12.51 °C ± 0.03°C while daily maximum water temperature averaged 17.60°C ± 0.10°C. Box-and-whisker plots allowed comparison of 5 min temperature data between sites and confirmed the homogeneity between the recorded temperatures as medians of water temperature as well as minimum and maximum temperature did not differ significantly between sites (Figure 14.5). Every box-and-whisker plot shows the presence of outliers related to the low-flow period in mid June.

To determine the effect of hydrological conditions and stream thermal capacity on spatial stream temperature variability, water temperature was studied explicitly for the extended low-flow period from 18 June 2010 to 28 June 2010 (mean daily discharge: 0.07 m^3 s^{-1}) and the high-flow period occurring from 15 July 2010 to 25 July 2010 (mean daily discharge: 0.77 m^3 s^{-1}; Figure 14.5). Averaged over all temperature loggers, mean daily water temperature during the low-discharge period was 16.51°C ± 0.10 °C (Table 14.2). Mean daily water temperature measured over the high-flow period was considerably lower at 13.49 °C ±0.08 °C than under low flows. Similar to mean daily temperature, mean daily maximum and minimum temperature was higher under low flows, at 21.25 °C and 12.63 °C, respectively, compared to the high-flow period when mean daily maximum and minimum water temperature were 15.18 °C and 12.06 °C, respectively. Standard deviation of minimum daily temperature was 0.03 °C and 0.04 °C under low-flow and high-flow conditions, respectively, and thus as low as for mean daily water temperature. The spatial variability of daily maximum temperature during high flows was similarly low (Std = 0.06 °C) but considerably higher under low flows (Std = 0.25 °C). Box-and-whisker plots of

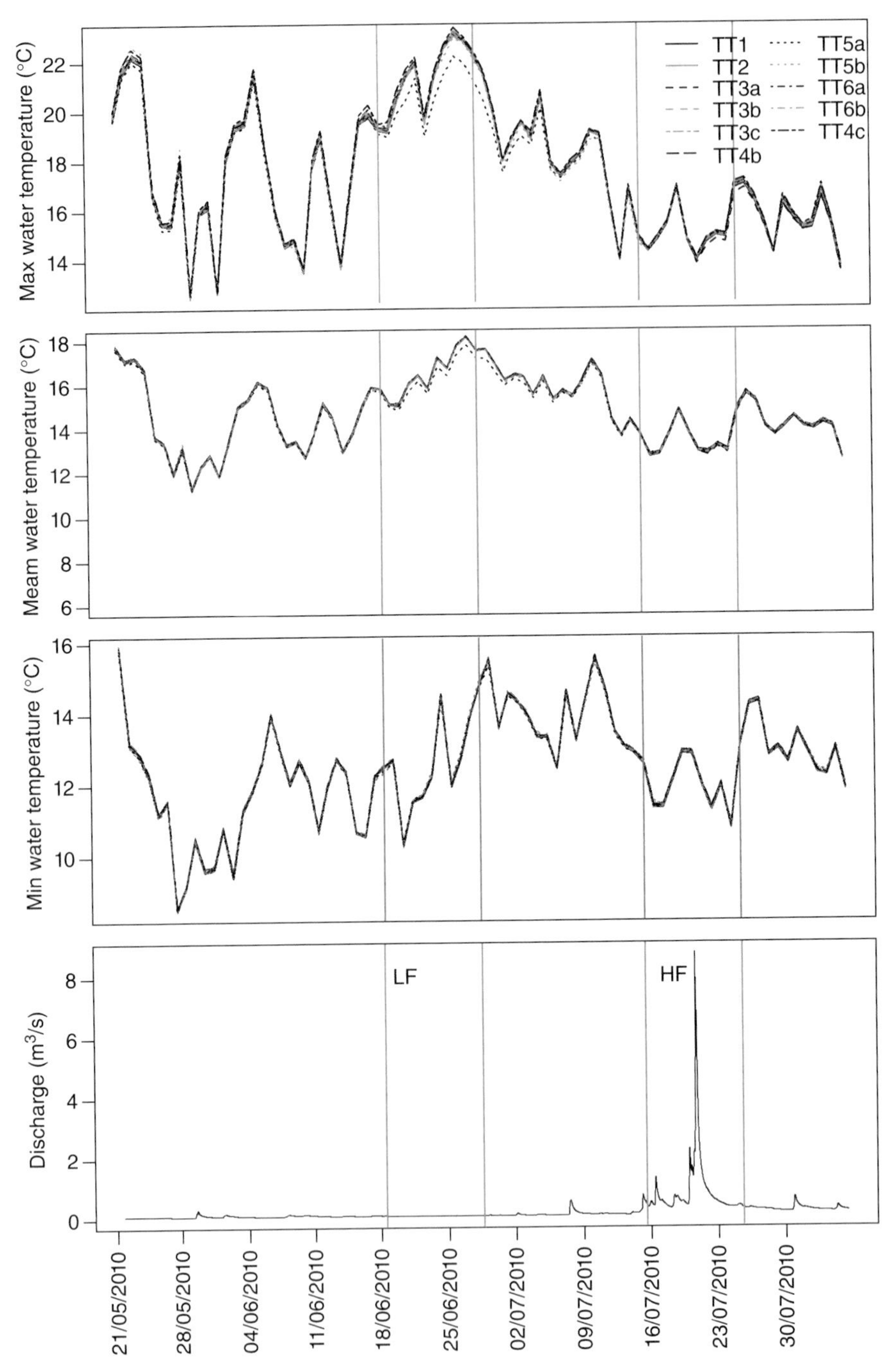

Figure 14.4 Daily maximum, mean and minimum water temperature measured with Tinytag loggers (TT) at different positions within the channel and river discharge over the study period. (High- (HF) and low-flow (LF) periods are denoted between vertical lines).

Table 14.2 Daily mean, minimum and maximum water temperature (°C) averaged over different periods (with standard deviation in parentheses).

Period	Mean (°C)	Min (°C)	Max (°C)
study period	14.81 (±0.05)	12.51 (±0.03)	17.60 (±0.10)
low flow	16.51 (±0.10)	12.63 (±0.03)	21.25 (±0.25)
high flow	13.49 (±0.08)	12.06 (±0.04)	15.18 (±0.06)

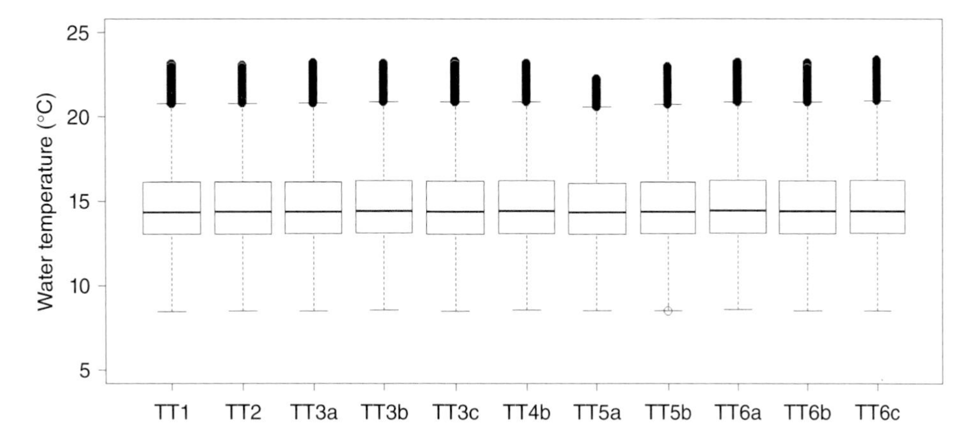

Figure 14.5 Box-and-whisker plots of water temperature measured with Tinytag loggers (TT) at different positions within the channel over the study period in summer 2010.

water temperature based on the recorded 5 min data confirmed the spatial homogeneity of stream temperature for low- and high- discharge conditions as temperature medians did not show any significant deviations among the sites (Figure 14.6 a, b). However, box-and-whisker plots revealed that maximum temperature at position 5a was at the minimum 0.65 °C lower than temperature at other sites during the low-flow period while no differences were apparent under high flows.

Averaged over all positions, daily ranges of stream temperature were considerably lower during the high-flow (3.13 °C) compared to the low-flow (8.62 °C) period despite relatively consistent air temperatures, which becomes clearly visible when comparing the respective box-and-whisker plots in Figure 14. 6 a and b. Furthermore, averaged over all sites, daily water temperature, particularly daily maximum and daily ranges, were higher relatively during the low-discharge than high-discharge period (Figure 14.6 a, b; Table 14.2).

On a diurnal basis, water temperature along the channel was most similar during the night and morning (Figure 14.7). From midday onwards, temperature measured at position 5a did not increase as strongly as the temperature registered at the other channel positions and daily maximum values at this position remained below those of the remaining loggers. The maximum divergence, accounting for up to 0.26 °C between temperature at position 5a and the remaining sites, occurred around 15:00. During this time period, a temporary small divergence

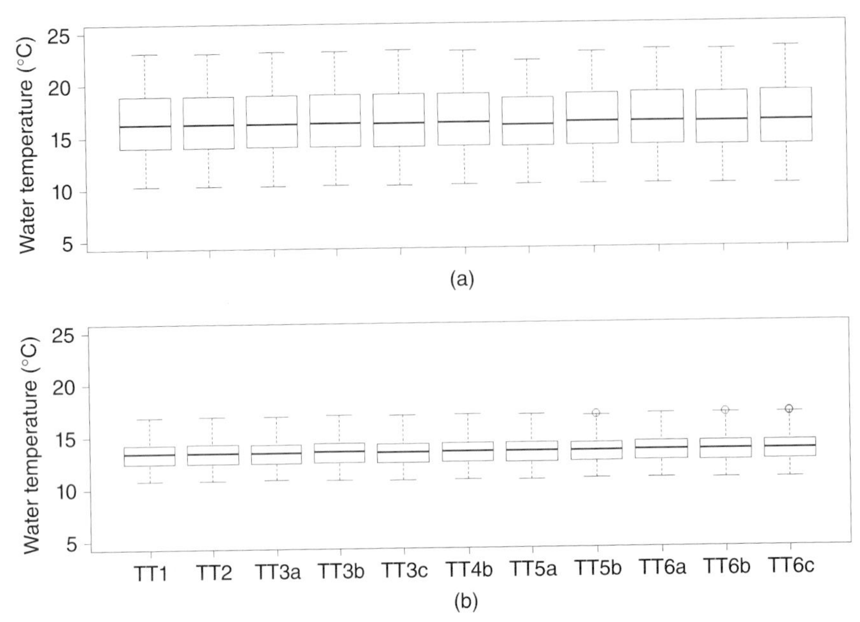

Figure 14.6 Box-and-whisker plots of water temperature measured with Tinytag loggers (TT) at different positions during the low- (a) and high-flow (b) period in summer 2010.

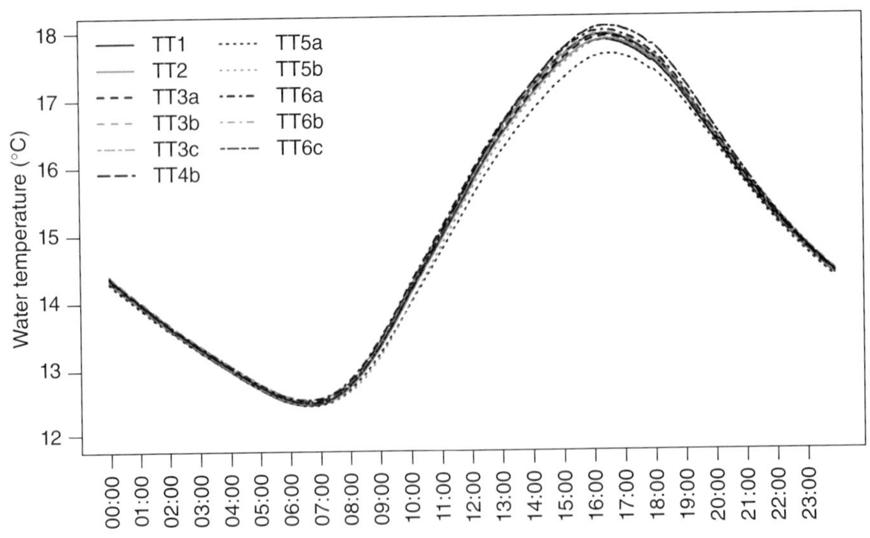

Figure 14.7 Mean diurnal cycles of water temperature measured with Tinytag loggers (TT) at different channel positions over the study period in summer 2010.

between the other temperature loggers was also visible. Towards the evening and throughout the night, water temperature at the different sites was identical.

Thermal imaging of stream channel

Thermal pictures were taken on two days with similar meteorological conditions

(a) (b) (c)

Figure 14.8 (a) Visual (top) and corresponding infrared image (bottom) taken on 21 May 2010 15:02 just downstream of site 6a. (b) Visual (top) and corresponding infrared image (bottom) taken on 16 June 2010 15:09 downstream of site 6c. (c) Visual (top) and corresponding infrared image (bottom) taken on 16 June 2010 15:18 at cross-section 4a/b. (Note: White housings on visual images contain water temperature loggers, and show monitoring position within channel. Channel width is about 4 m and flow is from right to left. Vantage point and scale of visual and infrared pictures are not exactly the same.) *(See colour plate section for colour figure.)*

(see above). Figure 14.8a shows the reach immediately downstream of position 6a. The picture demonstrates that water temperature was uniform across and along the channel (ranging between 20 °C and 21 °C). Light blue patches in the top-left indicate slightly lower water temperature, but values were still almost 20 °C. Flashy bright green and orange patches at the right riverbank were related to shaded riverbank structures and do not represent stream water. The thermal image displays clearly that channel water flowing along the riverbank or around in stream vegetation had the same temperature as water flowing in the middle of the channel.

However, at a riffle section where the stream was relatively shallow with larger boulders and cobbles protruding from the water surface, the water surface appeared slightly cooler (by 0.5–1.0 °C) compared with other areas within the channel (Figure 14.8b). This section was located about 5 m downstream of logger position 6c. At this site, channel width is about 4 m and water level was about 0.13 m at the time of the IR measurements. Since the in situ monitoring of stream temperature did not include this riffle sequence, the IR monitored temperature variability could not be confirmed by point measurements. However, in general, water surface temperature measured via IR thermometry corresponded well with water column temperature measured by the mercury-in-glass thermometer.

Figure 14.8c gives an impression of difficulties that are faced in association with thermal IR imaging. For logistical reasons, IR pictures had to be taken in the early afternoon during the summer time. Therefore shading effects, mainly arising from the right (south-side) riverbank, were recorded as apparent temperature differences at the water surface. Furthermore, direct insolation of the channel resulted in strong reflectance of solar radiation from the water surface and appeared as a virtual variation in stream temperature as seen at the top-left of Figure 14.8c.

Discussion

This chapter yields insight into the spatial distribution of water temperature within a small upland stream and examines the potential impact of channel morphology, groundwater influence and hydrological flow conditions on microthermal patterns. Stream temperature was measured in situ at multiple sites (including different morphological channel features) over a 10-week summer period to reveal uniform thermal patterns with differences in mean daily water temperature across monitoring locations of < 0.1 °C (cf. 0.2 °C accuracy of measurements). Hence, channel water temperature within the study reach appears largely insensitive to morphological channel features such as channel incision, channel width–depth ratio and riparian vegetation. It is likely that distinct spatial temperature patterns within the study reach were prevented by high-flow velocities and the associated high turbulence within the water column. Spatial variability was low irrespective of prevailing flow conditions. However, daily maximum temperature at logger position 5a was up to 0.65 °C lower compared with other sites during the low-flow period, facilitated by the reduced thermal capacity of the water column associated with low flows. Diurnal analysis of water temperature confirmed this finding in that, compared with other sites, recorded temperature at position 5a was lower at the diurnal temperature peak in the afternoon. Since this logger was positioned about 0.4 m away from the south-facing channel bank, which rises approximately 1 m above the channel surface at this section, the slightly reduced temperature maxima at this site are likely to be related to the reduced solar insolation during midday/afternoon associated with shading from the channel bank. Although temperature divergence at position 5a was relatively small, temperature variation is unlikely to have been caused by an instrument error. All temperature loggers were cross-calibrated in advance, the observed temperature deviations occurred temporarily and were limited to maximum water temperature at the respective position; this is suggestive of a real shading effect.

Thermal IR images of different sections within the study reach confirmed the generally low spatial heterogeneity of water temperature within the channel and did not show any considerable cross-sectional or longitudinal temperature gradients. However, pictures taken of a riffle section about 5 m downstream of logger position 6c identified some small patches of water column that appeared to be 0.5 °C to 1.0 °C cooler than the surrounding water column. The slightly lower water temperature at the riffle section may have resulted from the local upwelling of relatively cooler groundwater, as reported in a previous study wherein ground-based thermal IR imagery was considered a valuable and promising method to detect local groundwater inflow into small streams (Schuetz and Weiler, 2011). The upwelling of groundwater associated with riffles is consistent with the findings of other studies, which reported that riffles exhibit complex thermal behaviour and may cause local alterations of groundwater–surface water interactions (Evans and Petts, 1997; Hannah *et al.*, 2009; Krause *et al.*, 2011). Nevertheless, the exchange between channel water and groundwater was apparently limited to this section of the study reach, as in situ measurements and IR images of other stream sections, including also another riffle sequence, did not show any further coldwater patches. The minor importance of groundwater influx at the bed and the dominant influence of air temperature was indicated by the high correlation between air and stream temperature ($r > 0.7$) and conforms to the outcomes of a parallel study, which was conducted on temporal stream temperature dynamics of the Afon Llwyd stream (Gangi, 2010). Accordingly, temporal temperature dynamics within the stream bed and water column suggested that within the study reach, the stream is surface water- rather than groundwater-dominated.

Comparison of small-scale stream temperature variability with previous work is hampered by the fact that studies with a focus on local, micro-scale temperature variations over distances of centimetres to metres are scarce. In contrast, more research on longitudinal stream temperature distribution at the reach-scale exists and the factors and processes that control the spatial variability at the catchment and reach-scale have been proposed (Torgersen *et al.*, 2001; Malcolm *et al.*, 2004; Loheide and Gorelick, 2006; Cristea and Burges, 2009). For instance, a conceptual model provided by Malcolm *et al.* (2004) outlined that the catchment topography and the channel geometry (e.g., channel incision, orientation and width) exert substantial control on the thermal regime of running waters at the reach-scale. However, at the micro-scale these landscape factors are considered rather constant and may not be important in controlling thermal heterogeneity. Clark *et al.* (1999) showed that water depth and shading by riparian vegetation and river banks yielded considerable (up to 7 °C) lateral temperature gradients due to differences in water heat capacity and incoming solar radiation, respectively. However, their study was focused on groundwater dominated streams, which had generally lower stream gradients (0.3–0.5%) and a greater channel width (>7 m) compared with the Welsh upland stream examined herein, whose average channel gradient and channel width are 0.6% and 4–5 m, respectively. Given the relatively steep channel gradient, stream flow within the study reach is quite fast flowing and turbulent yielding a strong mixing of the water column. Hence, morphological channel features in the study reach showed no considerable effect on the spatial distribution of water temperature. Furthermore, in contrast to the channel examined by Clark *et al.* (1999), water depth across the channel was relatively uniform and this may prevent strong lateral temperature gradients. At the reach-scale, Malcolm *et al.* (2004) found that spatial variability of stream temperature was most apparent during summer months when stream temperature is relatively high. Due to reduced thermal capacity at lower flow depths, and the enhanced incoming solar radiation during this season, lateral temperature contrasts at the micro-scale are expected to be pronounced in summer. It can be reasoned consequently that spatial temperature variability, which was examined in summer and found to be low, is unlikely to be enhanced or more pronounced at another time of the year.

Conclusions

This research has revealed limited microthermal variability within a Welsh upland stream. Neither significant lateral or longitudinal temperature gradients nor thermal stratification occurred during the 10-week monitoring of stream temperature.

Small local temperature anomalies were limited to a coldwater patch associated with possible upwelling groundwater at a riffle tail and to local reductions in daily maximum temperature due to shading from the channel bank. This study is in contrast to previous research on microthermal variability that reported lateral temperature gradients of up to 7°C related to solar heating of shallow channel margin zones and considerable temperature variations within channels associated with sandbars or periphyton.

There is evidence from the present study to suggest that the occurrence of spatial temperature patterns was prevented through the generally high flow velocities, hence strong vertical mixing of the water column, owing to the steep gradient of the channel. Furthermore, the absence of floating vegetation and the consistently low water depth across the channel prevented local temperature variations and thermal stratification. Examination of hydrological conditions showed that daily temperature range and absolute water temperature across the study reach were enhanced consistently at low flows compared with high flows, but variability of daily mean water temperature within the study reach was not affected by changes in flow. However, low flows enhanced local divergence of maximum water temperature.

Thermal IR imaging provided a useful tool to monitor temperature distribution within the channel in a spatially continuous way even though shading from the river bank and reflectance of solar radiation from the water surface may hamper the detection of temperature variation. Comparison of previous small-scale stream temperature studies with our results indicates that microthermal variability is not comparable between different streams and highlights the need for further small-scale studies exploring the microthermal impact of hydrological conditions and channel features within different stream types. Future studies should combine techniques such as ground-based IR thermography and detailed in situ logging of water temperature as this allows one to monitor stream temperature in a spatially as well as temporally continuous way. Furthermore, there remains a demand for long-term research on spatial temperature distribution at the micro-scale that provides understanding of potential temporal changes in thermal variability within streams.

Acknowledgements

This research was support by a University of Birmingham Dean's Pilot Research Grant. The Environment Agency provided additional meteorological data for the study reach. Mark Robinson, Centre for Ecology and Hydrology, make available Plynlimon flume stage data. We are grateful to the Mr Tudor for allowing site access. Richard Johnson, University of Birmingham, gave technical assistance during fieldwork. Anne Ankcorn and Kevin Burkhill, University of Birmingham, helped with cartography.

David Hannah would like to thanks Geoff Petts and Angela Gurnell for initial encouragement to move from investigating the surface energy balance over snow and glaciers to the heat budget processes and thermal dynamics of rivers and streams. Your guidance has in this, and other regards, always been most insightful and highly valued.

References

Anderson, J.M. and S.B. Wilson. 1984. The physical basis of current infrared remote-sensing techniques and the interpretation of data from aerial surveys. *International Journal of Remote Sensing* 5: 1–18.

Bradley, C., A. Clay, N.J. Clifford, J. et al. 2010 Variations in saturated and unsaturated water movement through an upland floodplain wetland, mid-Wales, UK. *Journal of Hydrology* 393: 349–361.

Brown, L.E. and D.M. Hannah. 2008. Spatial heterogeneity of water temperature across an alpine river basin. *Hydrological Processes* 22: 954–967.

Cardenas, M.B., J.W. Harvey, A.I. Packman and D.T. Scott. 2008. Ground-based thermography of fluvial systems at low and high discharge reveals potential complex thermal heterogeneity driven by flow variation and bioroughness. *Hydrological Processes* 22: 980–986.

Carling, P.A., H.G. Orr and M.S. Glaister. 1994. Preliminary observations and significance of dead zone flow structures for solute and fine particle dynamics. In: *Mixing and Transport in the Environment*, 139–157. John Wiley and Sons.

Carrivick J., L.E. Brown, D.M. Hannah and A.G. Turner 2012. Numerical modelling of spatio-temporal thermal heterogeneity in a complex river system. *Journal of Hydrology* 414–415: 491–502.

Clark, E., B.W. Webb and M. Ladle. 1999. Microthermal gradients and ecological implications in Dorset rivers. *Hydrological Processes* 13: 423–438.

Coutant, C. 1977. Compilation of temperature preference data. *Journal of the Fisheries Research Board of Canada* 34: 739–745.

Cristea, N.C. and S.J. Burges. 2009. Use of thermal infrared imagery to complement monitoring and modeling of spatial stream temperature. *Journal of Hydrologic Engineering* 14: 1080–1090.

Ebersole, J., W. Liss and C.A. Frissell. 2003. Cold water patches in warm streams: physicochemical characteristics and the influence of shading. *Journal of the American Water Resources Association* 39: 355–368.

Ebersole, J.L., W.J. Liss and C.A. Frissell. 2001. Relationship between stream temperature, thermal refugia and rainbow trout Oncorhynchus mykiss abundance in arid-land streams in the northwestern United States. *Ecology of Freshwater Fish* 10: 1–10.

Emery, J.C. 2003. Characteristics and controls of gravel-bed riffle-pool sequences for habitat assessment and river rehabilitation design. *Doctor of Philosophy*. University of Birmingham, Birmingham.

Evans, E.C. and G.E. Petts. 1997. Hyporheic temperature patterns within riffles. *Hydrological Sciences Journal* 42: 199.

Gangi L., 2010 *unpublished Water temperature and heat exchange dynamics of a Welsh upland stream*. MSc Thesis. University of Freiburg.

Hannah, D.M., I.A. Malcolm and C. Bradley. 2009. Seasonal hyporheic temperature dynamics over riffle bedforms. *Hydrological Processes* 23: 2178–2194.

Hawkins, C.P., J.N. Hogue, L.M. Decker and J.W. Feminella. 1997. Channel Morphology, Water Temperature, and Assemblage Structure of Stream Insects. *Journal of the North American Benthological Society* 16: 728–749.

Hynes, H.B.N. 1970. *Ecology of Running Waters*. University of Toronto, Toronto.

Imholt C, Soulsby C, Malcolm IA, et al. 2013. Influence of scale on thermal characteristics in a large montane river basin. *River Research and Applications* 29: 403–419.

Kirby, C., M.D. Newson, and K. Gilman. 1991. Plynlimon research: The first two decades. IH Report 109. Available at: http://www.ceh.ac.uk/products/publications/documents/ih109plynlimonresearch.pdf\ignorespaces[Accessed\ignorespaces28/11/13].

Krause S., D.M. Hannah and T. Blume 2011. Interstitial pore-water temperature dynamics across a pool-riffle-pool sequence. *Ecohydrology* 4: 549–563.

Krause, S., T. Blume, and N.J. Cassidy 2012. Investigating patterns and controls of groundwater up-welling in a lowland river by combining Fibre-optic Distributed Temperature Sensing with observations of vertical hydraulic gradients. *Hydrology and Earth System Sciences* 16(6): 1775–1792.

Loheide II, S.P. and Gorelick S.M. 2006. Quantifying stream–aquifer interactions through the analysis of remotely sensed thermographic profiles and in situ temperature histories. *Environmental Science & Technology 2006* 40 (10): 3336–3341 DOI: 10.1021/es0522074

Madej, M., C. Currens, V. Ozaki, J. Yee, and D. Anderson. 2006. Assessing possible thermal rearing restrictions for juvenile coho salmon Oncorhynchus kisutch through thermal infrared imaging and in-stream monitoring, Redwood Creek, California. *Canadian Journal of Fisheries and Aquatic Sciences* 63: 1384–1396.

Malcolm I.A., D.M. Hannah, M.J Donaghy et al. 2004. The influence of riparian woodland on the spatial and temporal variability of stream water temperatures in an upland salmon stream. *Hydrology & Earth Systems Science* 8: 449–459

Neal, C. et al. 1997. The occurrence of groundwater in the Lower Palaeozoic rocks of upland Central Wales. *Hydrology & Earth Systems Sciences* 1: 3–18.

Peterson, J.T. and C.F. Rabeni. 1996. Natural thermal refugia for temperate warmwater stream fishes. *North American Journal of Fisheries Management* 16: 738–746.

Petts, G.E. 2000. A perspective on the abiotic processes sustaining the ecological integrity of running waters. *Hydrobiologia* 422: 15–27.

Rutherford, J.C., J.B. Macaskill and B.L. Williams. 1993. Natural water temperature variations in the lower Waikato River, New Zealand. *New Zealand Journal of Marine and Freshwater Research* 27: 71–85.

Schuetz, T. and M. Weiler 2011. Quantification of localized groundwater inflow into streams using ground-based infrared thermography. *Geophysical Research Letters* 38: L03401, doi:10.1029/2010GL046198.

Selker, J. S., L. Thevenaz, H. Huwald, A. et al. 2006. Distributed fiber-optic temperature sensing for hydrologic systems. *Water Resources Research* 42(12): W12202.

Torgersen, C.E., D. Price, H. Li, and B. Mcintosh. 1999. Multiscale thermal refugia and stream habitat associations of Chinook Salmon in northeastern Oregon. *Ecological Applications* 9: 301–319.

Torgersen, C.E., R.N. Faux, B.A. Mcintosh, et al. 2001. Airborne thermal remote sensing for water temperature assessment in rivers and streams. *Remote Sensing of Environment* 76: 386–398.

Webb, B.W., D.M. Hannah, R.D. Moore, et al. 2008. Recent advances in stream and river temperature research. *Hydrological Processes* 22: 902–918.

CHAPTER 15

River resource management and the effects of changing landscapes and climate

James A. Gore[1], James Banning[1] and Andrew F. Casper[2]
[1] *College of Natural and Health Sciences, Department of Biology, University of Tampa, Tampa, USA*
[2] *Illinois River Biological Station, University of Illinois, Havana, USA*

Introduction

The evolution of criteria to characterise river discharge regimes that maintain biological integrity (variously termed minimum flows, conservation flows, environmental flows or instream flows) has evolved considerably over the past four decades. These techniques can range from a purely legal stance, such as the concept of riparian rights or prior appropriation, both of which grant primary standing for water consumption, to the most recent concepts in ecohydrology which mandate the restoration of a 'natural state' (Poff *et al.*, 1997). Indeed, it is widely acknowledged that the natural flow regime has shaped the evolution of riverine communities and the biological processes that support them (Naiman *et al.*, 2002; Poff *et al.*, 1997; Lytle and Poff, 2004). That is, individual rivers will have a characteristic flow regime and an associated aquatic biota that demand a unique set of strategies to ensure the future integrity of the system. How these organisms and communities respond to alterations of some or all of the characteristics of the flow regime remains to be explored. However, an understanding of the resilience of the lotic ecosystem and its unique flow regime is a critical element in developing an effective management strategy (Lytle and Poff, 2004).

One of the greatest problems in river resources management has been the conflict between economic development and the legal and scientific definition of 'beneficial' use of water resources. That is, what are the mechanisms that can be used to not only predict the impact of modified flows on lotic biota but, whether any gains or losses in biota can be equated with some sort of economic index which allows assessment of their impact on other uses of the resource. Whether this ecohydrolgical assessment focuses upon instream communities, riparian communities or anthropogenic users, its assumptions and exceptions need to be considered. A number of intermediate philosophies have been introduced over the past two decades to attempt to address ecohydrological management issues. These philosophies range from an engineering approach focused upon water for human consumption and a modicum of water for ecosystem requirements, to more recent proposals for techniques that advocate a

River Science: Research and Management for the 21st Century, First Edition.
Edited by David J. Gilvear, Malcolm T. Greenwood, Martin C. Thoms and Paul J. Wood.

management programme based on 'naturalised' flows (see, for example, Richter *et al.*, 1997).

Estimating the flow allocation assumed to conserve the biological integrity of an ecosystem has historically employed best professional judgement (Tennant, 1976) or hydrographic approaches that initially relied upon an estimated flow that was exceeded at least 70% of the time (Gore, 1989). These methods were followed by attempts to link hydrographic and hydraulic conditions to the habitat requirements of key biota (often game fish but sometimes endangered species) (Gore, 1989). Basing flow conservation on this approach assumes that gauged hydrographic data and the species(s) selected for modelling also reflect flows that support other aquatic life in that system (Gore and Nestler, 1988). A more recent alternative for a flow allocation when there is little knowledge of specific requirements is to recreate (model) the pre-development flow duration curve or the 'naturalised hydrograph'. This naturalised hydrograph prescription assumes a simple ecological idea; that organisms and communities occupying any particular river have evolved and adapted their lifecycles to long-term, pre-development flow conditions in that river (Richter *et al.*, 1996, Stanford *et al.*, 1996, Poff *et al.*, 1997). Thus, with limited knowledge of specific biological flow requirements, the best alternative is to recreate the hydrographic conditions under which the communities originally existed. To accomplish this objective, a 'building-block' model, termed the 'Range of Variability Approach' (RVA) (Richter *et al.*, 1996, 1997) has been specifically designed as a river management strategy that attempts to reconstruct the natural hydrograph. This analysis requires a minimum of 20 years of daily streamflow records available for the analysis. RVA uses a statistical examination of the 20+ year dataset to establish management targets for hydrological parameters most likely to influence ecological conditions (Richter *et al.*, 1996; Olden and Poff, 2003; Monk *et al.*, 2007). Intermediate models rely upon hydrographic information for decision processes but attempt to incorporate some amount of biological information, usually gained through consultation with professional fisheries biologists and aquatic ecologists. In some respects, these intermediate models are similar to expert systems. The most popular of these are a group of computer models often referred to as wetted perimeter methods (Gore and Meade, 2008). Finally, linked models include those models that tie open channel hydraulics with measured elements of fish or macroinvertebrate behaviour. The most widely used example of this model is the Physical Habitat Simulation (PHABSIM) (Bovee, 1982; Nestler *et al.*, 1989). PHABSIM is the model most frequently used within the procedure called the Instream Flow Incremental Methodology (IFIM) (Bovee *et al.*, 1998; Lamouroux and Capra, 2002; Munson and Delfino, 2007; Petts, 2008; Peñas *et al.*, 2013).

Regardless of the type of approach to defining environmental flows in riverine ecosystems, five elements must be considered before an adequate decision can be made. These are: (i) the goal – such as restoring or maintaining a level of ecosystem structure; (ii) the resource – target fish species or physical conditions; (iii) the unit of analysis – how achievement of the goal is measured, such as a certain discharge or discharge regime; (iv) the benchmark time period – an arbitrary number of years of hydrographic record; and (v) the protection statistics – one or more critical flow variable, such as mean monthly flow, mean daily or, in some instances, mean weekly flow (Beecher, 1990).

Developing a management strategy for riverine resources now also requires consideration of both economic constraints to baseline data for decisions and multi-decadal shifts in precipitation and weather patterns. For example, with reduced expenditures on historical record keeping (witness the continual closure of the system of national and international gauging records), it becomes more difficult to accurately assess changes in long-term weather patterns and their resulting influence on habitat availability; thus, resulting in management strategy that may shift through time in order to best replicate the natural flow condition. In this chapter, we discuss a successful management strategy to address some of these concerns, the potential impact of decadal (or longer) shifts in weather conditions, and alternatives for estimating long-term flow patterns when gauging records are interrupted, terminated or they have never existed.

Multi-decadal shifts in weather pattern

A number of oceanic temperature anomalies that affect continental precipitation and runoff patterns at a regional scale have been recognised (Pekarova *et al.*, 2003; McCabe and Wolock, 2008; Millman *et al.*, 2008). However, only in recent decades has the connection between these anomalies and inland river hydrology begun to be appreciated (Whited *et al.*, 2007; Durance and Ormerod, 2007; Nunn *et al.*, 2007). In North America, one of the more influential anomalies is the Atlantic Multi-decadal Oscillation (AMO), a sea surface temperature oscillation that occurs in the North Atlantic between 0° and 70° N (Figure 15.1) (Maxwell *et al.*, 2013). Historically it has been assumed that annual variation in rainfall and thus streamflow is a more or less random event, Enfield *et al.* (2001) concluded that a long-term oscillation in rainfall (approximately 60 to 80 years) is evident, although this pattern may be affected by other short-term (e.g., El Nino – 6 years) or long-term cycles (*e.g.*, Pacific Decadal Oscillation – PDO, McCabe *et al.*, 2004; Johnson *et al.*, 2013). Moreover Mantua *et al.* (1997) noted that that there is a statistical relationship between El Nino (ENSO), the PDO and the AMO, meaning that they are not actually independent events as once assumed. Although Enfield *et al.* (2001) made these conclusions based on analysis of flow data from the Kissimmee River in Florida and the Mississippi, the Southwest Florida Water Management District (SWFWMD), ecological evaluation section (2004) and Kelly and Gore (2008) have concluded that there is also a similar multi-decadal shift in flow patterns of rivers of Florida and the southeastern United States. In the last decade, the AMO has been a popular topic of interest in the realms of environmental flow studies, habitat and ecological restoration and regulatory committees (Gaiser *et al.*, 2009; Munson and Delfino, 2007; Johnson *et al.*, 2013; Maxwell *et al.*, 2013). So, whether this oscillation does or does not exist is of more than academic interest.

Response of river hydrographs to oceanic multi-decadal shifts in weather pattern

Beecher (1990) correctly identified a 'baseline' (or benchmark) time period was a necessary element when developing

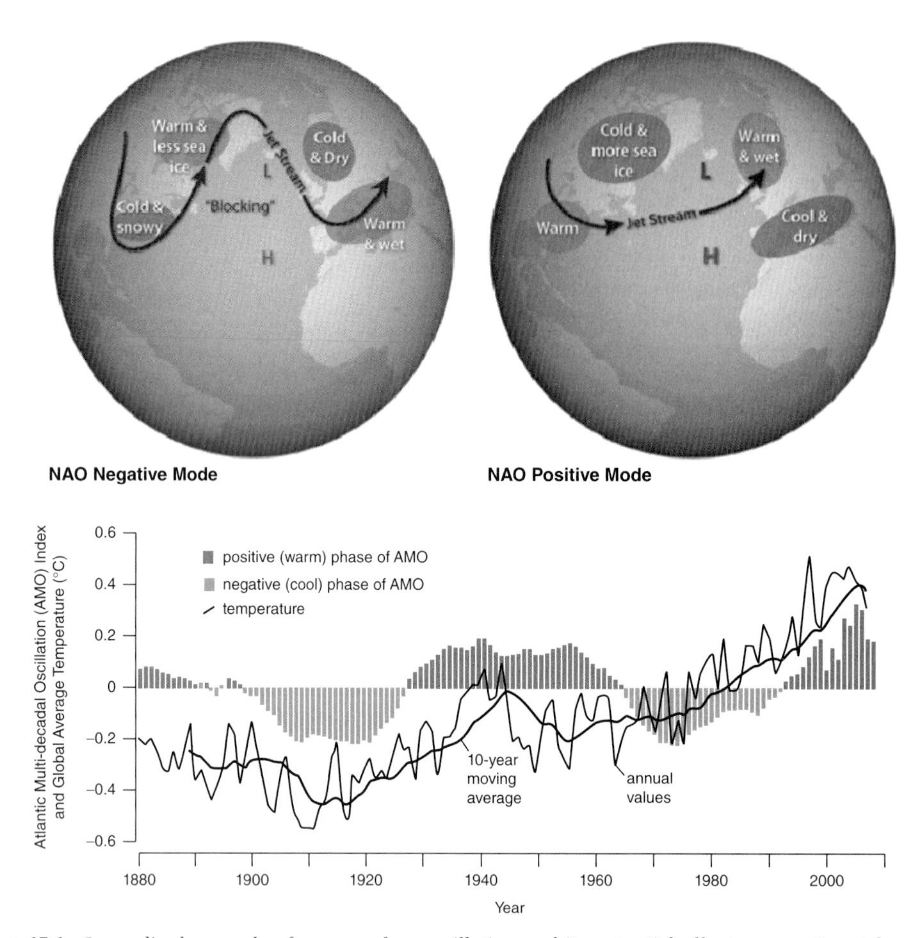

Figure 15.1 Generalised example of a sea surface oscillation and its potential effects on continental weather patterns (top). AMO index showing the warm (above line bars) and cool periods (below line bars) with global average temperature superimposed (bottom). Top graphic adapted from AIRMAP by Ned Gardiner and David Herring, NOAA. Bottom graphic created by Michon Scott, National Snow and Ice Data Center. *(See colour plate section for colour figure)*.

environmental flow criteria. However, Kelly and Gore (2008) suggest that care should be taken when selecting the baseline period since an ignorance of multi-decadal variations in flow could lead to either unreasonably high or low predictions. For example, although 20 years or so of record is generally considered sufficient for a baseline period (e.g., Richter *et al.*, 1996) if the entire 20-year baseline period falls in a different portion of the oscillation than the current one, warm instead of cool, then any deviations from the baseline hydrograph cannot be unambiguously linked to anthropogenic activities. Thus, Kelly and Gore (2008) further suggested that it may be appropriate to have at least two baseline periods, one based on a 'wet' period and one based on a 'dry' period (Figure 15.2).

There is little doubt that lotic species are adapted to a natural flow regime (Poff *et al.*, 1997; Lytle and Poff, 2004). Indeed,

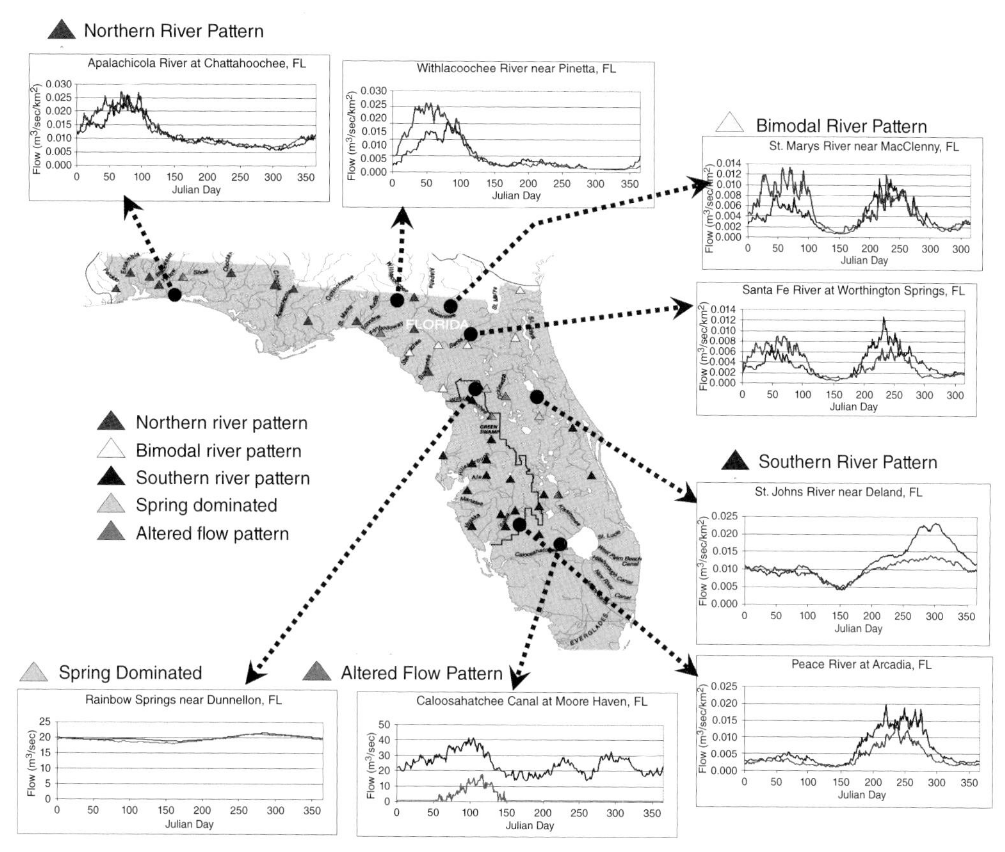

Figure 15.2 Examples of geographic differences in how the AMO can influence both the volume and, independently, the seasonality of seasonal baseline flow rates in the rivers of Florida. (Graphics adapted from M. Kelly and J.A. Gore, 2008.) Blue lines show the warm AMO period (1940–69) while green lines indicate the cooler AMO period (1970–99). Note that the effect of the AMO is different, and even reversed, depending on the geographic location of the river basin. Data from the USGS National Water Information service. *(See colour plate section for colour figure.)*

many of our current regulated flow management strategies are based upon that very assumption. However, a critical management question yet to be tested is the ability of the assembled communities to adapt to alteration in the frequency, timing and duration of low-flow extremes (Lytle and Poff, 2004). More importantly, the ability to predict the pattern of community changes and, to a greater extent, the resilience of communities will become important factors in effectively managing systems under the threat of anthropogenic modifications and potential climate changes in the future.

The building-block approach

The Florida legislature has directed the Florida Department of Environmental Protection (FDEP) and the state's five water management districts (WMDs) to develop ecologically defensible minimum flows and levels that would protect the resources of the state from 'significant harm' due to water withdrawals (Section 373.042, Florida Statutes). It is essential for the development of minimum flows and levels (MFLs) that temporal and spatial flow trends are understood in terms of natural and anthropogenic effects. The Southwest Florida Water Management District (SWFWMD) approached the development of environmental flows as an opportunity to examine the relationship between the potential effects of the AMO (Munson *et al.*, 2005; Kelly and Gore, 2008), potential changes in lotic community structure (e.g., biodiversity) and the management decisions that might alter flow allocations in the future.

There is a considerable body of knowledge that indicates that virtually every habitat parameter and life-stage requirement of lotic biota are linked, either directly or indirectly, to changes in hydrological or hydraulic conditions (Heede and Rinne, 1990; Gore *et al.*, 2008). Gore *et al.* (2008) have provided a recent review of these relationships for invertebrates, and similar reviews have been provided by Heede and Rinne (1990) for fish species. With the development of the field of ecohydraulics, it is sufficient to say that variation in the year-class strength of fish species (Bonvechio and Allen, 2005), fish community structure (Pyron and Lauer, 2004; Sheldon, 2011; Caiola *et al.*, 2014) and large-scale and small-scale changes in benthic macroinvertebrate composition (Statzner *et al.*, 1988; Cowell *et al.*, 2004; Brooks *et al.*, 2005; Statzner, 2008; Zigler *et al.*, 2008; Hussain and Pandit, 2012; Sanz-Ronda *et al.*, 2014) can all be attributed to changes in hydrologic and hydraulic conditions. In turn, recent work shows how hydrologic and hydraulic conditions in rivers are strongly influenced by oscillations like the AMO (Gaiser *et al.*, 2009; Domisch *et al.*, 2013; Johnson *et al.*, 2013; Sheldon and Burd, 2013; Maxwell *et al.*, 2013) and we can link global scale climate to river biotic response. Thus, the approach to modelling the response of river resources to water abstraction should also consider climatological oscillations and variability.

The approach used by SWFWMD to propose environmental flows (MFLs) is consistent with the building-block approach proposed by Postel and Richter (2003). Most of the SWFWMD jurisdictional rivers follow the Southern River Pattern (SRP) of Florida (Kelly and Gore, 2008), and we concentrated on those rivers for this research. Three distinct flow periods were evident in hydrographs of median daily flows for each river (Figure 15.3). Lowest flows (block 1) occurred during late spring and summer, approximately Julian days 110–175, and were considered to be the dry season. Highest flows occurred during an approximate 125-day period (block 3) that immediately followed the dry season. This is the period when the floodplain is most likely to be inundated. The remaining days constituted an intermediate or medium flow period (block 2). The approach to setting minimum flows and levels for all SWFWMD rivers is habitat based and assumes that an overriding consideration should be established for each block. During high-flow periods (block 3), the primary concern is maintaining adequate floodplain connections as potential spawning and nursery habitat, as well as soil nutrient regeneration. HEC-RAS (Hydrologic Engineering Center – River Analysis System) – developed by the US Army Corps of Engineers – is a modelling system commonly used to estimate the inundation characteristics of various habitats at transects placed across the rivers. The greatest potential impacts to lotic communities are likely to occur during the annual low-flow period, block 1, and the water management district opted for the modelling approach (PHABSIM) to assess potential changes in available habitat under these conditions.

PHABSIM requires a transect of measurements of depth, mean water column

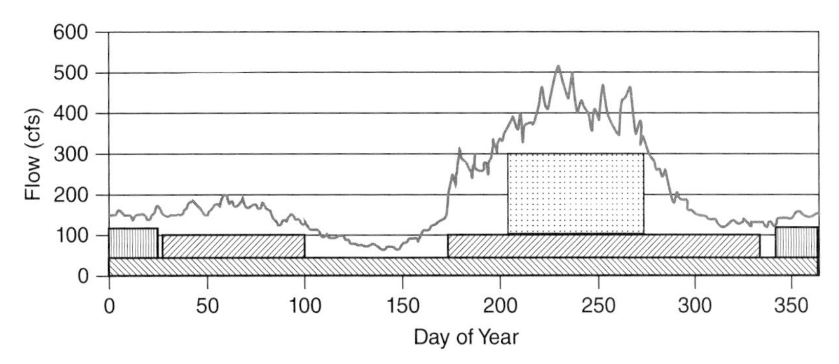

Figure 15.3 Building-block approach to restoring a hydrograph similar to the historical record. The line depicts a theoretical median flow hydrograph based upon 30 years of daily flow records. Block 1 (with left hash-marks) represents the wetted perimeter. Block 2 (with right hash-marks) represents the additional flow required by physical habitat requirements (PHABSIM). Block 3 (with stippling) represents additional flow to maintain annual inundation of riparian floodplains. Block 4 (with vertical lines) represents additions for special biological/physicochemical conditions.

velocity, evaluation of substrate and cover, slope, and water surface elevations for a series of discharges located at one or more hydrologically typical stream reaches (Bovee *et al.*, 1998). As an example, cross-sections were examined at nine sites along the length of Myakka River while six cross-sections were selected for the Alafia; three representing broad-shallow reaches of the upper river and three representing the incised downstream reaches of the river. Following accepted practices for PHABSIM analysis (Milhous *et al.*, 1989), we chose a suite of target biota including popular game fish, largemouth bass (*Micropterus salmoides*), bluegill sunfish (*Lepomis macrochirus*), spotted sunfish (*Lepomis punctatus*) and macroinvertebrate community diversity (sensu Gore *et al.*, 2001). Macroinvertebrate data from the northern Withlacoochee were provided by Florida Wildlife Conservation Commission, and we developed habitat suitability criteria according techniques described by Gore and Judy (1981).

Predictions of habitat available (expressed as weighted usable area, WUA) over a range of typical discharges were produced for each species' life-stage as well as for benthic community diversity. These predictions of habitat change were analysed over benchmark time periods using time-series analysis (Bovee *et al.*, 1998). As shown by Kelly and Gore (2008), rivers in Florida with a SRP generally exhibited higher flows during the 1940 to 1969 period compared with flows during the 1970 to 1999 period. Therefore, for each river system, we chose to break the time series analysis into wet periods (1940 to 1969) and dry periods (1970 to 1999) in order to elucidate any differences in habitat availability and/or changes in potential targets for management.

Since the differences in mean daily and mean monthly flows between wet and dry periods can vary by an order of magnitude during some seasons, we conducted a species sensitivity analysis using PHABSIM simulations of each time period and the effect of mean monthly flow reductions of 10, 20, 30 and 40%. This allowed us to determine how wet and dry periods would affect habitat availability in relation to MFL statutes; such regulations requiring that no 'significant harm' be brought to the riverine ecosystem. 'Significant harm' remains undefined in statutory documentation meaning that there is no numerical criterion with which to make comparisons. Gore *et al.* (2002) recommended a habitat loss of 15% or more from current or undisturbed conditions as an appropriate indicator; however, this is not a universally accepted breakpoint and the criteria used by other states in the USA can vary from habitat losses of 10 to 33% (see recommendations by Dunbar *et al.* 1998 and Jowett 1997). Of particular interest were the potential changes in habitat available for target species and communities as a result of the multi-decadal shifts in hydrologic conditions alone. For example, we concentrated on conservation flow recommendations for the block 1 and block 2 hydrologic conditions (sensu Postel and Richter, 2003, Figure 15.3, during dry and wet multi-decadal periods on the Myakka and (northern) Withlacoochee rivers. We followed Gore *et al.* (2002) and defined significant threat of population loss to be the point at which the species, functional group or life-stage lost more than 15% of available habitat in any given month. Some biota were predicted to experience an increase in available habitat while others a significant loss of habitat for the same flow reduction. It is assumed that 'habitat gain' has the potential to modify community dynamics, if flows are maintained over a long period of time. For species that were predicted to experience a significant habitat loss, the life-stage or species that had the greatest loss with the lowest percentage of flow reduction was considered to be the indicator of change in community composition.

In an example using benthic invertebrates (Figure 15.4) from the Myakka River, the time-series analysis predicted changes in the habitat available as the percentage of flow reduction increased, though primarily during the winter dry season. However, the relative proportions of habitat increase or reduction did not appear to be sufficient to conclude that there was sensitivity to either 'dry-period' or 'wet-period' conditions (Figures 15.1 and 15.2). In the case of the habitat available to juvenile spotted sunfish, considered to be the most sensitive life-stage and most likely to experience a reduction in survival, the 15% critical habitat loss threshold was exceeded during the dry season of both phases of the AMO. However, the threshold was crossed during more months with greater flow reductions during the cold, wet (1940–69) AMO cycle than the warm, dry cycle (1970–99). Thus, unlike benthic invertebrate diversity, juvenile spotted sunfish increased in relative abundance during the month of April during the wet AMO periods (Figure 15.5). For any reduction in flow greater than 10%, juveniles were predicted to lose more than 15% of their habitat. However, during the dry AMO periods, available habitat for juvenile spotted sunfish was predicted to increases for all levels of flow reduction; exhibiting modest increases of up to 5%. This result suggests that statutory changes in flow allocations could be based upon predicted changes in fish community structure (as indicated by the potential success of certain target species).

When tabulated, the shift in sensitivity can be used to indicate the species or life-stage most sensitive during each combination of month and dry or wet AMO phase (Table 15.1). With the exception of two months, different life-stages or functional groups were most sensitive. This set of observations explains anecdotal comments by local fisherman that, when they were young, some 30 or more years ago, a different fish (either largemouth bass or spotted sunfish) dominated the river systems.

Similar shifts in the sensitivity to overall changes in flow pattern between the

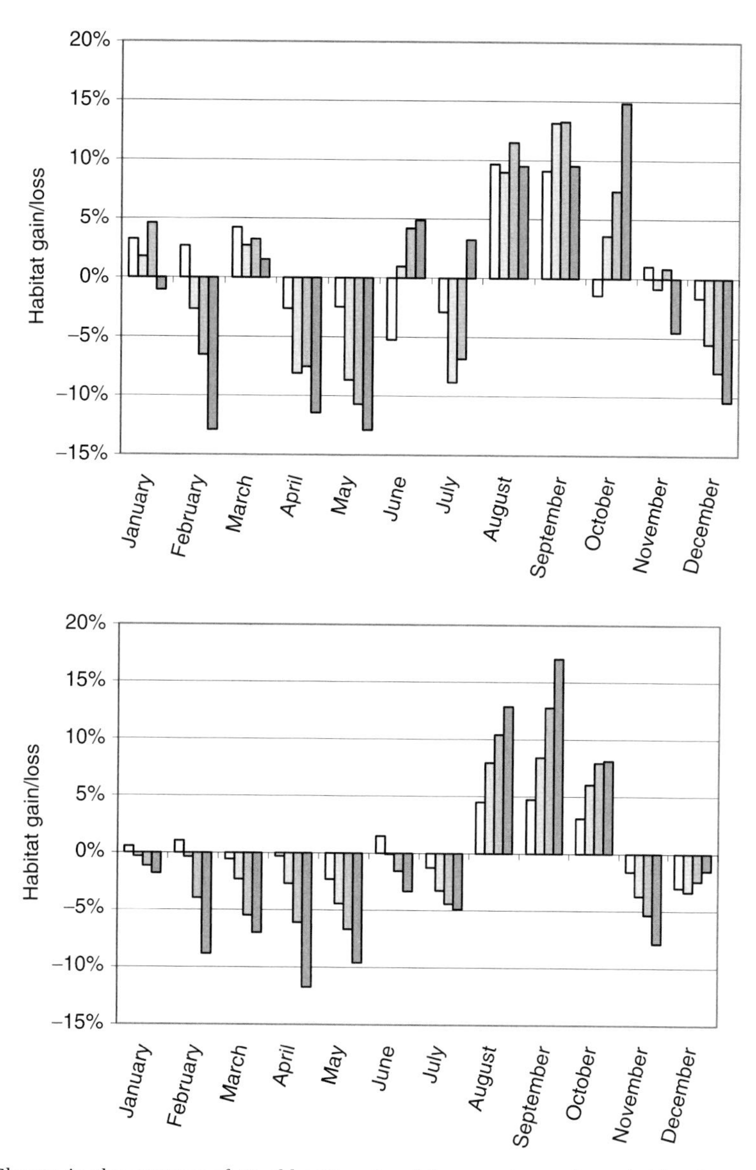

Figure 15.4 Change in the amount of Myakka River benthic macroinvertebrate habitat in response to four water abstraction scenarios (10, 20, 30 and 40%; increasing grey scale) during a recent warm phase (top panel) and cool phase (bottom panel) of the Atlantic Multi-decadal Oscillation.

wet and dry periods of AMO cycles were observed for biota in the northern Withlacoochee River. However, although the target fish species did not experience significant habitat gains or losses between phases, some species of benthic macroinvertebrates did, especially the Chironomidae (the non-biting midges). The most dramatic of these shifts in habitat availability were predicted for *Psuedochironomus* sp. (shallow–moderate flow, tube dwellers) and *Cricotopus bicinctus* (shallow-to-deep, slow flow, tube dwellers);

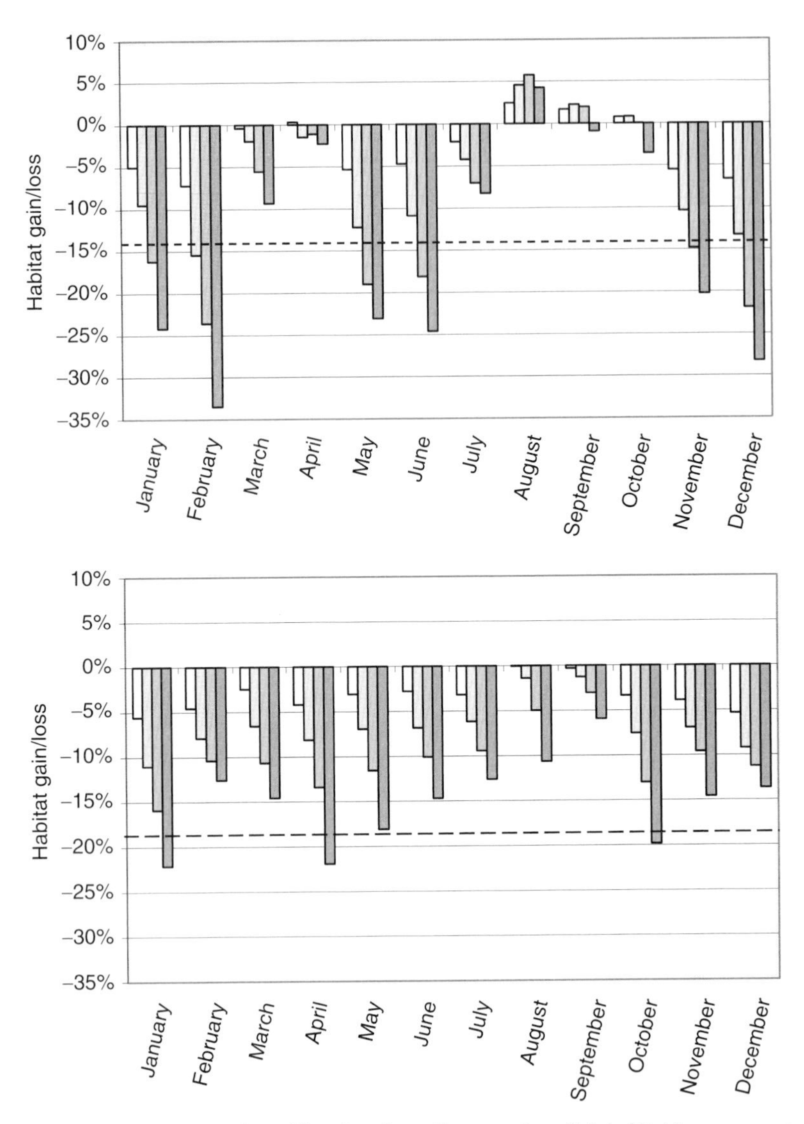

Figure 15.5 Change in the amount of Myakka River juvenile spotted sunfish habitat in response to four water abstraction scenarios (10, 20, 30 and 40%; increasing grey scale) during a recent warm phase (top panel) and cool phase (bottom panel) of the Atlantic Multi-decadal Oscillation.

two species with contrasting habitat requirements. For example, with *Psuedochironomus* sp., the magnitude of habitat lost at greater flow reductions was higher and the duration of loss extended two months longer during the dry season periods than the wet season periods (Figure 15.6 and 15.7).

Similarly, *Cricotopus bicinctus* was also predicted to experience significant changes in habitat availability during the 30-year wet and dry periods. A tabulation of sensitivities reveals that these two genera are equally likely to lose habitat and experience reduced population success during dry periods, while *Psuedochironomus*, a species that

Table 15.1 Comparisons of dominant life-stages in the Myakka River between wet and dry periods as influenced by the Atlantic Multi-decadal Oscillation.

Month	Dry hydrograph Most sensitive life-stage (>15% habitat loss)	Wet hydrograph Most sensitive life-stage (>15% habitat loss)
January	adult largemouth bass	adult spotted sunfish
February	adult largemouth bass	adult spotted sunfish
March	adult largemouth bass	adult spotted sunfish
April	adult largemouth bass	juvenile largemouth bass
May	spawning largemouth bass	juvenile largemouth bass
June	juvenile largemouth bass	adult spotted sunfish
July	adult spotted sunfish	no sensitive life-stages
August	no sensitive life-stages	benthic macroinvertebrates
September	adult spotted sunfish	adult spotted sunfish
October	adult largemouth bass	adult spotted sunfish
November	adult largemouth bass	adult spotted sunfish
December	adult largemouth bass	adult largemouth bass

favours dry periods, is predicted to be most impaired during wet intervals (Table 15.2). These results indicate that the dramatic changes in community structure, function and energy flow associated with wetter and dryer periods are as much or more than those associated with anthropogenic reductions in flows. As for the largemouth bass and juvenile spotted sunfish-based criteria (Table 15.1), choice of target species will affect the regulatory trigger for MFLs depending largely on which phase of the AMO is occuring. The implication is that a robust MFL starategy calls for determination of any confounding AMO phase effects and use of a target species and criteria appropriate to the specific phase occuring.

Creating hydrographic information in ungauged catchments

Clearly, knowledge of the effect of oceanic oscillations like the AMO on hydrographs is important for water resource/river flow and habitat management. However, a large number of the rivers and streams that are likely to be exploited for water abstractions in the future are ungauged or without long-term discharge time series (Sivapalan 2003; Sivapalan *et al.*, 2003). Fortunately, a variety of hydrologic models can be used to establish the necessary discharge–habitat relationship; however, even these require sufficient records of the river-specific hydrograph to produce reasonably accurate estimation of expected conditions (Gore and Mead, 2008). Due to the specific data requirements, the absence of a network of long-term gauging stations, such as those maintained by the USGS, USEPA or water utilities, is a major constraint upon development and application of river and reach specific minimum flow standards. Unfortunately, a large proportion of both new water resource demands and critical fish habitats throughout the world are in ungauged watersheds or river reaches (Smakhtin, 2001; Tharme, 2003). This creates a significant management conundrum because the

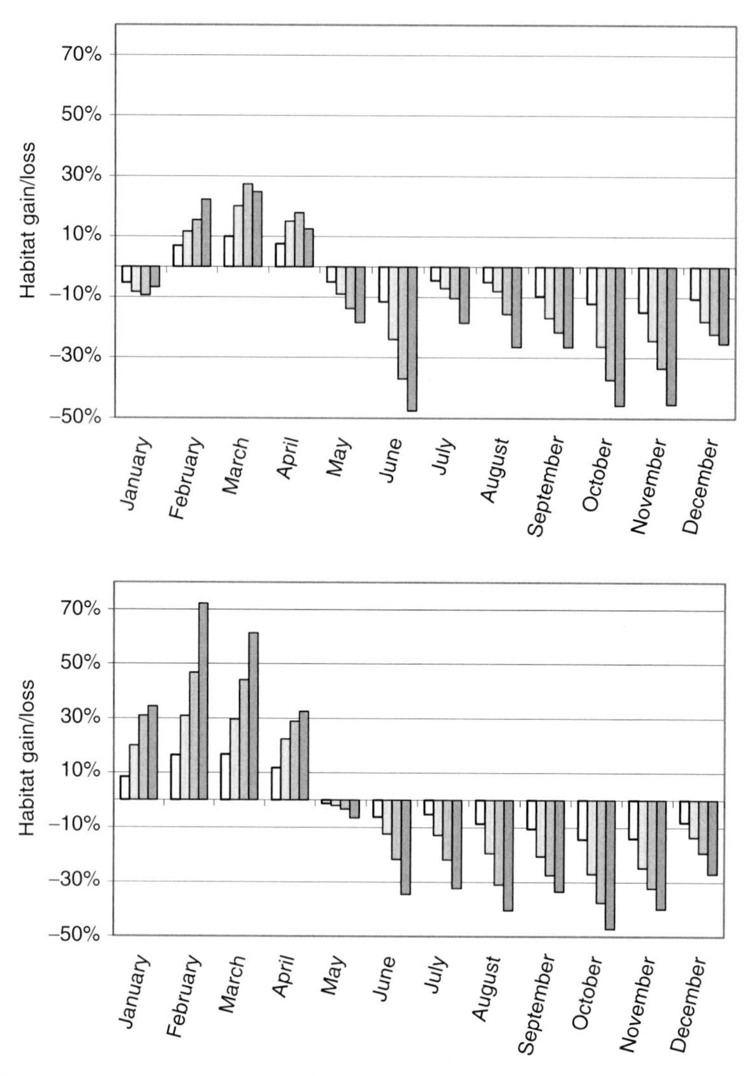

Figure 15.6 Change in the amount of Northern Withlacoochee River *Pseudochironomus* sp. habitat in response to four water abstraction scenarios (10, 20, 30, 40%; increasing grey scale) during a recent warm phase (top panel) and cool phase (bottom panel) of the Atlantic Multi-decadal Oscillation.

most widely approved, applied and successful tools for establishing minimum flows (hydrologic habitat modelling) are limited to systems where gauge data already exist (Gore and Mead, 2008). At the same time the priority systems for future assessment, planning and management are those that do not have gauging stations or historical hydrographs; that is, at least 20 years of hydrographic records. Clearly, new tools are required for these formerly-low-priority, ungauged systems.

One potential solution makes use of GIS-based models to estimate the local stream conditions, including discharge hydrographs of watershed topography, precipitation and subsequent runoff (Kokkonen, 2003; Martin *et al.*, 2005; Cheng

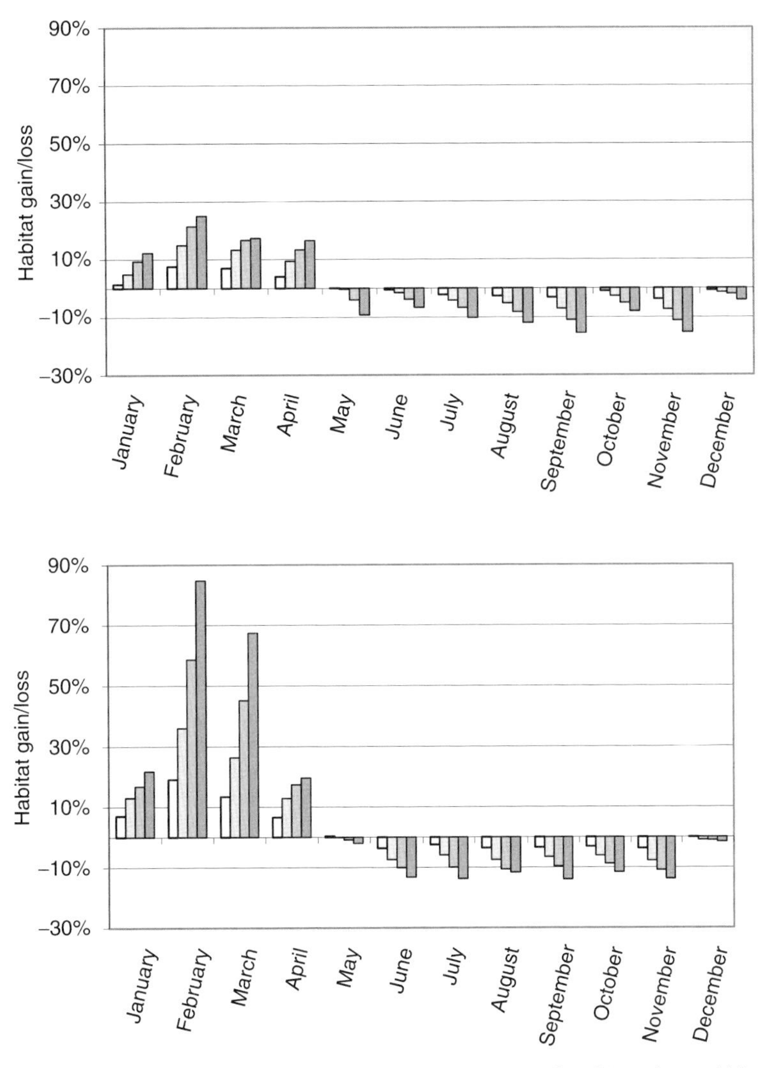

Figure 15.7 Change in the amount of Northern Withlacoochee River benthic *Crictopus bicinctus* habitat in response to four water abstraction scenarios (10, 20, 30 and 40%; increasing grey scale) during a recent warm phase (top panel) and cool phase (bottom panel) of the Atlantic Multi-decadal Oscillation.

et al., 2006; Casper *et al.*, 2011). One of the most widely used of these watershed yield models is the Soil and Water Assessment Tool (SWAT, Neitsch *et al.*, 2001). SWAT was selected for this study because it is spatially explicit and widely available in the public domain and is already in widespread use for water resource management (Reiser *et al.*, 1989; Gore and Mead, 2008). SWAT has the ability to produce watershed specific instream discharge hydrographs using GIS-formatted soil properties, land cover/use, meteorological data and digital elevation models of topography (DEMs). These SWAT-generated hydrographs have the potential to replace gauge records as the input for instream habitat availability models in ungauged river reaches, limited only

Table 15.2 Comparison of dominant Chironomids and the impact of streamflow reduction during wet and dry periods. Numbers in parentheses are the flow reductions determined to cause 'significant' harm.

	1940–69 (Wet AMO)	1970–99 (Dry AMO)
Jan	*Pseudochironomus* sp. (-10)	*Pseudochironomus* sp. (-20)
Feb	*Pseudochironomus* sp. (-20)	*Pseudochironomus* sp. (-20)
	Cricotopus bicinctus (-20)	
Mar	*Pseudochironomus* sp. (-20)	*Pseudochironomus* sp. (-20)
Apr	*Pseudochironomus* sp. (-20)	*Pseudochironomus* sp. (-20)
	Cricotopus bicinctus (-20)	*Cricotopus bicinctus* (-20)
May	*Pseudochironomus* sp. (-20)	*Pseudochironomus* sp. (-20)
	Cricotopus bicinctus (-20)	*Cricotopus bicinctus* (-20)
Jun	*Pseudochironomus* sp. (-10)	*Pseudochironomus* sp. (-20)
Jul	*Pseudochironomus* sp. (-20)	*Pseudochironomus* sp. (-20)
	Cricotopus bicinctus (-20)	
Aug	*Pseudochironomus* sp. (-20)	*Pseudochironomus* sp. (-20)
	Cricotopus bicinctus (-20)	
Sep	*Pseudochironomus* sp. (-20)	*Pseudochironomus* sp. (-10)
Oct	*Pseudochironomus* sp. (-10)	*Pseudochironomus* sp. (-10)
Nov	*Pseudochironomus* sp. (-10)	*Pseudochironomus* sp. (-10)
Dec	*Pseudochironomus* sp. (-20)	*Pseudochironomus* sp. (-20)
	Tvetenia (-20)	

by the length and completeness of precipitation and land use records. However, before this approach can be used more widely, we need validation studies to be confident that GIS-based hydrograph predictions are reliable and accurate. One known limitation to the accuracy of water yield predicted by a SWAT model is the effect of the DEM topographic resolution (Vieux and Needham, 1993; Vieux, 1993; Usery *et al.*, 2004). In the United States, digital elevation models (DEMs) are publicly or commercially available at several different grid resolutions, typically at 300, 90 and 30 metre sampling grids (in Europe they are available at 500, 50 and 10 metres). Choice of resolution with which to simulate watershed topography has strong effects on the accuracy of resulting modelled hydrographs, thus they ultimately have the potential to introduce ecologically significant error into the hydraulic habitat modelling involved in the MFL process (Kelly and Gore, 2008; Casper *et al.*, 2011). Development and use of spatially explicit models are needed in order to provide protection from over-allocation in undeveloped and/or ungauged systems. However, certain caveats and assumptions must be explicit when model output is to take the place of directly measured river basin hydrographs: precipitation records should be of sufficient length to accurately portray one if not both phases of an oscillation such as the AMO, assumptions about changes in land use characteristics during the period modelled need to be explicitly stated, and topographical data such as Digital Elevation Models (DEM) or LIght Detection And Ranging (LIDAR) need to have appropriate resolution and basin coverage to accurately reflect dynamics (timing, volumes) of overland flow to the river in question.

Conclusions

In the search for fish habitat conservation policies and supporting tools, there are several confounding challenges. First among these is disentangling the effects of anthropogenic water abstraction from that of shifts in the hydrologic regime due to natural rainfall–runoff cycles and oscillations like the AMO. These oscillations are primarily oceanic and operate on multi-year or multi-decadal time scales. Thus, it becomes for policy makers, water managers and conservation scientists to connect decisions to continental patterns of river hydrology and predictions of water allocation. However there is ample evidence accumulating and a consensus of opinion forming that these global multi-decadal oscillations play a major role in driving the availability of aquatic habitat and thus structure and function in many rivers across the globe.

Acknowledgements

We thank Dr Gary Warren, Florida Wildlife Conservation Commission, for sharing macroinvertebrate data to supplement some of our analysis.

References

Beecher, H.A. 1990. Standards for instream flows. *Rivers* 1: 97–109.

Bonvechio, T.F. and M.S. Allen. 2015 Relations between hydrological variables and year-class strength of sportfish in eight Florida waterbodies. *Hydrobiologia* 532: 193–207.

Bovee, K.D. 1982. A guide to stream habitat analysis using the instream flow incremental methodology. Instream Flow Info. Paper 12, U.S. Fish and Wildlife Service, FWS/OBS-82/26.

Bovee, K.D., B.L. Lamb, J.M. Batholow, *et al.* 1998. Stream habitat analysis using the instream flow incremental methodology. U.S. Geological Survey, Biological Resources Division, USGS/BRD-1998-004.

Brooks, A.J., T. Haeusler, I. Reinfelds and S. Williams. 2005. Hydraulic microhabitats and the distribution of macroinvertebrate assemblages in riffles. *Freshwater Biology* 50: 331–344.

Caiola, N., C. Ibáñez, J. Verdú and A. Munné. 2014. Effects of flow regulation on the establishment of alien fish species: a community structure approach to biological validation of environmental flows. *Ecological Indicators* 45: 598–604.

Casper, A.F., B. Dixon, J. Earls and J.A. Gore. 2011. Linking a spatially explicit watershed model (SWAT) with an in-stream fish habitat model (PHABSIM): A case study of setting minimum flows and levels in a low gradient, sub-tropical river. *River Research and Applications* 27(3): 269–282.

Cheng Q.M., C. Ko, Y.H., Yuan *et al.* 2006. GIS modeling for predicting river runoff volume in ungauged drainages in the Greater Toronto Area, Canada. *Computers & Geosciences* 32(8): 1108–1119.

Cowell, B.C, A.H. Remley and D.M. Lynch. 2004. Seasonal changes in the distribution and abundance of benthic invertebrates in six headwater streams in central Florida. *Hydrobiologia* 522: 99–115.

Domisch, S., M. Keummerlen, S.C. Jähnig and P. Haase. 2013. Choice of study area and predictors affect habitat suitability projections, but not the performance of species distribution models of stream biota. *Ecological Modeling* 257: 1–10.

Dunbar, M.J., A. Gustard, M.C. Acreman and C.R. Elliott. 1998. Overseas Approaches to Setting River Flow Objectives. *Institute of Hydrology, Technical Report W6-161*, Oxon, England.

Durance I. and S.J. Ormerod. 2007. Climate change effects on upland stream macroinvertebrates over a 25-year period. *Global Change Biology* 13(5): 942–957. DOI: 10.1111/j.1365-2486.2007.01340.x.

Enfield, D., A. Mestas-Nunez and P. Trimble. 2001. The Atlantic Multidecadal Oscillation and its relation to rainfall and river flows in the continental U.S. *Geophysical Research Letters* 28: 2077–2080.

Gaiser, E., N.D. Deyrup, R.W. Bachmann *et al.* 2009. Multidecadal climate oscillations detected in a transparency record from a subtropical Florida lake. *Limnology and Oceanography* 54: 2228–2232.

Gore, J.A. 1989. Case histories of instream flow analyses for permitting and environmental impact assessments in the United States. *South African Journal of Aquatic Sciences* 15: 194–208.

Gore, J.A. and R.D. Judy, Jr., 1981. Predictive models of benthic macroinvertebrate density for use in instream flow studies and regulated flow management. *Canadian Journal of Fisheries and Aquatic Science* 38: 1363–1370.

Gore, J.A. and J.M. Nestler. 1988. Instream flow studies in perspective. *Regulated Rivers* 2: 93–101.

Gore, J.A. and J. Mead. 2008. The benefits and dangers of ecohydrological models to water resource management decisions. In: D. Harper, M. Zalewski, S.E. Jorgensen and N. Pacini (eds) *Ecohydrology: Processes, Models and Case Studies. An Approach to the Sustainable Management of Water Resources*. CABI Publishers, London: pp. 112–137.

Gore, J.A., J.B. Layzer, and J. Mead. 2001. Macroinvertebrate instream flow studies after 20 years: a role in stream and river restoration. *Regulated Rivers* 17: 527–542.

Gore, J.A. C. Dahm and C. Klimas. 2002. *A review of 'Upper Peace River: an analysis of minimum flows and levels'*. Southwest Florida Water Management District, Brooksville, FL, available at [http://www.swfwmd.state.fl.us/documents/reports/files/peace%20river%20review2.pdf]

Gore, J.A., J. Mead., T. Pencczak, *et al.* 2008. Aquatic fauna. In: D. Harper, M. Zalewski, S.E. Jorgensen, and N. Pacini (eds) *Ecohydrology: Processes, Models and Case Studies. An Approach to the Sustainable Management of Water Resources*. CABI Publishing, London: pp. 62–87.

Heede, B.H. and J.N. Rinne. 1990. Hydrodynamic and fluvial morphologic processes: Implications for fisheries management and research. *North American Journal of Fisheries Management* 10: 249–268.

Hussain, Q.A. and A.K. Pandit. 2012. Macroinvertebrates in streams: A review of some ecological factors. *International Journal of Fisheries and Aquaculture* 4: 114–123.

Johnson, N.T., C.J. Martinez, G.A. Kiker and S. Leitman, S. 2013. S. Pacific and Atlantic sea surface temperature influences on streamflow in the Apalachicola-Chattahoochee-Flint river basin. *Journal of Hydrology* DOI 10.1016/j.jhydrol.2013.03.005.

Jowett, I.G. 1997. Instream flow methods: A comparison of approaches. *Regulated Rivers: Research & Management* 13: 115–127.

Kelly, M.N. and J.A. Gore. 2008. Florida river flow patterns and the Atlantic Multidecadal Oscillation. *River Research and Applications* 24: 598–616.

Kokkonen T.S., Jakeman A.J., Young P.C. and H.J. Koivusalo. 2003. Predicting daily flows in ungauged catchments: model regionalization from catchment descriptors at the Coweeta Hydrologic Laboratory, North Carolina. *Hydrological Processes* 17: 2219.

Lamouroux, N. and H. Capra. 2002. Simple predictions of instream habitat model outputs for target fish populations. *Freshwater Biology* 47: 1543–1556.

Lytle, D.A. and N.L. Poff. 2004. Adaptation to natural flow regimes. *Trends in Ecology and Evolution* 19: 94–100.

Mantua, N.J., S.R. Hare, Y. Zhang *et al.* 1997. A Pacific interdecadal climate oscillation with impacts on salmon production. *Bulletin of the American Meteorogical Society* 78: 1069–1079.

Martin, P.H., E.J. Lebouf, J.P. Dobbins *et al.* 2005. Interfacing GIS with water resource models. A state-of-the-art review. *Journal of the American Water Resources Association* 41: 1471–1487.

Maxwell, J.T., P.A. Knapp and J.T. Ortegren 2013. Influence of the Multidecadal Oscillation on tupelo honey production from AD 1800 to 2010. *Agriculture and Forest Meteorology* 174–175: 129–134.

McCabe, G.J. and D.M. Wolock. 2008. Joint variability of global runoff and global sea surface temperatures. *Journal of Hydrometeorology* 9: 816. DOI: 10.1175/2008JHM943.1.

McCabe, G., M. Palecki and J. Betancourt. 2004. Pacific and Atlantic Ocean influences on multidecadal drought frequency in the United States. *Proceedings of the National Academy of Sciences* 101: 4136–4141.

Milhous, R.T., M.A. Updike and D.M. Schneider. 1989. Physical Habitat Simulation System Reference Manual – Version II. Instream Flow Information Paper No. 26. U.S. Fish and Wildlife Service. Biological Report 89(16).

Milliman, J.D., K.L. Farnsworth, P.D. Jones, *et al.* 2008. Climatic and anthropogenic factors affecting river discharge to the global ocean, 1951–2000. *Global and Planetary Change* 62: 187–194.

Monk, W.A., P.J. Wood, D.M. Hannah and D.A. Wilson. 2007. Selection of river flow indices for the assessment of hydroecological change. *River Research and Applications* 23: 113–122.

Munson, A.B. and J.J. Delfino. 2007. Minimum wet-season flows and levels in southwest Florida rivers. *Journal of the American Water Resources Association* 43: 522–532.

Munson, A.B., J.J. Delfino and D.A. Leeper. 2005. Determining minimum flows and levels: The Florida experience. *Journal of the American Water Resources Association* 41: 1–10.

Naiman, R.J., S.E. Bunn, C. Nilsson, *et al.* 2002. Legitimizing fluvial ecosystems as users of water: an overview. *Environmental Management* 30: 455–467.

Neitsch, S.L., J.G. Arnold, J.R. Kiniry and J.R. Williams (2001). *SWAT 2000 user's manual.* Blackland Research Center, Temple Texas.

Nestler, J.M., R.T. Milhous and J.B. Layzer. 1989. Instream habitat modeling techniques. In: J.A. Gore and G.E. Petts (eds) *Alternatives in Regulated River Management*, CRC Press, Boca Raton, FL: pp. 295–315 .

Nunn, A.D., J.P. Harvey, J.R. Britton, *et al.* 2007. Fish, climate and the Gulf Stream: the influence of abiotic factors on the recruitment success of cyprinid fishes in lowland rivers. *Freshwater Biology* 52(8):1576–1586. DOI: 10.1111/j.1365-2427.2007.01789.x.

Olden, J.D. and N.L. Poff. 2003. Redundancy and the choice of hydrologic indices for characterizing streamflow regimes. *River Research and Applications* 19: 101–121.

Pekarova, P., P. Miklanek and J. Pekar. 2003. Spatial and temporal runoff oscillation analysis of the main rivers of the world during the 19th–20th centuries. *Journal of Hydrology* 274: 62–79.

Peñas, F.J., J.A. Juanes, M. Álvarez-Cabria, *et al.* 2013. Integration of hydrological and habitat simulation methods to define minimum environmental flows at the basin scale. *Water and Environment Journal* 2013: 1–9.

Petts, G.E. 2009. Instream-flow science for sustainable river management. *Journal of the American Water Resources Association* 45: 1071–1086.

Poff, N.L., J.D. Allan, M.B. Bain *et al.* 1997. The natural flow regime. *BioScience* 47: 769–784.

Postel, S. and B. Richter. 2003. *Rivers for Life: Managing Water for People and Nature.* Island Press, Washington, DC.

Pyron, M. and T.E. Lauer. 2004. Hydrological variation and fish assemblage structure in the middle Wabash River. *Hydrobiologia* 525: 203–213.

Reiser D.W., T.A. Wesche and C. Estes. 1989. Status of instream flow legislation and practices in North America. *Fisheries* 14(2): 22–29.

Richter, B.D., J.V. Baumgartner, J. Powell and D.P. Braun. 1996. A method for assessing hydrologic alteration within ecosystems. *Conservation Biology* 10: 1163–1174.

Richter, B.D., J.V. Baumgartner, R. Wiggington and D.P. Braun. 1997. How much water does a river need? *Freshwater Biology* 37: 231–249.

Sanz-Ronda, F.J., S. López-Sáenz, R. San-Martín and A. Palau-Ibars. 2014. Physical habitat of zebra mussel (*Dreissena polymorpha*) in the lower Ebro River (Northeastern Spain): influence of hydraulic parameters in their distribution. *Hydrobiologia* DOI 10.1007/s10750-013-1638-y.

Sheldon, A.L. 2011. Comparative habitat use by grazing fishes in a Bornean stream. *Environmental Biology of Fish* DOI 10.1007/s10641-011-9849-4.

Sheldon, J.E. and A.B. Burd. 2013. Alternating effects of climate drivers on Altamaha River discharge to coastal Georgia, USA. *Estuaries and Coasts* DOI 10.1007/s12237-013-9715-z.

Sivapalan, M. 2003. Prediction in ungauged basins: a grand challenge for theoretical hydrology. *Hydrological Processes* 17(15): 3163–3170. DOI: 10.1002/hyp.5155

Sivapalan, M., K. Takeuchi, S.W. Franks, *et al.* 2003. IAHS decade on predictions in ungauged basins (PUB), 2003 – 2012: Shaping an exciting future for the hydrological sciences, *Hydrological Sciences Journal* 48(6): 857–880, doi:10.1623/hysj.48.6.857.51421.

Smakhtin, V.U. 2001. Low flow hydrology: a review. *Journal of Hydrology* 240(3): 147–186.

Southwest Florida Water Management District (SWFWMD). Ecological Evaluation Section. 2004. *Florida River Flow Patterns and the Atlantic Multidecadal Oscillation.* SWFWMD, Brooksville, FL.

Stanford, J.A., J.V. Ward, W.J. Liss, *et al.* 1996. A general protocol for restoration of regulated rivers. *Regulated Rivers* 12: 391–413.

Statzner, B. 2008. How views about flow adaptations of benthic stream invertebrates changed over the last century. *International Review of Hydrobiology* 93: 593–605.

Statzner, B., J.A. Gore and V.H. Resh. 1988. Hydraulic stream ecology: observed patterns and potential applications. *Journal of the North American Benthological Society* 7: 307–360.

Stenseth, N.C., G. Ottersen, J.W. Hurrell *et al.* 2003. Studying climate effects on ecology through the use of climate indices: the North Atlantic Oscillation, El Nino Southern Oscillation and beyond. *Proceedings of the Royal Society of London series B.* 270: 2087–2096.

Tennant, D.L. 1976. Instream flow regimes for fish, wildlife, recreation and related environmental resources. *Fisheries* 1: 6–10.

Tharme, R.E. 2003. A global perspective on environmental flow assessment: emerging trends in the development and application of environmental flow methodologies for rivers. *River Research and Applications* 19(5–6): 397–441.

Usery E.L., M.P. Finn, D.J. Scheidt *et al.* 2004. Geospatial data resampling and resolution effects on watershed modeling: A case study using the agricultural non-point source pollution model. *Journal of Geographic Systems* 6: 289–306.

Vieux, B.E. 1993. DEM aggregation and smoothing effects on surface runoff modeling. *Journal of Computing in Civil Engineering* 7(3): 310–338.

Vieux, B.E. and S. Needham. 1993. Nonpoint-pollution model sensitivity to Grid-cell size. *Journal of Water Resources Planning and Management* 119(2): 141–157.

Whited, D.C., M.S. Lorang, M.J. Harner, M.J. , *et al.* 2007. Climate, hydrologic disturbance, and succession: Drivers of floodplain pattern. *Ecology* 88(4): 940–953.

Zalewski, M., G.A. Janauer and G. Jolánkai (Eds) 1997. Ecohydrology: A new paradigm for the sustainable use of aquatic resources. *International Hydrological Programme, IHP-V, Technical Documents in Hydrology*, No. 7, UNESCO, Paris.

Zigler, S.J., T.J. Newton, J.J. Steuer, *et al.* 2008. Importance of physical and hydraulic characteristics to unionid mussels: a retrospective analysis in a reach of large river. *Hydrobiologia* 598: 343–360.

CHAPTER 16

River restoration: from site-specific rehabilitation design towards ecosystem-based approaches

Jenny Mant[1], Andy Large[2] and Malcolm Newson[3]
[1] *The River Restoration Centre, Cranfield University, Cranfield, UK*
[2] *School of Geography, Politics and Sociology, Newcastle University, Newcastle upon Tyne, UK*
[3] *Tyne Rivers Trust, Corbridge, Northumberland, UK*

Introduction

In the context of practical river restoration (with the UK being the main focus of this chapter), we aim to explore how the quest for an integrated catchment approach (i.e., one that recognises the existence of ecosystems and their role in supporting flora and fauna, providing services to human societies, and regulating the human environment), and one based on natural river process principles, has developed. We also aim to identify trajectories of change in thinking and illustrate how far we have moved from small-scale opportunist restoration, with limited indicators of success, to arguably more integrated and ambitious catchment-scale approaches that can deliver a range of services for the environment and for society.

Such discussion is timely, since despite inevitable barriers there is clear appetite, enthusiasm and opportunities, both at the local and national levels, to instigate catchment-scale restoration. Within this chapter we piece together the last 25 years of practical and scientific elements of river restoration, explore what have been the key drivers for change, and ask challenging questions about where gaps still remain in terms of policy, research and practice.

Trajectories of change

The context of river restoration worldwide is one of past damage, both to morphology and to flow, with resultant ecological degradation. Freshwater conservationists campaigned throughout the 1970s and 1980s to 'mend the morphology' or, at least, slow the rate and extent of damage. Despite publications such as *Fluvial Hydrosystems* (Petts and Amoros, 1996) the hydro-ecological concept took time to be accepted, especially in Europe (see Petts, 1979; Richter *et al.*, 1996; King and Brown, 2010, for more details). This in part was because dam-building activities reduced significantly in Europe during the late twentieth and early twenty-first century compared to other places in the world. Furthermore, it was generally easier to explain how the removal of morphological

River Science: Research and Management for the 21st Century, First Edition.
Edited by David J. Gilvear, Malcolm T. Greenwood, Martin C. Thoms and Paul J. Wood.

complexity had resulted in a reduction of ecological and aesthetic value rather than, for example, explain the importance of how a range of discharges was critical for the lifecycle requirements of fish and macroinvertebrates. Despite this, a number of landmark contributions to interdisciplinary research aimed at better river conservation and management outcomes (e.g., Petts and Calow, 1996; Petts *et al.*, 1989). These highlighted both the benefits of such approaches and the need to understand the relationship between 'services' the ecosystem provides based largely on the hydrogeomorphic character of the river (Thorpe *et al.*, 2010). Over time, this has resulted in an evidence base that has begun to demonstrate that it is the combination of morphology and hydrology that determines how effective a river can be in terms of supporting a range of wildlife (e.g., Newson, 2010a). The concept of 'hydromorphology' is now embedded within the EU Water Framework Directive along water and sediment quality issues: collective quality improvement of all these elements is now seen as critically important to improving ecological habitat and restoration success. But, despite this, questions remain about how far this understanding has translated into 'best practice' river restoration that has as its outcome enhanced catchment-scale connectivity. How far this has been achieved and what the future might hold is discussed in this chapter.

Restoration beginnings in Britain and Europe: EU-LIFE and the River Restoration Project

In September 1990, the then Nature Conservancy Council in Great Britain held an international conference entitled *River Conservation and Management*. The conference was held at York University, attracted 337 delegates from 29 countries (Boon, 2012), and its emphasis reflected the fact that the subject of river conservation and management, including restoration, was then at an early stage of evolution. A serendipitous outcome of the York conference was sparked by a paper entitled 'The struggle to conserve one Czech river', delivered by Nadia Johanisova (cited in Holmes and Janes, 2012). Johanisova described how, even in the 1980s, initiatives existed (within Europe but outside the UK) where local communities were undertaking practical river restoration schemes under their own volition. This led to a number of conference delegates, championed by the late Nigel Holmes, to meet to determine what of a practical nature could be done to improve river management in the UK. The outcome was the founding of the River Restoration Project (RRP) in 1993, which led directly to the formation of a national River Restoration Centre (RRC). This novel idea gained significant impetus via European Union-LIFE funding along with financial support from local authorities, river management and other organisations. The practical outcome was three comparative restoration schemes in Europe with their shared objective being development of formal and agreed generic methods for improving the ecological and aesthetic value of degraded rivers. One of these schemes was on the River Brede in Denmark; the other two were located in the UK and addressed both urban and rural settings through the well-documented, both in terms of design and evaluation; the River Skerne in north east England and the River Cole in southern England (e.g., Holmes and Nielson, 1998; Murphy and Vivash, 1998; Biggs *et al.*, 1998; Åberg, 2010).

The rivers restored within this EU-funded study had all undergone significant human-induced modification. The aspiration was to demonstrate, through the implementation of a variety of techniques, that a range of services could be addressed and improved including, for example, recreational amenity, water quality, fisheries, wildlife habitat and flood risk (Vivash *et al.*, 1998). The project had a key objective of serving as a catalyst, through which knowledge could be gained and expertise shared, to benefit all who might subsequently undertake river restoration work. As the aim of this demonstration project was promotion of further river restoration work, sites were chosen where degraded river reaches could be improved with low risk of failure (Åberg and Tapsell, 2012). In terms of geomorphology and stream power calculations, for example, these sites fell well below the 'dynamic adjustment' threshold outlined by Brookes 1987, and therefore, once modified, needed the implementation of restoration techniques to 'kick start' any natural processes.

Early restoration designs were developed with input from a mix of academics and practitioners from a range of disciplines. Although channel restoration guides and concepts had already been compiled and implemented in other countries, especially in the USA as discussed by Palmer *et al.* 2005 and in Germany (Kern, 1992, 1994), no reach-scale river rehabilitation had been carried out prior to this venture in either the UK or Denmark. As a result, the designs, it could be argued, were cautious, underpinned by river engineering principles, and completed in opportunistic locations where there was a single landowner with a positive attitude to restoration principles. Certainly, compared with more recent guiding principles on river restoration, the projects were not specifically set against a reference reach within the catchment or process-based principles as discussed, for example, by Beechie *et al.* 2010; Roni and Beechie (2013) and the RRC's online *Manual of River Restoration Techniques* (2014). Historic maps were consulted to provide background about the river prior to its degradation, but there was no translation of the restoration concept into what is now termed in the EU Water Framework as 'hydromorphological quality' (i.e., the structure, evolution and dynamic morphology of a hydrological system over years, decades and centuries), or 'favourable condition/conservation status in the context of optimal habitat condition for a specific designation', as defined by Jones (2002). This is not to denigrate what proved to be very worthwhile and high-profile activities (such was the prestige that the then UK Prime Minister, John Major, attended the official RRP launch). Rather, it reflects the prevailing issues of the time in that whilst academic attention was directed at understanding river catchment processes in the context of reaches of erosion, deposition and transport for management purposes, this information was not cascaded down and embedded in the more practical elements of best practice interventions and appropriate use of restoration techniques. The general consensus was that restoration required a close approximation to a structural and functional return to some pre-disturbance condition (Cairns, 1991; Downs and Gregory 2004). Defining such pre-disturbance condition often became the hardest thing to accurately define, especially in systems that had been altered in terms of their morphology and hydrological regime. The outcome was that many approaches ended up as piecemeal, site-specific eco-engineering projects designed as sediment transfer reaches and hence often unable to recover through natural physical processes.

Like many schemes (e.g., Moerke and Lamberti, 2004), the River Skerne's long-term ecological recovery has been hampered by water quality. In the River Cole, barriers to sediment transport (including low stream power) have reduced the natural development of riffles (Åberg, 2010). Pederson *et al.* (2007) observed that, whilst wetland bird species abundance and diversity had increased on the rehabilitated Danish River Skjern, an unintended consequence was increased predation on migrating smolts of salmon and trout. Despite this, the authors argue that these inspirational 1990s projects did do exactly what they intended to do: act as a catalyst for future change in attitudes towards restoration of our water courses. Since then, despite there being a range of legal environmental mandates in place for many countries (Beechie *et al.*, 2010), criticism has often been voiced, particularly during the late 1990s and early 2000s (e.g., Ormerod, 2004; Palmer *et al.*, 2005), about the lack of frameworks and strategies that focus on integrated delivery at the catchment scale. Within the same time period, Petts *et al.* (2002) stressed that it is essential that rivers, particularly urban ones, are perceived as the central focus in terms of connecting different catchment areas and thus contributing to social cohesion, recreation, navigation, flood management and nature conservation. It is these wider benefits and attributes that are now beginning to be enveloped in the concept of 'ecosystem services' (Palmer *et al.*, 2014; Gilvear *et al.*, 2013; Large and Gilvear, 2015).

Delivery of restoration

A workshop on river rehabilitation at Loughborough University, England, in the early 1990s raised a number of specific questions regarding early attempts at river restoration. Much discussion at the time centred on definitions of restoration (interpreted as strategic approaches towards full structural and functional return to a pre-disturbance state (as discussed above) and as opposed to the less-ambitious and often opportunistic rehabilitation efforts prevalent at the time). Even at this early point in the UK's river restoration history, there was a recognition that schemes needed to be holistic and entail at least an element of 'catchment consciousness' and, wherever possible, concentrate on self-sustaining regimes so that disturbance and natural subsidies of energy would foster natural succession and rejuvenation. Yet, demonstrating how to do this effectively has not been a straightforward path towards success.

Towards catchment consciousness?

Initially, it was felt that to achieve 'catchment consciousness', restoration workers needed a shared vision (the *'leitbilt'* of Kern, 1992) of what was feasible in any particular case study. A recurring issue over the ensuing two decades, however, has been a consistent definition of this vision. The amount of freedom for restoration is often site specific, and constraints exist as to how far we can 'let nature rule'. Wohl (2004) highlights an enduring problem in that people interpret a 'healthy' river as a 'pretty' one with, for example, clear water and stable banks, as well as being 'active'. However, healthy systems often possess neither attribute and even decent-looking systems may be critically impaired due to activities such as flow regulation or long-term degradation of riffle and pool habitats.

More recently, stakeholders have been encouraged to become involved with restoration schemes (Newson, 2011). As such they expect, and indeed deserve, accessible information in order to properly judge risks and benefits. In Scotland, for example, a scheme run by the World Wildlife Fund (WWF) in the 1990s entitled 'Wild Rivers' struggled somewhat, due to negative public perception about the term 'wild'. Similarly, the series of damaging flood events in the UK since 2000 has made local communities more risk adverse to changes in how our river systems are managed. Understandably there is often a reticence for river and associated floodplain schemes aimed at 'slowing the flow' in the name of more sustainable flood management with wider environmental benefits. Such schemes are often perceived as increasing the uncertainty about how rivers work. Yet, at the same time, government agencies advocating societal benefits are actively encouraging the idea of a catchment-based approach to river restoration on the grounds that it will deliver positive and sustaining outcomes, and through transparent decision-making processes will deliver activities that improve the water environment (Defra, 2013). With such differing messages, and with the current UK emphasis on science with societal impact, river scientists need to be mindful not to generate 'evidence' without having a well-defined end user or policy dimension focus. Without this link between policy, science, practice and societal needs, the current optimism (and funding base) for catchment-based restoration could be short-lived.

Knowing what we are aiming for: a question of pre-disturbance or future resilience

Ensuring catchment characteristics are understood, and developing river restoration plans within this context, is now a recognised mainstream approach across the UK, in part promoted by the RRC (RRC, 2011). This reflects the situation elsewhere and especially in the United States (e.g., Roni and Beechie 2013). Whilst the aim of restoration efforts must always be long-term, sustainability and resilience questions remain as to what demonstrates success. An issue here is that whilst those funding the work often want very specific answers to questions, the research community traditionally deals in trends, not real numbers (e.g., Bradshaw, 1988, 1996). This divergence of approaches and attitudes is further complicated by the premise that river ecosystems are driven by natural, nonlinear succession, which complicates both the vision of what state a system should be restored towards and how to define outcomes. This debate brings us to the discussion about redefining what is meant by 'reference' conditions so that success can be confidently expressed to all interested parties. In early restoration efforts, a reference condition was sought that reflected any changes in restoration potential that may occur through time, defined by Kondolf and Downs (1996) as 'a relatively natural channel reach with a relatively high conservation value (presumed to be similar to the pre-disturbance state'. However, identifying such a pre-disturbance reference point that can determine 'normal' ecological communities and physical processes is often difficult: if inappropriately identified, it can hamper demonstration of success, as reported by Sear *et al.* (1998). Others favoured the idea that reference conditions for river restoration should be based on calculating the expected/predicted changes and improvements based on the understanding of current processes (Power *et al.*, 1998).

The need for a clear approach to reference condition thinking is demonstrated through the limited conclusions derived from completed restoration schemes. For example, De Waal *et al.* (1995) detailed 66 restoration schemes operating in Europe in the early 1990s; of these 68% were rural in character, 11% were purely urban, with the remainder being a combination of both. Since the completion of these schemes in the 1990s, several studies have been carried out seeking to demonstrate their success, yet the findings have been far from conclusive. Few of these studies consider reference conditions, and this together with the inevitable lack of baseline data collection (Downs and Kondolf, 2002) make necessary repeated research requests for long time series involving repeat monitoring to quantify trends (e.g., Tockner *et al.*, 1998; Roni *et al.*, 2002; Woolsey *et al.*, 2007). The quest to effectively quantify success was also hampered by the 'disappearance' of results and information gained from pre- and post-project appraisal into unpublished reports and the 'grey' literature (Wade *et al.*, 1998) along with the overall lack of project monitoring (Smith *et al.*, 2014).

As a result of the above, the evidence base about river ecological and physical processes stems from empirical observations that seek to identify repeatable patterns and trends over time. We now appreciate that there are predictable species–habitat relationships over short timescales, that wood plays a key role in rivers worldwide and that large rivers behave differently to smaller streams (e.g., Gupta, 2008). Crucially, we also now accept that heavily modified rivers are the 'norm'. As Geoff Petts concluded at the 2013 ISRS symposium in Beijing, the traditional three-reach classification model favoured since the 1960s often no longer applies: 'we've done away with the mid-reaches by damming upstream reaches and extending downstream reach types back upstream principally through channel modification' (Petts G.E., 2013, *personal communication*). System dynamics are often more redolent of transient states than those in equilibrium, and hydro-ecological models are limited by modified river conditions, simplified channels and degraded biological conditions compounded by alien and introduced species. Changes in the quality, quantity and seasonal availability of food for organisms, deterioration of water quality through temperature changes and excessive turbidity, habitat modification, water quantity and flow and reduced biotic interactions, as explained in NRC 1992, are the key recognised stresses on our river systems. Clearly, collectively, there has been much discussion about restoration success and the impacts of management on rivers, but questions remain as to where this leaves us in terms of understanding river restoration processes and projects. By the mid 1990s it was appreciated that monitoring and evaluation methods needed to be tailored to restoration objectives and targets to heighten this understanding, and to convince river managers that river restoration objectives could lead to a reduction in the need for regular maintenance. Case studies that could demonstrate this therefore became important.

Restoration models: towards natural rivers?

As indicated above, it is undeniable that, in their contemporary state, the vast majority of the planet's rivers are anything but 'pristine', or even near-natural, despite many rivers appealing to human aesthetic judgements in a landscape context. Worldwide, restoration scientists are no longer dealing with 'natural' rivers, defined by Newson and Large (2006) as 'those requiring minimum management interventions to support system resilience and protect a diversity of physical habitat'. While resilience and habitat diversity are not universal or perpetual 'ecosystem services', their value increases with the proportion of the channel network within the 'fluvial hydrosystem' (Petts and Amoros, 1996; Petts, 2000) exhibiting full interplay of unmanaged water and sediment fluxes with local boundary conditions. Since efforts are regularly hampered by lack of definition of what constitutes useful reference points or baselines typical of unmodified rivers to make real progress with river restoration, decision-making frameworks are needed that promote progressive and strategic actions. At the same time, they need to give confidence that actions are realistic and worthwhile (Mainstone and Holmes, 2010). The concept of 'shifting baseline syndrome' is useful here and is gaining credence in conservation ecology (Pauly, 1995; Papworth *et al.*, 2009). The premise here is that river restoration efforts are hampered by knowledge extinction occurring because younger generations are simply not aware of past river and catchment conditions ('generational amnesia'), as well as where individuals and communities forget their own experience of change. A pertinent example of shifting river baselines is provided by Walter and Merritts (2008), who describe how gravel-bedded mid-Atlantic streams in the USA are considered to have a characteristic meandering form bordered by self-forming, fine-grained floodplains – an ideal that guides and drives the contemporary multibillion-dollar US stream restoration industry. Analysis of stream deposits shows that before European settlement, these were instead small anabranching channels within extensive vegetated wetlands that accumulated little sediment but stored substantial organic carbon. Subsequent deposition of anything between 1 to 5 metres of slackwater sediments behind tens of thousands of seventeenth- to nineteenth-century mill dams buried these pre-settlement wetlands under fine-sediment fill terraces.

This further adds to the argument that using process science approaches to define what constitutes a 'natural' river is more appropriate than just using the more traditional empirical 'look and see' reference condition approach. Understanding and identifying the 'historical range of variability' (Brown *et al.*, 2013) in river attributes such as flow regime and planform that existed prior to intensive human alteration (Wohl, 2011) is important, and whilst this may be acknowledged within the academic arena, the challenge now is to ensure that this concept and how to apply it is widely disseminated to practitioners.

Delivery challenges: processes and practice

The opportunistic nature of early river restoration projects inferred a high degree of 'buy-in' for all parties involved. Everard and Capper (2004) recognised, however, the distinct difference between river protection via what they coined the 'common law of rivers' and statute law: whilst the first may provide a rich opportunity to achieve holistic outcomes. Especially where local stakeholders are engaged, the latter often either pay for or at least administer Directives. Rolling out the bigger agendas of the EU Habitats and Water Framework Directives has subsequently highlighted the issue of delivery barriers, not just in the context of government laws but also in the public perception that river restoration is merely 'turning the clock back'.

Public engagement for river restoration ideals

Early approaches to river restoration in the UK implied river engineering and a 'technocracy' of design, delivery and maintenance by agencies, consultancies and works gangs, often resulting in depriving riparian owners and communities of their former direct responsibilities. Public relations has now become integral to restoration projects and needs considerable care given the inherent uncertainty in river designs and outcomes (Newson and Clark, 2008). In England and Wales, conservationists devised more strategic approaches to evaluating in-river and riparian habitat features via the system of River Corridor Surveys (National Rivers Authority, 1992) and the River Habitat Survey (Environment Agency, 2003). Both of these approaches remain wonderful resources where, in the case of the Corridor Surveys, the documents have been preserved and, in the case of the Habitat Surveys, provide a baseline condition assessment. However, outside the game angling fraternity and conservation NGOs – notably the Royal Society for the Protection of Birds (RSPB) who commissioned handbooks and videos for knowledge transfer (Ward *et al.*, 1994; Jose *et al.*, 1999, RRC, 2014) – there was little public participation as there was little awareness of the pivotal role of rivers in British landscapes (Holmes and Raven, 2014). Another yawning gap in the early 1990s was the lack of a natural sciences equivalent of 'engineering design' outside academia. Three main components have characterised later restoration approaches:

- A greater *spatial awareness* of river processes than that characteristic of the civil engineering site approach – requiring walk-over surveys.
- A bigger emphasis on the *ecosystem health* of rivers and a move away from prevailing institutional preoccupations with human safety (from flooding and chemical pollution) – requiring a new attention to physical habitat.
- A realistic approach to the *uncertainties* of the 'new', non-engineering, sciences such as geomorphology compared with traditional public confidence in engineering solutions, which in fact are often over-designed and require expensive maintenance and are rarely performance tested or subject to evaluations of sustainability.

The third principle outlined here is vital in that political processes are already advancing the first two via, for example, River Corridor and Habitat Surveys together with critiques of both traditional 'land drainage' and the prominent role under law for chemical water quality. The EU-LIFE-supported river restoration project in the 1990s (see 'Restoration beginnings in Britain and Europe: EU-LIFE and the River Restoration Project' above), had its success boosted by the fact it brought together the most practically experienced of the campaigners for change and the least risk-averse members of the traditional engineering fraternity. Notably, the restoration designs had major technical inputs from geomorphologists and ecologists (Kronvang *et al.*, 1998). By forming effective partnerships and riding a wave of enthusiasm, restoration was successfully and sustainably delivered to significant reaches of the rural River Cole and urban River Skerne (Åberg and Tapsell, 2012).

More importantly, these lessons were not lost a decade later on the burgeoning third sector in river projects in the United Kingdom – the Rivers Trusts. These began to appear during the mid 1990s and now number nearly 50 across the UK. A Rivers Trust is readily able to form partnerships without impeding the important campaigning benefit of angling and other interest groups. It can justifiably expect to gain charitable status, which confers several important benefits not always available to public bodies or vested interest groups. Of vital importance is the public benefit, which can easily be demonstrated for larger catchments or groups of catchments, but may need greater consideration for smaller rivers and catchments where a few riparian interests may be perceived as deriving disproportionate benefits, for example commercial angling interests.

Typically, a Trust starts with a Board of Trustees overseeing and freely offering their time and a wealth of knowledge covering important aspects of the Trust's activity, including legal, business and accounting, fisheries, agriculture, tourism and education. As the Trust develops, the demands on time become more onerous and, funds permitting, the Board usually takes the (significant) step of appointing a small team of professionals, often beginning with a scientist or educationalist. This team works closely with the local community, including river owners and land managers. This interaction is central to success, healthy growth and sustainability (Newson, 2012). Whilst the Tweed Foundation is the 'doyen' of the Rivers Trusts, its mainly rural landscape has involved it less in river restoration than the Trusts in England. The EU-LIFE project discussed above occurred without a Rivers Trust playing a central role, but the 'movement' was burgeoning and, together with the RRC, provided momentum to carry on the work of river improvement. The roles of the two organisations have over time become clear. The RRC has become the strategic hub for gathering and disseminating best practice river restoration information through various media and training. This hub role provides an essential support mechanism for all organisations, local communities and associated Rivers Trust involved with river restoration at the local/site level.

Restoring habitat and processes: entering the catchment

Early river restoration efforts were heavily channel-focused; 20 years later many would query the economic and financial wisdom of river restoration without 'catchment consciousness' (Newson, 2010a, 2000b), and Defra in England have recently espoused a 'Catchment Based Approach' (Defra, 2013) within the River Basin Districts of the WFD – often amalgamating several WFD water-bodies. Scale arguments are vital. An early focus and an obvious target for catchment-scale campaigns has been rural diffuse pollution and the role of land users in 'aggravating' sediment yields beyond the natural fluxes (Newson, 2010c). Acidification, eutrophication and siltation remain important targets for regulatory and voluntary Best Practices at the catchment scale. Academic attention has begun to focus on the role of the hydrological connectivity introduced by development and management of catchments, from farm/forest drainage to roads and tracks (Bracken and Croke, 2007). A landscape-scale approach, within and beyond the catchment-scale, is becoming fashionable. Ironically, catchment connectivity is being embraced at the very time a 'disconnected rivers' approach to fisheries improvement has been supported by those in favour of increasing the commitment to the generation of hydropower by using old weirs or building new ones (e.g., Edgeworth, 2011). Here, the scientific tools for regulating from a base of evidence were ill-prepared; anglers and conservationists have resorted to protest campaigns to point out the irony of weir removal and weir building going side-by-side.

Rice *et al.* (2010) have put forward a '2020 vision' for river management, emphasising that fluvial geomorphology needs to sit alongside stream hydrology and hydraulic engineering as a key element of an integrated, interdisciplinary scientific approach. To this must be added ecology. While it is true that the last 25 years of river restoration has seen an upscaling of this integrated agenda from in-channel schemes towards the catchment scale, it is important to note that this progress has attendant geopolitical aspects – the pressure is on to have more done, in more places, affecting more people. Despite this, scientific underpinning is patchy. The Defra-funded and Environment Agency-administered catchment restoration fund, covering 42 projects across England and Wales, may well start to highlight successes and issues in a more strategic way over the coming few years. Building on its recognised independency, the RRC has encouraged the catchment restoration fund teams to think carefully about realistic monitoring that should collectively help to answer a series of pressing river restoration questions about habitat enhancement, water quality and community involvement and the potential benefits of up-scaling where possible using a riverine ecosystem approach or 'synthesis' (Thorp *et al.*, 2006). Table 16.1 summarises the potential benefits of upscaling. While there are key potential gains to be achieved by moving from more traditional channel-oriented schemes to larger, more holistic landscape-scale approaches, such step-ups are rare due to the drop-off in predictability of outcome with increasing scale of approach. Without government financial research funding for larger spatial-scale approaches however, the knowledge base on catchment connectivity will remain relatively intangible.

Rivers: local delivery and demands for an evidence base

Despite this move to widening management towards the catchment/river ecosystem scale, the severe floods of winter 2014 and 2015 in England, and in particular the resultant adverse public and political response to existing management paradigms, created apprehension in the river restoration community in the UK. Misinformation about the policy balance between flood risk and ecosystem management was rife, notably in the press and in Parliament. The public was encouraged to consider that 30 years of technical, practical and regulatory progress had been achieved by some sort of public deceit about the true hazards of 'natural' rivers. Prominent academics wrote to the national and local press about calls for renewed 'dredging' of channels. The professional responses (e.g., CIWEM, 2014) have highlighted the need to incorporate stakeholder perceptions of river processes and forms; these however are often couched in the form of 'myths' about flooding, erosion, deposition and habitat. In reality, almost the whole agenda of restoration implementation may now require renegotiation in the context of the 2014 flooding in the southwest and southeast of England and revitalised debates around the issue (Newson, 2014a, b). Practical experiences of the new framework for restoring protected rivers ('Catchment Restoration Strategies' under the EU Habitats Directive, and in the UK Defra's 'Catchment Restoration Fund') have shown the constant necessity for river scientists to deliver the evidence base to counter these myths and misapprehensions within stakeholder groups and the Cumbria floods

Table 16.1 Potential and limitations of conventional restoration compared with landscape-scale approaches (adapted from Temmerman *et al.*, 2013).

Restoration objective/ecosystem service enhancement	Traditional engineering-based restoration	Ecosystem-based restoration/ 'catchment consciousness'
Natural habitat	To reverse degradation	Landscape-based conservation or restoration
Sediment issues	Remove sediment trapped by dams, weirs etc. and hence seen as a problem.	Sustainable solutions to sediment as seen in context of river's dynamic equilibrium. Sediment addition along with woody debris.
Damage from extreme weather events exacerbated by wetland reclamation, land drainage and groundwater depletion	Wetland reclamation addresses site-specific issues	Catchment approach to flood attenuation, expanding water storage and friction
Cost-benefit appraisal	Moderate to high	Consistently high due to added ecosystem service benefits
Water quality	Doesn't address degradation by organic matter accumulation and toxic algal growth	Improved and sustained by nutrient and contaminant cycling in restored wetlands in river corridors
Climate mitigation through carbon sequestration	Very little	Wetlands and marshes important sinks
Fisheries and aquaculture production	Low habitat for fish and other biota; often a lack of sediment movement	Improved: more habitat creation for fish and other biota due to inclusion of floodplain refugia etc.
Human recreation potential	Negative perceptions of engineered restoration solutions	Positive perception of natural landscape
Required space	Moderate	High, therefore not applicable near large urban areas
Potential for using natural subsidies of energy and materials	Low to moderate	Relatively high due to natural dynamics and variability
Implementation and experience	Substantial, but lack of monitoring/empirical evidence of success in the past	Limited so far. More research urgently needed
Social and political acceptance	Widely accepted	So far only accepted in limited areas
Health hazards (other than flooding)	None	Wetlands with stagnant water a potential hazard

of 2015 and communities. It is often vital, for example, to separate out the component options in catchment-scale restoration before particularising to project reaches: for example 'processes', 'dimensions' and 'forms' (PDF model: Figure 16.1). Because dimensions (including both channel and corridor widths, but also inundated areas) are often the most contentious elements of restoration design, the most convincing negotiating stance for affected stakeholders is often to stress the major aim as restoring

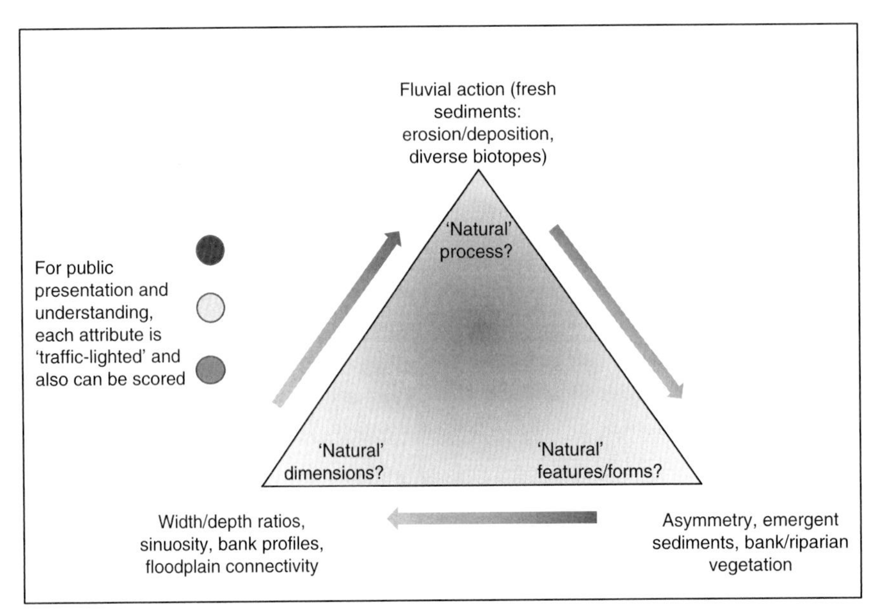

Figure 16.1 PDF (process-dimensions-form) model, outlining how stakeholders can access ecological assessments of river habitat. 'Quantity' issues are towards lower left, 'quality' towards upper right.

process. In reality the typical approach is to restore *form* (e.g., using gravel seeding or large woody debris) and while this can result in improved habitat *quality*, it may be of reduced *quantity*.

While negotiating strategies such as these may be anathema to scientists, enduring myths about river processes also necessitate an alternate, more easily-understood platform for traditional empirical evidence. Amongst the most regularly encountered myths from stakeholder consultation exercises are:

(1) Channel deposition causes floods by reducing conveyance (held for *every case!*).
(2) Fast flows 'directed' by bars cause bank erosion (observations mainly at low flow).
(3) 'Big stones stay where they're put' (often used as a defence for 'rip-rap').
(4) Obstructive weirs should 'just be taken out' (often the anglers' view).

Readers of this volume will immediately see the errors in these statements, but as a profession we must refute them confidently, using appropriately illustrated, secure evidence (Newson, 2014a, b).

Science supporting practice

In a 25-year review period it is inevitable that far more than policy and political contexts have radically changed. The need for integrated 'river science' has become clear, with Defra (2011) recognising that the natural world, its biodiversity and its ecosystems are critically important to our wellbeing and economic prosperity. Practical river restorers are now in a position where they can now call up a much more impressive raison d'etre than in the 1990s, when acts of faith and enthusiasm were more common. Perhaps more impressive, however, has been the growth of survey, sensor and computing power, which has developed the available databases and allowed their incorporation in powerful predictive models.

Scientific progress towards restoration strategies

River science has progressed over the last half-century via a series of conceptual advances in our understanding (Figure 16.2). In many ways, the engaged, publically supported movement for more 'healthy' rivers has been much patchier. On the one hand, there has been the dimension of 'know your river' (ranging from knowledge through regular use to a more formal citizen science ethos), which provides the holistic picture and the evidence base often lacking in regulatory institutions. On the other hand, finances for works, however low key and 'green', are hard to come by, and the conditions for funding often occur serendipitously. *'Have a drawer full of projects and wait your chance'* was the advice provided by the Board of one UK Rivers Trust to its Director!

Whilst opportunism retains a key role in promoting river improvements, two factors have spurred the entry of the Trusts (and, before them, the RRC) into mainline delivery of restored river habitats and the realisation that more scientifically-underpinned proof of concept is essential to future proof river restoration success. One was the acceptance by regulators throughout the UK that a third sector – catchment stakeholders – could play a vital *and competent* role in rehabilitation and restoration. The other was the arrival of the European Directives, most notably

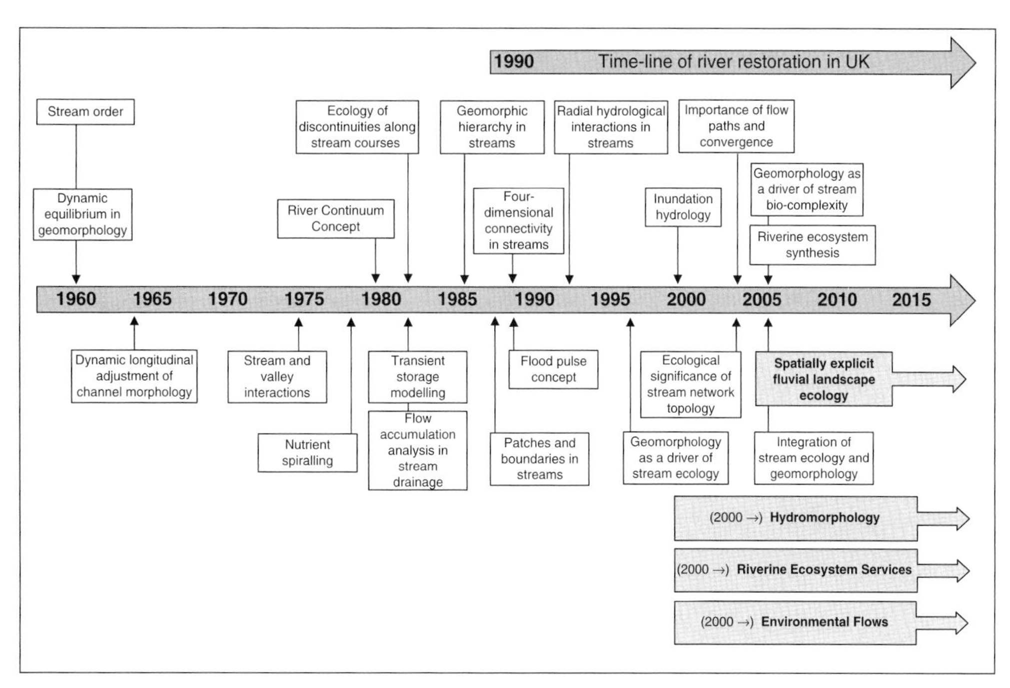

Figure 16.2 Key developments in stream science over the past half century as identified by Poole (2010) and key areas predicted to define river restoration approaches in the future (grey scale) emphasising the new 'triple bottom line' of hydromorphology, ecosystem services and environmental flows, set in the context of spatially-explicit fluvial landscape ecology. Poole 2010. Reproduced with permission from The Society for Freshwater Science.

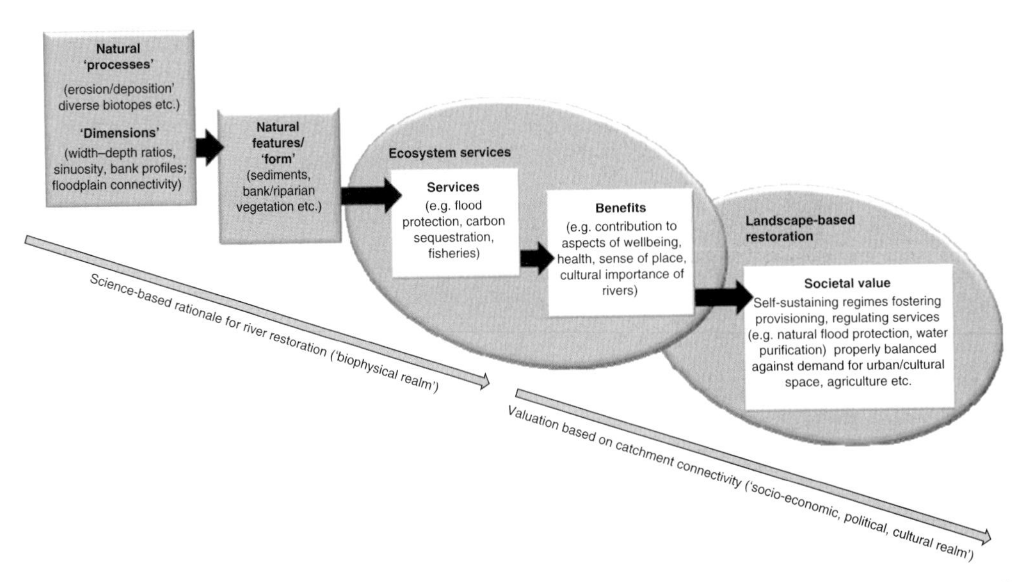

Figure 16.3 The cascade model of Haines-Young and Potschin 2010, adapted to illustrate the need for both science-based underpinning to restoration activity as well as publically-accessible valuation of what connected 'natural' river ecosystems can do for us. Haines-Young and Potschin 2010. Reproduced with permission from Cambridge University Press.

the Water Framework Directive and the earlier Habitats Directive, underpinned by the ecosystem concept of the freshwater environment (Figure 16.3). There has been much discussion about the role and relevance of EU policy (see Newson and Large, 2006; Newson, 2012; Wharton and Gilvear, 2006) but its significant legacy has been to raise the status of hydrologically-defined river physical habitat or 'hydromorphology' as the new body of knowledge forming the basis for river channel restoration. Hydromorphology adds the vital components of flow hydraulics and habitat sensitivity to traditional fluvial geomorphology but as yet has not been sufficiently supported by interdisciplinary academic research to the point where a ready-to-go set of tools is available to use in river restoration (Vaughan *et al.*, 2007); nor indeed a demonstration of the degree of impact of delivering river restoration at a point in a catchment – even though research has arguably been quicker on the catchment dynamics aspect of river restoration. Nevertheless, encouraged by the willingness of flood risk professionals to work with natural processes (Environment Agency, 2014), we are now on a rapid learning curve with progress being made on:

- Understanding the patch dynamics of river habitats, created by interactions of flow and substrate (e.g., Newson and Newson, 2000).
- Paying more attention to the uniqueness and complexity of individual river sites (e.g., the RRC, 2011).
- Including, at last, the impact of our 'disconnected rivers' (e.g., Wohl, 2004) and our modified flows on the operation of river networks as hydromorphological systems.

The impacts of Britain's heavy (two-thirds) reliance on surface water supplies, manifested through a dense network of headwater impoundments and modified flow regimes, has made it necessary to classify many impacted rivers under the WFD as 'heavily modified'. Although mitigation of adverse impacts on the flow regime is necessary for these water bodies, there are clear economic implications. Recent guidance (UKTAG, 2013), based on ecological indicators of damage caused by artificial flow regimes, has marked a significant advance in evidence gathering in hydromorphology, and the recommended 'building-block' approach championed in both South Africa and Australia (King and Louw, 1998; Arthington and Pusey, 2003) is now recommended in England and Wales as a consultative process designed to help agree practical measures to mitigate 'heavy' modifications to flows. One of the principal shortcomings of many rehabilitation schemes was lack of monitoring, for example Skinner and Bruce-Burgess's (2005) post-project appraisal system for assessing morphological recovery, or the RRC's PRAGMO toolkit (River Restoration Centre, 2011), remain under-used. Finally, channel engineering has not been discounted as a means to influence flows downstream of reservoirs: the proposal by the Environment Agency in England and Wales (EA, 2013) to use charges from abstractors to fund channel restoration that accepts modified flows but restores ecological features again represents a trade-off between habitat quality and habitat quantity. The context for this is contentious given the drive for flow modification to create renewable energy from hydropower.

Technical innovation and river restoration

Recalling the strategic development of early river restoration projects, such as the River Restoration Centre's (RRC) *LIFE* Project, there was a huge burden of proof on the geomorphology and ecology protagonists as they entered a world dominated by 'precise' engineering design and hard engineering. In the 1990s, the UK had no standardised geomorphological survey along the lines of the USGS (and in 2016 the UK still has no sediment load gauging stations). The technique of River Habitat Survey was yet to make its impact and the notion of geomorphological assessments, such as Fluvial Audit, was unknown. Research sponsored by the Environment Agency in England and Wales and by SEPA, the Scottish Environmental Protection Agency, led to a suite of scale- and cost-sensitive assessments to facilitate river management (Williamson *et al.*, 2015). Evolving techniques included Catchment Baseline assessment, Fluvial Audit, Dynamics Assessment and Environmental Channel Design (Sear *et al.*, 2010) and, in Scotland, the Morphological Impact Assessment System (MImAS: SNIFFER, 2006). The progression of the assessment base has been helped by the arrival of *inter alia*:

- Available remote sensing and time-lined imagery (including virtual globes like *Google Earth*).
- Remote survey techniques such as LiDAR, making the 'z' dimension over extended reaches much easier to determine.
- Robust field dGPS instruments.
- Robust field tablet computers for form filling.
- Software for digital analysis of bed sediment size.
- Compilation and collation of results via peer groups and processes initiated by RRC, SEPA (for MImAS, a screening tool for all river development pressures in Scotland) and the Environment Agency (for Fluvial Audits in England and Wales).

Advances in information technology facilitated both Fluvial Audit as a catchment-scale, resource-intensive prelude to restoration, and MImAS as a screening tool for all river development pressures in Scotland. The first Fluvial Audit was conducted in 1988 (Newson and Bathurst, 1990). The authors were, at the time, among a group of fewer than a dozen UK geomorphologists doing practical, 'applied' field assessments. The early River Habitat Surveys (http://www.riverhabitatsurvey.org/) carried out by the UK National Rivers Authority (1992) did not consider a remote sensing component (or even a measurement of local channel slope); by the time the method evolved into 'geoRHS' (Branson *et al.*, 2005) it was inevitable that a desk study component using readily available remote sensing and digitised 'old maps' was the logical start point. RHS was subsequently adapted as the standard morphological survey method for calibrating river assessment methods across Europe for Water Framework Directive implementation (Furse *et al.*, 2006), and data gathered using the method also formed a key information base for SERCON (System for Evaluating Rivers for Conservation: Wilkinson *et al.*, 1998). Subsequent research sought to establish remote sensing as the prime river assessment tool (see reviews by Gilvear *et al.*, 1999, Gilvear and Bryant, 2003 and Gilvear *et al.*, 2004) but problems remain with, for example, river bank profiles and channel features under dense riparian tree cover. There is scope for a wide range of imagery platforms (Carbonneau and Piégay, 2012) using, for example, terrestrial laser scanning and aimed at either ex-channel (e.g., Gilvear *et al.*, 2008) or in-channel (e.g., Milan *et al.*, 2011) habitat-scale assessment, but this is restricted to smaller spatial scales than airborne LiDAR and, as with the airborne remote-sensing platform, dataset size and management rapidly become an issue, as do the specialist skills required to manipulate the data itself.

Looking forward: future challenges to the evidence base and monitoring outcomes

The major uncertainty for current river restoration remains future climate change (a topic beyond the scope and remit of this chapter). At present, in the UK, almost all proposed and operational river restoration projects make an overt or assumed claim of future-proofing by increasing resilience; however, specific contextual evidence has not yet been gathered (a 'Catch 22') though we have grounds from theory and 'obvious' hydrological arguments about the merits of increasing both ecological connectivity and water storage. In his keynote address as outgoing President of the International Society for River Science at the 3rd Biennial Symposium in Beijing in August 2015, Geoff Petts reflected back on the time elapsed since the publication of *Impounded Rivers: Perspectives for Ecological Management* in 1984. At the time of that landmark publication, much of the river science of the time was concerned with issues of time, rates of change, trajectories and feedbacks with the underlying assumption being that there was 'equilibrium' and that every action had a complex response as systems tried to regain equilibrium. In his conclusions section in *Impounded Rivers,* Petts (1984) called for consideration of the long-term consequences of river regulation and a forward-looking 'anticipatory' approach to river management based upon a recognition that the full range of natural flows has a role in sustaining river systems. Even 30 years ago, Petts highlighted the pressing need to encourage morphological diversity and to maximise degree of flow variation. We now have evidence about the diversity of 'hydraulic biotopes' needed to support 'healthy' rivers in terms of biodiversity (Newson and Newson, 2000) – biotopes can now be assessed from aerial photography or from drones flown in tree-covered reaches. There is a prime strategic position here for the river restoration community to stress the need to share water between human needs and ecosystem services.

While there is no shortage of opportunity, there remains a need for more synergy between the research and planning arenas and, importantly, people living in catchments. If we manage this we can potentially achieve cost-effective benefits for all at the catchment scale (we're not there yet, but we are steadily getting there and bodies like the Rivers Trusts are key in this regard). Since 2011, the UK National Ecosystems Assessment (UKNEA) has further encouraged wider-ranging, multi-stakeholder, cross-disciplinary approaches (Defra, 2007). The UKNEA aims to provide a comprehensive picture of past, present and possible future trends in ecosystem services and their values (UK National Ecosystem Assessment, 2011) with one of the key objectives being to identify and understand what has driven change observed in the natural environment since 1950, and what are the implications for ecosystems services, both over the intervening 60 years and as we move deeper into the twenty-first century. As such, it provides a way to minimise the issue of 'shifting baseline syndrome' and to better define pragmatic reference points for river restoration. We still need better ways of expressing what nature does for us, and there remains a pressing need to ensure that the ecosystem services concept has tools not only to define benefits of services but also policy mechanisms to encourage its use. The concept does, however, resonate with river scientists as it emphasises the need for understanding both ecosystem structure and function, while at the same time concentrating attention on the value of rivers to modern human society.

As we move forward into the twenty-first century, a number of factors give rise to optimism for the future of river restoration efforts in the UK. Restoration remains a science but one given momentum by a series of external drivers. Momentum is given by the drive towards better flood risk management and the current paradigm of 'working with natural processes' in the UK (EA, in press). It is also driven by conceptual shifts; consideration of ecosystem services and 'what nature can do for us' now finds resonance in public valuation of landscapes, and restoration efforts going forwards will be driven not just by environmental policy but also community health and wellbeing aspirations. Other incentives will still come into play here (e.g., European Union Common Agricultural Policy reforms, agri-environmental schemes) but ultimately, to be sustainable, river restoration has to be seen as not just for rivers, but part of a wider, catchment-based, green-blue water approach to landscape conservation and management. Such discussion is timely, since despite inevitable barriers, there is clear appetite, enthusiasm and there are opportunities, both at the local and national levels, to instigate catchment-scale restoration.

Within this chapter we have pieced together the last 25 years of practical and scientific elements of river restoration in UK, explored what have been the key drivers for change, and asked challenging questions about where gaps still remain in terms of policy, research and practice. Our conclusions are, therefore:

- River restoration in the UK originally set out to address a legacy of morphological damage at locations driven by opportunity; it was intended as a catalyst and succeeded. There was little question of cost-effectiveness.
- As catchments became the standard UK river management unit and river ecosystems and 'services' derived from natural processes gained 'official' recognition, the scientific agenda and challenges evolved in a context where 'catchment consciousness' also grew through Rivers Trusts and their supporters.
- A subsequent challenge has become the identification of a reference (or 'natural') condition as a basis for restoration design and ecosystem service maximisation, as well as to what degree human modification of morphology and flows is allowable before process becomes impaired.
- These complications have spilled over into other demands on, for example, engagement, messaging and monitoring. The larger scale of operation is benefiting from IT, but monitoring to check our resilience assumptions remains weak.
- Going forward, the rapidly-evolving Ecosystems Services model, supported by a regulatory emphasis on hydromorphology, prioritises a balanced restoration effort focusing on flow.

References

Åberg EU. 2010. At the confluence of people, environment and ecology: Assessing river rehabilitation in Japan and the UK. Unpublished PhD Thesis, School of Geography, University of Leeds.

Åberg EU, Tapsell S. 2012. Rehabilitation of the River Skerne and the River Cole, England: A long-term public perspective. In: Boon P, Raven P. (Eds), *River Conservation and Management*, Wiley, Chichester: 249–259.

Arthington AH, Pusey BJ. 2003. Flow restoration and protection in Australian rivers. *River Research and Applications* 19: 377–395.

Beechie TJ, Sear DA, Olden JD, et al. 2010. Process-based principles for restoring river ecosystems. *BioScience* 60: 209–222.

Biggs J, Cornfield A, Gron P, et al. 1998. Restoration of the rivers Brede, Cole and Skerne: a joint Danish and British EU-Life demonstration project, V-short-term impacts on the conservation value of aquatic macroinvertebrate and macrophyte assemblages. *Aquatic Conservation-Marine and Freshwater Ecosystems* 8(1):241–255.

Boon, PJ. 2012. Revisiting the case for river restoration. In: Boon P, Raven P. (Eds), *River Conservation and Management*, Wiley, Chichester: 3–14.

Bracken LJ, Croke J. 2007. The concept of hydrological connectivity and its contribution to understanding runoff-dominated geomorphic systems. *Hydrological Processes* 21: 1749–1763.

Bradshaw AD. 1988. Alternate endpoints for restoration. In Cairns JJ. (Ed.), *Rehabilitating Damaged Ecosystems*. Volume 2, CRC Press, Boca Raton: 69–85.

Bradshaw AD. 1996. Underlying principles of restoration. *Canadian Journal of Fisheries and Aquatics Sciences* 53: 3–9.

Branson J, Hill C, Hornby DD, Newson MD, Sear DA. 2005. *A refined geomorphological and floodplain component River Habitat Survey (GeoRHS)*. Environment Agency/Defra R&D Technical Report SC020024/TR, Bristol.

Brookes A. 1987. River channel adjustment downstream from channelization works in England and Wales. *Earth Surface Processes and Landforms* 12: 337–351

Brown AG, Tooth S, Chiverrell RC, et al. 2013. The Anthropocene: is there a geomorphological case? *Earth Surface processes and Landforms* 38: 431–434.

Cairns JJ. 1991. The status of the theoretical and applied science of restoration ecology. *The Environmental Professional* 13: 186–294.

Carbonneau PE, Piégay H. (Eds). 2012. *Fluvial Remote Sensing for Science and Management*, Wiley-Blackwell, Chichester.

CIWEM. 2014. *Floods and dredging – a reality check*, CIWEM, London.

Defra. 2007. *Securing a healthy natural environment: An action plan for embedding an ecosystems approach*, Defra Publications, London.

Defra. 2011. *The natural choice: securing the value of nature*. Defra Publications, London.

Defra. 2013. *Catchment based approach: improving the quality of our water environment*. Department for Environment, Food & Rural Affairs Report PB13934 Defra Publications, London.

Downs PW, Kondolf GM. 2002. Post-project appraisals in adaptive management of river channel restoration. *Environmental Management* 29: 477–496.

Downs PW, Gregory, KJ. 2004. *River Channel Management: Towards Sustainable Catchment Hydrosystems*. Routledge, Oxford.

Edgeworth M. 2011. *Fluid Pasts: Archaeology of Flow*. Bristol Classical Press, Bristol.

Environment Agency. 2003. *River Habitat Survey Manual in Britain and Ireland - Field Survey Guidance Manual*. Environment Agency, Bristol.

Environment Agency. 2013. *Consultation on the proposal to use Environmental Improvement Unit Charge funds to implement hydromorphological measures*. Environment Agency, Bristol.

Environment Agency. 2014. *Working with natural processes to reduce flood risk – research and development framework*, Project Summary SC130004. Environment Agency, Bristol.

Everard M, Capper K. 2004. Common Law and conservation. *Environmental Law and Management* 16: 31–35.

Furse MT, Hering D, Brabec K, et al. (Eds). 2006. The ecological status of European rivers: evaluation and intercalibration of assessment methods. *Hydrobiologia* 566: 1–555.

Gilvear D, Bryant R. 2003. Analysis of aerial photography and other remotely sensed data. In: Kondolf GM, Piégay H. (Eds), *Tools in Fluvial Geomorphology*. Wiley, Chichester: 135–170.

Gilvear DJ, Bryant R, Hardy T. 1999. Remote sensing of channel morphology and in-stream fluvial process. *Progress in Environmental Science* 1: 257–284

Gilvear DJ, Davids C, Tyler AN. 2004. The use of remotely sensed data to detect channel hydromorphology: River Tummel, Scotland. *River Research and Applications* 20: 795–811.

Gilvear DJ, Sutherland P, Higgins T. 2008. An assessment of the use of remote sensing to map habitat features important to sustaining lamprey populations. *Aquatic Conservation: Marine and Freshwater Ecosystems* 18: 807–818.

Gilvear DJ, Spray CJ, Casas-Mulet R. 2013. River rehabilitation for the delivery of multiple ecosystem services at the river network scale. *Journal of Environmental Management* 126: 30–43.

Gupta A. 2008. *Large Rivers: Geomorphology and Management*. Wiley- Blackwell, Chichester.

Haines-Young RH, Potschin M. 2010. The links between biodiversity, ecosystem services and human well-being. In: Raffaelli D, Frid C. (Eds), *Ecosystem Ecology: A New Synthesis*. BES Ecological reviews Series, Cambridge University Press, Cambridge: 110–139.

Holmes NTH, Nielson MB. 1998. Restoration of the Rivers Breda, Cole and Skerne: a joint Danish and British EU-LIFE demonstration project, I – setting up and delivery of the project. *Aquatic Conservation: Marine and Freshwater Ecosystems* 8: 185–196.

Holmes NTH, Janes M. 2012. The history, development, role and future of River Restoration Centres. In: Boon P, Raven P. (Eds), *River Conservation and Management*. Wiley: Chichester: 285–293.

Holmes NTH, Raven P. 2014. *Rivers*. British Wildlife Publishing, Oxford.

Jones W, (2002) *EC habitats directive: favourable conservation status*. Joint nature conservation committee, JNCC 02 D07

Jose PV, Joyce C, Wade PM. 1999. *European Wet Grassland: Guidelines for management and restoration*. Royal Society for the Protection of Birds, Sandy, Bedfordshire.

Kern K. 1992. Rehabilitation of streams in south-west Germany. In Boon PJ, Calow P, Petts GE. (Eds), *River Conservation and Management*. Wiley, Chichester: 321–335.

Kern, K. 1994. Restoration of lowland rivers: the German experience. In Carling PA, Petts GE. (Eds), *Lowland Floodplain Rivers: Geomorphological Perspectives*. Wiley, Chichester: 279–297.

King J, Brown C. 2010. Integrated basin flow assessments: concepts and method development in Africa and South-east Asia. *Freshwater Biology* 55: 127–146.

King J, Louw D. 1998. Instream flow assessments for regulated rivers in South Africa using the Building Block Methodology. *Aquatic Ecosystem Health and Management* 1: 109–124.

Kronvang B, Svendsen LM, Brookes A, et al. 1998. Restoration of the Rivers Brede, Cole and Skerne: a joint Danish and British EU-LIFE demonstration project. 3. Channel morphology, hydrodynamics and transport of sediment and nutrients. *Aquatic*

Conservation: Marine and Freshwater Ecosystems 8: 209–222.

Kondolf GM, Downs PW. 1996. Catchment approach to channel restoration. In Brooks A, Shields FD Jr, (Eds), *River Channel Restoration*. Wiley, Chichester.

Large ARG, Gilvear DJ. 2014. Using *Google Earth*, a virtual-globe imaging platform, for ecosystem services-based river assessment. *River Research and Applications* 31: 406–421.

Mainstone CP, Holmes NTH. 2010. Embedding a strategic approach to river restoration in operational management processes - experiences in England. *Aquatic Conservation: Marine and Freshwater Ecosystems* 20: S82–S95.

Milan DJ, Heritage GL, Large ARG, Fuller IC. 2011. Filtering spatial error from DEMs: Implications for morphological change estimation. *Geomorphology* 125: 160–171.

Moerke AH, Lamberti GA. 2004. Restoring stream ecosystems: lessons from a midwestern state. *Restoration Ecology* 12: 327–334.

Murphy D, Vivash R. 1998. *Revetment techniques used on the River Skerne restoration project*. Environment Agency R & D Technical Report, Environment Agency, Bristol

National Rivers Authority. 1992. River Corridor Surveys. *Conservation Technical Handbook 1*. National Rivers Authority, Bristol.

Newson MD. 2010a. 'Catchment consciousness' – will mantra, metric or mania best protect, restore and manage habitats? In Kemp P. (Ed.), *Salmonid Fisheries: Freshwater Habitat Management*. Wiley-Blackwell, Oxford: 28–54.

Newson MD. 2010b. *Land Water and Development: Sustainable and Adaptive Management of Rivers*. 3rd Edition. Routledge, London.

Newson MD. 2010c. *Aggravated Erosion and Freshwater Habitat Siltation: Definition, Identification, Remedial Actions. A practical guidebook for Natural England policy and advisory staff*. Tyne Rivers Trust, Corbridge, UK.

Newson MD. 2011. Rivers in Trust: stakeholders and the delivery of the EU Water Framework Directive. *Water Management* 164: 433–440.

Newson, MD. 2012. From channel to catchment: a twenty year journey for river management in England and Wales. In Boon P, Raven P. (Eds), *River Conservation and Management*. Wiley, Chichester: 17–27.

Newson MD. 2014a. Making hydromorphological processes accessible to stakeholder audiences. *Proceedings of the Annual Conference of the British Society for Geomorphology*, University of Manchester.

Newson MD 2014b. *Aide Memoire for Channel Management in a Catchment Context*. Tyne Rivers Trust, Corbridge UK (available at www.tyneriverstrust.org).

Newson MD, Bathurst JC. 1990. *Sediment movement in gravel bed rivers*. University of Newcastle upon Tyne, Department of Geography, Unpublished Seminar Paper Series, No. 58.

Newson MD, Newson CL. 2000. Geomorphology, ecology and river channel habitat: mesoscale approaches to basin-scale challenges. *Progress in Physical Geography* 24: 195–217.

Newson MD, Large ARG. 2006. 'Natural'rivers,' hydromorphological quality'and river restoration: a challenging new agenda for applied fluvial geomorphology. *Earth Surface Processes and Landforms* 31: 1606–1624.

Newson MD, Clark MJ. 2008. Uncertainty and sustainable management of restored rivers. In: Darby S, Sear D. (Eds), *River Restoration: Managing the Uncertainty in Restoring Physical Habitat*. Wiley, Chichester: 290–301.

NRC (1992) *Restoration of Aquatic Ecosystems*. National Academy Press, Washington, DC.

Ormerod S.J. 2004. A golden age of river restoration science? *Aquatic Conservation: Marine and Freshwater Ecosystems* 14: 543–549.

Palmer MA, Bernhardt ES, Allan JD, et al. 2005. Standards for ecologically successful river restoration. *Journal of Applied Ecology* 42: 208–217.

Palmer MA, Menninger HL, Bernhardt E. 2010. River restoration, habitat heterogeneity and biodiversity: a failure of theory or practice? *Freshwater Biology* 55: S205–S222.

Palmer MA, Filoso S, Fanelli, RM. 2014. From ecosystems to ecosystem services: Stream restoration as ecological engineering. *Ecological Engineering* 65: 62–70.

Papworth SK, Rist J, Coad L, Milner-Gulland EJ. 2009. Evidence for shifting baseline syndrome in conservation. *Conservation Letters* 2: 93–100.

Pauly D. 1995. Anecdotes and the shifting baseline syndrome of fisheries. *Trends in Ecology and Evolution* 10: 430.

Pedersen ML, Andersen JM, Nielsen K, Linnemann M. 2007. Restoration of Skjern River and its valley: Project description and general ecological changes in the project area. *Ecological Engineering* 30: 131–144.

Petts GE. 1979. Complex response of river channel morphology subsequent to reservoir construction. *Progress in Physical Geography* 3: 329–362.

Petts GE. 1984. *Impounded Rivers: Perspectives for Ecological Management*. Wiley, Chichester.

Petts GE. 2000. A perspective on the abiotic processes sustaining the ecological integrity of running waters. *Hydrobiologia* 422: 15–27.

Petts GE, Moller H, Roux AL. (Eds), 1989. *Historical Change of Large Alluvial Rivers: Western Europe*. Wiley, Chichester.

Petts GE, Amoros C. 1996. *Fluvial Hydrosystems*. Chapman and Hall, London.

Petts GE, Calow P. 1996. *River Restoration*. Wiley-Blackwell, London.

Petts GE, Heathcote J, Martin D. (Eds). 2002. *Urban Rivers: Our Inheritance and Future*. IWA Publishing, London.

Poole GC. 2010. Stream hydrogeomorphology as a physical science basis for advances in stream ecology. *Journal of the North American Benthological Society* 29: 12–25.

Power ME, Stout RJ, Cushing CE, et al. 1998. Biotic and abiotic controls in river and stream communities. *Journal of the North American Benthological Society* 7: 456–479

Rice SP, Lancaster J, Kemp P. 2010. Experimentation and the interface of fluvial geomorphology, stream ecology and hydraulic engineering and the development of an effective interdisciplinary river science. *Earth Surfaces Processes and Landforms* 35: 64–77.

Richter BD, Baumgartner JV, Powell J, Braun DP. 1996. A method for assessing hydrologic alteration within ecosystems. *Conservation Biology* 10: 1163–1174.

RRC. 2011. *Practical River Restoration Appraisal Guidance for Monitoring Options (PRAGMO)*. River Restoration Centre, Cranfield, Bedfordshire.

RRC. 2014. *Manual of River Restoration Techniques http://www.therrc.co.uk/manual-river-restoration -techniques*. River Restoration Centre, Cranfield, Bedfordshire.

Roni P, Beechie TJ, Bilby RE, et al. 2002. A review of stream restoration techniques and a hierarchical strategy for prioritizing restoration in Pacific Northwest watersheds. *North American Journal of Fisheries Management* 22: 1–20.

Roni P, Beechie, T. 2013. *Stream and Watershed Restoration: A Guide to Restoring Riverine Process and Habitats*. Wiley-Blackwell, Chichester.

Sear DA, Briggs A, Brookes A. 1998. A preliminary analysis of the morphological adjustment within and downstream of a lowland river subject to river restoration. *Aquatic Conservation: Marine and Freshwater Ecosystems* 8: 167–184.

Sear DA, Newson MD, Thorne CR. 2010. *Guidebook of Applied Fluvial Geomorphology*. Thomas Telford, London.

Skinner KS, Bruce-Burgess L. 2005. Strategic and project level river restoration protocols — key components for meeting the requirements of the Water Framework Directive (WFD). *Water and Environment Journal* 19: 135–142.

Smith B, Clifford NJ, Mant J. 2014. The changing nature of river restoration. *Wires Water* 1: 249–261

SNIFFER. 2006. *Trialling of MImAS and proposed Environmental Standards*. Scotland & Northern Ireland Forum for Environmental Research Report WFD49, Edinburgh.

Temmerman S, Meire P, Bouma TJ, et al. 2013. Ecosystem-based coastal defence in the face of global change. *Nature* 504: 79–83.

Thorp JH, Thoms MC, Delong MD. 2006. The riverine ecosystem synthesis: biocomplexity in river networks across space and time. *River Research and Applications* 22: 123–147.

Thorp JH, Flotemersch JE, Delong MD, et al. 2010. Linking ecosystem services, rehabilitation and river hydrogeomorphology. *BioScience* 60: 67–74.

Tockner K, Schiemer F, Ward JV. 1998. Conservation by restoration: the management concept for a river-floodplain system on the Danube River in Austria. *Aquatic Conservation: Marine and Freshwater Ecosystems* 8: 71–86.

UK National Ecosystem Assessment. 2011. *The UK National Ecosystem Assessment: Synthesis of the Key Findings*. UNEP-WCMC, Cambridge.

UKTAG. 2013. *River flow for good ecological potential*. Draft Recommendations.

Vaughan IP, Diamond M, Gurnell AM, et al. 2007. Integrating ecology with hydromorphology: a priority for river science and management. *Aquatic Conservation: Marine and Freshwater Ecosystems* 19: 113–125.

Vivash R, Ottosen O, Janes M, Sorensen HV. 1998. Restoration of the Rivers Breda, Cole and Skerne: a joint Danish and British EU-LIFE demonstration project, II – The river restoration works and other related practical aspects. *Aquatic Conservation: Marine and Freshwater Ecosystems* 8: 197–208.

Waal LC de, Large ARG, Gippel CJ, Wade PM. 1995. River and floodplain rehabilitation in Western

Europe: opportunities and constraints. *Archiv für Hydrobgiologie Supplement 101 Large Rivers* 9: 1–15.

Wade, PM, Large ARG, de Waal LC. 1998. Rehabilitation of degraded river habitat: an introduction. In: de Waal LC, Large ARG and Wade PM (Eds), *Rehabilitation of Rivers: Principles and Implementation*. John Wiley & Sons ltd, Chichester: pp. 1–12 .

Walter RC, Merritts DJ. 2008. Natural streams and the legacy of water-powered mills. *Science* 319: 299–304.

Ward D, Holmes NTH, Jose P. 1994. *The New Rivers and Wildlife Handbook*. RSPB, NRA & The Wildlife Trusts, Sandy, Bedfordshire, UK.

Wharton G, Gilvear DJ. (2006) River restoration in the UK: meeting the dual needs of the European Union Water Framework Directive and flood defence? *International Journal of River Basin Management* 4: 1-12.

Wilkinson J, Martin J, Boon PJ, Holmes NTH. 1998. Convergence of field survey protocols for SERCON (System for Evaluating Rivers for Conservation) and RHS (River Habitat Survey). *Aquatic Conservation, Marine and Freshwater Ecosystems* 8: 579–596.

Williamson, P, Ogunyoye, F, Dennis, I, et al. 2015. *Channel Management Handbook: Working with Natural Processes*. Environment Agency, Bristol.

Wohl EE. 2004. *Disconnected Rivers: Linking Rivers to Landscapes*. Yale University Press, London.

Wohl EE. 2011. What should these rivers look like? Historical range of variability and human impacts in the Colorado Front Range, USA. *Earth Surface Processes and Landforms* 36: 1378–1390.

Woolsey S, Capelli F, Gonser TOM, et al. 2007. A strategy to assess river restoration success. *Freshwater Biology* 52: 752–769.

CHAPTER 17

Ecosystem services of streams and rivers

J. Alan Yeakley[1], David Ervin[2], Heejun Chang[3], Elise F. Granek[1], Veronica Dujon[4], Vivek Shandas[5] and Darrell Brown[6]

[1] *Department of Environmental Science and Management, School of the Environment, Portland State University, Portland, USA*

[2] *Department of Economics, Department of Environmental Science and Management, and Institute for Sustainable Solutions, Portland State University, Portland, USA*

[3] *Department of Geography, Portland State University, Portland, USA*

[4] *Department of Sociology, Portland State University, Portland, USA*

[5] *Toulan School of Urban Studies and Planning, Portland State University, Portland, USA*

[6] *School of Business, Portland State University, Portland, USA*

Introduction

River ecosystems function across a large range of spatial and temporal scales, from small stream reaches and intermittent waterways to large river basins at the continental scale (Vannote *et al.*, 1980). As such, the ecosystem services they provide span comparable scales varying both spatially and temporally in terms of the goods and services they provide human societies. In addition to considerations of scale, the ecosystem services provided by streams and rivers span broad categories of human societal needs, including resource provisioning such as drinking water and fisheries, regulating services such as flood mitigation and water filtration, supporting services such as maintaining riparian wildlife habitat and biodiversity, and cultural values such as recreation and aesthetics. Sustainable management of river ecosystems relies on achieving a balance among most if not all these broad categories of ecosystem services, and maintaining a balance among riverine ecosystem services frequently requires tradeoffs among services. This chapter provides an overview of the current understanding regarding ecosystem services of streams and rivers, with a focus on the effects of scale and consideration of tradeoffs.

Ecosystem services have been defined as the conditions and processes through which natural ecosystems, and the species that make them up, sustain and fulfil human life (Daily, 1997). The ecosystem services concept has proliferated in the scientific literature over the past two decades (e.g., Costanza *et al.*, 1997; Loomis *et al.*, 2000; Costanza, 2008; Crossman *et al.*, 2010; Ervin *et al.*, 2012; Costanza *et al.*, 2014) (Figure 17.1). The Millennium Ecosystem Assessment identified four broad categories of ecosystem services, including *provisioning services* (i.e., material or energetic outputs from ecosystems, including food, water and other resources); *regulating services*

River Science: Research and Management for the 21st Century, First Edition.
Edited by David J. Gilvear, Malcolm T. Greenwood, Martin C. Thoms and Paul J. Wood.

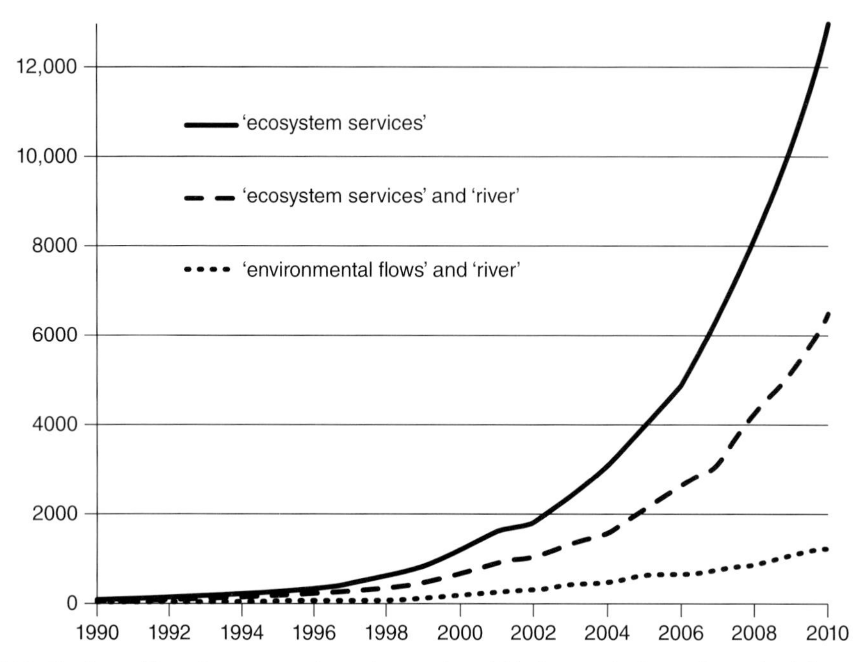

Figure 17.1 Citations of key phrases over time. The number of hits by year is shown for each of three searches using Google Scholar: (a) 'ecosystem services', (b) 'ecosystem services' and 'river', (c) 'environmental flows' and 'river'.

(i.e., factors that the ambient environment provides such as flood control); *supporting services* (i.e., maintenance of life supporting conditions such as nutrient cycling, primary productivity and habitat); and *cultural services* (i.e., non-material uses, such as spiritual and recreational benefits) (Millennium Ecosystem Assessment, 2005; Costanza *et al.*, 2011a; Milcu *et al.*, 2013). Recent studies have categorised ecosystem services somewhat differently, by combining regulating and supporting services into a single category of *regulating and maintenance services* (Haines-Young and Postchin, 2010). Additionally, focus has shifted from the functionality of the services to the products or benefits of those services (Haines-Young and Postchin, 2010).

The utility of the ecosystem services concept has been the subject of much debate, which has included an examination of the difference between intrinsic vs. instrumental ecosystem values. An *intrinsic value* of an organism or an ecosystem pertains to the existence of that entity, as an end in itself, and is essentially biocentric rather than anthropocentric (Leopold, 1949; O'Neill, 1992; Vilkka, 1997; Winter, 2007). Intrinsic values can be difficult to apply to conservation decisions in natural ecosystems such as river basins (Maguire and Justus, 2008) because they include intangible components of ecosystems that are very difficult to measure quantitatively. Conservation scholars argue for qualitative characterisation (Vucetich *et al.*, 2015). As our general understanding of connections between and within ecosystems remains incomplete, it is important that conservation managers carefully consider isolated, presumably economically irrelevant, or difficult to quantify ecosystems or their properties before deciding that they are not relevant to human welfare (Farber *et al.*, 2002).

In contrast, *instrumental value* describes properties of the entity that are a means to a valuable end as perceived by humans (Farber *et al.*, 2002; Reyers *et al.*, 2012). Instrumental values are ascribed to a wide array of ecosystem services ranging from provisioning, such as for human consumption, to non-use properties including spiritual values associated with rivers such as the Ganges River in India, a holy site to Hinduism. While ecosystem services valuation approaches have often been criticised for not addressing intrinsic values of ecosystems that go beyond human use and perception (McCauley, 2006; Minteer, 2009), there can yet be 'win–win' outcomes where the consideration of instrumental values also addresses intrinsic properties of ecosystems (Reyers *et al.*, 2012; Ervin *et al.*, 2014). Such win–win combinations are more likely to emerge from stream and river management if both intrinsic and instrumental values enter institutional decision-making processes. As an example, the restoration of riparian areas both increases habitat quality and biodiversity across a large variety of taxa (including species not considered economically relevant) as well as providing regulating and cultural ecosystem services, such as flood control and recreation.

The geographic scale of rivers and streams and the ecosystem services associated with them pose unique management challenges. These water bodies can flow across lands that are managed or governed by a number of different governance structures within a country, or across countries, triggering the need for collaborative approaches to management (Raadgever *et al.*, 2008; Dugan *et al.*, 2010; Ziv *et al.*, 2012). In addition to conflicts over water availability, the type of use at different points along a river can be contested. Upstream uses for industrial or conservation purposes, for example, affect downstream users, and in a similar fashion, downstream use in the form of urban development that, for example, impedes returning migratory fish can impact upstream fishing potential (Yeakley *et al.*, 2014). In some scenarios these tensions can be addressed through negotiated multinational, state or regional level policies (e.g., the Columbia River in Canada and USA). In others the problems remain intractable (e.g., the Jordan River in the Middle East).

A key challenge in the management of complex ecosystems such as streams and rivers is thus maintaining critical supporting and regulating ecosystem services to sustain the functionality of the river, while also serving the provisioning and cultural ecosystem service demands placed upon the ecosystem by human societies (Wilson and Carpenter, 1999; Vörösmarty *et al.*, 2010). Ecosystem services provided by streams and rivers vary spatially depending on geologic and climatic setting, landscape location and river size (i.e., stream order). Further, the communities that benefit or are affected by changes in the river system are an essential part of developing win–win management options that enable society to gain benefit from these systems. The institutions that oversee the management of rivers, such as the US Army Corps of Engineers in the USA or the Environment Agency in the UK, often attempt to manage competing stakeholder interests in terms of who benefits and who loses when specific decisions impact the quantity or quality of water within the system. As a result, we need approaches that both characterise the varying types of ecosystem services and recognise the distribution of ecosystem services (and disservices) accruing from watershed management.

In this chapter we examine how ecosystem services generally vary with these factors, with an emphasis on stream order, a measure of stream size and location (Strahler, 1957). We also discuss the critical resources needed to sustain the stream ecosystems in the form of the minimal stream flow, or discharge, necessary to maintain stream ecological and hydrologic function, and how that consideration relates to the ecosystem services concept. Although our emphasis is on the hydrological system, we include domains of institutional management of ecosystem services where relevant.

River ecosystems

Many rivers run along a continuous gradient of physical conditions, from headwaters to the mouth, which has been described as the River Continuum Concept (Vannote *et al.*, 1980). Characteristics of stream hydrology, water quality and aquatic ecology all vary in somewhat predictable ways along the river continuum gradient. Large-scale drivers of river continua include both climatic and geological factors, while small-scale controls include riparian characteristics as well as geomorphic factors (e.g., valley form) and the nature of the hydrological input (e.g., rain vs. snow vs. groundwater sources) (Naiman *et al.*, 1992). For effective management of healthy riverine ecosystems, it is important to understand the varying controls and their scales of influence, as well as how they affect ecosystem response variables such as biodiversity, productivity and biogeochemical cycles in rivers and streams (Naiman *et al.*, 1992).

Anthropogenic impacts on stream ecosystems have a long history (Vitousek *et al.*, 1997) and range from minor alterations to stream banks and watershed hydrologic routing, to major reconfigurations of stream networks, imposition of dams and, in the case of smaller streams, outright elimination of the surface expression of the stream course by encasing it within pipes (Petts, 1984; Hughes and Dunham, 2014; Yeakley, 2014). Anthropogenic alterations to streams occur for primarily economic reasons, such as for transport of goods or extraction of resources such as hydro-electric power or fish. These alterations can affect both a physical modification of the stream channel as well as a hydrological modification of the water quantity and/or quality of the stream. While occasionally these conversions may result in an enhancement of fish habitat, these alterations more typically constitute a diminishment or replacement of a number of regulating and maintenance ecosystem services with provisioning services (Table 17.1). Often these replacement activities narrow the suite of ecosystem services to a small overall range of benefits provided by the river ecosystem. By narrowing the range of ecosystem services and by reducing the effectiveness of the maintenance and regulating services (for example, channelising a river corridor for boat navigation that results in a reduced flood and pollutant uptake mitigation capacity due to removal of riverine wetlands and multiple stream and slough channels), the resilience of the stream ecosystem for dissipation and reduction of impacts is decreased (Palmer *et al.*, 2005).

A recent development in ecosystem management that seeks to maintain or restore the resilience of river ecosystems is the concept of *environmental flows* (Petts, 1996; Poff *et al.*, 1997; Arthington *et al.*, 2006; Petts, 2009; Arthington *et al.*, 2010). Managing rivers to maintain critical minimum environmental flows, which has also been referred to as ELOHA (i.e., the ecological limits of

Table 17.1 Ecosystem Services vs. Stream/River Size. Note that gradient bar indicates where a given ecosystem service is expected to be most emphasised or realised along a river continuum, with darker shades showing higher emphasis and lighter shades showing less emphasis. Stream order relates to stream size, from headwater streams (low order) to large rivers (high order).

Category of Ecosystem Service	Specific Ecosystem Service	Low order streams (1st-2nd)	Mid-order streams & rivers (3rd-5th)	High order rivers and floodplains (6th and up)
Provisioning	Consumptive use water: domestic			
Provisioning	Consumptive use water: agricultural			
Provisioning	Consumptive use water: industrial			
Provisioning	Non-consumptive use water: transportation/navigation			
Provisioning	Abiotic materials (e.g. gravel, aggregate)			
Provisioning	Aquatic organisms: food			
Provisioning	Non-consumptive use water: power generation			
Provisioning	Consumptive use water: drinking water			
Provisioning	Aquatic organisms: medicines			
Regulating	Erosion control thru flood control infrastructure			
Regulating	Buffering of flood flows			
Regulating	Erosion control thru land water interactions			
Regulating	Role in climate regulation			
Regulating	Maintenance of water quality			
Supporting	Role in maintenance of floodplain fertility			
Supporting	Role in primary production			
Supporting	Role in food webs and predator/prey relationships			
Supporting	Role in habitat maintenance			
Supporting	Maintenance of genetic resources (e.g. wild salmonids)			
Supporting	Role in nutrient cycling			
Cultural	Recreation from fishing as a sport			
Cultural	Tourism (river viewing)			
Cultural	Aesthetic, heritage			
Cultural	Recreation from river rafting			
Cultural	Recreation from hiking			
Cultural	Existence (personal satisfaction from free flowing rivers)			
Cultural	Traditional foods			

hydrologic alteration), is a strategy to help preserve or restore regulating and maintenance ecosystem services and thus maintain or improve the ecosystem resilience of the river system (Poff *et al.*, 2010; Poff and Matthews, 2013). Additionally, management of rivers to better mimic the natural flow regime addresses not just instrumental but also intrinsic ecosystem values. As seen in Figure 17.1, the growth of the concept of environmental flows in the scientific literature has substantially increased in step with the explosion of the ecosystem services concept over the past 20 years. Figure 17.1 demonstrates the increased consideration of environmental flows, and therefore of intrinsic values in river ecosystems, by river scientists and managers, even as management efforts have focused more on the entire suite of ecosystem services from rivers including provisioning services, that is, more traditional economic considerations of river management.

Spatial considerations of ecosystem services in rivers

Riverine ecosystem services are unevenly distributed over a river basin (Bastian *et al.*, 2012; Bagstad *et al.*, 2013). Typically, upstream forest areas provide more regulating and supporting ecosystem services while downstream areas of the watershed, particularly urbanised areas, receive the benefits of such ecosystem services from upstream areas. Larger streams and rivers tend to have relatively lower levels of regulating and supporting services and more greatly emphasise provisioning services (Table 17.1). In this regard, it is worthwhile distinguishing the areas where ecosystem services are generated from the areas where ecosystem services are supported or consumed. Syrbe and Walz (2012) identified these areas as 'service provisioning areas' (SPAs) and 'service benefiting areas' (SBAs), respectively. The intervening connecting areas between the SPAs and SBAs are referred to as 'service connecting areas' (SCAs). SCAs could be natural or human-constructed. For example, a natural spring at the foothill of a mountain stream can transport groundwater from upstream areas, while a mountain reservoir and the associated drinking water pipe network can connect to downstream urban water consumers.

A river course provides an illustrative example of how SPAs, SCAs and SBAs are positioned and connected to each other across different spatial and temporal scales. First, riparian areas and upstream areas serve as SPAs given that those areas produce high water yield and purify water by retaining high amounts of sediment and nutrients, and are expected to continue supplying high levels of services in a changing climate (Hoyer and Chang, 2014). By sustaining adequate flow, these riparian and upstream areas also provide diverse cultural ecosystem services to downstream users, such as river rafting and sport fishing (Gutierrez and Alonso, 2013). SCAs in a river system can be identified through how water-dependent ecosystem services are transported or transformed. Namely, SCAs spatially and functionally connect SPAs and SBAs (Syrbe and Walz, 2012). Stream order (Strahler, 1957) can be a useful concept for understanding such connections since mid-order streams can serve as SCAs, connecting lower order to higher order streams in a river (Table 17.1). Downstream areas and large river deltas or any major population centres in watersheds serve as SBAs, utilising purified water for residential use, industrial consumption and agriculture. SBAs may also use the water delivered from SPAs for recreation, transportation and wildlife viewing. The development of deltas in SBAs relies on the sediment supply from upstream SPAs.

Humans often attempt to overcome spatial and temporal mismatches between SPAs and SBAs by increasing SCAs. Some additional SCAs can be created by such anthropogenically-made infrastructure as dams, irrigation canals and groundwater storage tanks. For example, in the interior part of the Columbia River Basin, USA, to temporally maximise SBAs, water managers pump water from the main stem Columbia River in winter when excessive water is available, store the water in aquifer recovery storage, and use the stored water for irrigation during the summer months when the volume of the river discharge is reduced (Chang *et al.*, 2013). The 390 km long aqueduct from the Colorado River to the Los Angeles metropolitan area, USA, illustrates how humans connect SPAs to SBAs over vast spatial scales.

While such infrastructure may increase or reallocate provisioning ecosystem services, that same infrastructure can diminish regulating or cultural ecosystem services. As the natural flow regime is disturbed due to the construction of dams or irrigation channels, native species, habitat and water quality (e.g., temperature, turbidity, suspended solids and nutrient concentrations) may be affected. In the Willamette River basin, Oregon, USA, 13 major dams were created to regulate winter floods and provide hydropower generation, but the temperature of stream water released from the bottom of these reservoirs is not ideal for salmonid species (Rounds, 2010). Water from the Colorado River that is diverted to the city of Los Angeles via the state-wide aqueduct causes the river to run dry before it reaches the ocean; thus, the lower segments of the river have diminished levels of regulating and provisioning services, habitat provisioning and biodiversity (Postel, 2000; Zamora *et al.*, 2013). In some instances, infrastructure can create some synergistic effects, increasing multiple ecosystem services. In the lower Tualatin River basin, Oregon, USA, for example, an upstream reservoir not only provides water supply to growing municipalities but also offers some lake-oriented recreational opportunities otherwise unavailable under a natural flow regime. At a reach scale, adding dead wood to restore stream channels can increase the quantity and quality of selected ecosystem services in temperate forests (Acuna *et al.*, 2013).

Tradeoffs and benefits in riverine ecosystem services management

Riverine ecosystems and the biodiversity they embody provide a rich array of benefits to humans and nature. Prominent examples include renewable flows of fresh water, production of fish and wildlife, aesthetic, recreational and spiritual experiences, and maintenance of a genetic library of aquatic biodiversity. To capture the full range and character of these instrumental and intrinsic values, rigorous quantitative and qualitative methods are required (Ervin *et al.*, 2014). For those benefits that can be quantified with acceptable precision, some may be valued in monetary terms through the use of market prices, for example, freshwater for use in agricultural or domestic uses. Others may be analysed with quantitative non-monetary methods and metrics, for example using choice experiments or contingent valuation survey of riverine recreation areas (Loomis,

2012), and still others can only be assessed with qualitative methods, such as cultural services (Daniel *et al.*, 2012), which can constitute human preference for specific aesthetics of river systems or spiritual values that are embedded in traditional cultures.

Rivers are utilised as a major source of fresh water for human consumptive and agricultural uses as well as a key resource for energy production. This diversity of uses leads to a common management scenario whereby rivers are often managed to maximise the provisioning of one or a few ecosystem services of instrumental value. However, prioritisation of one service can result in large tradeoffs whereby other services may be concomitantly reduced or lost (Bullock *et al.*, 2011). When rivers are managed primarily for irrigation, water levels may become too low to support high-quality fish habitat or too warm to support native fish species and their prey (Lemly *et al.*, 2000). In the San Joaquin River, California, USA, the diversion of river water for agriculture has led to the loss of Chinook salmon runs and other native aquatic species (Moore *et al.*, 1996). Another frequent tradeoff in river ecosystems exists between 'soft use' recreation and 'hard use' extractive industries (Ruiz-Frau *et al.*, 2013). In a number of locales, mining and logging industries have led to the exclusion of recreational users from certain river systems. However, some studies find that recreation, in the form of fishing, camping and wildlife viewing, can contribute a higher economic value to the region and provide a more sustainable alternative to the extractive industries (Ziv *et al.*, 2012).

Another frequently observed tradeoff in rivers around the world is between indigenous and/or local cultural uses and economic extractive uses (Rosenberg *et al.*, 2000; Pringle *et al.*, 2000). A common example of this conflict is the construction of dams that negatively impact indigenous uses and species of riverine and riparian habitats (Table 17.2). For example, the Elwah Klallams lived in the Elwah River watershed and believed that they were brought into existence on the river, formed from the dirt scooped out of deep holes in the rocks (Crane, 2011). Their complex dependence, like other traditional cultures, relied on the material outcomes from the river, which also provided the ceremonial, cultural and spiritual resources necessary for their survival (US Department of the Interior, 1994). The damming of the Elwah River, which began in September, 1910, and ended in December, 1913, provided one primary ecosystem service: hydropower (Aldwell, 1950), to the exclusion of many other services, including cultural, supporting and regulating. With all these tradeoffs come winners and losers across different scales (in the form of individuals, businesses, tribal and non-tribal communities, species and ecosystems).

On the other hand, there are examples of management that, by focusing on a particular ecosystem service, lead to beneficial outcomes in the provisioning of other, non-target services. In several locales, maintenance of river reaches specifically managed for recreational fishing can lead to increases in habitat, water quality and flow, and biodiversity (Granek *et al.*, 2008). Managing rivers for drinking water can increase protection of riparian habitats and biodiversity as well as water quality (e.g., Bull Run, Oregon, USA; Hetchy Hetch, California, USA). Also, managing rivers for cultural ecosystem services, such as for traditional food and medicine sources of Native American Tribes in North America, can provide further benefits for the functionality of the river ecosystem

Table 17.2 Dam impacts on native species and habitats. Shown are examples of rivers on which dams have threatened or will threaten indigenous uses as well as native habitat and some of the key taxa threatened. Note that this list is not exhaustive, either of the dams listed or of the taxa affected by any given dam, but is provided to give a sense of the wide-ranging impacts that dams can have on ecosystem services and biodiversity.

River (Location)	Indigenous uses	Riparian habitat	Native/endemic taxa affected
Elwah[1] (Washington, USA)	Fisheries; cultural value; transportation	Floodplain temperate rain forest	Fish (e.g., Chinook, coho, chum, pink and sockeye salmon; steelhead/rainbow trout; cutthroat and bull trout; native char; lamprey).
Yaqui (Sonora, Mexico)[2]	Drinking water, irrigation, ceremonial uses	Riparian forest	Fish (e.g., Yaqui chub, Yaqui topminnow, Yaqui catfish, Taqui beautiful shiner, Mexican stoneroller, Yaqui sucker); Amphibians (e.g., Chiricahua leopard frog).
Changuinola (Panama)[3]	Water for cooking, drinking; subsistence fishery; transportation; agriculture	Tropical forest	Fish (e.g., freshwater eel, clingfish, grunt, mullet, goby); Macroinvertebrates; Shrimp (palaemonid and atyid).
Bayano (Panama)[4]	Agriculture; residence; subsistence fishing and hunting	Tropical forest; caves	Bats; Birds (e.g., macaw, heron, parakeet, harpy eagle); Fish (e.g., snook, trogon, tarpon) Mammals (e.g., white faced monkey, jaguar); Plants.
Xingu (Brazil)[5]	Agriculture; cultural value; transportation; subsistence fishery	Flooded forests of the Amazon	Birds; Fish (endemic and migratory); Mammals (e.g., white-cheeked spider monkey; black-bearded saki monkey); Plants; Reptiles.

(continued overleaf)

Table 17.2 *(Continued)*

River (Location)	Indigenous uses	Riparian habitat	Native/endemic taxa affected
Yangtze (China)[6]	Agriculture; cultural value; subsistence fishery, transportation	Floodplain; subtropical forests	Amphibians; Fish (e.g. 44 endemics threatened by the Three Gorges Dam, including Yangtze sturgeon); Plants (including Pinaceae, Magnoliacae, Lauraceae, Ranuculaceae families); Mammals; Reptiles.
Mekong (Vietnam)[7]	Agriculture; cultural value; subsistence fishery, transportation	Floodplain; tropical rainforest	Amphibians; Birds (e.g. river lapwing, pranticole, stork, adjutant, ibis, tern, wagtail); Fish (e.g., giant catfish, whitefish, blackfish, carp, freshwater herring); Mammals (e.g. otter, Irrawaddy dolphin); Mollusks; Reptiles (e.g. turtle, Siamese crocodile).

[1](Duda *et al.*, 2011; note that dam was removed during 2011–2014),
[2](Contreras-Balderas *et al.*, 2003; Minckley and Marsh, 2009),
[3](McCLarney *et al.*, 2010),
[4](Goodland, 1977; Dalle and Potvin, 2004),
[5](Rylands, *et al.* 1997; Borges and Carvalhaes 2000; Camargo *et al.*, 2004; Heckenberger *et al.*, 2007),
[6](Dudgeon, 2000; Park *et al.*, 2003; An *et al.*, 2007; Xie *et al.*, 2012),
[7](Dudgeon, 2000; Dugan *et al.*, 2010; Costanza *et al.*, 2011b)

(Flanagan and Laituri, 2004). Other cultural services, such as tourism and white-water rafting, can also result in maintenance of a 'wilder' state of the river ecosystem and thus enhance the natural ecological and hydrological character of the river.

An illustrative example of both tradeoffs and benefits in river-based ecosystem services can be seen in the recent removal of the Condit Dam on the White Salmon River in south central Washington state, USA. The Condit Dam, located 5 km upstream of the confluence of the White Salmon and Columbia rivers, was built in 1913 and stood as the only human-made obstruction on the 71 km river (Figure 17.2). The dam obstructed some 53 km of salmonid (*Oncorhynchus* spp.) habitat (Allen and Connolly 2011), and in 1994 the US Fish and Wildlife Service and NOAA Fisheries imposed fish passage conditions on Federal Energy Regulatory Commission license renewal for the dam. PacificCorp, the owner of the dam, completed a cost–benefit analysis for providing fish passage and determined that decommissioning the dam was the most cost-effective option. Subsequently, on 26 October 2011, the dam was breached in what, at the time, was the largest dam breach in US history, and the river resumed unimpeded flow from the headwaters to the mouth (National Geographic, 2011).

The removal of the Condit Dam constituted a tradeoff in ecosystem services, with both gains and losses resulting. The most obvious loss was in the provisioning service of electric power generation, as the Condit Dam provided some 80,000 MWh of electric power, serving 7000 homes per year (PacificCorp, 2014). An additional loss in ecosystem services was in the cultural service of recreational fishing and lakeside home ownership and water supply, with the elimination of the reservoir behind the dam,

Figure 17.2 Condit Dam on the White Salmon River, Washington State, USA. This dam measured 38 m high and 144 m wide, and provided 80,000 MWh of power generation per year. The dam also resulted in the loss of 53 km of salmonid habitat. Breach of the dam occurred on October 26, 2011, and it was subsequently removed. (Photo by A. Yeakley.). (*See colour plate section for colour figure*).

Figure 17.3 Capture and relocation of Chinook salmon on the lower White Salmon River. Prior to removal of the Condit Dam, personnel from the US Fish and Wildlife Service captured some 679 fall Chinook salmon adults, and relocated them upstream of the dam. (Photo by A. Yeakley.). (*See colour plate section for colour figure*).

which also lowered the groundwater table in lakeside properties. Gains in ecosystem services included the return of salmonid habitat to most of the river, increased access by local indigenous tribes to a culturally important fishery, and increased recreational activity in the form of white-water rafting. In fact, with careful efforts by the US Fish and Wildlife Service to relocate an estimated 679 fall Chinook salmon (*Oncorhynchus tshawytscha*) upstream of the dam prior to the breach (Figure 17.3), salmon spawning and reestablishment occurred throughout the White Salmon River within months following dam removal (National Geographic, 2011). Additional regulating and supporting ecosystem services were enhanced, such as increased water quality (eventually, following redistribution of the sediment released from the dam) and riparian habitat. Substantial cultural ecosystem services were enhanced, most notably for the many Native American Tribes whose rights to fish in their 'usual and accustomed fishing places' on the Columbia River and its tributaries were restored in the White Salmon River. These rights included fishing both for salmonids and for Pacific lamprey (*Lampetra tridentata*), a species on which the Yakama Nation place a high cultural value. Additional cultural ecosystem services were gained in general recreational fishing for salmonids and white-water rafting. Overall, the removal of the Condit Dam restored the natural hydrologic conditions to the White Salmon River, and the dam removal process was triggered by concerns for broader ecosystem services than just electric power generation. The removal was enabled by the prioritisation of fish passage restoration, which resulted in an economic decision to remove the dam by PacificCorps.

Management: minimum standards for critical service flows

A very promising approach that creates many potential win–win scenarios in river management focuses on environmental flows (i.e., ELOHA, the ecological limits of hydrological alteration). Environmental flows are a growing consideration by stream and river ecosystem managers as an approach to maintaining critical ecosystem functionality and sustainability. Environmental flows relate both to intrinsic and to instrumental ecosystem values in the form of regulating and maintenance services. The concept of environmental flows is also justified in the economics literature (Ojeda *et al.*, 2008). Environmental scientists and economists recognise that some natural resources have minimum levels below which the resource base cannot renew itself through natural processes or is irreplaceable. Minimum viable populations of fish species and unique riverine systems are examples. If these natural resources subject to irreversibility are deemed critical to the continued functioning and sustainability of particular riverine ecosystems, it may be economically prudent to establish a *safe minimum standard* (SMS) below which resource levels are not permitted to fall (Toman, 1994; Castle *et al.*, 1996). The SMS argument accords with a worldview that rejects the extremes of perfect substitutability or no substitutability for natural resources. Instead, it acknowledges uncertainty as the dominant condition describing the relationship of natural to anthropogenically-made resources in the distant future. The nature of the substitution relationship between anthropogenically-made and natural capital is unknown, and perhaps unknowable. There exists an implicit recognition of the evolutionary processes both in economic and ecological spheres. The rationale for imposing a safe minimum standard for critical ecological resources argues that with low information but high potential asymmetry in the loss function, cost–benefit analysis should give way to a greater presumption in favour of species preservation unless society judges that the cost of preservation is 'intolerable' (Toman, 1994). In effect, using the SMS approach can be seen as a social compact for expressing agreed-upon moral sentiments in the face of high ecological uncertainty and the potential loss asymmetry (Toman, 1994). Although these arguments hold a high common sense quotient, the application of the SMS is hampered by uncertainty in knowing the loss function and forecasting the substitutability of human-made resources for nature in the future. For those reasons, analysts recommend using an adaptive management approach to allow time for learning and adjustments to evolving social and natural conditions (Castle *et al.*, 1996).

Summary and future prospects

Ecosystem services are provided by riverine ecosystems in the form of provisioning, supporting, regulating and cultural benefits to both ecosystems and the human societies that depend upon them. River ecosystems vary greatly in scale, from headwater streams to vast river deltas, and the relative importance of various types of ecosystem services changes dramatically with spatial and temporal scales.

Ecosystem service provisioning areas and service benefiting areas are not necessarily co-located, so humans may choose to increase service-connecting areas through infrastructure. Given the diversity and complexity of river ecosystems, it is a significant challenge to synthesise the many ecosystem services that river ecosystems provide to develop clear insights to inform better management of rivers. Numerous tradeoffs exist in the management of ecosystem services from rivers, some of which include: navigation vs. water storage; recreation vs. industrial usage; irrigation vs. fish habitat; and cultural services vs. hydropower production. Over-emphasis on any one ecosystem service (e.g., provisioning of hydropower) can cause the reduction or even elimination of other valuable ecosystem services (e.g., provisioning of native fisheries), and hence it is critical to achieve a balanced approach that considers multiple ecosystem services in any management for river ecosystems across different scales.

Although questions that face riverine management may extend beyond the application of an ecosystem services approach, this concept can provide a mechanism for broadening the perspective of linking humans to river systems. The future of river science may require clearer articulation of the complex problems facing specific river systems and the development of novel methods for integrating the suite of ecosystem services into natural resource management paradigms. One promising approach, for example, lies in the concept of environmental flows, which posits minimal flow regimes for the recovery and/or continued maintenance of critical ecosystem services such as supporting healthy conditions for aquatic species in rivers. This approach is most appropriate when the threat of irreversible damage exists for critical services, and parallels the literature on creating safe minimum standards for such resource systems.

As the supply and demand of riverine ecosystem services continue to shift due to changing climatic, demographic and land use conditions, research on riverine ecosystems that examines the dynamic feedbacks in the coupled human and natural systems that produce different services is vitally needed. A useful tool may be a hierarchical approach combining qualitative and quantitative models to address how changing river governance affects the management of riverine ecosystem services and thus the provision of specific ecosystem services across different scales. Equally important is having an understanding of how different stakeholders may perceive these changes and potentially affect policy decisions. Incorporating experiential knowledge provided by stakeholders can improve the long-term management of rivers and their ecosystem services.

We conclude that, in the management of river ecosystems, there is a paramount need to analyse the full range of ecosystem services and their values, both instrumental and intrinsic, across different scales, and that environmental flow minima are useful criteria in helping achieve a balanced approach to managing riverine ecosystems and the services they provide human societies.

Acknowledgements

This work was funded in part by NSF grants #0966376, #1026629 and #0948983 as well as by internal support from Portland State University. We thank our many graduate students who have joined us in the classroom, most notably Jennifer Carman, Kenneth Lyons and Lauren Senkyr. We also thank Robert Costanza for providing helpful comments on an earlier version of this manuscript. Finally, we thank Paul Wood as well as two anonymous reviewers for their thorough comments and suggestions that significantly improved this manuscript.

References

Acuna, V., Diez, J.R., Flores, L. *et al.* (2013) Does it make economic sense to restore rivers for their ecosystem services? *Journal of Applied Ecology* 50: 988–997.

Aldwell, T.T. (1950) *Conquering the Last Frontier*. Seattle: Artcraft Engraving and Electrotype Co.

Allen, M.B. and Connolly, P.J. (2011) *Composition and Relative Abundance of Fish Species in the Lower White Salmon River, Washington, Prior to the Removal of Condit Dam*. USGS Open-File Report: 2011-1087.

An, S., Li, H., Guan, B., Zhou, C. *et al.* (2007) China's natural wetlands: past problems, current status, and future challenges. *AMBIO: A Journal of the Human Environment* 36: 335–342.

Arthington, A.H., Bunn, S.E., Poff, N.L. and Naiman, R.J. (2006) The challenge of providing environmental flow rules to sustain river ecosystems. *Ecological Applications* 16 (4): 1311–1318.

Arthington, A. H., Naiman, R.J., McClain, M.E. and Nilsson, C. (2010) Preserving the biodiversity and ecological services of rivers: new challenges and research opportunities. *Freshwater Biology* 55: 1–16.

Bagstad, K.J., Semmens, D.J. and Winthrop, R. (2013) Comparing approaches to spatially explicit ecosystem service modeling: a case study from the San Pedro River, Arizona. *Ecosystem Services* 5: 40–50.

Bastian, O., Grunewald, K. and Syrbe, R-U. (2012) Space and time aspects of ecosystem services, using the example of the EU Water Framework Directive. *International Journal of Biodiversity Science, Ecosystem Services & Management* 8: 5–16.

Borges, S.H. and Carvalhaes, A. (2000) Bird species of black water inundation forests in the Jaú National Park (Amazonas state, Brazil): their contribution to regional species richness. *Biodiversity & Conservation* 9: 201–214.

Bullock, J.M., Aronson, J., Newton, A.C. *et al.* (2011) Restoration of ecosystem services and biodiversity: conflicts and opportunities. *Trends in Ecology and Evolution* 26: 541–549.

Camargo M., Giarrizzo, T. and Isaac V. (2004) Review of the geographic distribution of fish fauna of the Xingu river basin, Brazil. *Ecotropica* 10: 123–147.

Castle, E., Berrens, R. and Polasky, S. (1996) Economics of sustainability. *Natural Resources Journal* 36: 715–730.

Chang, H., Jung, I., Strecker, A. *et al.* (2013) Water supply, demand, and quality indicators of the spatial distribution of water resources vulnerability in the Columbia River Basin, USA. *Atmosphere-Ocean* 51: 339–356.

Contreras-Balderas, S., Almada-Villela, P., de Lourdes Lozano-Vilano, M. and García-Ramírez, M.E. (2003) Freshwater fish at risk or extinct in Mexico. *Reviews in Fish Biology and Fisheries* 12: 241–251.

Costanza, R. (2008) Ecosystem services: multiple classification systems are needed. *Biological Conservation* 141: 350–352.

Costanza, R., d'Arge, R., de Groot, R. *et al.* (1997) The value of the world's ecosystem services and natural capital. *Nature* 387: 253–260.

Costanza, R. Kubieszewski, I., Ervin, D. *et al.* (2011a) Valuing ecological systems and their services. *F1000 Biology Reports 2011*, 3:14 (doi:10.3410/B3-14).

Costanza, R., Kubiszewski, I., Paquet, P. *et al.* (2011b) Planning Approaches for Hydropower Development in the Lower Mekong Basin. Institute for Sustainable Solutions, Portland State University, Portland, OR, USA. http://web.pdx.edu/~kub/publicfiles/Mekong/LMB_Report_FullReport.pdf [12 December 2015].

Costanza, R., de Groot, R., Sutton, P. *et al.* (2014) Changes in the global value of ecosystem services. *Global Environmental Change* 26: 152–158.

Crane, J. (2011) *Finding the River: An Environmental History of the Elwha*. Corvallis: Oregon State University Press.

Crossman, N.D., Connor, J.D., Bryan, B.A. *et al.* (2010) Reconfiguring an irrigation landscape to improve provision of ecosystem services. *Ecological Economics* 69: 1031–1042.

Daily, G. (1997) *Nature's Services: Societal Dependence on Natural Ecosystems*. Washington DC: Island Press.

Dalle, S.P. and Potvin, C. (2004) Conservation of useful plants: an evaluation of local priorities from two indigenous communities in eastern Panama. *Economic Botany* 58: 38–57.

Daniel, T.C., Muhar, A., Arnberger, A. *et al.* (2012) Contributions of cultural services to the ecosystem services agenda. *Proceedings of the National Academy of Sciences* 109: 8812-8819.

Duda, J.J., Warrick, J.A. and Magirl, C.S. (eds) (2011) Coastal habitats of the Elwha River, Washington—Biological and physical patterns and processes prior to dam removal: U.S. Geological Survey Scientific Investigations Report 2011–5120, 264 p.

Dudgeon, D. (2000) Large-scale hydrological changes in tropical Asia: prospects for riverine biodiversity. *BioScience* 50: 793–806.

Dugan, P.J., Barlow, C., Agostinho, A.A. *et al.* (2010) Fish migration, dams, and loss of ecosystem services in the Mekong Basin. *Ambio* 39: 344–348.

Ervin, D., Brown, D., Chang, H. *et al.* (2012) Managing ecosystem services supporting urbanizing areas. *Solutions* 6: 74–86.

Ervin, D., Vickerman, S., Ngawhika, S. *et al.* (2014) *Principles to Guide Assessments of Ecosystem Service Values*, Portland, Oregon: Cascadia Ecosystem Services Partnership, Institute for Sustainable Solutions, Portland State University.

Farber, S.C., Costanza, R. and Wilson, M.A. (2002) Economic and ecological concepts for valuing ecosystem services. *Ecological Economics* 41: 375–392.

Flanagan, C. and Laituri, M. (2004) Local cultural knowledge and water resource management: The Wind River Indian Reservation. *Environmental Management* 33: 262–270.

Goodland, R. (1977) Panamanian development and the global environment. *Oikos* 29: 195–208.

Granek, E.F., Madin, E.M.P., Brown, M.A. *et al.* (2008) Engaging recreational fishers in management and conservation: global case studies. *Conservation Biology* 22: 1125–1134.

Gutierrez, M. and Alonso, M.L.S. (2013) Which are, what is their status and what can we expect from ecosystem services provided by Spanish rivers and riparian areas? *Biodiversity and Conservation* 22: 2469–2503.

Haines-Young, R. and Potschin, M. (2010) The links between biodiversity, ecosystem services and human well-being. In D. Raffaelli, and C. Frid (Hg.) (eds) *Ecosystem Ecology: A New Synthesis. BES Ecological Reviews Series*. Cambridge: Cambridge University Press (iE), pp. 110–139.

Heckenberger, M.J., Russell, J.C., Toney, J.R. and Schmidt, M.J. (2007) The legacy of cultural landscapes in the Brazilian Amazon: implications for biodiversity. *Philosophical Transactions of the Royal Society B: Biological Sciences*. 362: 197–208.

Hoyer, W. and Chang, H. (2014) Assessment of freshwater ecosystem services in the Tualatin and Yamhill basins under climate change and urbanization. *Applied Geography* 53: 402–416. DOI: 10.1016/j.apgeog.2014.06.023

Hughes, R.M. and Dunham, S. (2014) Fish passage through urban and rural- residential areas. In Yeakley, J.A., Maas-Hebner, K.G. and Hughes, R.M. (eds) *Wild Salmonids in the Urbanizing Pacific Northwest*, New York: Springer, pp. 93–100.

Lemly, A.D., Kingsford, R.T. and Thompson, J.R. (2000) Irrigated agriculture and wildlife conservation: conflict on a global scale. *Environmental Management* 25: 485–512.

Leopold, Aldo. (1949) *A Sand County Almanac*. New York: Oxford University Press.

Loomis, J.B. (2012) Comparing households' total economic values and recreation value of instream flow in an urban river. *Journal of Environmental Economics and Policy* 1: 5–17.

Loomis, J., Kent, P., Strange, L., *et al.* (2000) Measuring the total economic value of restoring ecosystem services in an impaired river basin: results from a contingent valuation survey. *Ecological Economics* 33: 103–117.

Maguire L.A. and Justus, J. (2008) Why intrinsic value is a poor basis for conservation decisions. *BioScience* 58: 910–911.

McCauley, D.J. (2006) Selling out on nature. *Nature* 443: 27–28.

McLarney, W., Mafla, M., Arias, A. and Bouchonnet, D. (2010) The threat to biodiversity and ecosystem function of proposed hydroelectric dams in the LA Amistad world heritage site, Panama and Costa Rica, from proposed hydroelectric dams. Technical report. UNESCO World Heritage Committee. Asociación ANAI, Sabanilla de Montes de Oca, Costa Rica. Available: http://ww.w.biologicaldiversity.org/programs/international/pdfs/UNESCO_PDF.pdf [30 July 2014]

Milcu, A.I., Hanspach, D., Abson, D. and Fischer, J. (2013) Cultural ecosystem services: a literature review and prospects for future research. *Ecology and Society* 18(3): 44.

Millennium Ecosystem Assessment. (2005) *Ecosystems and Human Well-being: Synthesis*. Washington, DC: Island Press.

Minckley, W.L. and Marsh, P.C. (2009) *Inland Fishes of the Greater Southwest: Chronicle of a Vanishing Biota*. Tuscon, AZ: University of Arizona Press.

Minteer, B.A. (ed.) (2009). *Nature in Common? Environmental Ethics and the Contested Foundations of Environmental Policy*. Philadelphia: Temple University Press.

Moore, M. R., Mulville, A. and Weinberg, M. (1996). Water allocation in the American West: Endangered fish versus irrigated agriculture. *Natural Resources Journal* 36: 319–358.

Naiman, R.J., Beechie, T.J., Benda, L.E., *et al.* (1992) Fundamental elements of ecologically healthy watersheds in the Pacific Northwest coastal ecoregion. In Naiman, R. (ed.) *Watershed Management*, New York: Springer, pp. 127–188.

National Geographic. (2011) *Spectacular Dam "Removal:" More Revealed*, [Online], Available: http://video.nationalgeographic.com/video/news/us-condit-dam-salmon?source=searchvideo [12 December 2015].

O'Neill, J. (1992) The varieties of intrinsic value. *Monist* 75: 119–137.

Ojeda, M.I., Mayer, A.S. and Solomon, B.D. (2008) Economic valuation of environmental services sustained by water flows in the Yaqui River Delta. *Ecological Economics* 65: 155–166.

PacificCorp. (2014) *Condit Dam*, [Online], Available: http://www.pacificorp.com/about/newsroom/mr/cdmr.html [8 Mar 2014].

Palmer, M.A., Bernhardt, E.S., Allan, J.D., *et al.* (2005) Standards for ecologically successful river restoration. *Journal of Applied Ecology* 42: 208–217.

Park, Y-S., Chang, J., Lek, S. *et al.* (2003) Conservation strategies for endemic fish species threatened by the Three Gorges Dam. *Conservation Biology* 17: 1748–1758.

Petts, G.E. (1984) *Impounded Rivers: Perspectives for Ecological Management*. Chichester: John Wiley.

Petts, G.E. (1996) Water allocation to protect instream flows. *Regulated Rivers - Research & Management* 12: 353–365.

Petts, G.E. (2009) Instream Flow Science for Sustainable River Management. *Journal of the American Water Resources Association* 45:1071–1086.

Poff, N.L., Allan J.D., Bain M.B. *et al.* (1997) The natural flow regime: a paradigm for river conservation and restoration. *BioScience* 47: 769–784.

Poff, N.L., Richter, B.D., Arthington, A.H. *et al.* (2010) The Ecological Limits of Hydrologic Alteration (ELOHA): A new framework for developing regional environmental flow standards. freshwater biology. *Freshwater Biology* 55: 147–170.

Poff, N.L. and Matthews, J.H. (2013) Environmental flows in the anthropocene: past progress and future prospects. *Current Opinion in Environmental Sustainability* 5: 667–675.

Postel, S.L. (2000) Entering an era of water scarcity: The challenges ahead. *Ecological Applications* 10: 941–948.

Pringle, C.M., Freeman, M.C. and Freeman, B.J. (2000) Regional effects of hydrologic alterations on riverine macrobiota in the New World: tropical-temperate comparisons. *BioScience* 50: 807–823.

Raadgever, G.T., Mostert, E., Kranz, N. *et al.* (2008) Assessing management regimes in transboundary river basins: do they support adaptive management? *Ecology and Society* 13(1): 14. [online]

Reyers, B., Polasky, S., Tallis, H. *et al.* (2012) Finding common ground for biodiversity and ecosystem services. *BioScience* 62: 503–507.

Rosenberg, D.M., McCully, P. and Pringle, C.M. (2000) Global-scale environmental effects of hydrological alterations: introduction. *BioScience* 50: 746–751.

Rounds, S. (2010) Thermal Effects of Dams in the Willamette River Basin, *Oregon. USGS Scientific Investigations Report* 2010-5153.

Ruiz-Frau, A., Hinz, H., Edwards-Jones, G. and Kaiser, M.J. (2013) Spatially explicit economic assessment of cultural ecosystem services: Non-extractive recreational uses of the coastal environment related to marine biodiversity. *Marine Policy* 38: 90–98.

Rylands, A.B., Mittermeier, R.A. and Rodriguez-Luna, E. (1997) Conservation of neotropical primates: threatened species and an analysis of primate diversity by country and region. *Folia Primatologica* 68: 134–160.

Strahler, A.N. (1957) Quantitative analysis of watershed geomorphology. *Transactions of the American Geophysical Union* 38: 913–920.

Syrbe, R-U. and Walz, U. (2012) Spatial indicators for the assessment of ecosystem services: providing, benefiting and connecting areas and landscape metrics. *Ecological Indicators* 21: 80–88.

Toman, M. (1994) Economics and "sustainability:" balancing tradeoffs and imperatives. *Land Economics* 70: 399–413.

U.S. Department of the Interior (1994) *The Elwha Report: Restoration of the Elwha River Ecosystem & Native Anadromous Fisheries*. A Report Submitted to Congress Pursuant to PL 102-495.

Vannote, R.L., Minshall, G.W., Cummins, K.W. *et al.* (1980) The river continuum concept. *Canadian Journal of Fisheries and Aquatic Sciences* 37(1): 130–137.

Vilkka, L. (1997) *The Intrinsic Value of Nature*. Amsterdam: Rodopi.

Vitousek, P.M., Mooney, H.A., Lubchenco, J. and Melillo, J.M. (1997) Human domination of Earth's ecosystems. *Science* 277(5325): 494–499.

Vörösmarty, C.J., McIntyre, P.B., Gessner, M.O. *et al.* (2010) Global threats to human water security and river biodiversity. *Nature* 467: 555–561.

Vucetich, J., Bruskotter, J. and Nelson, M. (2015) Evaluating whether nature's intrinsic value is an axiom of or anathema to conservation. *Conservation Biology* 29(2): 321–332. (doi:10.1111/cobi.12464)

Wilson, M.A. and Carpenter, S.R. (1999) Economic valuation of freshwater ecosystem services in the United States: 1971-1997. *Ecological Applications* 9: 772–783.

Winter, C. (2007) The intrinsic, instrumental and spiritual values of natural area visitors and the general public: a comparative study. *Journal of Sustainable Tourism* 15: 599–614.

Xie, Z. (2003) Characteristics and conservation priority of threatened plants in the Yangtze valley. *Biodiversity and Conservation* 12: 65–72.

Yeakley J.A. (2014) Urban hydrology in the Pacific Northwest. In Yeakley, J.A., Maas-Hebner, K.G. and Hughes, R.M. (eds) *Wild Salmonids in the Urbanizing Pacific Northwest*, New York: Springer, pp. 59–74.

Yeakley, J.A., Maas-Hebner, K.G. and Hughes, R.M. (eds) (2014) *Wild Salmonids in the Urbanizing Pacific Northwest*. New York: Springer.

Zamora, H.A., Nelson, S.M., Flessa, K.W. and Nomura, R. (2013) Post-dam sediment dynamics and processes in the Colorado River estuary: Implications for habitat restoration. *Ecological Engineering* 59: 134–143.

Ziv, G., Baran, E., Nam, S. *et al.* (2012) Trading-off fish biodiversity, food security, and hydropower in the Mekong River Basin. *Proceedings of the National Academy of Sciences* 109: 5609–5614.

CHAPTER 18

Managing rivers in a changing climate

Robert L. Wilby
Department of Geography, Loughborough University, Loughborough, UK

Introduction

Managers have long recognised the sensitivity of river systems to climate variation, but the practical challenges of designing and operating long-lived infrastructure for future, non-stationary climate conditions remain largely unresolved (Němec and Schaake, 1982). Nonetheless, representations of the hydrologic cycle have been integral to General Circulation Models (GCMs) since the earliest climate simulations (e.g., Manabe *et al.*, 1965). By the 1970s, GCMs were beginning to imitate distributions of arid zones, tropical rain belts and patterns of continental runoff (Manabe and Holloway, 1975). By the 1980s they were being used to explore changing distributions of summer soil moisture deficit as a consequence of increased atmospheric concentrations of CO_2 (Manabe *et al.*, 1981). Other early research assessed the direct physiological impact of higher CO_2 concentrations on stomatal resistance of plants and hence catchment-scale water balance (Aston, 1984).

The idea that regional hydrologic impacts of global climate change can be evaluated using a water-balance model fed by GCM output was first proposed by Gleick (1986). His seminal study of the Sacramento River Basin established a methodological framework that has changed little in the ensuing 25 years (Gleick, 1987). Advances in computing power and proliferation of models merely enable more thorough quantification of uncertainty in hydrological impacts, typically approached from a 'top-down' perspective (Figure 18.1). This is legitimate scientific enquiry, but there is growing resistance to the view that higher resolution impacts modelling necessarily adds value to river management under climate change uncertainty (Stakhiv, 2011). Others are more optimistic and assert that societal demands for policy-relevant climate predictions should stimulate investment in climate services, including higher resolution climate and impacts modelling (Shukla *et al.*, 2010).

This chapter takes a middle path, namely that GCM outputs have utility for some aspects of river management but a smarter approach is needed when applying climate risk information in a decision-making context (Brown and Wilby, 2012; Poff *et al.*, 2015; Wilby *et al.*, 2009). This view is shared by others, as evidenced by reports published

River Science: Research and Management for the 21st Century, First Edition.
Edited by David J. Gilvear, Malcolm T. Greenwood, Martin C. Thoms and Paul J. Wood.

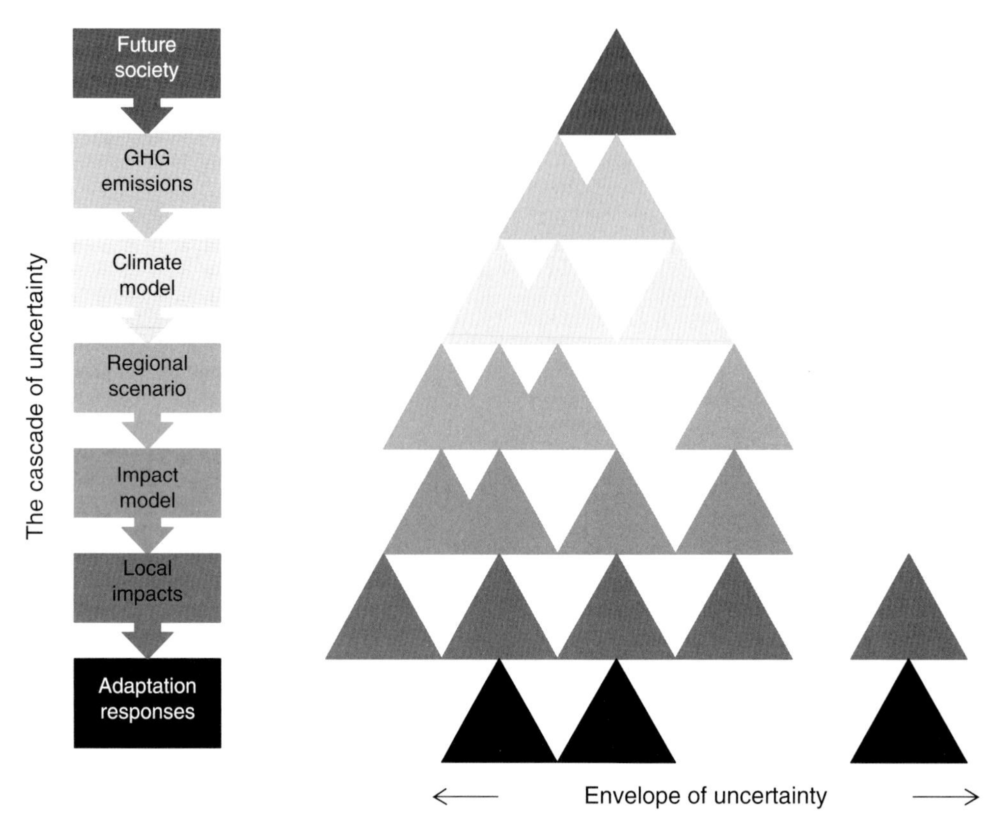

Figure 18.1 The cascade of uncertainty proceeds from different socio-economic and demographic pathways, their translation into concentrations of atmospheric greenhouse gas (GHG) concentrations, expressed climate outcomes in global and regional models, translation into local impacts on human and natural systems, and implied adaptation responses. The number of triangles at each level symbolises the growing number of permutations and hence the expanding envelope of uncertainty. For example, even relatively reliable hydrological models yield very different results depending on the methods (and data) used for calibration. Missing triangles represent incomplete knowledge or sampling of uncertainty. Adapted from Wilby and Dessai (2010). *(See colour plate section for colour figure.)*

by the World Bank Independent Evaluation Group (IEG, 2012) and the Organisation for Economic Co-operation and Development (2013). The chapter begins with a set of principles for testing the credibility of GCM output in hydrologic applications. These are necessary but not sufficient for determining whether GCM outputs provide so-called 'actionable information' (Kerr, 2011). Five examples are then given of strategies for adapting freshwater systems based on varying levels of reliance on climate model information. These include the conventional scenario-led risk assessment (introduced above), through scenario-neutral safety margins and sensitivity analyses, and adaptive management of whole systems. The strengths and weaknesses of each approach are explored in relation to typical water resource management activities, such as demand profiling, estimation of source yields and achievement of in-stream environmental objectives. Opportunities for further research and development are outlined in the concluding section.

Testing climate models for hydrological applications

Table 18.1 lays out five guiding principles for rigorous, impartial evaluation of GCM output from the point of view of hydrological applications. First, it should be recognised that observed hydro-meteorological data used to benchmark model 'skill' are never perfect, since they are often lacking in homogeneity, spatial coverage, longevity and sampling of extreme events. Second, for fair comparison GCM output should be evaluated against equivalent gridded (and sometimes interpolated) data rather than point observations. Third, greatest insight is gained when testing the physical realism of processes within the GCM, rather than comparing spatial patterns or time series (often aggregated to global scales). Fourth, climate model uncertainty must be assessed in the context of other uncertainties (such as imperfect hydrological models, or other long-term pressures on river catchments including land use change). Finally, ability to reproduce the present hydro-climate is no guarantee of future proficiency, although real-time model verification should become increasingly feasible within seamless prediction systems used for seasonal and decadal outlooks (Hurrell *et al.*, 2009; Shapiro *et al.*, 2010).

Some contest that GCMs are not yet 'ready for prime time' in hydrologic design and operation (Kundzewicz and Stakhiv, 2010). For example, when subjected to careful evaluation, GCMs used in the Intergovernmental Panel on Climate Change (IPCC) Fourth Assessment Report (AR4) did not even conserve mass – the majority generated more global annual mean precipitation than evaporation (Liepert and Previdi, 2012). Bias correction methods can adjust GCM output such that the local observed and modelled climates match. However, without realistic variability in water balance terms at the macro scale it is implausible that basin-scale hydrological simulations can be used in the type of deterministic assessment traditionally applied by engineers.

Compared with other components (such as the method of downscaling or hydrological modelling, Figure 18.1), the GCM generally contributes most to the cascade of hydrological uncertainty (Wilby and Harris, 2006). This arises from partial or simplistic representations (parameterisation) of important features such as clouds, from sensitivity to the conditions used to initiate GCM experiments, and to the natural variability within the model. Uncertainties from model processes/parameters and internal variability are relatively large over the next couple of decades, but are overtaken by the

Table 18.1 Five guiding principles for climate model evaluation. Source: Wilby (2010).

1. Quantify the uncertainty in the observed data used for model evaluation (homogeneity, confidence intervals, outliers)
2. Compare like with like (grid to grid, scale to scale)
3. Select indicators of performance relevant to the intended applications (extremes, low-frequency variability)
4. Evaluate climate models relative to other components of hydrological uncertainty (impact model, weighting)
5. Test combined climate, downscaling and hydrological model skill using near-term applications (seasonal forecasts)

uncertainty from socio-economic scenarios that diverge from mid-century onwards (Hawkins and Sutton, 2010). In other words, the mix of uncertainty in climate projections varies within the time horizons used to manage rivers, design and operate new water infrastructure (i.e., 10 to 80 years).

Additional, non-climatic uncertainties accrue in hydrologic projections. For example, future runoff scenarios derived from regional climates forced only by greenhouse gases do not reflect feedbacks from land-surface changes (including rapid urbanisation), major water transfers and withdrawals that may also be locally significant. In some water scarce regions concerns about future climate change may distract from the immediate imperative of water insecurity heighted by rapid economic growth and/or demographic change. Moreover, some of the most stressed river systems also reside in some of the most data sparse regions of the world. Hence, lack of basic information on key variables (such as the temperature lapse rate in snowmelt-dominated catchments) may be a significant impediment to river flow forecasting and management even under present climate conditions (e.g., Dobler *et al.*, 2012).

From the previous discussion the reader might be left wondering whether there is any merit in using GCM output to guide adaptation. Regional changes in precipitation may be highly uncertain, but all GCMs agree that air temperatures and sea levels will rise for decades to come even if emissions are curbed. Furthermore,the physical reasoning behind the Clausius–Clapeyron law warns us that a warmer atmosphere is more likely to deliver heavy precipitation events. At the very least these narratives (i.e., higher temperatures, sea level rise and more extreme rainfall) sketch the direction of travel. The remainder of this chapter considers whether GCMs can provide any further insight for adaptation beyond these simple statements.

A typology of climate model use in water-management decisions

Conventional risk-management frameworks assume that climate models can provide 'hard' values for design purposes (step 3 in Table 18.2). According to this conventional approach, the climate scenario always leads the decision-making process. However, the process can be inverted: climate models can be used in other ways to stress test adaptation options and inform design decisions. For example, Wilby (2012) showed that recent climate assessments undertaken by the World Bank Group could be classified by scenario method (Table 18.3) and approach to adaptation options analysis (Table 18.4). The same twin-axis typology is now illustrated using case studies of long-term river management and planning in the UK.

Three types of scenario analysis may be defined (Table 18.3):

(a) *qualitative* – simple narratives such as hotter/drier, or reference to high-level primary sources such as the Assessment Reports of the IPCC;

(b) *sensitivity test* – arbitrary ranges of climatic (and non-climatic) change factors that are used to adjust the data inputs for hydrological impacts modelling;

(c) *scenario-led* – conventional top-down approach to quantify outcomes arising from various combinations of greenhouse gas emissions, climate model, regional downscaling and hydrological impacts model uncertainty.

Four types of adaptation options analysis may be defined (Table 18.4):

Table 18.2 A risk-based decision framework for 'climate smart' investments. Adapted from the World Bank (2009).

Step	Activities
1	Identify problem, objectives, performance criteria and rules for decision-making • Define decision problem and objectives for exposure units/receptors and timeframe • Establish 'success' or 'performance' criteria and associated thresholds of tolerable risk • Identify rules for decision-making that will be applied to evaluate options (e.g., cost–benefit analysis)
2	Assess risks • Identify the climate and non-climate variables that could influence the potential outcomes • Identify the alternative future states or circumstances that may occur and associated impacts
3	Identify and evaluate options to manage risk • Identify (any) potential adaptation options to meet 'success' criteria (within thresholds of tolerable risk) • Evaluate adaptation options according to degree of uncertainty and established decision rules

(1) *low regret* – largely qualitative (often common-sense) appraisals of measures that should realise benefits under present and future climate conditions;

(2) *adaptively managed* – flexible operations, forecasting, or innovative use of existing infrastructure to meet emergent climate trends and/or changes in climate variability;

(3) *precautionary principle* – safety margins that are incorporated within infrastructure designs and operations to manage climate risk and uncertainty;

(4) *cost–benefit* – appraisal of multiple options (sometimes monetised) under climatic and non-climatic scenarios, including robust decision-making techniques that emphasise 'satisficing' rather than optimisation.

According to this system, a study that is heavily reliant on regional climate downscaling and yields non-specific recommendations (such as improved weather observations or water conservation) would be classified as scenario-led, low-regret adaptation.

Case studies

The following examples show five different ways in which GCM outputs are used in scenario and adaptation options appraisal for water-resource management. Although the cases refer mainly to the UK, the examples demonstrate the extent to which these methodologies are practicable in a broader sense.

Scenario-led (risk assessment only)

One meta-analysis of the downscaling literature found that the most common application is climate risk assessment for water resources (Wilby and Dawson, 2013). Of all the downscaling publications in the last decade, more than 40% have addressed some aspect of flood, drought or water quality, and the majority refer to North America, Europe or Australasia. The typical study adopts the same conceptual framework used by Gleick (1986) over a quarter of a century ago. That is, it takes GCM output to drive a water balance model. The main technical advance during the intervening period

Table 18.3 Typology of climate assessment by scenario method (ordered by burden of effort).

Strategy	Description	Advantages	Disadvantages	Examples
Qualitative	Refers to primary sources (such as IPCC) and/or applies simple climate narratives (such as hotter, warmer, drier or earlier).	Quick/cheap to implement and auditable when based on peer-reviewed/ public sources. May place greater emphasis on understanding vulnerabilities.	Perceived lack of due diligence or rigour in exploring location-specific factors.	Policy advocacy such as water conservation and source protection.
Sensitivity test	Application of arbitrary climatic (and non-climatic) change factors to the inputs driving model(s) of the system(s) of interest.	Informed by but not limited to the ranges specified by climate model and downscaled scenarios. Useful for stress-testing existing water systems to identify 'weak links', limits to performance, and to develop contingency plans.	Typically performed using one or two climate variables, problematic to envisage multi-dimensional pressures. Can be computationally demanding and presupposes existence of a functionally realistic impact model.	Appraising individual schemes or portfolios of adaptation options identified by stakeholders, including mixes of hard and soft measures to improve resilience.
Scenario-led	Applies top-down approach to quantify outcomes arising from combinations of emissions, climate model, downscaling, and impact model uncertainty.	Useful for preliminary climate risk screening when system vulnerabilities are not well understood. Can be performed alongside assessment of non-climate pressures. Range of risks may be so wide as to logically invoke robust solutions.	Computationally demanding with uncertainty in projected risks/impacts that can be so large as to be nonsensical to decision-makers, thereby delaying action, or undermining scientific credibility.	Raising awareness of high-level risks, identifying water storage and distribution system thresholds, characterising main sources of uncertainty.

Table 18.4 Typology of climate assessment by adaptation options analysis.

Strategy	Description	Advantages	Disadvantages	Examples
Low regret	Largely qualitative appraisal of scenario-neutral measures that should realise benefits under present climate variability as well as future climate change.	Draws on common-sense/ win–win/ low regret solutions that may already be known. Can be quick and easy to apply.	Measures may be easier to identify in theory than in practice. Often presupposes that strong governance systems are in place. Solutions may be generic or incur opportunity costs.	Reducing vulnerability to flood risk by controlling floodplain development.
Adaptively managed	Flexible operations, forecasting, or innovative use of existing infrastructure to meet emergent climate trends and/or changes in variability.	Decision trigger points can be related to stakeholder coping ranges. Evaluation of climate detection times and design of observing networks. Some investment decisions can be delayed until risks are better characterised.	Depends on long-term resourcing of monitoring systems (which may not necessarily be cheaper than a one-off intervention). There may also be adjustment costs when switching between adaptation actions.	Modified operating rules, water abstraction licensing, and water quality management.
Precautionary principle	Well-established safety margin approach for managing risk and uncertainty.	Climate safety factors can be explicitly linked to climate model output, enshrined within professional guidelines and periodically updated.	Possibility of mal-adaptation and large opportunity costs. Very low confidence in the underpinning science of hydrological extremes (especially heavy rainfall and peak river flows).	Designing levee, coastal defence, urban drainage system, and spillways.
Cost-benefit	Monetisation of adaptation options under climatic and non-climatic scenarios. Includes robust decision-making with emphasis on 'satisficing' rather than determining optimal solutions.	High economic discount rate or social/development imperatives may render climate risks as irrelevant to the decision.	Data may not be readily available to derive cost-functions, and/or the form of the relationship may change in the future.	Expected revenues from hydropower from new or rehabilitated infrastructure.

has been a switch from equilibrium (i.e., present, doubled or quadrupled CO_2 concentration) GCM experiments to transient GCM runs forced by prescribed emissions scenarios for the twenty-first century.

Scenario-led risk assessments help to raise awareness of water sector vulnerabilities (e.g., Watts *et al.*, 2015). In general, results for glacierised basins are consistent with observational evidence of earlier, more rapid spring melt, and lower summer flows in high-latitude regions, mediated by river basin properties (such as area, elevation range, extent of snow and ice cover). For example, Akhtar *et al.* (2008) applied a single climate change scenario from the PRECIS regional climate model to three river basins (the Hunza, Gilgit and Astore) in the Hindukush–Karakorum–Himalaya region under three prescribed stages of glacier coverage (100, 50 and 0%). PRECIS projected an annual mean temperature rise of 4.8 °C and precipitation change of +19% by the 2080s. Under this scenario, discharge generally increases for the case of the 100 and 50% glacier scenarios, but decreases for the 0% case, demonstrating the combined influence of climate forcing and catchment properties.

Water utility plans in England and Wales set out strategies for maintaining the balance between supply and demand. The plans have a time horizon of 25 years and are reviewed by the regulator every five years. Utilities have been required to undertake climate risk assessment as part of the planning process. Plans published since 1999 have been informed by national climate change scenarios in conjunction with methodologies developed jointly by the Environment Agency and water industry. Ahead of the 2009 plans, climate change factors were provided for river flows at the basin scale (Figure 18.2). The flow factor ranges captured uncertainty from a small ensemble of climate models combined with hydrological model uncertainty for the 2020s.

One estimate suggested that the aggregate loss of deployable output for England could be 3%, but the reduction for individual water resource zones could be as large as 50% (Charlton and Arnell, 2011). The 2009 UK Climate Projections (UKCP09) enabled sampling of an even greater range of uncertainty in regional climate projections. For example, by the 2050s the change in summer rainfall over parts of southern England could span −40 to +20% depending on the emissions scenario and climate model ensemble member (Figure 18.3). Overall, low flows could diminish in summer by as much as 70% in some catchments (Cloke *et al.*, 2010).

Thanks to these (and subsequent) national assessments the notion of 'warmer wetter winters' and 'hotter drier summers' has become firmly established within the psyche of the UK water industry. However, there are ensemble members that show the opposite tendency: drier winters and wetter summers. Hence, the pervasive message from UKCP09 is one of immense uncertainty about the future climate and associated water-resource impacts. This realisation alone brings to an end the notion that new projects and operating procedures can be designed with a single future in mind. Rather, the focus is now less about optimising outcomes and more about 'satisficing' (Stakhiv, 2011). These solutions are more robust to the acknowledged climate (and non-climate) uncertainties.

Scenario-led, low-regret adaptations

As noted above, robust decision-making approaches identify options that perform satisfactorily (but not necessarily optimally) across a wide range of plausible futures

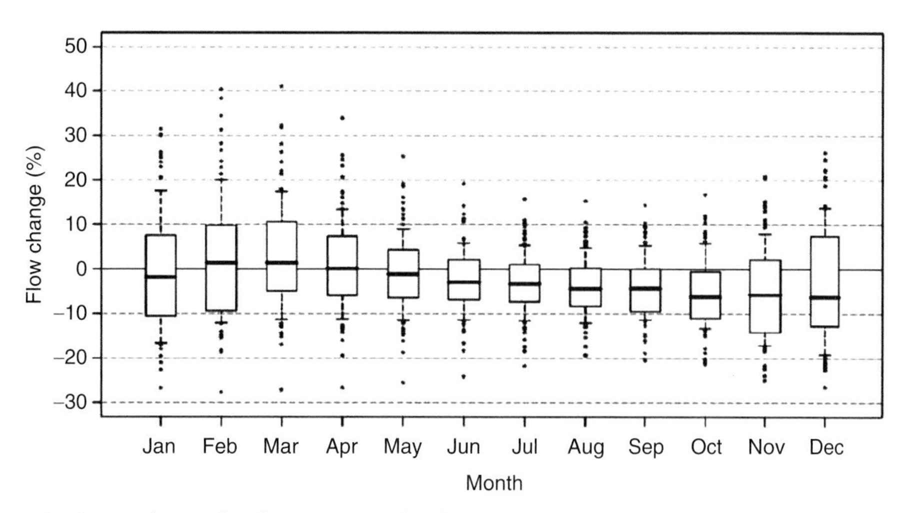

Figure 18.2 Climate change flow factors (2020s) for the River Itchen, UK. Source: Portsmouth Water (2014).

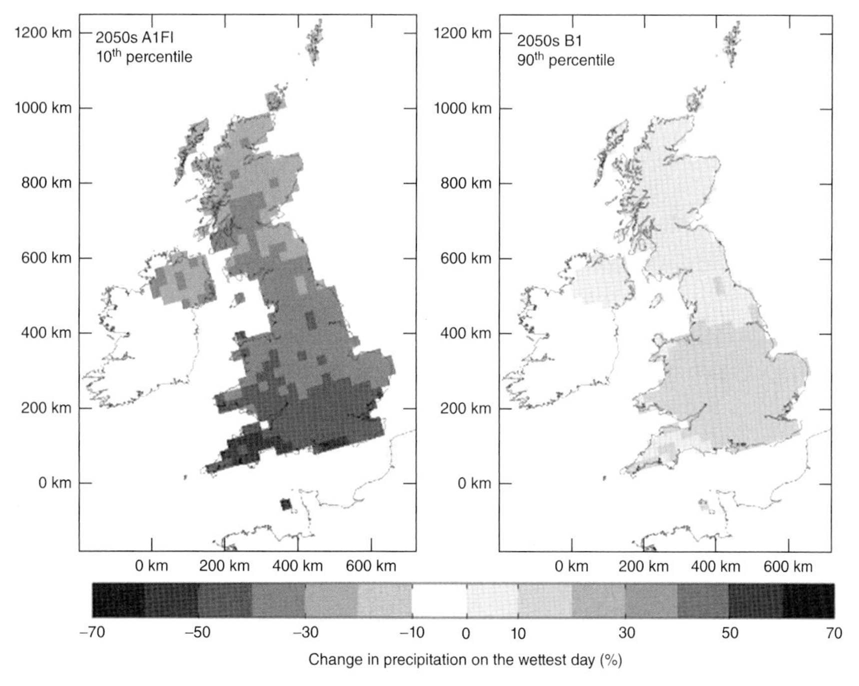

Figure 18.3 UKCP09 changes in summer rainfall under high emissions (left panel: A1FI, 10th percentile ensemble member) and low emissions (right panel: B1, 90th percentile ensemble member) by the 2050s. *(See colour plate section for colour figure.)*

(Lempert *et al.*, 2006). *In extremis* it is possible to envisage options without recourse to any climate model information but definitions of 'satisfactory' will reflect changing societal values. For example, under a business as usual water use scenario for east Devon, in southern England, and a very large set of climate model projections, water

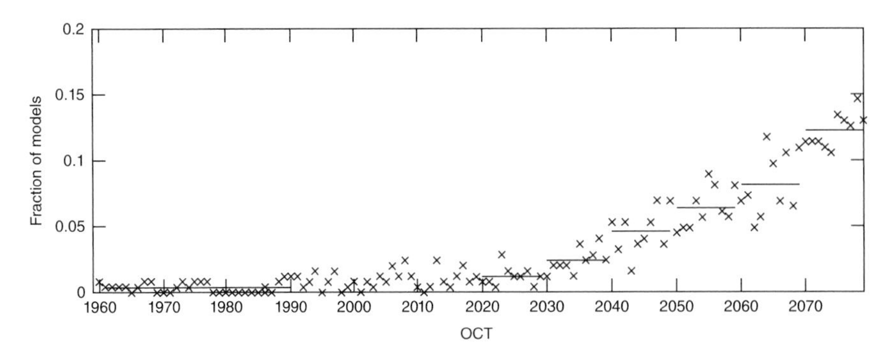

Figure 18.4 Outcome of a business as usual scenario for east Devon: the fraction of ClimatePrediction.net (CP.net) projections failing to meet average water demand in October under SRES A1B. Source: Lopez *et al* (2009).

supply failures could increase from near zero historic frequencies to an annual likelihood of about 5% by the 2040s (Figure 18.4). 'Do nothing' is always an option and, in this case, the regulator, water utility and customers would be accepting lower levels of service. In practice, this could mean periodic bans on non-essential water use to conserve stocks whilst delaying investment (see 'Scenario-led, adaptively managed').

'Scenario-neutral' or 'no regret' strategies should yield benefits regardless of climate change. In practice, there are always opportunity costs, tradeoffs, or externalities associated with adaptation actions so it is preferable to label such interventions as 'low regret' (Wilby and Dessai, 2010; Prudhomme *et al.*, 2010a). These measures should address present water management priorities as well as keeping open, or maximising, options for adaptation in the future. For example, protecting water sources from contamination is a sound strategy under any climate scenario. Likewise, long-term monitoring of environmental quality is necessary for estimating the sustainable resource and for benchmarking changing conditions or the outcome of management decisions.

Water-demand management is widely regarded as a low-regret adaptation option (Table 18.5) because all the measures make sense regardless of the very uncertain outlook for climatic and non-climatic drivers of water availability. For example, in the case of east Devon, the single-year water supply failure rate is more than halved by demand management measures. However, supply-side options tend to be preferred more than demand-side measures by water utilities because bulk yields from new sources can be quantified more easily than the aggregate water savings made by a large number of individual customers. Nonetheless, all of the options listed in Table 18.5 make sense under a qualitative (narrative) scenario of 'increased water scarcity'.

Sensitivity test, precautionary principle

Safety margins are an established method for accommodating uncertainties in infrastructure design. This is typically accomplished by expanding the capacity of reservoirs, irrigation systems and urban drainage networks, or by adding height to levees, coastal defences and platform levels. The safety margin (or 'head room') can be incorporated

Table 18.5 Examples of water demand management options. Adapted from EA (2009a).

- Convert permanent abstraction licenses to *time-limited* status, to provide flexibility to respond to climate variability and change.
- Accept a *reduction in the reliability of supply* as an option for resolving future deficits.
- Increase levels of *metering* with suitable tariffs to improve water and economic efficiency whilst protecting vulnerable groups.
- Support *water neutrality* where new development is planned and require developers to produce water cycle studies for proposed housing developments.
- Identify regulatory and voluntary *water efficiency* standards for non-household buildings.
- Set targets for *leakage control* that better reflect the costs to society and the environment.
- Introduce further *incentives* for the purchase and fitting of water efficient equipment and appliances.

at the design stage for new investments, during maintenance and rehabilitation cycles, or when retro-fitting existing systems. The critical question is: how large should the safety factor be to achieve a specified level of protection or performance throughout the design life of the infrastructure?

Governments have addressed the question in different ways. For example, the Australian State of Queensland (2010) assumes a 5% increase in rainfall intensity per degree of global warming (expected to be 2 °C by 2050, 3 °C by 2070 and 4 °C by 2100) when assessing peak flows. Climate change factors are applied to rainfall amounts whilst historic flood levels with probability 0.5% and 0.2% are scaled to 1% and 0.5% respectively by the 2050s. Other governments base their climate change factors more explicitly on model projections for heavy precipitation over the region of interest. In Germany, different factors have been used depending on the flood return period (Ihringer, 2004). The allowance for England and Wales once assumed a 20% increase to all peak flows (Reynard *et al.*, 2004) but the revised advice now has upper and lower bounds, given by region (EA, 2011). In New South Wales, Australia the recommended sensitivity analysis is based on increases in extreme rainfall and flood volumes of 10 to 30% (NSW DECC, 2007).

The advice for flood management in England requires that a sensitivity analysis be performed across a range of change (for example, extreme rainfall and flood flows) that could occur over the lifetime of the plan (EA, 2011). The central climate change factor is used for investment appraisals with the upper and lower bounds used to test the extent to which options can adapt. In practice, the uncertainty in peak flows can be very large. For example, in Eastern England (Anglian region) the lower, central and upper change factors for river flood flows are −10, +15 and +40% respectively by the 2050s, and −5, +25 and +70% by the 2080s.

Clearly, the more precautionary the allowance for climate change the greater the range of uncertainty that can be managed. On the other hand, economic viability and technical feasibility of the scheme may be brought into question. For example, based on available scientific evidence, a 50% safety margin on 20-year flood heights would be sufficient to manage climate risk in the majority of UK catchments to the 2080s (Figure 18.5). However, with pressure on flood defence budgets, such a

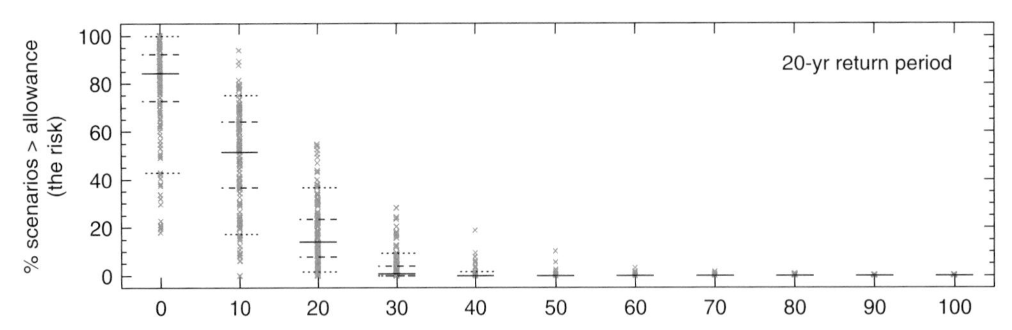

Figure 18.5 Percentage of model runs exceeding a given 20-year flood safety margin (%) based on 155 UK catchments and 16 AR4 GCMs (2080s, A1B emissions scenario). Each cross for a given allowance shows the results for one catchment. The 50th, 30th and 70th, and 10th and 90th percentiles (solid, dashed and dotted lines respectively) are shown for each ensemble. Source: Prudhomme *et al.* (2010b).

precautionary margin could imply higher levels of investment in fewer locations.

Sensitivity test, cost–benefit

The literature abounds with examples of hydrological sensitivity (or stress) testing based on prescribed temperature, precipitation and basin conditions. The climate 'change factors' may be arbitrary, originate from palaeo-climate data, or from climate models. The stress test can be performed as part of a vulnerability assessment for contingency planning, or to evaluate the outcome of different management decisions.

For example, Abu-Taleb (2000) calculated the annual water deficit for Jordan, taking into account projected water use by all sectors for specified precipitation and temperature changes to 2020. Under a temperature rise of 4 °C and 20% decrease in precipitation, the projected deficit would be 1020 mm^3/year, compared with a deficit of 408 mm^3/year under the no climate change scenario. Several options for reducing the deficit were tested, including: water pricing, conservation measures, water distribution network rehabilitation, enforcement of metering, billing and revenue collection, and reallocation through volumetric constraints. Taken together these measures could realise water savings of up to 566 mm^3. However, even an optimal combination of strategies would not produce enough water savings to offset the anticipated deficits. Under the best case scenario (no change in temperature, precipitation +20%) Jordan would still need to invest in water conservation measures.

There are relatively few examples of sensitivity analysis combined with adaptation options appraisal. Lopez *et al.* (2009) evaluated the frequency of years with water supply failures in East Devon under business as usual compared with three adaptation scenarios: water-demand management, increased storage capacity of a reservoir, and a combination of demand reduction and capacity expansion. The outcomes were modelled using industry standard water balance and distribution models, driven by precipitation and temperature changes from the ClimatePrediction.net (CP.net) climate model ensemble (Figure 18.6). The preferred adaptation strategy to reduce risk of supply interruptions would involve a mix of water-demand reduction measures with increased storage (shown as Demand+Reservoir in Figure 18.6).

Using the same approach, Whitehead *et al.* (2006) and Karamouz *et al.* (2010) evaluate

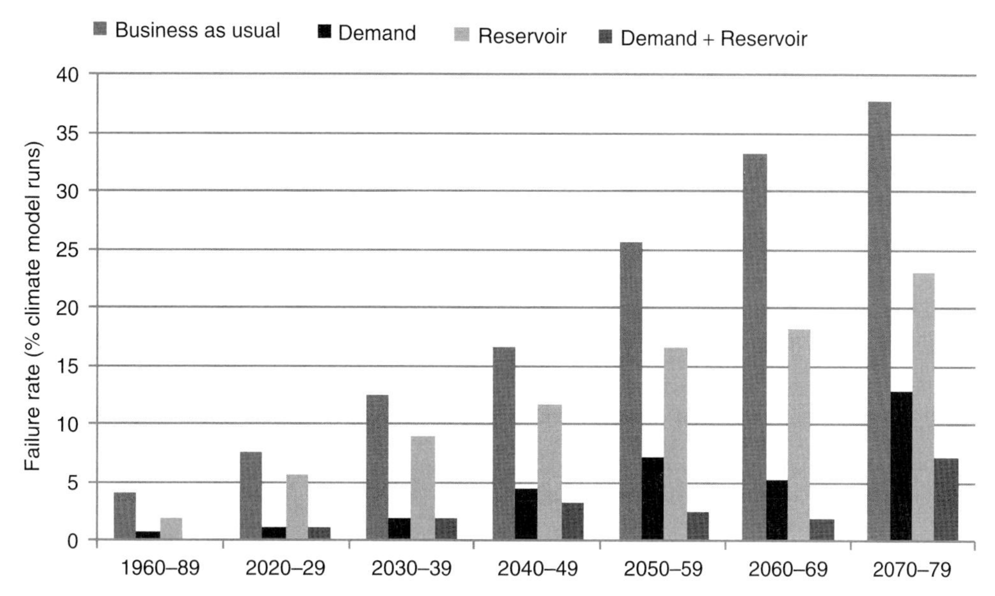

Figure 18.6 Percentage of model runs with single year water supply failure under various strategies for a water supply zone in East Devon under SRES A1B emissions. Adapted from Lopez *et al.* (2009). *(See colour plate section for colour figure.)*

the effectiveness of different control strategies for eutrophication using downscaled climate scenarios. For instance, Whitehead *et al.* (2006) show the relative reduction in simulated river nitrate concentrations to 2100 assuming lower fertiliser application rates, more stringent emission controls for nitrogen and/or wetland creation. Similarly, Henriques and Spraggs (2011) show how the population at risk from water supply failures due to widespread flooding (under a +20% peak flow scenario) can be reduced by increasing the resistance of individual assets and installing new links to improve overall distribution network resilience. Such studies are not using climate model output in a deterministic sense but rather as plausible scenarios with which to stress test existing systems and performance of planned adaptations. Other examples of the approach include management plans for the Upper Great Lakes in North America (Brown *et al.*, 2011) and the Niger Basin (Brown, 2010).

Scenario-led, adaptively managed

Adaptive management involves taking action to manage climate risks through predetermined interventions when particular trigger conditions or coping conditions are exceeded (e.g., Brown *et al.*, 2011; EA, 2009b). As such, the approach depends critically on good awareness of system vulnerabilities and routine surveillance of changing patterns of risk. The arrival of the trigger point may be delayed by improved forecasting and contingency planning, or by modifying control rules for existing water infrastructure to better cope with evolving conditions.

Despite heavy reliance on routine monitoring and review, adaptive management can still benefit from climate model analysis. For example, by exploring the relationship

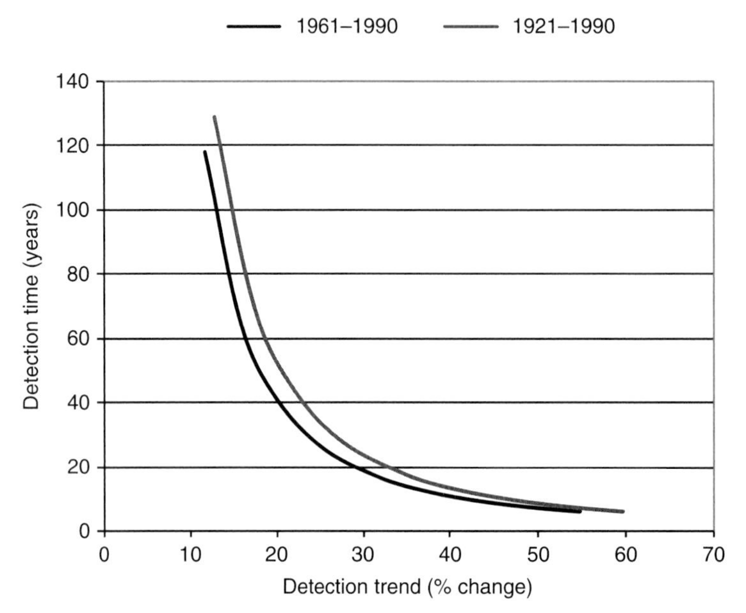

Figure 18.7 Detection times (years from 1990) for summer low flows in the River Itchen, southern England using two different periods to estimate natural variability (lower variability: 1961–90 and higher variability: 1921–90). The detection time for this river is relatively short because the chalk groundwater system dampens much of the inter-annual variability observed in less permeable catchments. Source: Wilby (2006).

between a projected climate trend and natural variability (i.e., signal-to-noise ratio), it is possible to determine the length of time required for detection at a specified level of confidence (Figure 18.7). In regions where the climate change signal is expected to be weak relative to inter-annual variability, detection may not be possible for many decades or even during the twenty-first century. Although the signal may not be statistically significant for many years, the practical significance of small changes in river flow could be felt sooner. There could also be abrupt changes in climate that are not well-resolved by climate models. Nonetheless, detection studies provide some insight to the scope for delaying investments, or for designing networks of observing systems at sentinel locations (Murphy *et al.*, 2013).

Non-structural measures may be deployed to reduce vulnerability to climate variability and change within coping ranges. As noted before, 'soft measures' include water conservation, better irrigation scheduling and demand management (Table 18.5). Where the legal and regulatory context allow, smarter water licensing (permit) arrangements for abstractions may also afford greater protection to the environment whilst safeguarding public water supplies (Figure 18.8).

Well-maintained and designed infrastructure can accommodate a degree of climate variability and change. For example, model studies in North America indicate that adaptive reservoir management can sustain levels of performance for water supply, energy production and environmental flows even under future droughts (Georgakakos *et al.*, 2012; Li *et al.*, 2010; Watts *et al.*, 2011).

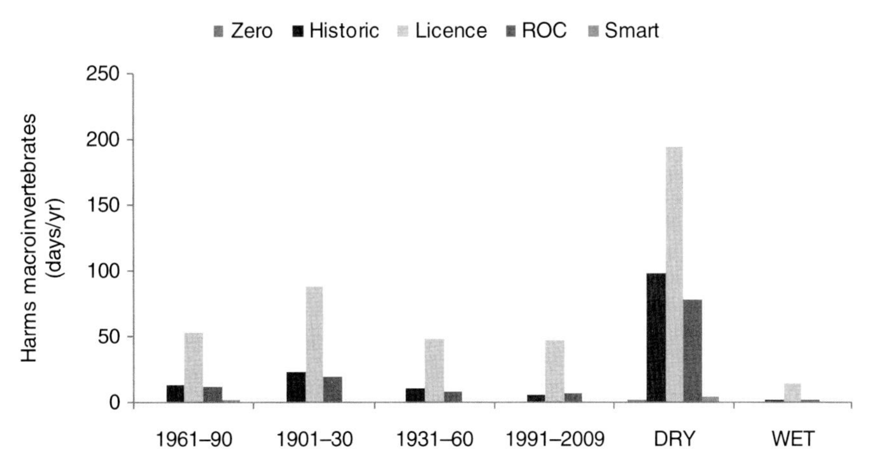

Figure 18.8 Mean annual frequency of low flows (<224 ml/d) that are harmful to macroinvertebrate communities in the River Itchen under various abstraction license conditions (Zero, Historic, Licence, Existing Review of Consents (ROC), Smart), precipitation variability (1961–90, 1901–30, 1931–60, 1991–2009) and climate change projections for the 2020s (DRY and WET). The Smart license takes water from the environment when it is least harmful, and imposes hands-off flow conditions under very dry conditions. Source: Wilby *et al.* (2011). *(See colour plate section for colour figure.)*

However, depending on the climate change scenario, increasing reservoir storage for one use (e.g., flood control) may require tradeoffs against other allocations (e.g., for hydropower production, irrigation and instream flow targets) in multi-purpose systems (Payne *et al.*, 2004; Gosh *et al.*, 2010; Raje and Mujumdar, 2010). The adaptation pathway and choice of rule curves also depend on decision-maker attitudes to risk and weight attached to different climate scenarios (Brekke *et al.*, 2009), as well as the anatomy of the water resource system in the case of multi-reservoir configurations (Eum and Simonovic, 2010). Once again, the role of climate models (when deployed) is to bind the range of conditions tested.

Concluding remarks

Water managers are divided as to when (or even whether) climate models will be capable of producing 'actionable information' for adaptation and development planning. Nonetheless, the top-down, scenario-led framework has been the dominant scientific paradigm for at least 25 years. This approach has improved understanding of the extent and character of climate uncertainty. The work has also raised awareness of risks but it has done little to inform adaptation in practice. Indeed, it can be argued that the single most important legacy of climate modelling to decision-makers has been to convince them that flexible, robust and imaginative measures are needed to adapt water resource systems to climate change. Some contest that this has always been the case.

This chapter shows that, despite uncertainty, climate model outputs can add value to a range of water management decisions. However, it is always helpful to ask 'how far can the plan proceed without using climate models?' Since climate change is expected to amplify non-climatic pressures, the first step should be to identify and implement the 'low-regret' (i.e., scenario-neutral) measures that are already an established part of best practice (see Stakhiv, 2011). Improved forecasting systems, inter-agency cooperation, soil and water conservation, contingency planning, and other enabling activities go a long way to reducing vulnerabilities to climate variability and change regardless of the outlook (Wilby and Keenan, 2012). Ideally, the management plan will recognise the broader catchment context (especially for transboundary river systems) and future tradeoffs between water use for domestic and commercial supply, agriculture and energy production. In many situations, non-climatic or indirect climate impacts may be the dominant driver of change in freshwaters for the foreseeable future (Vörösmarty *et al.*, 2010).

Perhaps the greatest value of climate model products lies in their heuristic use. This runs counter to the fashion for climate 'prediction' (e.g., Hurrell *et al.*, 2009; Shapiro *et al.*, 2010) but the term 'scenario' better reflects present capabilities and uncertainties. With scenarios it is possible to bound sensitivity tests of adaptation portfolios, identify system thresholds and fix weak links, explore the timing of investments, revise operating rules, or develop smarter permitting regimes.

The most problematic application remains the climate change safety margin because of the very low confidence in extreme precipitation and river flows generated by climate models. Nonetheless, risk-based design standards are needed for new infrastructure and government agencies are beginning to provide engineers with look-up tables based on available scientific evidence. Ultimately, there will be a tradeoff between the cost of the scheme and the allowance for uncertainty that reflects the degree of risk aversion. However, through legislated periodic review, it is possible to revisit the evolving scientific knowledge underpinning climate change safety margins. The guiding principle here should be one of transparency about the assumptions and evidence employed.

Existing water infrastructure could also be gradually upgraded during scheduled maintenance. This is to protect present assets and maintain levels of performance in the future (unless the decision is taken to accept declining levels of service). Elements with short lifetimes (~10 years for pumps, telemetering devices and detentions ponds) can be optimised with less attention to climate resilience than long-lived assets (~80 years for concrete sewer and pipe replacement) (Arnbjerg-Nielsen, 2011). Post-disaster reconstruction or routine replacement provides further opportunities to 'build back better' (i.e., incorporate higher specification designs or materials for vulnerable assets). However, adaptively managed systems may not necessarily be a cheaper strategy for managing uncertain climate risks than incorporating safety margins up-front. Again, scenario analysis can help elucidate the circumstances under which the former option might be preferable.

Finally, some are calling for a new era of field- and model-based research to test the effectiveness of adaptation interventions at river habitat (Everall *et al.*, 2012), river reach (Orr *et al.*, 2015), to river catchment scales (Wilby *et al.*, 2010). Such calls echo those in the 1980s for field campaigns to 'solve' acid rain. There is also a case for Research Councils to allocate more resources for the opportunistic study of the causes and consequences of extreme weather events. This all depends on wider enabling conditions, such as strong institutional memory and governance, long-term monitoring and reporting systems, freedom of access to data and analytical tools, technical capacities in public and private sectors, and bridging organisations between scientific and stakeholder communities. However, these initiatives will only bear fruit if comparable efforts are made to translate latest research findings into 'actionable' guidance for river managers.

Acknowledgements

This chapter draws on a paper presented at the workshop *'Including long-term climate change in hydrologic design'* held at the World Bank, Washington, DC, USA, 21 November 2011. It is also fitting to repeat the heartfelt thanks to Geoff Petts from the front-piece of my PhD thesis. This is because Geoff has lectured, tutored, inspired, enthused, financed and nudged me through everything from polo mint experiments in rivers to understanding(?) multi-dimensional hyperspace!

References

Abu-Taleb, M.F. 2000. Impacts of global climate change scenarios on water supply and demand in Jordan. *Water International*, **25**, 457–463.

Akhtar, M., Ahmad, N. and Booij, M.J. 2008. The impact of climate change on the water resources of Hindukush-Karakorum-Himalaya region under different glacier coverage scenarios. *Journal of Hydrology*, **355**, 148–163.

Arnbjerg-Nielsen, K. 2011. Past, present, and future design of urban drainage systems with focus on Danish experiences. *Water Science and Technology*, **63**, 527–535.

Aston, A.R. 1984. The effect of doubling atmospheric CO2 on streamflow: A simulation. *Journal of Hydrology*, **67**, 273–280.

Brekke, L.D., Maurer, E.P., Anderson, J.D., et al. 2009. Assessing reservoir operations risk under climate change. *Water Resources Research*, **45**, W04411.

Brown, C. 2010. Decision-scaling for robust planning and policy under climate uncertainty. *World Resources Report*. Washington DC. Available online from: http://www.worldresourcesreport.org/files/wrr/papers/wrr&uscore;brown&uscore;uncertainty.pdf

Brown, C. and Wilby, R.L. 2012. An alternate approach to assessing climate risks. *Eos*, **93**, 401–402.

Brown, C., Werick, W., Leger, W., et al. 2011. A decision-analytic approach to managing climate risks: application to the Upper Great Lakes. *Journal of the American Water Resources Association* **47**, 524–534.

Charlton, M.B. and Arnell, N.W. 2011. Adapting to climate change impacts in water resources in England – An assessment of draft Water Resources Management Plans. *Global Environmental Change*, **21**, 238–248.

Cloke, H.L., Jeffers, C., Wetterhall, F., et al. 2010. Climate impacts on river flow: projections for the Medway catchment, UK, with UKCP09 and CATCHMOD. *Hydrological Processes*, **24**, 3476–3489.

Dobler, C., Hagemann, S., Wilby, R.L. and Stötter, J. 2012. Quantifying different sources of uncertainty in hydrological projections at the catchment scale. *Hydrology and Earth Systems Science*, **16**, 4343–4360.

Environment Agency (EA), 2009a. *Water for people and the environment: Water resources strategy for England and Wales*. Rio House, Bristol.

Environment Agency (EA), 2009b. *TE2100 plan consultation document*. London: Thames Barrier.

Environment Agency (EA), 2011. *Adapting to climate change: Advice for flood and coastal erosion risk management authorities*. Environment Agency, Bristol.

Eum, H.I. and Simonovic, S.P. 2010. Integrated reservoir management system for adaptation to climate change: The Nakdong River basin in Korea. *Water Resources Management*, **24**, 3397–3417.

Everall, N.C., Farmer, A., Heath, A.F., et al. 2012. Ecological benefits of creating messy rivers. *Area*, **44**, 470–478.

Georgakakos, A.P., Yao, H., Kistenmacher, M., et al. 2012. Value of adaptive water resources management in Northern California under climatic variability and change: Reservoir management. *Journal of Hydrology*, **412–413**, 34–46.

Gleick, P.H. 1986. Methods for evaluating the regional hydrologic impacts of global climatic changes. *Journal of Hydrology*, **88**, 97–116.

Gleick, P.H. 1987. Regional hydrologic consequences of increases in atmospheric CO2 and other trace gases. *Climatic Change*, **10**, 137–161.

Gosh, S., Raje, D. and Mujumdar, P.P. 2010. Mahanadi streamflow: climate change impact assessment and adaptive strategies. *Current Science*, **98**, 1084–1091.

Hawkins, E. and Sutton, R. 2010. The potential to narrow uncertainty in projections of regional precipitation change. *Climate Dynamics*, **37**, 407–418.

Henriques, C. and Spraggs, G. 2011. Alleviating the flood risk of critical water supply sites: asset and system resilience. *Journal of Water Supply Research and Technology-Aqua*, **60**, 61–68.

Hurrell, J., Meehl, G.A., Bader, D., et al. 2009. A unified modeling approach to climate system prediction. *Bulletin of the American Meteorological Society*, **90**, 1819–1832.

Ihringer, J. 2004. Ergebnisse von Klimaszenarien und Hochwasserstatistik. *KLIWA Bericht 4*, S.153–168. KLIWA Symposium, Würzburg, München.

Karamouz, M., Taheriyoun, M., Baghvand, A., et al. 2010. Optimization of watershed control strategies for reservoir eutrophication management. *Journal of Irrigation and Drainage Engineering*, **136**, 847–861.

Kerr, R.A. 2011. Time to adapt to a warming world, but where's the science? *Science*, **334**, 1052–1053.

Kundzewicz, Z.W. and Stakhiv, E.Z. 2010. Are climate models 'ready for prime time' in water resources management applications, or is more research needed? *Hydrological Sciences Journal*, **55**, 1085–1089.

Lempert, R.J., Groves, D.G., Popper, S.W. and Bankes, S.C. 2006. A general analytic method for generating robust strategies and narrative scenarios. *Management Science*, **52**, 514–554.

Li, L.H., Xu, H.G., Chen, X. and Simonovic, S.P. 2010. Streamflow forecast and reservoir operation performance assessment under climate change. *Water Resources Management*, **24**, 83–104.

Liepert, B.G. and Previdi, M. 2012. Inter-model variability and biases of the global water cycle in CMIP3 coupled climate models. *Environmental Research Letters*, **7**, doi:10.1088/1748-9326/7/1/014006.

Lopez, A., Fung, F., New, M., et al. 2009. From climate model ensembles to climate change impacts: A case study of water resource management in the South West of England. *Water Resources Research*, **45**, W08419.

Manabe, S. and Holloway, J.L. Jr., 1975. The seasonal variation of the hydrologic cycle as simulated by a Global Model of the Atmosphere. *Journal of Geophysical Research*, **80**, 1617–1649.

Manabe, S., Smagorinsky, J. and Strickler, R.F. 1965. Simulated climatology of a General Circulation Model with a hydrologic cycle. *Monthly Weather Review*, **93**, 769–798.

Manabe, S., Wetherald, R.T. and Stouffer, R.J. 1981. Summer dryness due to an increase of atmospheric CO2 concentration. *Climatic Change*, **3**, 347–386.

Murphy, C., Harrigan, S., Hall, J. and Wilby, R.L. 2013. Assessing climate driven trends in mean- and high- river flows from a network of reference stations in Ireland. *Hydrological Sciences Journal*, **58**, 755–772.

Němec, J. and Schaake, J. 1982. Sensitivity of water resource systems to climate variation. *Hydrological Sciences Journal*, **27**, 327–343.

New South Wales Department of Environment and Climate Change (NSW DECC), 2007. *Floodplain Risk Management Guideline – Practical Consideration of Climate Change*. New South Wales Government, Sydney, 14pp.

Organisation for Economic Co-operation and Development (OECD), 2013. *Water and climate change adaptation: Policies to navigate unchartered waters*. OECD Studies on Water, OECD Publishing, Paris, 228pp.

Orr, H., Johnson, M., Wilby, R.L., et al. 2015. Warming rivers: What else do managers need to know? *WIREs Water*, **2**, 55–64.

Payne, J.T., Wood, A.W., Hamlet, A.F., et al. 2004. Mitigating the effects of climate change on the water resources of the Columbia River basin. *Climatic Change*, **62**, 233–256.

Poff, N.L., Brown, C.M., Grantham, T.E., et al. 2015. *Sustainable water management under future uncertainty with eco-engineering decision scaling*. doi: 10.1038/nclimate2765

Portsmouth Water, 2014. *Final Water Resource Management Plan 2014*. Havant, UK.

Prudhomme, C., Wilby, R.L., Crooks, S., et al. 2010a. Scenario-neutral approach to climate change impact studies: application to flood risk. *Journal of Hydrology*, **390**, 198–209.

Prudhomme, C., Wilby, R.L., Crooks, S., et al. 2010b. *Regionalisation of the impact of climate change on flood flows using a scenario-neutral approach: application in Great Britain*. Proceedings of the British Hydrological Society Third International Symposium, Newcastle, UK.

Raje, D. and Mujumdar, P.P. 2010. Reservoir performance under uncertainty in hydrologic impacts of climate change. *Advances in Water Resources*, **33**, 312–326.

Reynard, N., Crooks, S., Wilby, R.L. and Kay, A. 2004. Climate change and flood frequency in the

UK. *Proceedings of the 39th Defra Flood and Coastal Management Conference*, University of York, UK.

Shapiro, M., Shukla, J., Brunet, G. et al. 2010. An Earth-system prediction initiative for the twenty-first century. *Bulletin of the American Meteorological Society*, **91**, 1377–1388.

Shukla, J., Palmer, T.N., Hagedorn, R., et al. 2010. Toward a new generation of world climate research and computing facilities. *Bulletin of the American Meteorological Society*, **91**, 1407–1412.

Stakhiv, E.Z. 2011. Pragmatic approaches for water management under climate change uncertainty. *Journal of the American Water Resources Association*, **47**, 1183–1196.

State of Queensland, 2010. *Increasing Queenland's resilience to inland flooding in a changing climate: Final report of the Inland Flooding Study*. Department of Environment and Resource Management, Department of Infrastructure and Planning, and the Local Government Association of Queensland. Queensland Government, 16pp.

Vörösmarty, C.J., McIntryre, P.B., Gessner, M.O., et al. 2010. Global threats to human water security and river biodiversity. *Nature*, **467**, 555–561.

Watts, G., Battarbee, R., Bloomfield, J.P., et al. 2015. Climate change and water in the UK – past changes and future prospects. *Progress in Physical Geography*, **39**, 6–28.

Watts, R.J., Richter, B.D., Opperman, J.J. and Bowmer, K.H. 2011. Dam reoperation in an era of climate change. *Marine and Freshwater Research*, **62**, 321–327.

Whitehead, P.G., Wilby, R.L., Butterfield, D. and Wade, A.J. 2006. Impacts of climate change on nitrogen in a lowland chalk stream: An appraisal of adaptation strategies. *Science of the Total Environment*, **365**, 260–273.

Wilby, R.L. 2006. When and where might climate change be detectable in UK river flows? *Geophysical Research Letters*, **33**, L19407, doi:10.1029/2006GL027552.

Wilby, R.L. 2010. Evaluating climate model outputs for hydrological applications – opinion. *Hydrological Sciences Journal*, **55**, 1090–1093.

Wilby, R.L. 2012. *Utility of climate model scenarios for water policy, investment and operational decisions*. Report on behalf of the World Bank Independent Evaluation Group, Washington, 49pp.

Wilby, R.L. and Harris, I. 2006. A framework for assessing uncertainties in climate change impacts: low flow scenarios for the River Thames, UK. *Water Resources Research*, **42**, W02419.

Wilby, R.L. and Dessai, S. 2010. Robust adaptation to climate change. *Weather*, **65**, 180–185.

Wilby, R.L. and Keenan, R. 2012. Adapting to flood risk under climate change. *Progress in Physical Geography*, **36**, 349–379.

Wilby, R.L. and Dawson, C.W. 2013. The Statistical DownScaling Model (SDSM): Insights from one decade of application. *International Journal of Climatology, International Journal of Climatology*, **33**, 1707–1719.

Wilby, R.L., Troni, J., Biot, Y., et al. 2009. A review of climate risk information for adaptation and development planning. *International Journal of Climatology*, **29**, 1193–1215.

Wilby, R.L., Orr, H., Watts, G., et al. 2010. Evidence needed to manage freshwater ecosystems in a changing climate: turning adaptation principles into practice. *Science of the Total Environment*, **408**, 4150–4164.

Wilby, R.L., Fenn, C.R., Wood, P.J., et al. 2011. Smart licensing and environmental flows: Modeling framework and sensitivity testing. *Water Resources Research*, **47**, W12524.

World Bank, 2009. *Water and Climate Change: Understanding the Risks and Making Climate Smart Investment Decisions*. World Bank, Washington, 174pp.

World Bank Independent Evaluation Group, 2012. *Adapting to Climate Change: Assessing World Bank Group Experience*. World Bank Group, Washington DC, 193pp.

CHAPTER 19

Conclusion: The discipline of river science

David J. Gilvear[1], Malcolm T. Greenwood[2], Martin C. Thoms[3] and Paul J. Wood[2]

[1] *School of Geography, Earth and Environmental Sciences, Plymouth, UK*

[2] *Centre for Hydrological and Ecosystem Science, Department of Geography, Loughborough University, Loughborough, UK*

[3] *Riverine Landscapes Research Laboratory, Geography and Planning, University of New England, Australia*

The chapters contained in this volume and the nature of present-day river science, are the outcome of the research by many of the authors and their colleagues over the last 30 to 40 years. The chapters form a fitting tribute to their scientific endeavour, charting the historical development of river science and also highlighting where significant advances may be achieved in the future. During the 1970s and 1980s, the integrated and holistic understanding of what today we call 'river science', was not in the vocabulary of river managers or scientists. Since the start of the twenty-first century, river science has become a primary focus for many early career scientists and practitioners. A new generation of academics has emerged, describing themselves as 'Professors of River Science' or 'River Ecosystems', including a number who have contributed to this book. In marked contrast, during the 1970s and 1980s, academics with a primary focus on riverine systems were described as stream biologists, ecologists, hydrologists, chemists, fluvial geomorphologists or hydraulic engineers. Each of these separate disciplines evolved on parallel paths with limited, if any, cross-fertilisation or collaboration with others. Each of the disciplines developed their own concepts, theories and terminology independently, even when working to achieve common research goals on the same river system. Thus, in the UK, there has never been a single society for those working on rivers; stream ecologists typically being members of the Freshwater Biological Association, hydrologists, the British Hydrological Society, fluvial geomorphologists, the British Geomorphological Research Group (now British Society for Geomorphology) and hydraulic engineers, the Institute of Civil Engineers (e.g., Scottish Hydraulics Group). In the United States the Society for Freshwater Science, formerly North American Benthological Society (NABS), encompasses those working on both lotic and lentic systems, while engineers and a few geomorphologists work with the United States Corps of Engineers on river erosion and sediment issues. Similarly, in Australia, New Zealand, South Africa and other regions of the globe, most river scientists are members of either the limnological, geographical or engineering societies within their respective countries or region.

River Science: Research and Management for the 21st Century, First Edition.
Edited by David J. Gilvear, Malcolm T. Greenwood, Martin C. Thoms and Paul J. Wood.

As a result of the emergence of a systems based approach within science during the 1970s, the significance of research within other scientific areas became apparent within a range of academic subjects. There was inevitably a convergence of individual disciplines towards river science, although the term would not come into contemporary use until the start of the twenty-first century. Adoption and integration of concepts, terminology and techniques has, however, not been straightforward or without resistance. At the end of the twentieth century and at the start of the twenty-first, scientific coupling of research within established disciplines that were focusing on riverine environments occurred. Eco-hydrology (e.g., Baird and Wilby, 1999; Zalewski *et al.,* 1997), hydroecology (e.g., Wood *et al.,* 2007), eco-hydraulics (e.g. Nestler *et al.,* 2007), hydromorphology (e.g., Orr *et al.,* 2008), eco-geomorphology (e.g., Thoms and Parsons, 2002), and eco-hydromorphology (e.g., Vaughan *et al.,* 2009) all emerged as research foci. In essence, this was the coupling of two or more of the traditional disciplinary areas. Notable under-representation among the academic fields experiencing academic coupling in the area of river science were hydrochemistry and biogeochemistry; although this research gap is being rapidly addressed. This trend can be clearly tracked in the meta-analysis of papers published in the journal *River Research and Applications* (Thoms *et al.,* Chapter 1 this volume).

The ideas, concepts and approaches of river science reflect an interdisciplinary endeavour emerging from a variety of disciplines (see Thoms *et al.,* Chapter 1, Figure 1.1 this volume). Like all interdisciplinary activities, river science is challenged by working across spatial and temporal scales that link different disciplinary paradigms and conceptual tools (Dollar *et al.,* 2007; Delong and Thoms, Chapter 2 this volume). Disciplinary paradigms often lose their usefulness in an interdisciplinary domain and the development and acceptance of new approaches takes time. As a consequence, the focus of river science is evolving and expanding. As illustrated by the chapters contained in this volume, river science has expanded to directly engage with the human/social dimension of rivers and their landscapes; clearly illustrated in Figure 19.1. Thus rivers and their associated landscape are increasingly being viewed as social/ecological systems with cultural dimensions. There are few, if any, large natural river systems that are not influenced by people, nor their communities. The social and ecological nature of river systems recognises that they are truly interdependent and constantly co-evolving across spatial and temporal scales. Indeed, it is difficult to truly understand the dynamics of riverine systems and their ability to generate services without consideration of the anthropogenic dimension, both past and present. Focusing solely on ecological or biophysical elements as a basis for decision-making simplifies reality to the extent that the results become incomplete and the conclusions partial. Undertaking scientific research in isolation, with the social science dimension included later in the processes, overlooks essential feedbacks.

Many of the problems facing the world's riverine ecosystems cannot be solved using the same tools or approaches that partially created them. New approaches, such as resilience thinking, offer novel opportunities for the sustainable management of our global river ecosystems in general and specifically for determining water allocations. Resilience thinking provides an approach in which ecosystems, economies

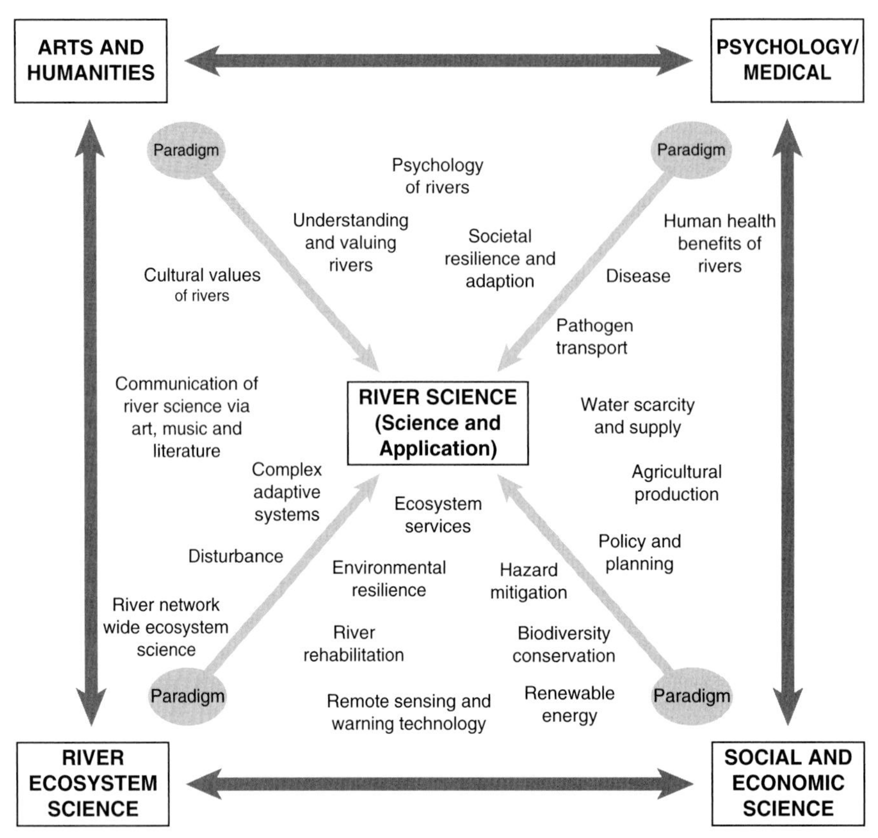

Figure 19.1 A conceptual diagram showing the likely evolution of river science moving forward with incorporation over time of disciplines from the Arts and Humanities, Psychology and Medicine, and Social and Economic Sciences alongside River Ecosystem Science. The arrows that flow towards the centre of the diagram, from their subject specific paradigm, are conceptual timelines converging on the river science of the future with an emphasis on both research and application. In 2-dimensional space a selection of sub-disciplines and fields of enquiry (shown in lower case font) that need to emerge over time are shown to illustrate the future development of river science as a subject. The locations in 2-dimensional space of these sub-disciplines and fields of enquiry relate to their subject content on the disciplinary continuum between the four corners (keystones) of the diagram.

and societies must be managed as integrated social-ecological systems (Wilson, 2012). It is a rapidly developing research area at the interface of science–management–policy and has three key building blocks:

- the concept of thresholds and alternate stable states at different scales;
- the concept of adaptive loops;
- the concepts of management and intervention in ecosystems that can incorporate uncertainty and variability (see Parsons *et al.*, Chapter 10 this volume).

In terms of an immediate future, the evolution of river science is clear – traditional disciplines will be increasingly integrated into interdisciplinary teams, in recognition that rivers function as a system of tightly integrated abiotic and biotic components (Nestler *et al.*, 2012). A conceptual model developed to guide management and restoration planning on the Mississippi River (Lubinski and Barko, 2003), for example, describes the river condition using five categories of variables called essential ecosystem characteristics: (i) hydrology and hydraulics, (ii) biogeochemical cycling, (iii) geomorphology, (iv) habitat and (v) population dynamics. Moreover, the model visualises the dynamics of the Mississippi River as interactions amongst and between the variables – the very essence of a river ecosystem!

River science of the future

There have undoubtedly been significant advances in our understanding of rivers over the last 30–40 years (e.g., Vannotte *et al.*, 1980; Townsend, 1989; Junk *et al.*, 1989) and despite the contemporary focus on common interdisciplinary goals (e.g., Benda *et al.*, 2004; Thorp *et al.* 2006) there are still areas where understanding needs to advance. A recent useful description of a river ecosystem elegantly illustrates the complexity of the challenges that river science faces (McCluney *et al.*, 2014). They define 'riverine macrosystems as hierarchical dynamic networks, influenced by strong directional connectivity and ones that integrate processes across multiple scales and broad distances through time' (p. 48). However, like the majority of the other concepts and theories identified above, this has an ecosystem focus and does not explicitly integrate societal values or socio-economic systems.

River science, like other hybrid disciplines, faces a number of challenges, such as the integration of approaches that have varying levels of high uncertainty at large and medium spatial scales (e.g., hydromorphology), together with those of low levels of uncertainty at small scales (e.g., traditional ecohydraulics). There is also an urgent need for a heightened empirical evidence base, and hence predictive capability, as to the extent to which the concepts articulated in the River Ecosystem Synthesis (Thorp *et al.*, 2006) underpin the functioning of different types of river system in different biogeographical regions. River science needs to acknowledge that in a functional sense, our ability to predict how human alterations to rivers modify biogeochemical processing, including carbon sequestration and biodiversity, is very limited. In this regard there are severe constraints in terms of our capability to measure the level to which ecosystem functioning is occurring in an individual river.

There is little doubt that new conceptual models will emerge that aid the process of understanding river ecosystems. These are likely to revolve around heterogeneity and multi-scalar perspectives (McCluney *et al.*, 2014). The means by which ecosystem services are transported from river reach to reach, and from headwaters to lowlands, and the degree to which these are dependent upon ecosystem functioning, requires greater conceptual understanding. Many of the relatively new interdisciplinary research fields, such as ecohydraulics and ecogeomorphology, are focusing on the same questions but at different scales, and as a result novel spatial and temporal frameworks for multi-disciplinary science are required. There may also be a need to return to and advance the application of basic physics, chemistry and biology to rivers, which has to a degree been lost in the drive for cross-disciplinary studies. In this regard, the dearth of mathematical under-pinning to conceptual frameworks is notable. New technologies, however, are helping in terms of monitoring processes and characterising riverscapes, at high temporal and spatial resolution, both on the ground and from the air (Gilvear *et al.*, Chapter 9 this volume). Even the conservation and management of our most threatened species is now being aided by new technologies, for example passive hydro-acoustics being used to monitor the distribution, population and behaviour of endangered species such as the freshwater dolphins of the rivers Ganges, Yangtze, Indus and Amazon (Marques *et al.*, 2013; Richman *et al.*, 2014).

Furthermore, one needs to think about the multiple components of the fluvial hydrosystem and the three-dimensional structure of the river ecosystem so as to consider where the greatest gaps in our knowledge exist. The role of the hyporheic zone in river functioning will surely be an area where further advances in understanding are to be made – at present it is very much a 'black box' (Krause *et al.*, 2011). Limited research on the air–water interface and gaseous losses from river systems has been undertaken with the exception of the thermal characteristics of rivers (Gangi *et al.*, Chapter 14 this volume; Stanford *et al.*, Chapter 13 this volume). The manner in which biogeochemical transformations along the river occur in relation to network pattern and structure also represents a challenging area where research advances are required.

River science has largely emerged from within the natural sciences, and from the onset those scientists involved were wishing to 'make a difference' in the way rivers are valued and managed by society. Currently, the ecosystem services concept is an important cross-fertilisation from the social sciences and is an important vehicle for linking environmental systems to people (Yeakley *et al.*, Chapter 17 this volume; Comino *et al.*, 2014; Large and Gilvear, 2014). This is perceived to be critical to the development of the field, since traditional single disciplines have failed to provide the science for the wise use of rivers and thus the need for an interdisciplinary approach. Moreover, it is evident that any attempt to restrict river science solely to the domain of natural scientists will lead to a failure in the sustainable use and management of the system. It is increasingly apparent that to make a real difference, river science needs to welcome and engage with the fields of sociology and economics to better understand the relationship between society at large, people and rivers; including how individuals make decisions regarding their engagement with rivers. Stakeholder involvement is increasingly acknowledged as crucial to decision making. However as Athrington *et al.* (2009) state 'how does one educate the public to fully appreciate the importance of the natural rhythm of rivers?' (p. 10). Building communities that value rivers can also bring benefits to the natural sciences in terms of crowd sourcing information and citizen science. The health of rivers, and the range and level of ecosystem services they provide, will ultimately be stronger if individuals have a catchment consciousness and see rivers as being at the heart of communities. If this is achieved, the final goal for river science is to articulate its knowledge to people living beside and working with the river. We hope that river science books in the future will be written with a balance of natural and social science, as wider recognition of its importance is accepted.

Making it happen

We hope that this volume demonstrates that rivers are not only complex ecosystems, but are in fact complex socio-economic–ecological systems. Given that the goal of river science is to develop resilient river systems that meet societal and ecosystem needs, they clearly need to be viewed and managed as such. However, as already demonstrated, because different scientific disciplines use different concepts and language to describe the system, progress has been slow to date in many areas. Ostrom (2009) argues that without a common framework to organise findings they are isolated and knowledge does not accumulate and develop with regard to complex systems with relationships at multiple levels. Ostrom (2009) proposes four sub-systems at the most basic level, representing resource units, resources systems, governance systems and users. In terms of river systems these equate to: (i) the catchment, the river network and river reaches; (ii) the flora and fauna of the river corridor network together with fluxes of water, nutrients and sediments; (iii) those responsible for river management and conservation and the rules for water governance and (iv) the users (e.g., fisherman, hydro-power generators, irrigators and canoeists). This framework seems attractive for river science if it is to move forward in dealing with the understanding of the social–ecological system of rivers across the globe.

Capacity building and education of a new generation of water scientists and policy makers is critical for the future of river science. There is the need to develop a passion in individuals for the subject at an early age through education and initiatives. This can initially be achieved through learning about river landforms and the plants, fish and other organisms that inhabit our local rivers. Rivers also need to be a focus for courses in higher education, but these will only flourish if rivers are seen as important and exciting, and crucially are adequately addressed in school/college syllabuses. A grass roots platform for river scientists needs to be created. This could take the form of a computer App similar to 'MineCraft', called 'RiverCraft' – its aim being to illustrate and model how river flows can be maintained and ecosystem collapse be prevented in the face of increasing anthropogenic pressure. River science in the future will be different from that of the past or present and will be richer for combining traditional approaches with the ever-increasing array of technologies available for interrogating and monitoring the world around us. Twenty-first century river science will hopefully see the health of the rivers of the world improve, their sustainable management secured and for society to realise the benefits of this endeavour for them individually, for society at large and for the natural environment.

References

Arthington AH, Naiman RJ, MClain ME and Nilsson C (2009). Preserving the biodiversity and ecological services of rivers: new challenges and research opportunities, *Freshwater Biology* 55: 1–16.

Baird AJ and Wilby RL (1999). *Eco-hydrology: Plants and Water in Terrestrial and Aquatic Environments*. Routledge, London.

Benda L, Poff NL, Miller D, *et al.* (2004). The network dynamics hypothesis: how channel networks structure riverine habitats. *BioScience* 54: 413–427.

Comino E, Bottero M, Pomarico S and Rosso M (2014). Exploring the environmental value of ecosystem services for a river basin through a spatial multicriteria analysis. *Land Use Policy* 36: 381–395.

Dollar ESJ, James CS, Rogers KH and Thoms MC (2007). A framework for interdisciplinary understanding of rivers as ecosystems. *Geomorphology* 89: 147.

Junk WJ, Bayley PB and Sparks RE (1989). The flood pulse concept in river floodplain systems, In: Dodge DP (ed.), *Proceedings of the International Large River Symposium. Canadian Special Publications Fisheries and Aquatic Science* 106: 110–127.

Krause S, Hannah DM, Fleckenstein JH, *et al.* (2011). Interdisciplinary perspectives on processes in the hyporheic zone. *Ecohydrology* 4: 481–499

Large ARG and Gilvear DJ (2014). Using Google Earth, a virtual imaging platform, for ecosystem services-based river assessment. *River Research and Applications* 31:406–421

Lubinski KS and Barko JW (2003). Upper Mississippi River-Illinois Waterway System Navigation Feasibility Study: Environmental Science Panel Report. US Army Corps of Engineers.

Marquest TA, Thomans L, Martin SW, *et al.* (2013). Estimating animal population density using passive acoustics. *Biological Reviews of the Cambridge Philosophical Society* 88: 287–309.

McCluney KE, Poff NL, Palmer MA, *et al.* (2014). Riverine macrosystems ecology: sensitivity, resistance, and resilience of whole river basins with human alterations. *Frontiers in Ecology and the Environment* 12: 48–58.

Nestler JM, Goodwin RA, Smith DL and Anderson JJ (2007). A mathematical and conceptual framework for hydraulics. In: Wood PJ, Hannah DM, Sadler JP (eds), *Hydroecology and Ecohydrology: Past, Present and Future*, John Wiley & Sons, Ltd: Chichester, pp 205–224.

Nestler JM, Pompeu PS, Goodwin RA, *et al.* (2012). The river machine: a template for fish movement and habitat, fluvial geomorphology, fluid dynamics and biogeochemical cycling. *River Research and Applications* 28: 490–503.

Orr HG, Large ARG, Newson MD and Walsh CL (2008). A predictive typology for characterising hydromorphology. *Geomorphology* 100**:** 32–40.

Ostrom E (2009). A generalising framework for analyzing sustainability of social-ecological systems. *Science* 324: 419–422.

Richman NI, Gibbons JM, Turvey ST, *et al.* (2014). To See or not to see: investigating detectability of ganges river dolphins using a combined visual-acoustic survey. *PLoS ONE* 9(5): e96811. doi:10.1371/journal.pone.0096811.

Thoms MC and Parsons M (2002). Ecogeomorphology: an interdisciplinary approach to river science. *International Association of Hydrological Sciences* 276**:** 113–119.

Thorp JH, Thoms MC and Delong MD (2006). The riverine ecosystem synthesis: biocomplexity in river networks across space and time. *River Research and Applications* 22: 123–147.

Townsend CR (1989). The patch dynamics concept of stream community ecology. *Journal of the North American Benthological Society* 8: 36–50.

Vannote RL, Minshall GW, Cummins KW, *et al.* (1980). The River Continuum Concept. *Canadian Journal of Fisheries and Aquatic Sciences* 37: 130–137.

Vaughan IP, Diamond M, Gurnell AM, *et al.* (2009). Integrating ecology with hydromorphology: a priority for river science and management. *Aquatic Conservation: Marine and Freshwater Ecosystems* 19: 113–125.

Wilson GA (2012). *Community Resilience and Environmental Transitions*. Earthscan/Routledge, London.

Wood PJ, Hannah DM and Sadler JP. (2007). *Hydroecology and Ecohydrology, Past, Present and Future*. John Wiley & Sons, Chichester.

Zalewski M, Janauer GA and Jolankai G. (1997). *Ecohydrology: A New Paradigm for the Sustainable Use of Aquatic Resources*. Unesco, Paris. IHP-V Technical Documents in Hydrology no. 7.

Index

Page numbers in *italic* refer to figures; those in **bold** to tables

River Science: Research and Management for the 21st Century, First Edition.
Edited by David J. Gilvear, Malcolm T. Greenwood, Martin C. Thoms and Paul J. Wood.